STUDENT'S SOLUTIONS MANUAL

GEX PUBLISHING SERVICES

ELEMENTARY AND INTERMEDIATE ALGEBRA

FOURTH EDITION

Tom Carson

Franklin Classical School

Bill E. Jordan

Seminole State College of Florida

PEARSON

Boston   Columbus   Indianapolis   New York   San Francisco   Upper Saddle River
Amsterdam   Cape Town   Dubai   London   Madrid   Milan   Munich   Paris   Montreal   Toronto
Delhi   Mexico City   São Paulo   Sydney   Hong Kong   Seoul   Singapore   Taipei   Tokyo

Reproduced by Pearson from electronic files supplied by the author.

ISBN-13: 978-0-321-92529-9
ISBN-10: 0-321-92529-7

 2 3 4 5 6 EBM 17 16 15 14

www.pearsonhighered.com

CONTENTS

Chapter 1

Foundations of Algebra

Exercise Set 1.1

1. {Sunday, Monday, Tuesday, Wednesday, Thursday, Friday, Saturday}

3. {a, e, i, o, u}   5. {5, 10, 15, 20, …}

7. {9, 11, 13, 15, …}   9. $\{-6, -5, -4, -3\}$

11. Rational because -4 and 5 are integers.

13. Rational because 9 is an integer and all integers are rational numbers.

15. Irrational because π cannot be written as a ratio of integers.

17. Rational because -0.21 can be expressed as $-\dfrac{21}{100}$, the ratio of two integers.

19. Rational because $0.\overline{6}$ can be expressed as the fraction $\dfrac{2}{3}$, the ratio of two integers.

21. True

23. True

25. False. All real numbers are either rational or irrational.

27. The number $-\dfrac{5}{6}$ is located $\dfrac{5}{6}$ of the way between 0 and -1, so we divide the space between 0 and -1 into 6 equal divisions and place a dot on the 5th mark to the left of 0.

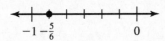

29. The number $2\dfrac{3}{8}$ is located $\dfrac{3}{8}$ of the way between 2 and 3, so we divide the space between 2 and 3 into 8 equal divisions and place a dot on the 3rd mark to the right of 2.

31. The number -3.5 is located $0.5 = \dfrac{5}{10}$ of the way between -3 and -4, so we divide the space between -3 and -4 into 10 equal divisions and place a dot on the 5th mark to the left of -3.

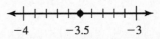

33. First divide the number line between 2 and 3 into tenths. The number 2.45 falls between 2.4 and 2.5 on the number line. Subdivide this section into hundredths and place a dot on the 5th mark to the right of 2.4.

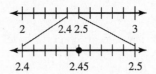

35. $|23| = 23$ because 23 is 23 units from 0 on a number line.

37. $|-2| = 2$ because -2 is 2 units from 0 on a number line.

39. $|-5.7| = 5.7$ because -5.7 is 5.7 units from 0 on a number line.

41. $\left|-3\dfrac{1}{8}\right| = 3\dfrac{1}{8}$ because $-3\dfrac{1}{8}$ is $3\dfrac{1}{8}$ units from 0 on a number line.

43. $|0| = 0$ because 0 is 0 units from 0 on a number line.

45. $9 > 3$ because 9 is farther to the right on a number line than 3.

47. $4 > -3$ because 4 is farther to the right on a number line than -3.

49. $-7 > -8$ because -7 is farther to the right on a number line than -8.

51. $-8 < 0$ because -8 is farther to the left on a number line than 0.

53. $6.2 > 4.5$ because 6.2 is farther to the right on a number line than 4.5.

55. $-4.3 > -7.6$ because -4.3 is farther to the right on a number line than -7.6.

57. $2\dfrac{2}{3} < 2\dfrac{3}{4}$ because $2\dfrac{2}{3}$ is farther to the left on a number line than $2\dfrac{3}{4}$.

59. $5.8 = |-5.8|$ because the absolute value of -5.8 is equal to 5.8.

61. $7.3 < |-8.7|$ because 7.3 is farther to the left on a number line than the absolute value of -8.7, which is equal to 8.7.

63. $|4.31| = |-4.31|$ because the absolute value of 4.31 and the absolute value of -4.31 are both equal to 4.31.

65. $\left|-6\dfrac{5}{8}\right| > 5\dfrac{3}{8}$ because the absolute value of $-6\dfrac{5}{8}$ is equal to $6\dfrac{5}{8}$, which is farther to the right on a number line than $5\dfrac{3}{8}$.

67. $|-4| > |-2|$ because the absolute value of -4 is 4, the absolute value of -2 is 2, and 4 is farther to the right on a number line than 2.

69. $|29.5| < |-29.7|$ because the absolute value of 29.5 is 29.5, the absolute value of -29.7 is 29.7, and 29.5 is farther to the left on a number line than 29.7.

71. $\left|-\dfrac{2}{3}\right| < \left|-\dfrac{4}{3}\right|$ because the absolute value of $-\dfrac{2}{3}$ is $\dfrac{2}{3}$, the absolute value of $-\dfrac{4}{3}$ is $\dfrac{4}{3}$, and $\dfrac{2}{3}$ is farther to the left on a number line than $\dfrac{4}{3}$.

73. $-4.7, -2.56, 5.4, \left|-7\dfrac{1}{2}\right|, |8.3|$

75. $-0.6, -0.44, 0, |-0.02|, 0.4, \left|1\dfrac{2}{3}\right|, 3\dfrac{1}{4}$

Exercise Set 1.2

1. $\dfrac{3}{10}$ 3. $\dfrac{1}{3}$ 5. $\dfrac{3}{4}$

7. $\dfrac{3}{8}$ 9. $\dfrac{15}{16}$

11. $\dfrac{3}{10} = \dfrac{?}{40} \Rightarrow \dfrac{3\cdot 4}{10\cdot 4} = \dfrac{12}{40}$
The missing number is 12.

13. $-\dfrac{7}{12} = \dfrac{?}{60} \Rightarrow -\dfrac{7\cdot 5}{12\cdot 5} = \dfrac{-35}{60}$
The missing number is -35.

15. $\dfrac{4}{7} = \dfrac{?}{28} \Rightarrow \dfrac{4\cdot 4}{7\cdot 4} = \dfrac{16}{28}$
The missing number is 16.

17. $\dfrac{8}{60} = \dfrac{2}{?} \Rightarrow \dfrac{8\div 4}{60\div 4} = \dfrac{2}{15}$
The missing number is 15.

19. The LCD of 5 and 3 is 15.
$\dfrac{4\cdot 3}{5\cdot 3} = \dfrac{12}{15}$ and $\dfrac{2\cdot 5}{3\cdot 5} = \dfrac{10}{15}$

21. The LCD of 9 and 6 is 18.
$\dfrac{4\cdot 2}{9\cdot 2} = \dfrac{8}{18}$ and $\dfrac{7\cdot 3}{6\cdot 3} = \dfrac{21}{18}$

23. The LCD of 18 and 24 is 72.
$-\dfrac{11\cdot 4}{18\cdot 4} = -\dfrac{44}{72}$ and $-\dfrac{17\cdot 3}{24\cdot 3} = -\dfrac{51}{72}$

25. The LCD of 16 and 36 is 144.
$-\dfrac{1\cdot 9}{16\cdot 9} = -\dfrac{9}{144}$ and $-\dfrac{15\cdot 4}{36\cdot 4} = -\dfrac{60}{144}$

27. $44 = 2\cdot 22$
$ = 2\cdot 2\cdot 11$

29. $36 = 2\cdot 18$
$ = 2\cdot 2\cdot 9$
$ = 2\cdot 2\cdot 3\cdot 3$

31. $64 = 2\cdot 32$
$ = 2\cdot 2\cdot 16$
$ = 2\cdot 2\cdot 2\cdot 8$
$ = 2\cdot 2\cdot 2\cdot 2\cdot 4$
$ = 2\cdot 2\cdot 2\cdot 2\cdot 2\cdot 2$

33. $250 = 10\cdot 25$
$ = 2\cdot 5\cdot 5\cdot 5$

35. $\dfrac{72}{90} = \dfrac{\cancel{2}\cdot 2\cdot 2\cdot \cancel{3}\cdot \cancel{3}}{\cancel{2}\cdot \cancel{3}\cdot \cancel{3}\cdot 5} = \dfrac{4}{5}$

37. $\dfrac{63}{99} = \dfrac{\cancel{3}\cdot \cancel{3}\cdot 7}{\cancel{3}\cdot \cancel{3}\cdot 11} = \dfrac{7}{11}$

39. $-\dfrac{78}{104} = -\dfrac{\cancel{2}\cdot 3\cdot \cancel{13}}{\cancel{2}\cdot 2\cdot 2\cdot \cancel{13}} = -\dfrac{3}{4}$

41. $-\dfrac{48}{90} = -\dfrac{\cancel{2}\cdot 2\cdot 2\cdot 2\cdot \cancel{3}}{\cancel{2}\cdot \cancel{3}\cdot 3\cdot 5} = -\dfrac{8}{15}$

43. Incorrect. You may divide out only factors, not addends.

45. Incorrect. The prime factorization of 240 should be $2\cdot 2\cdot 2\cdot 2\cdot 3\cdot 5$.

47. If the student scores 294 points out of 336, then the student's score is $\dfrac{294}{336}$ of the total points.

$\dfrac{294}{336} = \dfrac{\cancel{2}\cdot \cancel{3}\cdot \cancel{7}\cdot 7}{\cancel{2}\cdot 2\cdot 2\cdot 2\cdot \cancel{3}\cdot \cancel{7}} = \dfrac{7}{8}$

49. If 300 of the 575 employees have optional life insurance, then the fraction of the employees that have the optional life insurance is $\dfrac{300}{575}$

$\dfrac{300}{575} = \dfrac{2\cdot 2\cdot 3\cdot \cancel{5}\cdot \cancel{5}}{\cancel{5}\cdot \cancel{5}\cdot 23} = \dfrac{12}{23}$

51. There are $7\cdot 24 = 168$ hours in one week.

$\dfrac{40}{168} = \dfrac{\cancel{2}\cdot \cancel{2}\cdot \cancel{2}\cdot 5}{\cancel{2}\cdot \cancel{2}\cdot \cancel{2}\cdot 3\cdot 7} = \dfrac{5}{21}$

Carla spends $\dfrac{5}{21}$ of her week working.

53. $50+40+18+4 = 112$ hours

$\dfrac{112}{168} = \dfrac{\cancel{2}\cdot \cancel{2}\cdot \cancel{2}\cdot 2\cdot \cancel{7}}{\cancel{2}\cdot \cancel{2}\cdot \cancel{2}\cdot 3\cdot \cancel{7}} = \dfrac{2}{3}$

Carla spends $\dfrac{2}{3}$ of her week doing all of the listed activities combined.

55. $\dfrac{325}{1000} = \dfrac{\cancel{5}\cdot \cancel{5}\cdot 13}{2\cdot 2\cdot 2\cdot \cancel{5}\cdot \cancel{5}\cdot 5} = \dfrac{13}{40}$

57. $1000-325 = 675$ non-victims;

$\dfrac{675}{1000} = \dfrac{\cancel{5}\cdot \cancel{5}\cdot 3\cdot 3\cdot 3}{2\cdot 2\cdot 2\cdot \cancel{5}\cdot \cancel{5}\cdot 5} = \dfrac{27}{40}$

59. a) 1991

b) $\dfrac{177}{500}$

61. $\dfrac{154}{198} = \dfrac{\cancel{2}\cdot 7\cdot \cancel{11}}{\cancel{2}\cdot 9\cdot \cancel{11}} = \dfrac{7}{9}$

63. $\dfrac{30}{365} = \dfrac{2\cdot 3\cdot \cancel{5}}{\cancel{5}\cdot 73} = \dfrac{6}{73}$

65. $\dfrac{40}{60} = \dfrac{\cancel{2}\cdot \cancel{2}\cdot 2\cdot \cancel{5}}{\cancel{2}\cdot \cancel{2}\cdot 3\cdot \cancel{5}} = \dfrac{2}{3}$

67. $193+242 = 435$ total representatives.

$\dfrac{193}{435}$

69. $12+22+11 = 45$ atoms total

$\dfrac{12}{45} = \dfrac{2\cdot 2\cdot \cancel{3}}{\cancel{3}\cdot 3\cdot 5} = \dfrac{4}{15}$

Review Exercises

1. {Mercury, Venus, Earth, Mars}

2. It is a rational number because it can be written as a ratio of the integers 8 and 10.

3.

4. It is an expression because it has no = sign.

5. $|27| = 27$

6. $|-6| = 6$

Exercise Set 1.3

1. Commutative Property of Addition because the order of the addends is changed.

3. Additive identity because the sum of a number and 0 is that number.

5. Additive inverse because the sum of these opposites is 0.

7. Associative Property of Addition because the grouping is changed.

9. Commutative Property of Addition because the order of the addends is changed.

11. Additive inverse because the sum of the opposites 2.1 and −2.1 is 0.

13. $8+13 = 21$

15. $-6+(-12) = -18$

17. $-4+13 = 9$

19. $-14+5 = -9$

21. $27+(-13) = 14$

23. $-28+12 = -16$

25. $\dfrac{5}{8} + \dfrac{1}{8} = \dfrac{5+1}{8}$

$\qquad\qquad = \dfrac{6}{8}$

$\qquad\qquad = \dfrac{\cancel{2} \cdot 3}{\cancel{2} \cdot 2 \cdot 2}$

$\qquad\qquad = \dfrac{3}{4}$

27. $-\dfrac{4}{9} + \left(-\dfrac{2}{9}\right) = \dfrac{-4+(-2)}{9}$

$\qquad\qquad\qquad = -\dfrac{6}{9}$

$\qquad\qquad\qquad = -\dfrac{2 \cdot \cancel{3}}{3 \cdot \cancel{3}}$

$\qquad\qquad\qquad = -\dfrac{2}{3}$

29. $\dfrac{1}{6} + \left(-\dfrac{5}{6}\right) = \dfrac{1+(-5)}{6}$

$\qquad\qquad\qquad = -\dfrac{4}{6}$

$\qquad\qquad\qquad = -\dfrac{\cancel{2} \cdot 2}{\cancel{2} \cdot 3}$

$\qquad\qquad\qquad = -\dfrac{2}{3}$

31. The LCD of 4 and 6 is 12.

$\dfrac{3}{4} + \dfrac{1}{6} = \dfrac{3(3)}{4(3)} + \dfrac{1(2)}{6(2)}$

$\qquad\quad = \dfrac{9}{12} + \dfrac{2}{12}$

$\qquad\quad = \dfrac{9+2}{12}$

$\qquad\quad = \dfrac{11}{12}$

33. The LCD of 12 and 3 is 12.

$-\dfrac{1}{12} + \left(-\dfrac{2}{3}\right) = -\dfrac{1}{12} + \left(-\dfrac{2(4)}{3(4)}\right)$

$\qquad\qquad\qquad = -\dfrac{1}{12} + \left(-\dfrac{8}{12}\right)$

$\qquad\qquad\qquad = \dfrac{-1+(-8)}{12}$

$\qquad\qquad\qquad = -\dfrac{9}{12}$

$\qquad\qquad\qquad = -\dfrac{\cancel{3} \cdot 3}{\cancel{3} \cdot 4}$

$\qquad\qquad\qquad = -\dfrac{3}{4}$

35. The LCD of 6 and 21 is 42.

$-\dfrac{5}{6} + \dfrac{4}{21} = -\dfrac{5(7)}{6(7)} + \dfrac{4(2)}{21(2)}$

$\qquad\qquad = -\dfrac{35}{42} + \dfrac{8}{42}$

$\qquad\qquad = \dfrac{-35+8}{42}$

$\qquad\qquad = -\dfrac{27}{42}$

$\qquad\qquad = -\dfrac{\cancel{3} \cdot 3 \cdot 3}{2 \cdot \cancel{3} \cdot 7}$

$\qquad\qquad = -\dfrac{9}{14}$

37. $0.21 + 0.05 = 0.26$

39. $-0.18 + 6.7 = 6.52$

41. $-0.28 + (-4.1) = -4.38$

43. $-42 + |-14| = -42 + 14 = -28$

45. $|-2.4| + |-0.78| = 2.4 + 0.78 = 3.18$

47. The LCD of 8 and 6 is 24.

$\left|-\dfrac{3}{8}\right| + \left|\dfrac{5}{6}\right| = \dfrac{3}{8} + \dfrac{5}{6}$

$\qquad\qquad = \dfrac{3(3)}{8(3)} + \dfrac{5(4)}{6(4)}$

$\qquad\qquad = \dfrac{9}{24} + \dfrac{20}{24}$

$\qquad\qquad = \dfrac{29}{24}$

49. -5 because $5 + (-5) = 0$

51. 12 because $-12 + 12 = 0$

53. 0 because $0 + 0 = 0$

55. $-\dfrac{5}{6}$ because $\dfrac{5}{6} + \left(-\dfrac{5}{6}\right) = 0$

57. 0.29 because $-0.29 + 0.29 = 0$

59. x because $-x + x = 0$

61. $-\dfrac{m}{n}$ because $\dfrac{m}{n} + \left(-\dfrac{m}{n}\right) = 0$

63. $-(-2) = 2$

65. $-\left(-(-4)\right) = -(4) = -4$

67. $-|4| = -4$

69. $-|-12| = -(12) = -12$

71. $6 - 15 = 6 + (-15) = -9$

73. $-4 - 9 = -4 + (-9) = -13$

75. $4 - (-3) = 4 + 3 = 7$

77. $-8 - (-2) = -8 + 2 = -6$

79. $-\dfrac{1}{5} - \left(-\dfrac{1}{5}\right) = -\dfrac{1}{5} + \dfrac{1}{5} = 0$

81. The LCD of 10 and 5 is 10.
$$\dfrac{7}{10} - \left(-\dfrac{3}{5}\right) = \dfrac{7}{10} + \dfrac{3}{5}$$
$$= \dfrac{7}{10} + \dfrac{3(2)}{5(2)}$$
$$= \dfrac{7}{10} + \dfrac{6}{10}$$
$$= \dfrac{7+6}{10}$$
$$= \dfrac{13}{10}$$

83. The LCD of 5 and 7 is 35.
$$-\dfrac{4}{5} - \left(-\dfrac{2}{7}\right) = -\dfrac{4}{5} + \dfrac{2}{7}$$
$$= -\dfrac{4(7)}{5(7)} + \dfrac{2(5)}{7(5)}$$
$$= -\dfrac{28}{35} + \dfrac{10}{35}$$
$$= \dfrac{-28 + 10}{35}$$
$$= -\dfrac{18}{35}$$

85. $4.01 - 3.65 = 0.36$

87. $0.07 - 5.82 = 0.07 + (-5.82)$
$$= -5.75$$

89. $-6.1 - (-4.5) = -6.1 + 4.5$
$$= -1.6$$

91. $-|-4| - |-6| = -4 - 6$
$$= -4 + (-6)$$
$$= -10$$

93. $|6.2| - |-7.1| = 6.2 - 7.1$
$$= 6.2 + (-7.1)$$
$$= -0.9$$

95. $11,088 + 169 - 8128 - 93 - 993 - 7$
$$= 11,088 + 169 + (-8128) + (-93)$$
$$+ (-993) + (-7)$$
$$= 2036$$
The net income is $2,036,000,000.

97. $820.7 + 915.6 + (-2004.5) = -268.2 \, \text{N}.$
The negative indicates that the beam is moving downward.

99. $\$600.92 - \$574.97 = \$25.95$

101. $12,604.53 - (-48.59) = 12,604.53 + 48.59$
$$= 12,653.12$$

103. $-196 - (-208) = -196 + 208 = 12$

105. a) $18.6 - 18.8$
 b) -0.2
 c) The negative difference indicates that the mean composite score in 1989 was less than the score in 1986.

107. $\$67,790 - \$50,561 = \$17,229$

109. Doctorate
$\quad$ $\$144,720 - \$111,149 = \$33,571$

Review Exercises

1. {Washington, Adams, Jefferson, Madison}

2. $|-7| = 7$

3. $100 = 2 \cdot 50 = 2 \cdot 2 \cdot 25 = 2 \cdot 2 \cdot 5 \cdot 5$

4. $\dfrac{78}{91} = \dfrac{2 \cdot 3 \cdot \cancel{13}}{7 \cdot \cancel{13}} = \dfrac{6}{7}$

5. $\dfrac{40}{48} = \dfrac{?}{6}$
$\quad$ $\dfrac{40 \div 8}{48 \div 8} = \dfrac{5}{6}$

6. $-\dfrac{4}{5}? -\dfrac{8}{20}$
$\quad$ $-\dfrac{4 \cdot 4}{5 \cdot 4}? -\dfrac{8}{20}$
$\quad$ $-\dfrac{16}{20} < -\dfrac{8}{20}$

Exercise Set 1.4

1. Distributive Property of Multiplication over addition.

3. Multiplicative identity because the product of a number and 1 is the number.

5. Multiplicative Property of 0 because the product of a number and 0 is 0.

7. Commutative Property of Multiplication because the order of the factors is different.

9. Associative Property of Multiplication because the grouping of factors is different.

11. Commutative Property of Multiplication because the order of the factors is different.

13. $6(-3) = -18$

15. $(-10)5 = -50$

17. $(9)(-6) = -54$

19. $-4(-5) = 20$

21. $(-7)(-6) = 42$

23. $\dfrac{1}{2} \cdot \left(-\dfrac{1}{2}\right) = -\dfrac{1 \cdot 1}{2 \cdot 2} = -\dfrac{1}{4}$

25. $\left(-\dfrac{2}{3}\right)\left(-\dfrac{3}{4}\right) = \left(-\dfrac{\cancel{2}}{\cancel{3}}\right)\left(-\dfrac{\cancel{3}}{\cancel{2} \cdot 2}\right)$
$\quad$ $= \dfrac{1}{2}$

27. $\left(-\dfrac{5}{6}\right)\left(\dfrac{8}{15}\right) = -\dfrac{\cancel{8}}{\cancel{2} \cdot 3} \cdot \dfrac{\cancel{2} \cdot 2 \cdot 2}{3 \cdot \cancel{8}} = -\dfrac{4}{9}$

29. $\quad$ 3.8
$\quad$ $\underline{\times -10}$
$\quad$ -38.0

31. $\quad$ -1.6
$\quad$ $\underline{\times -4.2}$
$\quad$ $\quad 32$
$\quad$ $\underline{\quad 64}$
$\quad$ 6.72

33. $\quad$ 1.52
$\quad$ $\underline{\times -4.3}$
$\quad$ $\quad 456$
$\quad$ $\underline{\quad 608}$
$\quad$ -6.536

35. $6(-9)(-1) = 6(9)$
$\quad$ $= 54$

37. $6(-2)(4) = -12(4)$
$\quad$ $= -48$

39. $(-4)(-6)(-7) = (24)(-7)$
$\quad$ $= -168$

41. $12(-6)(-2)(-1) = -72(-2)(-1)$
$\quad$ $= 144(-1)$
$\quad$ $= -144$

43. $(-1)(-6)(-40)(-3)$
$\quad$ $= 6(-40)(-3)$
$\quad$ $= -240(-3)$
$\quad$ $= 720$

45. $(-2)(3)(-4)(-1)(2)$
$\quad$ $= -6(-4)(-1)(2)$
$\quad$ $= 24(-1)(2)$
$\quad$ $= -24(2)$
$\quad$ $= -48$

47. $\dfrac{3}{2}$ is the multiplicative inverse of $\dfrac{2}{3}$ because

$\dfrac{2}{3} \cdot \dfrac{3}{2} = 1$.

49. $-\dfrac{2}{5}$ is the multiplicative inverse of $-\dfrac{5}{2}$ because

$-\dfrac{5}{2} \cdot \left(-\dfrac{2}{5} \right) = 1$.

51. $-\dfrac{1}{5}$ is the multiplicative inverse of -5 because

$-5 \cdot \left(-\dfrac{1}{5} \right) = 1$.

53. There is no multiplicative inverse of 0.

55. $-6 \div 2 = -3$

57. $-56 \div (-4) = 14$

59. $\dfrac{-18}{9} = -2$

61. $\dfrac{-63}{-7} = 9$

63. $\dfrac{0}{-6} = 0$

65. $\dfrac{-8}{0}$ is undefined

67. $\dfrac{0}{0}$ is indeterminate

69. $6 \div \dfrac{2}{3} = \dfrac{6}{1} \cdot \dfrac{3}{2}$

$= \dfrac{\cancel{2} \cdot 3}{1} \cdot \dfrac{3}{\cancel{2}}$

$= 9$

71. $\dfrac{3}{5} \div \left(-\dfrac{9}{10} \right) = \dfrac{3}{5} \cdot \left(-\dfrac{10}{9} \right)$

$= \dfrac{\cancel{3}}{\cancel{5}} \cdot \left(-\dfrac{2 \cdot \cancel{5}}{\cancel{3} \cdot 3} \right)$

$= -\dfrac{2}{3}$

73. $-\dfrac{2}{7} \div \left(-\dfrac{8}{21} \right) = -\dfrac{2}{7} \cdot \left(-\dfrac{21}{8} \right)$

$= -\dfrac{\cancel{2}}{\cancel{7}} \cdot \left(-\dfrac{3 \cdot \cancel{7}}{\cancel{2} \cdot 2 \cdot 2} \right)$

$= \dfrac{3}{4}$

75. $-\dfrac{4}{9} \div \dfrac{10}{21} = -\dfrac{4}{9} \cdot \dfrac{21}{10}$

$= -\dfrac{\cancel{2} \cdot 2}{\cancel{3} \cdot 3} \cdot \dfrac{\cancel{3} \cdot 7}{\cancel{2} \cdot 5}$

$= -\dfrac{14}{15}$

77. $9.03 \div 4.3 = 90.3 \div 43$

$\begin{array}{r} 2.1 \\ 43\overline{)90.3} \\ \underline{86} \\ 43 \\ \underline{43} \\ 0 \end{array}$

79. $-36.72 \div (-0.4) = -367.2 \div (-4)$

$\begin{array}{r} 91.8 \\ 4\overline{)367.2} \\ \underline{36} \\ 07 \\ \underline{4} \\ 32 \\ \underline{32} \\ 0 \end{array}$

81. $-14 \div 0.3 = -140.0 \div 3$

$\begin{array}{r} -46.\overline{6} \\ 3\overline{)-140.0} \\ \underline{12} \\ 20 \\ \underline{18} \\ 20 \\ \underline{18} \\ 2 \end{array}$

83. $\dfrac{2}{3} \cdot \dfrac{3}{4} = \dfrac{\cancel{2}}{\cancel{3}} \cdot \dfrac{\cancel{3}}{\cancel{2} \cdot 2} = \dfrac{1}{2}$

$\dfrac{1}{2}$ of all respondents were women who agreed with the statement.

85. $3\frac{1}{2}\cdot(-2480)=\frac{7}{2}\cdot\left(-\frac{2480}{1}\right)$

$=\frac{7}{\cancel{2}}\cdot\left(-\frac{\cancel{2}\cdot1240}{1}\right)$

$=-\$8680$

Her debt in five years can be represented by $-\$8680$.

87. $-158,572.85\div(-258.75)\approx612.8$

Their debt is approximately 612.8 times greater.

89. $12.5(-32.2)=-402.5$ lb. The negative sign indicates that the force is downward.

91. $-1,658,181.8\div(-9.8)=169,202.2245$ kg

93. $-6.4\cdot8=-51.2$ V

95. $-120(-30)=3600$ W

Review Exercises

1. irrational

2.

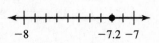

$\begin{array}{c}\\-8\qquad\qquad-7.2\ -7\end{array}$

3. $-\frac{2}{3}$

4. $|-6.8|=6.8$

5. $-7+(-15)=-22$

6. $-\frac{5}{8}-\left(-\frac{1}{6}\right)=-\frac{5}{8}+\frac{1}{6}$

$=-\frac{5\cdot3}{8\cdot3}+\frac{1\cdot4}{6\cdot4}$

$=\frac{-15+4}{24}$

$=-\frac{11}{24}$

Exercise Set 1.5

1. Base: 7; Exponent: 2; "seven squared"

3. Base: -5; Exponent: 3; "negative 5 cubed"

5. Base: 2; Exponent: 7 ; "additive inverse of two to the seventh power"

7. $3^4=3\cdot3\cdot3\cdot3=81$

9. $(-8)^2=(-8)(-8)=64$

11. $-8^2=-(8\cdot8)=-64$

13. $(-5)^3=(-5)(-5)(-5)=-125$

15. $-5^3=-5\cdot5\cdot5=-125$

17. $-(-2)^3=-((-2)\cdot(-2)\cdot(-2))$

$=-(-8)$

$=8$

19. $-(-1)^6=-((-1)\cdot(-1)\cdot(-1)\cdot(-1)\cdot(-1)\cdot(-1))$

$=-(1)$

$=-1$

21. $\left(-\frac{1}{5}\right)^2=\left(-\frac{1}{5}\right)\cdot\left(-\frac{1}{5}\right)=\frac{1}{25}$

23. $\left(-\frac{3}{4}\right)^3=\left(-\frac{3}{4}\right)\cdot\left(-\frac{3}{4}\right)\cdot\left(-\frac{3}{4}\right)=-\frac{27}{64}$

25. $(0.2)^3=(0.2)(0.2)(0.2)=0.008$

27. $(-4.1)^2=(-4.1)(-4.1)=16.81$

29. ±11

31. No real-number square roots exist.

33. ±14

35. ±16

37. $\sqrt{16}=4$

39. $\sqrt{144}=12$

41. $\sqrt{0.49}=0.7$

43. $\sqrt{-64}$ is not a real number.

45. $\sqrt{\frac{64}{81}}=\frac{\sqrt{64}}{\sqrt{81}}=\frac{8}{9}$

47. $\sqrt{\frac{50}{2}}=\sqrt{25}=5$

49. $5\cdot3+4=15+4$

$=19$

51. $8\div2-4=4-4$

$=0$

53. $8 + 4 \div 2 = 8 + 2$
$$= 10$$

55. $4 \cdot 6 - 7 \cdot 5 = 24 - 35$
$$= -11$$

57. $12 - 2^4 = 12 - 16$
$$= -4$$

59. $8 - 3(-4)^2 = 8 - 3(16)$
$$= 8 - 48$$
$$= -40$$

61. $4^2 - 36 \div 4(12 - 8) = 4^2 - 36 \div 4(4)$
$$= 16 - 36 \div 4(4)$$
$$= 16 - 9(4)$$
$$= 16 - 36$$
$$= -20$$

63. $-4 + 3(-1)^4 + 18 \div 3 \cdot 3 = -4 + 3(1) + 18 \div 3 \cdot 3$
$$= -4 + 3 + 18 \div 3 \cdot 3$$
$$= -4 + 3 + 6 \cdot 3$$
$$= -4 + 3 + 18$$
$$= -1 + 18$$
$$= 17$$

65. $-2^3 + 8 - 7(4 - 3) = -2^3 + 8 - 7(1)$
$$= -8 + 8 - 7(1)$$
$$= -8 + 8 - 7$$
$$= 0 - 7$$
$$= -7$$

67. $24 \div (-9 + 3)(3 + 4) = 24 \div (-6)(7)$
$$= -4(7)$$
$$= -28$$

69. $13.02 \div (-3.1) + 6^2 - \sqrt{25}$
$$= 13.02 \div (-3.1) + 36 - \sqrt{25}$$
$$= 13.02 \div (-3.1) + 36 - 5$$
$$= -4.2 + 36 - 5$$
$$= 31.8 - 5$$
$$= 26.8$$

71. $18.2 + 3.4\left[(5 + 9) \div 7 - 6^2\right]$
$$= 18.2 + 3.4\left(14 \div 7 - 6^2\right)$$
$$= 18.2 + 3.4\left(14 \div 7 - 36\right)$$
$$= 18.2 + 3.4\left(2 - 36\right)$$
$$= 18.2 + 3.4(-34)$$
$$= 18.2 - 115.6$$
$$= -97.4$$

73. $-4^3 - 3^2 + 4|7 - 12| = -4^3 - 3^2 + 4|-5|$
$$= -4^3 - 3^2 + 4(5)$$
$$= -64 - 9 + 4(5)$$
$$= -64 - 9 + 20$$
$$= -73 + 20$$
$$= -53$$

75. $-\dfrac{3}{4} \div \left(\dfrac{1}{8}\right) + \left(-\dfrac{2}{5}\right)(-3)(-4)$
$$= -\dfrac{3}{4} \cdot \dfrac{8}{1} + \left(-\dfrac{2}{5}\right)(-3)(-4)$$
$$= -6 + \left(-\dfrac{24}{5}\right)$$
$$= -\dfrac{30}{5} + \left(-\dfrac{24}{5}\right)$$
$$= -\dfrac{54}{5}$$
$$= -10\dfrac{4}{5}$$

77. $-36 \div (-2)(3) + \sqrt{169 - 25} + 12$
$$= -36 \div (-2)(3) + \sqrt{144} + 12$$
$$= -36 \div (-2)(3) + 12 + 12$$
$$= 18(3) + 12 + 12$$
$$= 54 + 12 + 12$$
$$= 78$$

79. $8 - 3\left[5 - (7 + 4)\right] - \sqrt{49} = 8 - 3\left[5 - (7 + 4)\right] - \sqrt{49}$
$$= 8 - 3(5 - 11) - \sqrt{49}$$
$$= 8 - 3(-6) - \sqrt{49}$$
$$= 8 - 3(-6) - 7$$
$$= 8 + 18 - 7$$
$$= 26 - 7$$
$$= 19$$

81. $\sqrt{71-35} - 3^2\left[6-(4-9)\right] + 4^3$

$= \sqrt{36} - 3^2\left[6-(-5)\right] + 4^3$

$= \sqrt{36} - 3^2(6+5) + 4^3$

$= \sqrt{36} - 3^2(11) + 4^3$

$= 6 - 9(11) + 64$

$= 6 - 99 + 64$

$= -93 + 64$

$= -29$

83. $\dfrac{9}{8}\cdot\left(-\dfrac{2}{3}\right) + \left(\dfrac{1}{5}-\dfrac{2}{3}\right) \div \sqrt{\dfrac{125}{5}}$

$\dfrac{9}{8}\cdot\left(-\dfrac{2}{3}\right) + \left(\dfrac{3}{15}-\dfrac{10}{15}\right) \div \sqrt{25}$

$\dfrac{9}{8}\cdot\left(-\dfrac{2}{3}\right) + \left(-\dfrac{7}{15}\right) \div \sqrt{25}$

$\dfrac{9}{8}\cdot\left(-\dfrac{2}{3}\right) + \left(-\dfrac{7}{15}\right) \div 5$

$-\dfrac{3}{4} + \left(-\dfrac{7}{15}\right)\cdot\dfrac{1}{5}$

$-\dfrac{3}{4} + \left(-\dfrac{7}{75}\right)$

$-\dfrac{225}{300} + \left(-\dfrac{28}{300}\right)$

$-\dfrac{253}{300}$

85. $\dfrac{2}{5} \div \left(-\dfrac{1}{10}\right)\cdot(-3) + \sqrt{64+36}$

$= \dfrac{2}{5} \div \left(-\dfrac{1}{10}\right)\cdot(-3) + \sqrt{100}$

$= \dfrac{2}{5}\cdot\left(-\dfrac{10}{1}\right)\cdot(-3) + 10$

$= -4(-3) + 10$

$= 12 + 10$

$= 22$

87. $-16\cdot\left(\dfrac{3}{4}\right) \div (-2) + \left|9 - 3(4+1)\right|$

$= -16\cdot\left(\dfrac{3}{4}\right) \div (-2) + \left|9 - 3(5)\right|$

$= -16\cdot\left(\dfrac{3}{4}\right) \div (-2) + \left|9 - 15\right|$

$= -16\cdot\left(\dfrac{3}{4}\right) \div (-2) + \left|-6\right|$

$= -16\cdot\left(\dfrac{3}{4}\right) \div (-2) + 6$

$= -\dfrac{16}{1}\cdot\dfrac{3}{4}\cdot\left(-\dfrac{1}{2}\right) + 6$

$= 6 + 6$

$= 12$

89. $\dfrac{7 - \left|4(5)-13\right|}{6^2 - 3(2-14)} = \dfrac{7 - \left|20-13\right|}{6^2 - 3(-12)}$

$= \dfrac{7 - \left|7\right|}{36 - 3(-12)}$

$= \dfrac{0}{36 - (-36)}$

$= \dfrac{0}{72}$

$= 0$

91. $\dfrac{4\left[5 - 8(2+1)\right]}{3 - 6 - (-4)^2} = \dfrac{4(5 - 8\cdot 3)}{3 - 6 - 16}$

$= \dfrac{4(5-24)}{-3-16}$

$= \dfrac{4(-19)}{-19}$

$= 4$

93. $\dfrac{5^3 - 3\left(4^3 - 41\right)}{19 - (3-10)^2 + 38} = \dfrac{5^3 - 3(64-41)}{19 - (-7)^2 + 38}$

$= \dfrac{5^3 - 3(23)}{19 - 49 + 38}$

$= \dfrac{125 - 3(23)}{-30 + 38}$

$= \dfrac{125 - 69}{8}$

$= \dfrac{56}{8}$

$= 7$

95. $\dfrac{4\left(5^2-10\right)+3}{\sqrt{25-16}-3}=\dfrac{4\left(25-10\right)+3}{\sqrt{9}-3}$

$=\dfrac{4\left(15\right)+3}{3-3}$

$=\dfrac{60+3}{0}$

Because the divisor is 0, the answer is undefined.

97. Associative Property of Multiplication. The multiplication was not performed from left to right.

99. Distributive Property. The parentheses were not simplified first.

101. Mistake: Multiplied before dividing.
Correction: $24\div4\cdot2-11=6\cdot2-11$
$=12-11$
$=1$

103. Mistake: Found the square roots of the subtrahend and minuend in square root of a difference.
Correction: $40\div2+\sqrt{25-9}=40\div2+\sqrt{16}$
$=40\div2+4$
$=20+4$
$=24$

105. To find the average, find the sum of her test scores and divide by 5, the number of scores.
$\dfrac{82+76+64+90+74}{5}=\dfrac{386}{5}=77.2$

107. To get an average of 90, he must have a total number of points equal to the 5 tests times 90, or 450 points. Since he already has $96+88+86+84=354$ points, subtract the points he has from the total points he needs to find the minimum last test score.
$450-354=96$

109. To find the average, find the sum of the average earnings in each month and divide by 12, the number of months in a year.
$577.98+578.29+583.42+583.05$
$+585.42+589.86+589.81+591.50$
$+592.85+593.87+596.23+598.60=7060.88$
$\dfrac{\$7060.88}{12}\approx\$588.41$

111. To find the average, find the sum of the stock prices and divide by 5, the number of days.
$\dfrac{74.46+74.76+74.70+73.05+74.74}{5}$
$=\dfrac{371.71}{5}$
$\approx\$74.34$

Review Exercises

1. $\{0, 1, 2, 3, 4, 5, 6, 7, 8, 9, 10\}$

2.

3. $-\dfrac{1}{9}$

4. It is an expression because it has no = sign.

5. $-4(3)(-2)=-12(-2)$
$=24$

6. $\dfrac{-15}{-5}=3$

Exercise Set 1.6

1. $4x$　　　3. $4x+16$　　　5. $7x-8$

7. $-4\div y^3$ or $\dfrac{-4}{y^3}$　　　　　9. $8p-4$

11. $\dfrac{m}{14}$　　　13. x^4+5　　　15. $7w-\dfrac{1}{5}$

17. $-3(n-2)$　　19. $(4+n)^5$　　21. $mn-5$

23. $4\div n-2$ or $\dfrac{4}{n}-2$　　25. $-27-(a+b)$

27. $0.6-4(y-2)$　　　29. $(p-q)-(m+n)$

31. $\sqrt{y}-mn$　　　　33. $n-3(n-6)$

35. Mistake: Order is incorrect.
Correct: $3t-17$

37. Mistake: Multiplied x by 9 instead of the sum of x and y.
Correct: $9(x+y)$

39. $w+5$　　　41. $3w$　　　43. $\dfrac{1}{2}d$

45. $42 - n$ 47. $t + \dfrac{1}{4}$ 49. $2w + 2l$

51. $\dfrac{1}{3}\pi r^2 h$ 53. mc^2

55. $\sqrt{\left(x_2 - x_1\right)^2 + \left(y_2 - y_1\right)^2}$

57. Mistake: Could be translated as $3(x+4)$.
    Correct: Four more than three times a number.

59. Mistake: Could be translated as $5x-1$.
    Correct: Five times the difference of x and one.

61. Mistake: Could be translated as $(n+5)(n-6)$.
    Correct: n plus the product of five and the difference of n and six.

63. One-half the product of the base and the height.

65. The product of the length, width, and height.

67. Product of the length and width added to the product of the length and height added to the product of the width and height, all doubled.

69. The ratio of the difference of y_2 and y_1 to the difference of x_2 and x_1.

Review Exercises

1. $2(5+2) = 2\cdot 5 + 2\cdot 2$
    $\qquad\quad = 10 + 4$
    $\qquad\quad = 14$

2. $4 - 8 + 6 = -4 + 6$
    $\qquad\qquad\;\; = 2$

3. $\dfrac{2}{3} - \dfrac{1}{4} = \dfrac{2(4)}{3(4)} - \dfrac{1(3)}{4(3)}$
    $\qquad\quad = \dfrac{8}{12} - \dfrac{3}{12}$
    $\qquad\quad = \dfrac{5}{12}$

4. $-6^2 \div 3 + 4(5-9) = -6^2 \div 3 + 4(-4)$
    $\qquad\qquad\qquad\quad = -36 \div 3 + 4(-4)$
    $\qquad\qquad\qquad\quad = -12 + (-16)$
    $\qquad\qquad\qquad\quad = -28$

5. $-4|3-8| + 2^5 = -4|-5| + 2^5$
    $\qquad\qquad\qquad = -4(5) + 2^5$
    $\qquad\qquad\qquad = -4(5) + 32$
    $\qquad\qquad\qquad = -20 + 32$
    $\qquad\qquad\qquad = 12$

6. $-3^2 = -3\cdot 3 = -9$

Exercise Set 1.7

1. Let $a = 2, b = 3$.
    $3(a+5) - 4b = 3(2+5) - 4(3)$
    $\qquad\qquad\quad = 3(7) - 4(3)$
    $\qquad\qquad\quad = 21 - 12$
    $\qquad\qquad\quad = 9$

3. Let $x = 1$.
    $4 - 0.2(x+3) = 4 - 0.2(1+3)$
    $\qquad\qquad\quad = 4 - 0.2(4)$
    $\qquad\qquad\quad = 4 - 0.8$
    $\qquad\qquad\quad = 3.2$

5. Let $p = -3$.
    $2p^2 - 3p - 4 = 2(-3)^2 - 3(-3) - 4$
    $\qquad\qquad\quad = 2(9) - 3(-3) - 4$
    $\qquad\qquad\quad = 18 - (-9) - 4$
    $\qquad\qquad\quad = 18 + 9 - 4$
    $\qquad\qquad\quad = 27 - 4$
    $\qquad\qquad\quad = 23$

7. Let $y = -\dfrac{1}{2}$.
    $2y^2 - 4y + 3 = 2\left(-\dfrac{1}{2}\right)^2 - 4\left(-\dfrac{1}{2}\right) + 3$
    $\qquad\qquad\quad = 2\left(\dfrac{1}{4}\right) - 4\left(-\dfrac{1}{2}\right) + 3$
    $\qquad\qquad\quad = \dfrac{1}{2} + 2 + 3$
    $\qquad\qquad\quad = \dfrac{1}{2} + \dfrac{4}{2} + \dfrac{6}{2}$
    $\qquad\qquad\quad = \dfrac{11}{2}$

9. Let $y = -2.3$.
$$8 - 2(y + 4) = 8 - 2(-2.3 + 4)$$
$$= 8 - 2(1.7)$$
$$= 8 - 3.4$$
$$= 4.6$$

11. Let $x = 5$, $y = -1$.
$$-|3x| + |4y^3| = -|3(5)| + |4(-1)^3|$$
$$= -|15| + |4(-1)|$$
$$= -15 + |-4|$$
$$= -15 + 4$$
$$= -11$$

13. Let $r = -3$, $q = 5$.
$$|2r^2 - 3q| = |2(-3)^2 - 3(5)|$$
$$= |2(9) - 3(5)|$$
$$= |18 - 15|$$
$$= |3|$$
$$= 3$$

15. Let $a = -1$, $b = -2$, $c = 16$.
$$\sqrt{c} + 4ab^2 = \sqrt{16} + 4(-1)(-2)^2$$
$$= 4 + 4(-1)(4)$$
$$= 4 - 16$$
$$= -12$$

17. Let $a = 25$, $b = 4$.
$$-5\sqrt{a} + 2\sqrt{b} = -5\sqrt{25} + 2\sqrt{4}$$
$$= -5(5) + 2(2)$$
$$= -25 + 4$$
$$= -21$$

19. Let $x = 2$, $y = 7$.
$$\frac{6x^3}{y - 8} = \frac{6(2)^3}{7 - 8} = \frac{6(8)}{-1} = \frac{48}{-1} = -48$$

21. Let $b = 3$, $c = 25$, $d = 144$.
$$\frac{15 - b^2}{2\sqrt{c + d}} = \frac{15 - (3)^2}{2\sqrt{25 + 144}}$$
$$= \frac{15 - 9}{2\sqrt{169}}$$
$$= \frac{6}{2(13)}$$
$$= \frac{6}{26}$$
$$= \frac{3}{13}$$

23. a) Let $a = 2$, $b = 1$, $c = -3$.
$$b^2 - 4ac = 1^2 - 4(2)(-3)$$
$$= 1 - 4(2)(-3)$$
$$= 1 + 24$$
$$= 25$$
   b) Let $a = 1$, $b = -2$, $c = 4$.
$$b^2 - 4ac = (-2)^2 - 4(1)(4)$$
$$= 4 - 4(1)(4)$$
$$= 4 - 16$$
$$= -12$$

25. a) Let $x_1 = 2$, $y_1 = 1$, $x_2 = 5$, $y_2 = 7$
$$\frac{y_2 - y_1}{x_2 - x_1} = \frac{7 - 1}{5 - 2} = \frac{6}{3} = 2$$
   b) Let $x_1 = -1$, $y_1 = 2$, $x_2 = -7$, $y_2 = -2$
$$\frac{y_2 - y_1}{x_2 - x_1} = \frac{-2 - 2}{-7 - (-1)} = \frac{-4}{-6} = \frac{2}{3}$$

27. If $y = -5$, we have $\dfrac{-7}{5 + (-5)} = \dfrac{-7}{0}$, which is undefined because the denominator is 0.

29. If $m = -1$, we have
$$\frac{6(-1)}{(-1 + 1)(-1 - 3)} = \frac{-6}{(0)(-4)} = \frac{-6}{0},$$ which is undefined. Also, if $m = 3$, we have
$$\frac{6(3)}{(3 + 1)(3 - 3)} = \frac{18}{(4)(0)} = \frac{18}{0},$$ which is undefined.

31. If $x = 0$, we have $\dfrac{6 + 0^2}{0} = \dfrac{6}{0}$, which is undefined because the denominator is 0.

33. If $x = \dfrac{2}{3}$, we have $\dfrac{\dfrac{2}{3}+1}{3\left(\dfrac{2}{3}\right)-2} = \dfrac{\dfrac{5}{3}}{2-2} = \dfrac{\dfrac{5}{3}}{0}$, which

is undefined because the denominator is 0.

35. $6(a+2) = 6 \cdot a + 6 \cdot 2$
$\qquad = 6a + 12$

37. $-8(4-3y) = -8 \cdot 4 - (-8) \cdot (3y)$
$\qquad = -32 + 24y$

39. $\dfrac{7}{8}\left(\dfrac{1}{2}c-16\right) = \dfrac{7}{8} \cdot \dfrac{1}{2}c + \dfrac{7}{8} \cdot \left(-\dfrac{16}{1}\right)$
$\qquad\qquad = \dfrac{7}{16}c - 14$

41. $0.2(3n-8) = 0.2 \cdot 3n + 0.2 \cdot (-8)$
$\qquad\qquad = 0.6n - 1.6$

43. -6 $\qquad$ 45. 1 $\qquad$ 47. -1

49. $-\dfrac{2}{3}$ $\qquad$ 51. $\dfrac{1}{5}$

53. $3y + 5y = 8y$

55. $6a - 11a = -5a$

57. $14x - 3x = 11x$

59. $-3r - 8r = -11r$

61. $6.3n - 8.2n = -1.9n$

63. $\dfrac{1}{2}b^2 - \dfrac{5}{6}b^2 = \dfrac{1(3)}{2(3)}b^2 - \dfrac{5}{6}b^2$
$\qquad\qquad = \dfrac{3}{6}b^2 - \dfrac{5}{6}b^2$
$\qquad\qquad = -\dfrac{2}{6}b^2$
$\qquad\qquad = -\dfrac{1}{3}b^2$

65. $-11c - 10c - 5c = -21c - 5c$
$\qquad\qquad\qquad = -26c$

67. $6x + 7 - x - 8 = 6x - x + 7 - 8$
$\qquad\qquad\qquad = 5x - 1$

69. $-8x + 3y - 7 - 2x + y + 6$
$\quad = -8x - 2x + 3y + y - 7 + 6$
$\quad = -10x + 4y - 1$

71. $-10m + 7n + 1 + m - 2n - 12 + y$
$\quad = -10m + m + 7n - 2n + y + 1 - 12$
$\quad = -9m + 5n + y - 11$

73. $1.5x + y - 2.8x + 0.3 - y - 0.7$
$\quad = 1.5x - 2.8x + y - y + 0.3 - 0.7$
$\quad = -1.3x - 0.4$

75. $\dfrac{1}{6}c + 3d + \dfrac{2}{3}c + \dfrac{1}{7} - \dfrac{1}{2}d$
$\quad = \dfrac{1}{6}c + \dfrac{2}{3}c + 3d - \dfrac{1}{2}d + \dfrac{1}{7}$
$\quad = \dfrac{1}{6}c + \dfrac{2(2)}{3(2)}c + \dfrac{3(2)}{1(2)}d - \dfrac{1}{2}d + \dfrac{1}{7}$
$\quad = \dfrac{1}{6}c + \dfrac{4}{6}c + \dfrac{6}{2}d - \dfrac{1}{2}d + \dfrac{1}{7}$
$\quad = \dfrac{5}{6}c + \dfrac{5}{2}d + \dfrac{1}{7}$

77. $7a + \dfrac{4}{5}b^2 - 5 - \dfrac{3}{4}a - \dfrac{2}{7}b^2 + 9$
$\quad = 7a - \dfrac{3}{4}a + \dfrac{4}{5}b^2 - \dfrac{2}{7}b^2 - 5 + 9$
$\quad = \dfrac{7(4)}{1(4)}a - \dfrac{3}{4}a + \dfrac{4(7)}{5(7)}b^2 - \dfrac{2(5)}{7(5)}b^2 - 5 + 9$
$\quad = \dfrac{28}{4}a - \dfrac{3}{4}a + \dfrac{28}{35}b^2 - \dfrac{10}{35}b^2 - 5 + 9$
$\quad = \dfrac{25}{4}a + \dfrac{18}{35}b^2 + 4$

79. a) $14 + (6n - 8n)$
   b) $14 - 2n$
   c) Let $n = -3$
   $\qquad 14 - 2n = 14 - 2(-3)$
   $\qquad\qquad\quad = 14 + 6$
   $\qquad\qquad\quad = 20$

Review Exercises

1. $-3,\ -2.5,\ 4.2,\ 4\dfrac{5}{8},\ |-6|$

2. $\dfrac{72}{420} = \dfrac{72 \div 12}{420 \div 12} = \dfrac{6}{35}$

3. $-8 + (-14) = -22$

4. $\dfrac{1}{2}(12.5)(8) = 50$

5. $4(-2) + 9 = -8 + 9 = 1$

6. $\dfrac{3}{4}\left(5\dfrac{1}{3}\right)-1=\dfrac{3}{4}\left(\dfrac{16}{3}\right)-1$

$\qquad\qquad\quad=\dfrac{\cancel{3}}{\cancel{4}}\left(\dfrac{\cancel{4}\cdot 4}{\cancel{3}}\right)-1$

$\qquad\qquad\quad=4-1$

$\qquad\qquad\quad=3$

Chapter 1 Review Exercises

1. Equations and inequalities

   Expressions

   Constants and Variables

2. {March, May}   3. {2, 4, 6, ...}

4.   Real Numbers

Rational numbers	Irrational numbers
Integers	
Whole numbers	
Natural numbers	

5. The number $-3\dfrac{2}{5}$ is located $\dfrac{2}{5}$ of the way between -3 and -4, so we divide the space between -3 and -4 into 5 equal divisions and place a dot on the 2nd mark to the left of -3.

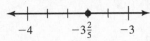

6. The number 8.2 is located $0.2=\dfrac{2}{10}$ of the way between 8 and 9, so we divide the space between 8 and 9 into 10 equal divisions and place a dot on the 2nd mark to the right of 8.

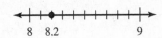

7. $\left|-6.3\right|=6.3$

8. $\left|2\dfrac{1}{6}\right|=2\dfrac{1}{6}$

9. $\left|-6.4\right|=6.4$

10. $-14<-9$

11. $-\dfrac{6}{7}=-\dfrac{?}{14}$

    $-\dfrac{6\cdot 2}{7\cdot 2}=-\dfrac{12}{14}$

    The missing number is 12.

12. $\dfrac{15}{30}=\dfrac{?}{10}$

    $\dfrac{15\div 3}{30\div 3}=\dfrac{5}{10}$

    The missing number is 5.

13. $\dfrac{5}{6}$ and $\dfrac{3}{8}$

    The LCD is 24.

    $\dfrac{5\cdot 4}{6\cdot 4}=\dfrac{20}{24}$ and $\dfrac{3\cdot 3}{8\cdot 3}=\dfrac{9}{24}$

14. $-\dfrac{7}{15}$ and $\dfrac{9}{10}$

    The LCD is 30.

    $-\dfrac{7\cdot 2}{15\cdot 2}=-\dfrac{14}{30}$ and $\dfrac{9\cdot 3}{10\cdot 3}=\dfrac{27}{30}$

15. $100=2\cdot 50=2\cdot 2\cdot 25=2\cdot 2\cdot 5\cdot 5$

16. $84=2\cdot 42=2\cdot 2\cdot 21=2\cdot 2\cdot 3\cdot 7$

17. $\dfrac{152}{200}=\dfrac{\cancel{2}\cdot\cancel{2}\cdot\cancel{2}\cdot 19}{\cancel{2}\cdot\cancel{2}\cdot\cancel{2}\cdot 5\cdot 5}=\dfrac{19}{25}$

18. $\dfrac{26}{39}=\dfrac{2\cdot\cancel{13}}{3\cdot\cancel{13}}=\dfrac{2}{3}$

19. Associative Property of Addition

20. Commutative Property of Addition

21. $-15+\left(-12\right)=-15-12=-27$

22. $-\dfrac{1}{4}+\dfrac{2}{3}=-\dfrac{1\cdot 3}{4\cdot 3}+\dfrac{2\cdot 4}{3\cdot 4}=-\dfrac{3}{12}+\dfrac{8}{12}=\dfrac{5}{12}$

23. $-8-\left(-4.2\right)=-8+4.2$

    $\qquad\qquad\qquad=-3.8$

24. $6-\left(-9.1\right)=6+9.1$

    $\qquad\qquad\quad=15.1$

25. $\left(-685.92\right)+\left(-45.80\right)+\left(250.00\right)+\left(-36.45\right)$

    $+\left(-12.92\right)+\left(32.68\right)+\left(-5.18\right)=-\$503.59$

26. a) $\$58.92-\$59.36=-\$0.44$

    It closed $0.44 lower at the end of the week.

b) $\$58.22 - \$58.70 = -\$0.48$
One would lose $0.48 per share.

27. Commutative Property of Multiplication

28. Distributive Property

29. Commutative Property of Multiplication

30. Associative Property of Multiplication

31. $-7(-13) = 91$

32. $6(-8) = -48$

33. $-\dfrac{3}{8}\left(\dfrac{5}{9}\right) = -\dfrac{3\cdot 5}{8\cdot 9} = -\dfrac{\cancel{3}\cdot 5}{2\cdot 2\cdot 2\cdot \cancel{3}\cdot 3} = -\dfrac{5}{24}$

34. $(-6.5)(-0.4) = 2.6$

35. $-25 \div 5 = -5$

36. $-30 \div (-3) = 10$

37. $-\dfrac{7}{12} \div \left(-\dfrac{3}{8}\right) = -\dfrac{7}{12}\cdot\left(-\dfrac{8}{3}\right)$

$\quad\quad = -\dfrac{7}{\cancel{2}\cdot\cancel{2}\cdot 3}\cdot\left(-\dfrac{\cancel{2}\cdot\cancel{2}\cdot 2}{3}\right)$

$\quad\quad = -\dfrac{7}{3}\cdot\left(-\dfrac{2}{3}\right)$

$\quad\quad = \dfrac{14}{9} \text{ or } 1\dfrac{5}{9}$

38. $30.66 \div (-7.3) = -4.2$

39. a) $200(59.36) = \$11,872$

b) First find the amount the 200 shares were worth on 6/15: $200(58.92) = \$11,784$. Now subtract the purchase price from the selling price to find the loss: $\$11,784 - \$11,872 = -\$88$, or a loss of $88.

40. You must divide the current loan balance by the original loan balance.
$(-14{,}896.08) \div (-2405.80) \approx 6.2$

41. $(-3)^2 = -3\cdot(-3) = 9$

42. $-6^2 = -(6\cdot 6) = -36$

43. ± 7

44. No real-number square roots exist.

45. $\sqrt{196} = 14$

46. $\sqrt{\dfrac{25}{81}} = \dfrac{\sqrt{25}}{\sqrt{81}} = \dfrac{5}{9}$

47. $(-2)^3(4)(-5) = -8(4)(-5)$
$\quad\quad = -32(-5)$
$\quad\quad = 160$

48. $3^5 \div 3^2 \div 3^2 \div 3 = 243 \div 9 \div 9 \div 3$
$\quad\quad = 27 \div 9 \div 3$
$\quad\quad = 3 \div 3$
$\quad\quad = 1$

49. $-\dfrac{1}{3} - \dfrac{3}{2} \div \dfrac{1}{6} = -\dfrac{1}{3} - \dfrac{3}{2}\cdot\dfrac{6}{1}$
$\quad\quad = -\dfrac{1}{3} - \dfrac{3}{\cancel{2}}\cdot\dfrac{\cancel{2}\cdot 3}{1}$
$\quad\quad = -\dfrac{1}{3} - \dfrac{9}{1}$
$\quad\quad = -\dfrac{1}{3} - \dfrac{9(3)}{1(3)}$
$\quad\quad = -\dfrac{1}{3} - \dfrac{27}{3}$
$\quad\quad = -\dfrac{28}{3}$

50. $4|-2| \div (-8) = 4(2) \div (-8)$
$\quad\quad = 8 \div (-8)$
$\quad\quad = -1$

51. $10 \div 5 + (-5)(-5) = 2 + 25$
$\quad\quad = 27$

52. $6 + \{3(4-5) + 2[6 + (-4)]\} = 6 + [3(-1) + 2(2)]$
$\quad\quad = 6 + (-3 + 4)$
$\quad\quad = 6 + 1$
$\quad\quad = 7$

53. $-1 - 3[4^2 - 3(-6)] = -1 - 3[16 - 3(-6)]$
$\quad\quad = -1 - 3(16 + 18)$
$\quad\quad = -1 - 3(34)$
$\quad\quad = -1 - 102$
$\quad\quad = -103$

54. $\dfrac{-6+3^2(8-10)}{2^3-16} = \dfrac{-6+3^2(-2)}{8-16}$

$= \dfrac{-6+9(-2)}{-8}$

$= \dfrac{-6-18}{-8}$

$= \dfrac{-24}{-8}$

$= 3$

55. $\dfrac{3(4-5)-4\cdot3+(11-20)}{(3-4)^3} = \dfrac{3(-1)-4\cdot3+(-9)}{(-1)^3}$

$= \dfrac{-3-12-9}{-1}$

$= \dfrac{-24}{-1}$

$= 24$

56. To find the average, add the monthly revenues and divide by 6, the number of months.

$\dfrac{\left(\begin{array}{c}45,320+38,250+61,400+42,500\\+74,680+62,800\end{array}\right)}{6}$

$= \dfrac{324,950}{6}$

$\approx \$54,158.33$

57. $14-2n$

58. $\dfrac{y}{7}$

59. $y-2(y+4)$

60. $7(m-n)$

61. Let $a=-3, b=-1, c=5$

$b^2-4ac = (-1)^2 - 4(-3)(5)$

$= 1-4(-3)(5)$

$= 1+60$

$= 61$

62. Let $a=2, b=6, c=-2$

$b^2-4ac = (6)^2 - 4(2)(-2)$

$= 36-4(2)(-2)$

$= 36+16$

$= 52$

63. Let $x=-3$, $y=-2$.

$-|4x|+|y^2| = -|4(-3)|+|(-2)^2|$

$= -|-12|+|4|$

$= -12+4$

$= -8$

64. Let $m=16$, $n=9$.

$\sqrt{m}+\sqrt{n} = \sqrt{16}+\sqrt{9}$

$= 4+3$

$= 7$

65. Let $m=16$, $n=9$.

$\sqrt{m+n} = \sqrt{16+9}$

$= \sqrt{25}$

$= 5$

66. Let $l=-4$, $n=16$.

$\dfrac{3l^2}{4-n} = \dfrac{3(-4)^2}{4-16} = \dfrac{3\cdot16}{-12} = \dfrac{48}{-12} = -4$

67. If $n=-6$, we have $\dfrac{-6}{-6+6} = \dfrac{-6}{0}$, which is undefined because the denominator is 0.

68. If $y=-3$, we have

$\dfrac{-3}{(-3-4)(-3+3)} = \dfrac{-3}{(-7)(0)} = \dfrac{-3}{0}$, which is undefined. If $y=4$, we have

$\dfrac{4}{(4-4)(4+3)} = \dfrac{4}{(0)(7)} = \dfrac{4}{0}$, which is undefined.

69. $5(x+6) = 5\cdot x + 5\cdot 6$

$= 5x+30$

70. $-3(5n-8) = -3\cdot5n - (-3)\cdot8$

$= -15n+24$

71. $\dfrac{1}{4}(8y+3) = \dfrac{1}{4}\cdot8y + \dfrac{1}{4}\cdot3$

$= 2y+\dfrac{3}{4}$

72. $0.6(4.5m-2.1) = 0.6\cdot4.5m - 0.6\cdot(2.1)$

$= 2.7m-1.26$

73. $6x+3y-9x-6y-15$

$= 6x-9x+3y-6y-15$

$= -3x-3y-15$

74. $5y^2 - 6y + 4y - y^2 + 9y$

    $= 5y^2 - y^2 - 6y + 4y + 9y$

    $= 4y^2 + 7y$

75. $-6xy + 9xy - xy = 2xy$

76. $8x^3 - 4x - 6x^2 - 10x^3 + 4x$

    $= 8x^3 - 10x^3 - 6x^2 - 4x + 4x$

    $= -2x^3 - 6x^2$

77. $-4m - 4m - 4n + 4n = -8m$

78. $14 - 6x + 4y - 8 - 10y - 8x$

    $= -6x - 8x + 4y - 10y + 14 - 8$

    $= -14x - 6y + 6$

Chapter 1 Practice Test

1. The number $-3\frac{2}{7}$ is located $\frac{2}{7}$ of the way between -3 and -4, so we divide the space between -3 and -4 into 7 equal divisions and place a dot on the 2nd mark to the left of -3.

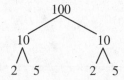

2. $|-3.67| = 3.67$ because -3.67 is 3.67 units from 0 on a number line.

3. Use a factor tree.

                100
          /   \
      10   10
   / \   / \
  2   5   2   5

    $100 = 2 \cdot 2 \cdot 5 \cdot 5$

4. $\dfrac{17}{51} = \dfrac{1 \cdot \cancel{17}}{3 \cdot \cancel{17}} = \dfrac{1}{3}$

5. $-\dfrac{6}{5} = -\dfrac{?}{10}$

    $-\dfrac{6 \cdot 2}{5 \cdot 2} = -\dfrac{12}{10}$

    The missing number is 12.

6. Distributive Property

7. Commutative Property of Addition because the order of the addends is changed.

8. $8 + (-4) = 4$

9. $\dfrac{7}{8} - \left(-\dfrac{5}{6}\right) = \dfrac{7}{8} + \dfrac{5}{6}$

    $= \dfrac{7 \cdot 3}{8 \cdot 3} + \dfrac{5 \cdot 4}{6 \cdot 4}$

    $= \dfrac{21}{24} + \dfrac{20}{24}$

    $= \dfrac{41}{24} = 1\dfrac{17}{24}$

10. $(-1.5)(-0.4) = 0.6$

11. $-\dfrac{5}{6} \div \dfrac{2}{3} = -\dfrac{5}{6} \cdot \dfrac{3}{2}$

    $= -\dfrac{5}{\cancel{6}_{2}} \cdot \dfrac{\cancel{3}^{1}}{2}$

    $= -\dfrac{5}{4}$

12. $(-4)^3 = (-4)(-4)(-4) = -64$

13. $-3^4 = -3 \cdot 3 \cdot 3 \cdot 3 = -81$

14. $-12 \div 3(6 - 2^2) = -12 \div 3(6 - 4)$

                    $= -12 \div 3(2)$

                    $= -4(2)$

                    $= -8$

15. $\dfrac{6^2 + 14}{(2 - 7)^2} = \dfrac{36 + 14}{(-5)^2}$

          $= \dfrac{50}{25}$

          $= 2$

16. $-8 \div |4 - 2| + 3^2 = -8 \div |2| + 3^2$

                     $= -8 \div 2 + 9$

                     $= -4 + 9$

                     $= 5$

17. $\sqrt{25 - 16} + \left[(14 + 2) - 3^2 \right] = \sqrt{9} + \left(16 - 3^2 \right)$

                                          $= 3 + (16 - 9)$

                                          $= 3 + 7$

                                          $= 10$

18. $2(m + n)$ 19. $3w - 5$

20. $-854.80 - (-1104.80) = -854.80 + 1104.80$
$$= \$250$$

21. To find the average, add the monthly snowfall measurements and divide by 6, the number of snowfalls. $\dfrac{6 + 10 + 4 + 3 + 2.5 + 8}{6} = \dfrac{33.5}{6}$
$$= 5.58\overline{3} \text{ inches}$$

22. $-\left|3(-1)^2 + 2 \cdot 4\right| = -\left|3 \cdot 1 + 2 \cdot 4\right|$
$$= -\left|3 + 8\right|$$
$$= -\left|11\right|$$
$$= -11$$

23. $\sqrt{64 - (-36)} = \sqrt{64 + 36}$
$$= \sqrt{100}$$
$$= 10$$

24. $-5(4y + 9) = -5 \cdot 4y + (-5) \cdot 9 = -20y - 45$

25. $3.5x - 8 + 2.1x + 9.3 = 3.5x + 2.1x - 8 + 9.3$
$$= 5.6x + 1.3$$

Chapter 2

Solving Linear Equations and Inequalities

Exercise Set 2.1

1. For $3x-5=41$, let $x=2$.

   $3(2)-5 \overset{?}{=} 41$

   $6-5 \overset{?}{=} 41$

   $1 \ne 41$

   No, $x=2$ is not a solution.

3. For $7y-1=y+3$, let $y=\dfrac{2}{3}$.

   $7\left(\dfrac{2}{3}\right)-1 \overset{?}{=} \left(\dfrac{2}{3}\right)+3$

   $\dfrac{14}{3}-\dfrac{3}{3} \overset{?}{=} \dfrac{2}{3}+\dfrac{9}{3}$

   $\dfrac{11}{3} = \dfrac{11}{3}$

   Yes, $y=\dfrac{2}{3}$ is a solution.

5. For $-4(x-5)+2=5(x-1)+3$, let $x=-2$.

   $-4(-2-5)+2 \overset{?}{=} 5(-2-1)+3$

   $-4(-7)+2 \overset{?}{=} 5(-3)+3$

   $28+2 \overset{?}{=} -15+3$

   $30 \ne -12$

   No, $x=-2$ is not a solution.

7. For $-\dfrac{1}{2}+\dfrac{1}{3}y=\dfrac{3}{2}$, let $y=6$.

   $-\dfrac{1}{2}+\dfrac{1}{3}\left(\dfrac{6}{1}\right) \overset{?}{=} \dfrac{3}{2}$

   $-\dfrac{1}{2}+2 \overset{?}{=} \dfrac{3}{2}$

   $-\dfrac{1}{2}+\dfrac{4}{2} \overset{?}{=} \dfrac{3}{2}$

   $\dfrac{3}{2} = \dfrac{3}{2}$

   Yes, $y=6$ is a solution.

9. For $4.3z-5.71=2.1z+0.07$, let $z=2.2$.

   $4.3(2.2)-5.71 \overset{?}{=} 2.1(2.2)+0.07$

   $9.46-5.71 \overset{?}{=} 4.62+0.07$

   $3.75 \ne 4.69$

   No, $z=2.2$ is not a solution.

11. For $b^2-4b=b^3+6$, let $b=-1$.

    $(-1)^2-4(-1) \overset{?}{=} (-1)^3+6$

    $1-4(-1) \overset{?}{=} -1+6$

    $1+4 \overset{?}{=} 5$

    $5 = 5$

    Yes, $b=-1$ is a solution.

13. For $\left|x^2-9\right|=-x+3$, let $x=3$.

    $\left|(3)^2-9\right| \overset{?}{=} -(3)+3$

    $\left|9-9\right| \overset{?}{=} 0$

    $0 = 0$

    Yes, $x=3$ is a solution.

15. For $\dfrac{4x-3}{x-4}=\sqrt{x+8}$, let $x=17$.

    $\dfrac{4(17)-3}{(17)-4} \overset{?}{=} \sqrt{(17)+8}$

    $\dfrac{68-3}{13} \overset{?}{=} \sqrt{25}$

    $\dfrac{65}{13} \overset{?}{=} 5$

    $5 = 5$

    Yes, $x=17$ is a solution.

17. a) We must find the perimeter of a rectangle.
       Let $l=20$ ft. and $w=13$ ft.
       $P=2l+2w$
       $P=2(20)+2(13)$
       $P=40+26$
       $P=66$ ft.
    b) $66 \div 12 = 5.5$
       Since you cannot buy half of a roll, she must purchase 6 rolls.
    c) $6(\$8.99)=\$53.94$

19. We must use the formula for the circumference of a circle. Let $r = 2$ km.

    $C = 2\pi r$

    $C = 2\pi(2)$

    $C \approx 12.57$ km

21. We must find the area of a rectangle. Let $l = 100$ ft. and $w = 70$ ft.

    $A = lw$

    $A = (100)(70)$

    $A = 7000$ ft.^{2}

23. Begin by finding the area (in square feet) of the wall. Let $l = 32.5$ ft. and $w = 12$ ft.

    $A = lw$

    $A = (32.5)(12)$

    $A = 390$ ft.^{2}

    Now multiply the area of the wall in square feet by the cost per square foot to paint the wall to obtain the total cost to have the wall painted.

    $390(\$2.50) = \$975$.

25. a) We must find the area of a rectangle. Let $l = 13$ ft. and $w = 6$ ft.

    $A = lw$

    $A = (13)(6)$

    $A = 78$ ft.^{2}

    b) $78 \div 10 = 7.8$

    Since you cannot buy part of a bale, she must purchase 8 bales of pine straw.

    c) $8(\$4.50) = \$36$

27. a) We must find the area of a triangle. Let $b = 26$ in. and $h = 20$ in.

    $A = \dfrac{1}{2}bh$

    $A = \dfrac{1}{2}(26)(20)$

    $A = 13(20)$

    $A = 260$ in.^{2}

    b) Find the cost of one sign:

    $260(\$0.75) = \$195$.

    Multiply the cost of one sign by 20 to get the total cost: $\$195(20) = \$3900$.

29. a) Begin by finding the area of the yard if the storage building were not there and the area of the storage building.

    Area of yard: $A = lw$

    $A = (82)(40)$

    $A = 3280$ ft.^{2}

    Area of building: $A = lw$

    $A = (8)(9)$

    $A = 72$ ft.^{2}

    Subtract the area of the building from the area of the yard: $3280 - 72 = 3208$ ft.^{2}

    b) Now multiply the area you just found by the fraction of the area to be covered with grass.

    $(3208)\left(\dfrac{2}{3}\right) = 2138\dfrac{2}{3}$ ft.^{2}

    c) Divide the area to be covered with grass by the number of square feet covered with one pallet: $2138\dfrac{2}{3} \div 504 \approx 4.24$. Since you cannot buy part of a pallet, they must purchase 5 pallets of grass sod.

    d) Multiply the number of pallets needed by the cost per pallet.

    $5(\$106) = \$530$

31. Begin by finding the area of the rectangle in square feet and area of one circle in square feet. Keep in mind that the radius is half of the diameter of a circle.

    Area of rectangle: $A = lw$

    $A = (8)(4)$

    $A = 32$ ft.^{2}

    Area of circle: $A = \pi r^2$

    $A = \pi(2)^2$

    $A \approx 12.57$ ft.^{2}

    Now multiply the area of the circle by 2 because there are two circles.

    $12.57 \cdot 2 = 25.14$ ft.^{2}

    Finally, subtract the area of the circles from the area of the rectangle.

    $32 - 25.14 \approx 6.9$ ft.^{2}

33. Begin by finding the area of the wall in square feet if the door were not there, as well as the area of the door in square feet.
Height of top part of wall: $9 - 5.5 = 3.5$ ft.
Area of wall:

$$A_1 = \frac{1}{2}h(a+b) \qquad A_2 = lw$$
$$A_1 = \frac{1}{2}(3.5)(8+12) \qquad A_2 = (12)(5.5)$$
$$A_1 = \frac{1}{2}(3.5)(20) \qquad A_2 = 66 \text{ ft.}^2$$
$$A_1 = 35 \text{ ft.}^2$$

Total wall area: $66 + 35 = 101$ sq. ft.
Area of door: $A = lw$

$$A = (7.5)(2.5)$$
$$A = 18.75 \text{ ft.}^2$$

Subtract the area of the door from the area of the wall.

$$101 - 18.75 = 82.25 \text{ ft.}^2$$

35. Use the formula for the volume of a box.
$V = lwh$

$$V = (2)(5)(4.75)$$
$$V = 47.5 \text{ m}^3$$

37. First find the radius.
$8 \div 2 = 4$ cm.
Then use the formula for the volume of a cylinder.

$$V = \pi r^2 h$$
$$V = \pi (4)^2 (20)$$
$$V = \pi (16)(20)$$
$$V \approx 1005.3 \text{ cm}^3$$

39. Use the formula for the volume of a pyramid.

$$V = \frac{1}{3}lwh$$
$$V = \frac{1}{3}(745)(745)(449)$$
$$V = 83,068,741.\overline{6} \text{ ft.}^3$$

or $83,068,741\frac{2}{3} \text{ ft.}^3$

41. Begin by finding the total drive time. Between 8:00 A.M. and 5:30 P.M. is 9 hours and 30 minutes. Takingout the two 15-minute breaks and one hour for lunch leaves a total drive time of 8 hours. We are looking for an average rate, so use the formula $r = \dfrac{d}{t}$.

$$r = \frac{d}{t}$$
$$r = \frac{516}{8}$$
$$r = 64.5 \text{ mph}$$

43. Begin by finding the total drive time. Between 10:00 A.M. and 4:00 P.M. is 6 hours. Taking out the three 15-minute stops (0.25 hour each) leaves a total drive time of 5.25 hours. We are looking for an average rate, so use the formula $r = \dfrac{d}{t}$.

$$r = \frac{d}{t}$$
$$r = \frac{285}{5.25}$$
$$r \approx 54.3 \text{ mph}$$

45. Since the flight begins in EST and ends in PST, you must add 3 hours to the difference between arrival and departure: $10 - 8.5 + 3 = 4.5$ hours
$d = rt$

$$d = \left(388\frac{1}{3}\right)(4.5)$$
$$d = 1747.5 \text{ miles}$$

47. $V = iR$
$V = (-6.5)(8)$
$V = -52$ V

49. $F = \dfrac{9}{5}(34) + 32$
$F = 61.2 + 32$
$F = 93.2° \text{ F}$

51. $F = \dfrac{9}{5}(-183) + 32$
$F = -329.4 + 32$
$F = -297.4° \text{ F}$

53. $C = \dfrac{5}{9}(-76 - 32)$
$C = \dfrac{5}{9}(-108)$
$C = -60° \text{ C}$

Review Exercises

1. $6(-3-3)+5(-3)=6(-6)+5(-3)$
    $$=-36+(-15)$$
    $$=-51$$

2. $6\left(2+3^2\right)-4(2+3)^2=6(2+9)-4(5)^2$
    $$=6(11)-4(25)$$
    $$=66-100$$
    $$=-34$$

3. $\dfrac{1}{3}(x+6)=\dfrac{1}{3}\cdot x+\dfrac{1}{3}\cdot 6$
    $$=\dfrac{1}{3}x+2$$

4. $-(4w-6)=-1(4w-6)$
    $$=(-1)\cdot 4w-(-1)\cdot 6$$
    $$=-4w+6$$

5. $5x-14+3x+9=5x+3x-14+9$
    $$=8x-5$$

6. $6.2y+7-1.5-0.8y=6.2y-0.8y+7-1.5$
    $$=5.4y+5.5$$

Exercise Set 2.2

1. Yes, because the variable terms contain a single variable and have an exponent of 1.

3. No, because the variable term has an exponent of 2.

5. No, because there are variable terms with exponents greater than 1.

7. Yes, because the variable terms contain a single variable and have an exponent of 1.

9. Yes, because the variable terms contain a single variable and have an exponent of 1.

11. No, because there are variable terms with exponents greater than 1.

13. Yes, because the variable term contains a single variable and has an exponent of 1.

15. Yes, because the variable terms contain a single variable and have an exponent of 1.

17. Solve the equation for x.
    $$\begin{aligned} x-7 &= 2 \\ \underline{+7}\quad &\underline{+7} \\ x+0 &= 9 \\ x &= 9 \end{aligned}$$
    Check: $9-7 \overset{?}{=} 2$
    $$2 = 2$$

19. Solve the equation for m.
    $$\begin{aligned} m-8 &= -3 \\ \underline{+8}\quad &\underline{+8} \\ m+0 &= 5 \\ m &= 5 \end{aligned}$$
    Check: $5-8 \overset{?}{=} -3$
    $$-3 = -3$$

21. Solve the equation for r.
    $$\begin{aligned} r-2 &= -9 \\ \underline{+2}\quad &\underline{+2} \\ r+0 &= -7 \\ r &= -7 \end{aligned}$$
    Check: $-7-2 \overset{?}{=} -9$
    $$-9 = -9$$

23. Solve the equation for x.
    $$\begin{aligned} x+7 &= 12 \\ \underline{-7}\quad &\underline{-7} \\ x+0 &= 5 \\ x &= 5 \end{aligned}$$
    Check: $5+7 \overset{?}{=} 12$
    $$12 = 12$$

25. Solve the equation for n.
    $$\begin{aligned} n+12 &= 2 \\ \underline{-12}\quad &\underline{-12} \\ n+0 &= -10 \\ n &= -10 \end{aligned}$$
    Check: $-10+12 \overset{?}{=} 2$
    $$2 = 2$$

27. Solve the equation for y.

$$-16 = y + 9$$

$$\underline{ -9 \qquad -9}$$

$$-25 = y + 0$$

$$-25 = y$$

Check: $-16 \stackrel{?}{=} -25 + 9$

$-16 = -16$

29. Solve the equation for m.

$$m + \frac{7}{8} = \frac{4}{5}$$

$$m + \frac{7}{8} - \frac{7}{8} = \frac{4}{5} - \frac{7}{8}$$

$$m + 0 = \frac{32}{40} - \frac{35}{40}$$

$$m = -\frac{3}{40}$$

Check: $-\frac{3}{40} + \frac{7}{8} \stackrel{?}{=} \frac{4}{5}$

$-\frac{3}{40} + \frac{35}{40} \stackrel{?}{=} \frac{4}{5}$

$\frac{32}{40} \stackrel{?}{=} \frac{4}{5}$

$\frac{4}{5} = \frac{4}{5}$

31. Solve the equation for y.

$$-\frac{5}{8} = y - \frac{1}{6}$$

$$-\frac{5}{8} + \frac{1}{6} = y - \frac{1}{6} + \frac{1}{6}$$

$$-\frac{15}{24} + \frac{4}{24} = y + 0$$

$$-\frac{11}{24} = y$$

Check: $-\frac{5}{8} \stackrel{?}{=} -\frac{11}{24} - \frac{1}{6}$

$-\frac{5}{8} \stackrel{?}{=} -\frac{11}{24} - \frac{4}{24}$

$-\frac{5}{8} \stackrel{?}{=} -\frac{15}{24}$

$-\frac{5}{8} = -\frac{5}{8}$

33. Solve the equation for y.

$$15.8 + y = 7.6$$

$$\underline{-15.8 \qquad\quad -15.8}$$

$$0 + y = -8.2$$

$$y = -8.2$$

Check: $15.8 + (-8.2) \stackrel{?}{=} 7.6$

$7.6 = 7.6$

35. Solve the equation for n.

$$-2.1 = n - 7.5 + 0.8$$

$$-2.1 = n - 6.7$$

$$\underline{+6.7 \qquad\quad +6.7}$$

$$4.6 = n + 0$$

$$4.6 = n$$

Check: $-2.1 \stackrel{?}{=} 4.6 - 7.5 + 0.8$

$-2.1 = -2.1$

37. Solve the equation for x.

$$6 - 18 = 7x - 3 - 6x$$

$$-12 = x - 3$$

$$\underline{+3 \qquad +3}$$

$$-9 = x + 0$$

$$-9 = x$$

Check: $6 - 18 \stackrel{?}{=} 7(-9) - 3 - 6(-9)$

$-12 \stackrel{?}{=} -63 - 3 + 54$

$-12 = -12$

39. Solve the equation for m.

$$5m = 4m + 7$$

$$\underline{-4m \quad -4m}$$

$$1m = 0 + 7$$

$$m = 7$$

Check: $5(7) \stackrel{?}{=} 4(7) + 7$

$35 \stackrel{?}{=} 28 + 7$

$35 = 35$

41. Solve the equation for x.

$$7x-9 = 6x+4$$

$$\underline{-6x \qquad -6x}$$

$$x-9 = 0+4$$

$$x-9 = 4$$

$$\underline{+9 \quad +9}$$

$$x+0 = 13$$

$$x = 13$$

Check: $7(13)-9 \overset{?}{=} 6(13)+4$

$$91-9 \overset{?}{=} 78+4$$

$$82 = 82$$

43. Solve the equation for y.

$$-2y-11 = -3y-5$$

$$\underline{+3y \qquad +3y}$$

$$y-11 = 0-5$$

$$y-11 = -5$$

$$\underline{+11 \quad +11}$$

$$y+0 = 6$$

$$y = 6$$

Check: $-2(6)-11 \overset{?}{=} -3(6)-5$

$$-12-11 \overset{?}{=} -18-5$$

$$-23 = -23$$

45. Solve the equation for x.

$$7x-2+x = 10x-1-x$$

$$8x-2 = 9x-1$$

$$\underline{-8x \qquad -8x}$$

$$0-2 = x-1$$

$$-2 = x-1$$

$$\underline{+1 \qquad +1}$$

$$-1 = x$$

Check: $7(-1)-2+(-1) \overset{?}{=} 10(-1)-1-(-1)$

$$-7-2+(-1) \overset{?}{=} -10-1+1$$

$$-9+(-1) \overset{?}{=} -11+1$$

$$-10 = -10$$

47. Solve the equation for n.

$$8n+7-12n = 4n+5-9n$$

$$-4n+7 = -5n+5$$

$$\underline{+5n \qquad +5n}$$

$$n+7 = 0+5$$

$$n+7 = 5$$

$$\underline{-7 \quad -7}$$

$$n+0 = -2$$

$$n = -2$$

Check: $8(-2)+7-12(-2) \overset{?}{=} 4(-2)+5-9(-2)$

$$-16+7+24 \overset{?}{=} -8+5+18$$

$$15 = 15$$

49. Solve the equation for a.

$$2.6+7a+5 = 8a-5.6$$

$$7a+7.6 = 8a-5.6$$

$$\underline{-7a \qquad -7a}$$

$$0+7.6 = a-5.6$$

$$7.6 = a-5.6$$

$$\underline{+5.6 \qquad +5.6}$$

$$13.2 = a+0$$

$$13.2 = a$$

Check: $2.6+7(13.2)+5 \overset{?}{=} 8(13.2)-5.6$

$$2.6+92.4+5 \overset{?}{=} 105.6-5.6$$

$$100 = 100$$

51. Solve the equation for h.

$$12-7(h+6)+8h = 14-8$$

$$12-7h-42+8h = 6$$

$$h-30 = 6$$

$$\underline{+30 \quad +30}$$

$$h+0 = 36$$

$$h = 36$$

Check: $12-7(36+6)+8(36) \overset{?}{=} 14-8$

$$12-7(42)+288 \overset{?}{=} 6$$

$$12-294+288 \overset{?}{=} 6$$

$$-282+288 \overset{?}{=} 6$$

$$6 = 6$$

53. Solve the equation for x.

$$5 - \frac{1}{2}(4x+6) = -3x - 8$$

$$5 - 2x - 3 = -3x - 8$$

$$2 - 2x = -3x - 8$$

$$\underline{\quad +3x \quad +3x \quad}$$

$$2 + x = \quad 0 - 8$$

$$2 + x = \quad -8$$

$$\underline{-2 \qquad -2}$$

$$0 + x = \quad -10$$

$$x = \quad -10$$

Check: $5 - \dfrac{1}{2}(4(-10)+6) \stackrel{?}{=} -3(-10)-8$

$$5 - \frac{1}{2}(-40+6) \stackrel{?}{=} 30 - 8$$

$$5 - \frac{1}{2}(-34) \stackrel{?}{=} 22$$

$$5 + 17 \stackrel{?}{=} 22$$

$$22 = 22$$

55. Solve the equation for y.

$$5y - (3y+7) = y + 9$$

$$5y - 3y - 7 = y + 9$$

$$2y - 7 = \quad y + 9$$

$$\underline{-y \qquad -y}$$

$$y - 7 = \quad 0 + 9$$

$$y - 7 = \quad 9$$

$$\underline{+7 \quad +7}$$

$$y + 0 = 16$$

$$y = 16$$

Check: $5(16) - (3(16)+7) \stackrel{?}{=} 16 + 9$

$$80 - (48+7) \stackrel{?}{=} 25$$

$$80 - 55 \stackrel{?}{=} 25$$

$$25 = 25$$

57. Solve the equation for x.

$$3(3x-5) - 2(4x-3) = 5 - 10$$

$$9x - 15 - 8x + 6 = -5$$

$$x - 9 = -5$$

$$\underline{+9 \quad +9}$$

$$x + 0 = \quad 4$$

$$x = \quad 4$$

Check: $3(3(4)-5) - 2(4(4)-3) \stackrel{?}{=} 5 - 10$

$$3(12-5) - 2(16-3) \stackrel{?}{=} -5$$

$$3(7) - 2(13) \stackrel{?}{=} -5$$

$$21 - 26 \stackrel{?}{=} -5$$

$$-5 = -5$$

59. Solve the equation for x.

$$1.4(2.5x - 4.5) - (6.2 + 2.5x) = -7.3 + 4.5$$

$$3.5x - 6.3 - 6.2 - 2.5x = -2.8$$

$$x - 12.5 = \quad -2.8$$

$$\underline{+12.5 \quad +12.5}$$

$$x + 0 = \quad 9.7$$

$$x = \quad 9.7$$

Check:

$1.4(2.5(9.7) - 4.5) - (6.2 + 2.5(9.7)) \stackrel{?}{=} -7.3 + 4.5$

$$1.4(24.25 - 4.5) - (6.2 + 24.25) \stackrel{?}{=} -2.8$$

$$1.4(19.75) - (30.45) \stackrel{?}{=} -2.8$$

$$27.65 - 30.45 \stackrel{?}{=} -2.8$$

$$-2.8 = -2.8$$

61. Solve the equation for y.

$$4 + 3y + y - 12 = 5y + 9 - y - 17$$

$$4y - 8 = 4y - 8$$

Because the linear equation is an identity, every real number is a solution.

63. Solve the equation for x.

$$7.3 - 0.2x + 1.3 - 0.6x = 12 - 0.8x - 3.6$$

$$-0.8x + 8.6 = -0.8x + 8.4$$

The expressions on each side of the equation have the same variable term but different constant terms, so the equation is a contradiction and has no solution.

65. Solve the equation for z.

$$20z + \frac{2}{3}(6z-48) - 9z = 14z + 0.2(50z-250) - 9z$$

$$20z + 4z - 32 - 9z = 14z + 10z - 50 - 9z$$

$$15z - 32 = 15z - 50$$

The expressions on each side of the equation have the same variable term but different constant terms, so the equation is a contradiction and has no solution.

67. Solve the equation for m.

$8(2m-4)+20(m-5)$
$\quad = 4m + 5(5m-20) - 32 + 7m$
$16m - 32 + 20m - 100$
$\quad = 4m + 25m - 100 - 32 + 7m$
$36m - 132 = 36m - 132$

Because the linear equation is an identity, every real number is a solution.

69. Let x represent the amount of money Latonia needs for the down payment.

$1947 + x = 2373$
$\underline{-1947 \qquad -1947}$
$0 + x = 426$
$\qquad x = 426$

She needs $426.

71. Let x be the distance Robert has already driven.

$16 + x = 42$
$\underline{-16 \qquad -16}$
$0 + x = 26$
$\qquad x = 26$

He has driven 26 miles.

73. Begin by finding the difference in her balance by subtracting the May 21st balance from the May 16th balance: $1741.62 - 1286.65 = 454.97$. This is the amount of change during that period. Now let x be the amount of the fourth check and set the total of the checks equal to the amount of change.
Translation:
$16.82 + 150.88 + 192.71 + x = 1741.62 - 1286.65$
$16.82 + 150.88 + 192.71 + x = 454.97$
$\qquad 360.41 + x = 454.97$
$\qquad \underline{-360.41 \qquad -360.41}$
$\qquad 0 + x = 94.56$
$\qquad x = \$94.56$

The fourth check was for $94.56.

75. Let x be the length of the missing side.
$12.4 + 27.2 + 16.3 + x = 67.2$
$\qquad 55.9 + x = 67.2$
$\qquad \underline{-55.9 \qquad -55.9}$
$\qquad 0 + x = 11.3$
$\qquad x = 11.3$

The length of the missing side is 11.3 cm.

77. Let x be the length of the missing side.
$x + 19 + 10 = 54$
$\qquad x + 29 = 54$
$\qquad \underline{-29 \qquad -29}$
$\qquad x + 0 = 25$
$\qquad x = 25$

The length of the missing side is 25 ft.

79. To find the total in sales, we must first multiply the quantity sold by the price per unit.
On 4/2: $1 \cdot \$900 = \$900$
On 4/9: $3(\$1500) = \$4500$
On 4/13: $2(\$1245) = \$2490$

Let x be the amount that John needs to sell to make the quota. Translate this situation into an equation.
$x + 900 + 3 \cdot 1500 + 2 \cdot 1245 = 8500$
$\qquad x + 900 + 4500 + 2490 = 8500$
$\qquad x + 7890 = 8500$
$\qquad \underline{-7890 \qquad -7890}$
$\qquad x + 0 = 610$
$\qquad x = 610$

John needs to sell $610 more. Yes, John will probably meet his quota because $610 is less than all of his prior weeks.

81. Let x be the fraction of the respondents having no opinion. Remember that the fractions are parts of a whole, so set the sum of the fractions equal to 1 whole.
$x + \dfrac{1}{3} + \dfrac{2}{5} = 1$
$x + \dfrac{5}{15} + \dfrac{6}{15} = 1$
$x + \dfrac{11}{15} = 1$
$x + \dfrac{11}{15} - \dfrac{11}{15} = 1 - \dfrac{11}{15}$
$x + 0 = \dfrac{15}{15} - \dfrac{11}{15}$
$x = \dfrac{4}{15}$

$\dfrac{4}{15}$ of the respondents had no opinion.

Review Exercises

1.   The multiplicative inverse of $-\dfrac{4}{5}$ is $-\dfrac{5}{4}$.

2. $4(-8.3) = -33.2$

3. $27 \div (-3) = -9$

4. $13 - 6(x+2) = 13 + (-6) \cdot x + (-6) \cdot 2$
$$= 13 - 6x - 12$$
$$= -6x + 1$$

5. $3x^2 - 6x + 5x^2 + x - 9 = 3x^2 + 5x^2 - 6x + x - 9$
$$= 8x^2 - 5x - 9$$

6. $12\left(\dfrac{1}{3}x - \dfrac{3}{4}\right) = 12\left(\dfrac{1}{3}x\right) - 12\left(\dfrac{3}{4}\right)$
$$= \dfrac{12}{3}x - \dfrac{36}{4}$$
$$= 4x - 9$$

Exercise Set 2.3

1.   Solve the equation for x.

$3x = 12$ Check: $3(4) \overset{?}{=} 12$

$\dfrac{3x}{3} = \dfrac{12}{3}$ $12 = 12$

$x = 4$

3.   Solve the equation for x.

$-6x = -18$ Check: $-6(3) \overset{?}{=} -18$

$\dfrac{-6x}{-6} = \dfrac{-18}{-6}$ $-18 = -18$

$x = 3$

5.   Solve the equation for n.

$\dfrac{n}{2} = 8$ Check: $\dfrac{16}{2} \overset{?}{=} 8$

$2\left(\dfrac{n}{2}\right) = 2(8)$ $8 = 8$

$n = 16$

7.   Solve the equation for y.

$$\dfrac{3}{4}y = -15$$

$$\dfrac{4}{3}\left(\dfrac{3}{4}y\right) = \dfrac{4}{\cancel{3}_{1}}\left(-\cancel{15}^{5}\right)$$

$$y = -20$$

Check: $\dfrac{3}{4}(-20) \overset{?}{=} -15$

$$\dfrac{-60}{4} \overset{?}{=} -15$$

$$-15 = -15$$

9.   Solve the equation for t.

$$-\dfrac{4}{7}t = -\dfrac{2}{3}$$

$$-\dfrac{7}{4}\left(-\dfrac{4}{7}t\right) = -\dfrac{7}{2\cancel{4}}\left(-\dfrac{\cancel{2}^{1}}{3}\right)$$

$$t = \dfrac{7}{6}$$

Check: $-\dfrac{^2\cancel{4}}{\cancel{7}}\left(\dfrac{\cancel{7}}{\cancel{6}_3}\right) \overset{?}{=} -\dfrac{2}{3}$

$$-\dfrac{2}{3} = -\dfrac{2}{3}$$

11.   Solve the equation for t.

$$-\dfrac{2}{5}t = \dfrac{8}{15}$$

$$-\dfrac{5}{2} \cdot \left(-\dfrac{2}{5}t\right) = -\dfrac{^1\cancel{5}}{_{1}\cancel{2}} \cdot \dfrac{\cancel{8}^4}{\cancel{15}_3}$$

$$t = -\dfrac{4}{3}$$

Check: $-\dfrac{2}{5}\left(-\dfrac{4}{3}\right) \overset{?}{=} \dfrac{8}{15}$

$$\dfrac{8}{15} = \dfrac{8}{15}$$

13.   Solve the equation for n.

$5n - 2n = 21$ Check: $5(7) - 2(7) \overset{?}{=} 21$

$3n = 21$ $35 - 14 \overset{?}{=} 21$

$\dfrac{3n}{3} = \dfrac{21}{3}$ $21 = 21$

$n = 7$

15. Solve the equation for x.

$$4x + 1 = 21$$
$$\underline{-1 \quad -1}$$
$$4x + 0 = 20$$
$$4x = 20$$
$$\frac{4x}{4} = \frac{20}{4}$$
$$x = 5$$

Check: $4(5) + 1 \overset{?}{=} 21$

$$20 + 1 \overset{?}{=} 21$$
$$21 = 21$$

17. Solve the equation for a.

$$4a - 3 = 17$$
$$\underline{+3 \quad +3}$$
$$4a + 0 = 20$$
$$4a = 20$$
$$\frac{4a}{4} = \frac{20}{4}$$
$$a = 5$$

Check: $4(5) - 3 \overset{?}{=} 17$

$$20 - 3 \overset{?}{=} 17$$
$$17 = 17$$

19. Solve the equation for x.

$$5x + 7 = -8$$
$$\underline{-7 \quad -7}$$
$$5x + 0 = -15$$
$$5x = -15$$
$$\frac{5x}{5} = -\frac{15}{5}$$
$$x = -3$$

Check: $5(-3) + 7 \overset{?}{=} -8$

$$-15 + 7 \overset{?}{=} -8$$
$$-8 = -8$$

21. Solve the equation for n.

$$9 - 2n = 12$$
$$\underline{-9 \qquad -9}$$
$$0 - 2n = 3$$

$$-2n = 3$$
$$\frac{-2n}{-2} = \frac{3}{-2}$$
$$n = -\frac{3}{2}$$

Check: $9 - \cancel{2}\left(-\dfrac{3}{\cancel{2}}\right) \overset{?}{=} 12$

$$9 - (-3) \overset{?}{=} 12$$
$$12 = 12$$

23. Solve the equation for x.

$$\frac{5}{8}x - 2 = 8$$
$$\underline{+2 \quad +2}$$
$$\frac{5}{8}x + 0 = 10$$
$$\frac{5}{8}x = 10$$
$$\frac{8}{5} \cdot \frac{5}{8}x = \frac{8}{\cancel{5}} \cdot \overset{2}{\cancel{10}}$$
$$x = 16$$

Check: $\dfrac{5}{\cancel{8}}\left(\overset{2}{\cancel{16}}\right) - 2 \overset{?}{=} 8$

$$10 - 2 \overset{?}{=} 8$$
$$8 = 8$$

25. Solve the equation for x.

$$2(x - 3) = -6$$
$$2x - 6 = -6$$
$$\underline{+6 \quad +6}$$
$$2x + 0 = 0$$
$$2x = 0$$
$$\frac{2x}{2} = \frac{0}{2}$$
$$x = 0$$

Check: $2(0 - 3) \overset{?}{=} -6$

$$2(-3) \overset{?}{=} -6$$
$$-6 = -6$$

27. Solve the equation for n.

$$3(n+7) = -6$$
$$3n+21 = -6$$
$$\underline{-21 \quad -21}$$
$$3n+0 = -27$$
$$3n = -27$$
$$\frac{3n}{3} = -\frac{27}{3}$$
$$n = -9$$

Check: $3(-9+7) \overset{?}{=} -6$

$$3(-2) \overset{?}{=} -6$$
$$-6 = -6$$

29. Solve the equation for a.

$$3a-2a+7+6a = -28$$
$$7a+7 = -28$$
$$\underline{-7 \quad -7}$$
$$7a+0 = -35$$
$$7a = -35$$
$$\frac{7a}{7} = \frac{-35}{7}$$
$$a = -5$$

Check: $3(-5)-2(-5)+7+6(-5) \overset{?}{=} -28$

$$-15+10+7-30 \overset{?}{=} -28$$
$$-5+7-30 \overset{?}{=} -28$$
$$2-30 \overset{?}{=} -28$$
$$-28 = -28$$

31. Solve the equation for b.

$$12b-8(2b-3) = 16$$
$$12b-16b+24 = 16$$
$$-4b+24 = 16$$
$$\underline{-24 \quad -24}$$
$$-4b+0 = -8$$

$$-4b = -8$$
$$\frac{-4b}{-4} = \frac{-8}{-4}$$
$$b = 2$$

Check: $12(2)-8(2\cdot2-3) \overset{?}{=} 16$

$$24-8(4-3) \overset{?}{=} 16$$
$$24-8(1) \overset{?}{=} 16$$
$$24-8 \overset{?}{=} 16$$
$$16 = 16$$

33. Solve the equation for y.

$$2(y-33)+3(y-2) = 8$$
$$2y-66+3y-6 = 8$$
$$5y-72 = 8$$
$$\underline{+72 \quad +72}$$
$$5y+0 = 80$$
$$5y = 80$$
$$\frac{5y}{5} = \frac{80}{5}$$
$$y = 16$$

Check: $2(16-33)+3(16-2) \overset{?}{=} 8$

$$2(-17)+3(14) \overset{?}{=} 8$$
$$-34+42 \overset{?}{=} 8$$
$$8 = 8$$

35. Solve the equation for x.

$$8x+5 = 2x+17$$
$$\underline{-2x \qquad -2x}$$
$$6x+5 = 0+17$$
$$6x+5 = 17$$
$$\underline{-5 \quad -5}$$
$$6x+0 = 12$$
$$6x = 12$$
$$\frac{6x}{6} = \frac{12}{6}$$
$$x = 2$$

Check: $8(2)+5 \overset{?}{=} 2(2)+17$

$$16+5 \overset{?}{=} 4+17$$
$$21 = 21$$

37. Solve the equation for y.

$$5y - 7 = 2y + 13$$
$$\underline{-2y \qquad -2y}$$
$$3y - 7 = 0 + 13$$
$$3y - 7 = 13$$
$$\underline{+7 \quad +7}$$
$$3y + 0 = 20$$
$$3y = 20$$
$$\frac{3y}{3} = \frac{20}{3}$$
$$y = \frac{20}{3}$$

Check: $5\left(\dfrac{20}{3}\right) - 7 \overset{?}{=} 2\left(\dfrac{20}{3}\right) + 13$

$$\frac{100}{3} - 7 \overset{?}{=} \frac{40}{3} + 13$$
$$\frac{100}{3} - \frac{21}{3} \overset{?}{=} \frac{40}{3} + \frac{39}{3}$$
$$\frac{79}{3} = \frac{79}{3}$$

39. Solve the equation for k.

$$9 - 4k = 15 - k$$
$$\underline{+k \qquad +k}$$
$$9 - 3k = 15 + 0$$
$$9 - 3k = 15$$
$$\underline{-9 \qquad -9}$$
$$0 - 3k = 6$$
$$-3k = 6$$
$$\frac{-3k}{-3} = \frac{6}{-3}$$
$$k = -2$$

Check: $9 - 4(-2) \overset{?}{=} 15 - (-2)$

$$9 + 8 \overset{?}{=} 15 + 2$$
$$17 = 17$$

41. Solve the equation for k.

$$4k + 5 = 7k - 7 + 9k$$
$$4k + 5 = 16k - 7$$
$$\underline{-16k \qquad -16k}$$
$$-12k + 5 = 0 - 7$$
$$-12k + 5 = -7$$
$$\underline{-5 \qquad -5}$$
$$-12k + 0 = -12$$
$$-12k = -12$$
$$\frac{-12k}{-12} = \frac{-12}{-12}$$
$$k = 1$$

Check: $4(1) + 5 \overset{?}{=} 7(1) - 7 + 9(1)$

$$4 + 5 \overset{?}{=} 7 - 7 + 9$$
$$9 = 9$$

43. Solve the equation for a.

$$3a - 4a + 9 = -12a + 43 - 6a$$
$$-a + 9 = -18a + 43$$
$$\underline{+18a \qquad +18a}$$
$$17a + 9 = 0 + 43$$
$$17a + 9 = 43$$
$$\underline{-9 \quad -9}$$
$$17a + 0 = 34$$
$$17a = 34$$
$$\frac{17a}{17} = \frac{34}{17}$$
$$a = 2$$

Check: $3(2) - 4(2) + 9 \overset{?}{=} -12(2) + 43 - 6(2)$

$$6 - 8 + 9 \overset{?}{=} -24 + 43 - 12$$
$$-2 + 9 \overset{?}{=} 19 - 12$$
$$7 = 7$$

45. Solve the equation for x.

$$9x + 12 - 3x - 2 = x + 8 + 5x$$
$$6x + 10 = 6x + 8$$

The expressions on each side of the equation have the same variable term but different constant terms, so the equation is a contradiction and has no solution.

47. Solve the equation for w.

$$14(w-2)+13 = 4w+5$$
$$14w-28+13 = 4w+5$$
$$14w-15 = 4w+5$$
$$\underline{-4w \qquad -4w}$$
$$10w-15 = 0+5$$
$$10w-15 = 5$$
$$\underline{+15 \quad +15}$$
$$10w+0 = 20$$
$$10w = 20$$
$$\frac{10w}{10} = \frac{20}{10}$$
$$w = 2$$

Check: $14(2-2)+13 \overset{?}{=} 4(2)+5$

$$14(0)+13 \overset{?}{=} 8+5$$
$$0+13 \overset{?}{=} 13$$
$$13 = 13$$

49. Solve the equation for n.

$$2n-(6n+5) = 9-(2n-1)$$
$$2n-6n-5 = 9-2n+1$$
$$-4n-5 = -2n+10$$
$$\underline{+2n \qquad +2n}$$
$$-2n-5 = 0+10$$
$$-2n-5 = 10$$
$$\underline{+5 \quad +5}$$
$$-2n+0 = 15$$
$$-2n = 15$$
$$\frac{-2n}{-2} = \frac{15}{-2}$$
$$n = -\frac{15}{2}$$

Check:

$$2\left(-\frac{15}{2}\right)-\left({}^{3}\cancel{6}\left(-\frac{15}{\cancel{2}}\right)+5\right) \overset{?}{=} 9-\left(2\left(-\frac{15}{2}\right)-1\right)$$
$$-15-(-45+5) \overset{?}{=} 9-(-15-1)$$
$$-15-(-40) \overset{?}{=} 9-(-16)$$
$$-15+40 \overset{?}{=} 9+16$$
$$25 = 25$$

51. Solve the equation for x.

$$4-(6x+5) = 7-2(3x+4)$$
$$4-6x-5 = 7-6x-8$$
$$-6x-1 = -6x-1$$

Because the linear equation is an identity, every real number is a solution.

53. Solve the equation for a.

$$5(a-1)-9(a-2) = -3(2a+1)-2$$
$$5a-5-9a+18 = -6a-3-2$$
$$-4a+13 = -6a-5$$
$$\underline{+6a \qquad +6a}$$
$$2a+13 = 0-5$$
$$2a+13 = -5$$
$$\underline{-13 \quad -13}$$
$$2a+0 = -18$$
$$2a = -18$$
$$\frac{2a}{2} = \frac{-18}{2}$$
$$a = -9$$

Check:

$$5(-9-1)-9(-9-2) \overset{?}{=} -3(2(-9)+1)-2$$
$$5(-10)-9(-11) \overset{?}{=} -3(-18+1)-2$$
$$-50+99 \overset{?}{=} -3(-17)-2$$
$$49 \overset{?}{=} 51-2$$
$$49 = 49$$

55. Solve the equation for x.

$$3(2x-5)-4x = 2(x-3)-12$$
$$6x-15-4x = 2x-6-12$$
$$2x-15 = 2x-18$$

The expressions on each side of the equation have the same variable term but different constant terms, so the equation is a contradiction and has no solution.

57. Solve the equation for n.

$$\frac{2}{3}n - 1 = \frac{1}{4}$$

$$12\left(\frac{2}{3}n - 1\right) = \overset{3}{\cancel{12}}\left(\frac{1}{\cancel{4}_1}\right)$$

$$\overset{4}{\cancel{12}} \cdot \frac{2}{\cancel{3}_1}n - 12 \cdot 1 = 3$$

$$8n - 12 = 3$$

$$\underline{+12 \quad +12}$$

$$8n + 0 = 15$$

$$8n = 15$$

$$\frac{8n}{8} = \frac{15}{8}$$

$$n = \frac{15}{8}$$

Check: $\dfrac{\overset{1}{\cancel{2}}}{\cancel{3}_1} \cdot \dfrac{\overset{5}{\cancel{15}}}{\cancel{8}_4} - 1 \overset{?}{=} \dfrac{1}{4}$

$$\frac{5}{4} - 1 \overset{?}{=} \frac{1}{4}$$

$$\frac{5}{4} - \frac{4}{4} \overset{?}{=} \frac{1}{4}$$

$$\frac{1}{4} = \frac{1}{4}$$

59. Solve the equation for n.

$$\frac{3}{4}n - 2 = -\frac{7}{8}n + 4$$

$$8\left(\frac{3}{4}n - 2\right) = 8\left(-\frac{7}{8}n + 4\right)$$

$$\overset{2}{\cancel{8}} \cdot \frac{3}{\cancel{4}_1}n - 8 \cdot 2 = \cancel{8} \cdot -\frac{7}{\cancel{8}}n + 32$$

$$6n - 16 = -7n + 32$$

$$\underline{+7n \qquad +7n}$$

$$13n - 16 = 0 + 32$$

$$13n - 16 = 32$$

$$\underline{+16 \quad +16}$$

$$13n + 0 = 48$$

$$13n = 48$$

$$\frac{13n}{13} = \frac{48}{13}$$

$$n = \frac{48}{13}$$

Check:

$$\frac{3}{\cancel{4}_1} \cdot \left(\frac{\cancel{48}^{12}}{13}\right) - 2 \overset{?}{=} -\frac{7}{\cancel{8}_1} \cdot \left(\frac{\cancel{48}^{6}}{13}\right) + 4$$

$$\frac{36}{13} - 2 \overset{?}{=} -\frac{42}{13} + 4$$

$$\frac{36}{13} - \frac{26}{13} \overset{?}{=} -\frac{42}{13} + \frac{52}{13}$$

$$\frac{10}{13} = \frac{10}{13}$$

61. Solve the equation for x.

$$\frac{3}{4}x + \frac{5}{6} = \frac{1}{4} + \frac{4}{3}x$$

$$12\left(\frac{3}{4}x + \frac{5}{6}\right) = 12\left(\frac{1}{4} + \frac{4}{3}x\right)$$

$$\overset{3}{\cancel{12}} \cdot \frac{3}{\cancel{4}}x + \overset{2}{\cancel{12}} \cdot \frac{5}{\cancel{6}} = \overset{3}{\cancel{12}} \cdot \frac{1}{\cancel{4}} + \overset{4}{\cancel{12}} \cdot \frac{4}{\cancel{3}}x$$

$$9x + 10 = 3 + 16x$$

$$\underline{-16x \qquad -16x}$$

$$-7x + 10 = 3 + 0$$

$$-7x + 10 = 3$$

$$\underline{-10 \quad -10}$$

$$-7x + 0 = -7$$

$$-7x = -7$$

$$\frac{-7x}{-7} = \frac{-7}{-7}$$

$$x = 1$$

Check: $\dfrac{3}{4}(1) + \dfrac{5}{6} \overset{?}{=} \dfrac{1}{4} + \dfrac{4}{3}(1)$

$$\frac{3}{4} + \frac{5}{6} \overset{?}{=} \frac{1}{4} + \frac{4}{3}$$

$$\frac{9}{12} + \frac{10}{12} \overset{?}{=} \frac{3}{12} + \frac{16}{12}$$

$$\frac{19}{12} = \frac{19}{12}$$

63. Solve the equation for y.

$$\frac{1}{2}(y+5) = \frac{3}{2} - y$$

$$2\left[\frac{1}{2}(y+5)\right] = 2\left[\frac{3}{2} - y\right]$$

$$\frac{\cancel{2}}{1} \cdot \frac{1}{\cancel{2}}(y+5) = \frac{\cancel{2}}{1} \cdot \frac{3}{\cancel{2}} - 2y$$

$$y+5 = 3 - 2y$$

$$\underline{+2y \qquad\qquad +2y}$$

$$3y+5 = 3+0$$

$$3y+5 = 3$$

$$\underline{-5 \qquad -5}$$

$$3y+0 = -2$$

$$3y = -2$$

$$\frac{3y}{3} = \frac{-2}{3}$$

$$y = -\frac{2}{3}$$

Check:

$$\frac{1}{2}\left(-\frac{2}{3}+5\right) \overset{?}{=} \frac{3}{2} - \left(-\frac{2}{3}\right)$$

$$\frac{1}{2}\left(-\frac{2}{3}+\frac{15}{3}\right) \overset{?}{=} \frac{3}{2} - \left(-\frac{2}{3}\right)$$

$$\frac{1}{2}\left(\frac{13}{3}\right) \overset{?}{=} \frac{3}{2}+\frac{2}{3}$$

$$\frac{1}{2}\left(\frac{13}{3}\right) \overset{?}{=} \frac{9}{6}+\frac{4}{6}$$

$$\frac{13}{6} = \frac{13}{6}$$

65. Solve the equation for m.

$$\frac{1}{6}(m-4) = \frac{2}{3}(m+1) - \frac{1}{3}m$$

$$6\left[\frac{1}{6}(m-4)\right] = 6\left[\frac{2}{3}(m+1) - \frac{1}{3}m\right]$$

$$\frac{\cancel{6}}{1} \cdot \frac{1}{\cancel{6}}(m-4) = \frac{\overset{2}{\cancel{6}}}{1} \cdot \frac{2}{\cancel{3}_1}(m+1) - \frac{\overset{2}{\cancel{6}}}{1} \cdot \frac{1}{\cancel{3}_1}m$$

$$m-4 = 4(m+1) - 2m$$

$$m-4 = 4m+4-2m$$

$$m-4 = 2m+4$$

$$\underline{-2m \qquad\quad -2m}$$

$$-m-4 = 0+4$$

$$-m-4 = 4$$

$$\underline{+4 \qquad +4}$$

$$-m+0 = 8$$

$$-m = 8$$

$$\frac{-m}{-1} = \frac{8}{-1}$$

$$m = -8$$

Check: $\dfrac{1}{6}(-8-4) \overset{?}{=} \dfrac{2}{3}(-8+1) - \dfrac{1}{3}(-8)$

$$\frac{1}{6}(-12) \overset{?}{=} \frac{2}{3}(-7) + \frac{8}{3}$$

$$-2 \overset{?}{=} -\frac{14}{3} + \frac{8}{3}$$

$$-2 \overset{?}{=} -\frac{6}{3}$$

$$-2 = -2$$

67. Solve the equation for u.

$$-2.9u + 3.6u = 6.3$$

$$10(-2.9u + 3.6u) = (6.3)10$$

$$-29u + 36u = 63$$

$$7u = 63$$

$$\frac{7u}{7} = \frac{63}{7}$$

$$u = 9$$

Check: $-2.9(9) + 3.6(9) \overset{?}{=} 6.3$

$$-26.1 + 32.4 \overset{?}{=} 6.3$$

$$6.3 = 6.3$$

69. Solve the equation for t.

$$0.2 - 1.3t - 0.8t = 1 - 2.1t - 0.8$$

$$0.2 - 2.1t = 0.2 - 2.1t$$

Because the linear equation is an identity, every real number is a solution.

71. Solve the equation for a.

$$0.3(a-18) = 0.9a$$
$$0.3a - 5.4 = 0.9a$$
$$10(0.3a - 5.4) = (0.9a)10$$
$$3a - 54 = 9a$$
$$\underline{-3a \qquad -3a}$$
$$0 - 54 = 6a$$
$$-54 = 6a$$
$$\frac{-54}{6} = \frac{6a}{6}$$
$$-9 = a$$

Check: $0.3(-9-18) \overset{?}{=} 0.9(-9)$

$$0.3(-27) \overset{?}{=} -8.1$$
$$-8.1 = -8.1$$

73. Solve the equation for x.

$$0.08(15) + 0.4x = 0.05(36 + 7x)$$
$$1.2 + 0.4x = 1.8 + 0.35x$$
$$100(1.2 + 0.4x) = 100(1.8 + 0.35x)$$
$$120 + 40x = 180 + 35x$$
$$\underline{-35x \qquad -35x}$$
$$120 + 5x = 180 + 0$$
$$120 + 5x = 180$$
$$\underline{-120 \qquad -120}$$
$$0 + 5x = 60$$
$$5x = 60$$
$$\frac{5x}{5} = \frac{60}{5}$$
$$x = 12$$

Check:

$$0.08(15) + 0.4(12) \overset{?}{=} 0.05(36 + 7(12))$$
$$1.2 + 4.8 \overset{?}{=} 0.05(36 + 84)$$
$$6 \overset{?}{=} 0.05(120)$$
$$6 = 6$$

75. Solve the equation for v.

$$9 - 0.2(v-8) = 0.3(6v-8)$$
$$10(9 - 0.2(v-8)) = 10(0.3(6v-8))$$
$$90 - 2(v-8) = 3(6v-8)$$
$$90 - 2v + 16 = 18v - 24$$
$$106 - 2v = 18v - 24$$
$$\underline{-18v \qquad -18v}$$
$$106 - 20v = 0 - 24$$
$$106 - 20v = -24$$
$$\underline{-106 \qquad -106}$$
$$0 - 20v = -130$$
$$-20v = -130$$
$$\frac{-20v}{-20} = \frac{-130}{-20}$$
$$v = 6.5$$

Check: $9 - 0.2(6.5 - 8) \overset{?}{=} 0.3(6(6.5) - 8)$

$$9 - 0.2(-1.5) \overset{?}{=} 0.3(39 - 8)$$
$$9 + 0.3 \overset{?}{=} 0.3(31)$$
$$9.3 = 9.3$$

77. Mistake: The minus sign was dropped in front of the 11. Correct solution:

$$6x - 11 = 8x - 15$$
$$\underline{-6x \qquad -6x}$$
$$-11 = 2x - 15$$
$$\underline{+15 \qquad +15}$$
$$4 = 2x$$
$$\frac{4}{2} = \frac{2x}{2}$$
$$2 = x$$

Correct answer: 2

79. Mistake: In the second line, the minus sign was not distributed to the 2. Correct solution:

$$4-(x+2)=2x+3(x-8)$$
$$4-x-2=2x+3x-24$$
$$2-x=5x-24$$
$$\underline{+x \quad +x}$$
$$2+0=6x-24$$
$$2=6x-24$$
$$\underline{+24 \qquad +24}$$
$$26=6x+0$$
$$\frac{26}{6}=\frac{6x}{6}$$
$$\frac{13}{3}=x$$

Correct answer: $\dfrac{13}{3}$

81. Substitute 140 for A and 10 for w. Then solve for l.

$$A=lw$$
$$140=l(10)$$
$$140=10l$$
$$\frac{140}{10}=\frac{10l}{10}$$
$$14=l$$

The length is 14 feet.

83. Substitute 45.6 for A and 4.8 for b. Then solve for h.

$$A=\frac{1}{2}bh$$
$$45.6=\frac{1}{2}(4.8)h$$
$$45.6=2.4h$$
$$\frac{45.6}{2.4}=\frac{2.4h}{2.4}$$
$$19=h$$

The side labeled h is 19 cm in length.

85. Substitute 212 for P and $w+2$ for l. Then solve for w.

$$P=2w+2l$$
$$212=2w+2(w+2)$$
$$212=2w+2w+4$$
$$212=4w+4$$
$$\underline{-4 \qquad -4}$$
$$208=4w+0$$
$$208=4w$$
$$\frac{208}{4}=\frac{4w}{4}$$
$$52=w$$

The dimensions of the field are $w=52$ ft. and $l=w+2=52+2=54$ ft.

87. Substitute 7500 for A, 75 for h, and 90 for a. Then solve for b.

$$A=\frac{1}{2}h(a+b)$$
$$7500=\frac{1}{2}(75)(90+b)$$
$$2\cdot 7500=2\cdot\frac{1}{2}(75)(90+b)$$
$$15{,}000=75(90+b)$$
$$\frac{15{,}000}{75}=\frac{75(90+b)}{75}$$
$$200=90+b$$
$$\underline{-90 \quad -90}$$
$$110=0+b$$

The side labeled b is 110 feet in length.

89. Substitute 106 for SA, 7.5 for l, and 4 for w. Then solve for h.

$$SA=2lw+2lh+2wh$$
$$106=2(7.5)(4)+2(7.5)h+2(4)h$$
$$106=60+15h+8h$$
$$106=60+23h$$
$$\underline{-60 \quad -60}$$
$$46=0+23h$$
$$46=23h$$
$$\frac{46}{23}=\frac{23h}{23}$$
$$2=h$$

The height is 2 inches.

91. Substitute 40,053.84 for C. Then solve for r.

$$C = 2\pi r$$

$$40,053.84 = 2\pi r$$

$$\frac{40,053.84}{2\pi} = \frac{2\pi r}{2\pi}$$

$$6374.8 \approx r$$

The equatorial radius is approximately 6374.8 km.

93. Substitute 60 for V and 2 for r. Then solve for h.

$$V = \pi r^2 h$$

$$60 = \pi (2)^2 h$$

$$60 = 4\pi h$$

$$\frac{60}{4\pi} = \frac{4\pi h}{4\pi}$$

$$4.8 \approx h$$

The height of the tank is approximately 4.8 meters.

95. Begin by sketching the figure as two rectangles, the one on the left (with width w and length 15 ft.) and the one on the right (with width 12 and length $w + 2$). The total area is found by adding the areas of each rectangle.

$$A = lw + lw$$

$$321 = (15)w + (w+2)(12)$$

$$321 = 15w + 12w + 24$$

$$321 = 27w + 24$$

$$\underline{-24 \qquad\qquad -24}$$

$$297 = 27w + 0$$

$$297 = 27w$$

$$\frac{297}{27} = \frac{27w}{27}$$

$$11 = w$$

Therefore, $w = 11$ ft., and $w + 2 = 11 + 2 = 13$ ft. Now make the necessary substitutions.

$$y = w + 12 \quad \text{and} \quad x = 15 - (w+2)$$
$$ = 11 + 12 \qquad\qquad = 15 - 13$$
$$ = 23 \text{ ft.} \qquad\qquad\;\; = 2 \text{ ft.}$$

97. Substitute -76 for V and 8 for R Then solve for i.

$$V = iR$$

$$-76 = i(8)$$

$$-76 = 8i$$

$$\frac{-76}{8} = \frac{8i}{8}$$

$$-9.5 \text{ A} = i$$

99. Substitute 1200 for m and 4800 for F. Then solve for a.

$$F = ma$$

$$4800 = 1200a$$

$$\frac{4800}{1200} = \frac{1200a}{1200}$$

$$4 \text{ m/sec.}^2 = a$$

101. Substitute -9265.9 for F and -9.8 for a. Then solve for m.

$$F = ma$$

$$-9265.9 = m(-9.8)$$

$$-9265.9 = -9.8m$$

$$\frac{-9265.9}{-9.8} = \frac{-9.8m}{-9.8}$$

$$945.5 \text{ kg} = m$$

103. Substitute 0.94 for w and 440 for f. Then solve for s.

$$w = \frac{s}{f}$$

$$0.94 = \frac{s}{440}$$

$$440 \cdot 0.94 = 440 \cdot \frac{s}{440}$$

$$413.6 \text{ m/sec.} = s$$

105. Since Pam only pays for minutes over 100, we must first find how many minutes she will pay for: $187 - 100 = 87$ minutes. Substitute 87 for m.

$$C = 0.3m + 25$$

$$C = 0.3(87) + 25$$

$$C = 26.1 + 25$$

$$C = 51.10$$

Pam's total phone cost is $51.10

107. Substitute 135 for d, 3 for t, and 20 for a. Then solve for v_i.

$$d = v_i t + \frac{1}{2} a t^2$$

$$135 = v_i (3) + \frac{1}{2} \cdot 20 \cdot 3^2$$

$$135 = 3v_i + \frac{1}{\cancel{2}} \cdot \overset{10}{\cancel{20}} \cdot 9$$

$$135 = 3v_i + 90$$

$$\underline{-90 \qquad\quad -90}$$

$$45 = 3v_i + 0$$

$$45 = 3v_i$$

$$\frac{45}{3} = \frac{3v_i}{3}$$

$$15 \text{ ft./sec.} = v_i$$

Review Exercises

1. $-6 + 3\left(5^2 + 2\right) - (6 - 1)$

$$= -6 + 3(25 + 2) - (6 - 1)$$
$$= -6 + 3(27) - 5$$
$$= -6 + 81 - 5$$
$$= 75 - 5$$
$$= 70$$

2. $-3.2x^2 + 0.4x + 7x^2 - 8.03x + 1.5$

$$= -3.2x^2 + 7x^2 + 0.4x - 8.03x + 1.5$$
$$= 3.8x^2 - 7.63x + 1.5$$

3. $-\dfrac{3}{4}(6x + 8) = -\dfrac{3}{\cancel{4}} \cdot \overset{3}{\cancel{6}} x - \dfrac{3}{4} \cdot 8$

$$= -\frac{9}{2}x - 6$$

4. Solve the equation for x.

$$6 + x = -14 + 2x$$

$$\underline{-x \qquad\qquad -x}$$

$$6 + 0 = -14 + x$$

$$6 = -14 + x$$

$$\underline{+14 \qquad +14}$$

$$20 = 0 + x$$

$$20 = x$$

Check: $6 + 20 \overset{?}{=} -14 + 2(20)$

$$26 \overset{?}{=} -14 + 40$$

$$26 = 26$$

5. Solve the equation for x.

$$-\frac{2}{5}x = 10$$

$$-\frac{5}{2}\left(-\frac{2}{5}x\right) = -\frac{5}{\cancel{2}}\left(\overset{5}{\cancel{10}}\right)$$

$$x = -25$$

Check: $-\dfrac{2}{5} \cdot (-25) \overset{?}{=} 10$

$$10 = 10$$

6. Solve the equation for x.

$$2x + 3(x - 3) = 10x - (3x - 11)$$

$$2x + 3x - 9 = 10x - 3x + 11$$

$$5x - 9 = 7x + 11$$

$$\underline{-5x \qquad\quad -5x}$$

$$0 - 9 = 2x + 11$$

$$-9 = 2x + 11$$

$$\underline{-11 \qquad\quad -11}$$

$$-20 = 2x + 0$$

$$-20 = 2x$$

$$\frac{-20}{2} = \frac{2x}{2}$$

$$-10 = x$$

Check:

$$2(-10) + 3(-10 - 3) \overset{?}{=} 10(-10) - (3(-10) - 11)$$

$$2(-10) + 3(-13) \overset{?}{=} 10(-10) - (-30 - 11)$$

$$-20 - 39 \overset{?}{=} 10(-10) - (-41)$$

$$-59 \overset{?}{=} -100 + 41$$

$$-59 = -59$$

Exercise Set 2.4

1. Solve for t.

$$t - 4u = v$$
$$\underline{+4u \quad +4u}$$
$$t + 0 = 4u + v$$
$$t = 4u + v$$

3. Solve for y.

$$-5y = x$$
$$\frac{-5y}{-5} = \frac{x}{-5}$$
$$y = -\frac{x}{5}$$

5. Solve for x.

$$2x - 3 = b$$
$$\underline{+3 \quad +3}$$
$$2x + 0 = b + 3$$
$$2x = b + 3$$
$$\frac{2x}{2} = \frac{b+3}{2}$$
$$x = \frac{b+3}{2}$$

7. Solve for m.

$$y = mx + b$$
$$\underline{-b \qquad -b}$$
$$y - b = mx + 0$$
$$y - b = mx$$
$$\frac{y-b}{x} = \frac{mx}{x}$$
$$\frac{y-b}{x} = m$$

9. Solve for y.

$$3x + 4y = 8$$
$$\underline{-3x \qquad -3x}$$
$$0 + 4y = 8 - 3x$$
$$4y = 8 - 3x$$
$$\frac{4y}{4} = \frac{8-3x}{4}$$
$$y = \frac{8-3x}{4}$$

11. Solve for Y.

$$\frac{mn}{4} - Y = f$$
$$\underline{+Y \quad +Y}$$
$$\frac{mn}{4} + 0 = f + Y$$
$$\frac{mn}{4} = f + Y$$
$$\underline{-f \quad -f}$$
$$\frac{mn}{4} - f = 0 + Y$$
$$\frac{mn}{4} - f = Y$$

13. Solve for c.

$$6(c + 2d) = m - np$$
$$6c + 12d = m - np$$
$$\underline{-12d \quad -12d}$$
$$6c + 0 = m - np - 12d$$
$$6c = m - np - 12d$$
$$\frac{6c}{6} = \frac{m - np - 12d}{6}$$
$$c = \frac{m - np - 12d}{6}$$

15. Solve for y.

$$\frac{x}{3} + \frac{y}{5} = 1$$
$$15\left(\frac{x}{3} + \frac{y}{5}\right) = 15(1)$$
$${}^5\cancel{15} \cdot \frac{x}{\cancel{3}} + {}^3\cancel{15} \cdot \frac{y}{\cancel{5}} = 15$$
$$5x + 3y = 15$$
$$\underline{-5x \qquad -5x}$$
$$0 + 3y = 15 - 5x$$
$$3y = 15 - 5x$$
$$\frac{3y}{3} = \frac{15 - 5x}{3}$$
$$y = \frac{15 - 5x}{3}$$

17. Solve for n.

$$\frac{3}{4} + 5a = \frac{n}{c}$$

$$c\left(\frac{3}{4} + 5a\right) = \frac{n}{c} \cdot c$$

$$c\left(\frac{3}{4} + 5a\right) = n$$

19. Solve for p.

$$t = \frac{A - P}{pr}$$

$$pr \cdot t = \cancel{pr} \cdot \frac{A - P}{\cancel{pr}}$$

$$prt = A - P$$

$$\frac{prt}{rt} = \frac{A - P}{rt}$$

$$p = \frac{A - P}{rt}$$

21. Solve for a.

$$P = a + b + c$$

$$\underline{-b \quad\quad -b}$$

$$P - b = a + 0 + c$$

$$P - b = a + c$$

$$\underline{\quad -c \quad -c}$$

$$P - b - c = a + 0$$

$$P - b - c = a$$

23. Solve for l.

$$A = lw$$

$$\frac{A}{w} = \frac{lw}{w}$$

$$\frac{A}{w} = l$$

25. Solve for d.

$$\frac{C}{d} = \pi$$

$$d \cdot \frac{C}{d} = d \cdot \pi$$

$$C = d\pi$$

$$\frac{C}{\pi} = \frac{d\pi}{\pi}$$

$$\frac{C}{\pi} = d$$

27. Solve for r^2.

$$V = \frac{1}{3}\pi r^2 h$$

$$3(V) = 3\left(\frac{1}{3}\pi r^2 h\right)$$

$$3V = \pi r^2 h$$

$$\frac{3V}{\pi h} = \frac{\pi r^2 h}{\pi h}$$

$$\frac{3V}{\pi h} = r^2$$

29. Solve for b.

$$A = \frac{1}{2}bh$$

$$2(A) = 2\left(\frac{1}{2}bh\right)$$

$$2A = bh$$

$$\frac{2A}{h} = \frac{bh}{h}$$

$$\frac{2A}{h} = b$$

31. Solve for w.

$$P = 2l + 2w$$

$$\underline{-2l \quad\quad -2l}$$

$$P - 2l = 0 + 2w$$

$$P - 2l = 2w$$

$$\frac{P - 2l}{2} = \frac{2w}{2}$$

$$\frac{P - 2l}{2} = w$$

33. Solve for l.

$$S = \frac{n}{2}(a + l)$$

$$2(S) = 2\left(\frac{n}{2}(a + l)\right)$$

$$2S = n(a + l)$$

$$2S = na + nl$$

$$\underline{-na \quad\quad -na}$$

$$2S - na = 0 + nl$$

$$2S - na = nl$$

$$\frac{2S - na}{n} = \frac{nl}{n}$$

$$\frac{2S - na}{n} = l$$

35. Solve for w.

$$V = h\left(\pi r^2 - lw\right)$$

$$V = \pi r^2 h - lwh$$

$$\underline{-\pi r^2 h \quad -\pi r^2 h}$$

$$V - \pi r^2 h = 0 - lwh$$

$$V - \pi r^2 h = -lwh$$

$$\frac{V - \pi r^2 h}{-lh} = \frac{-lwh}{-lh}$$

$$\frac{\pi r^2 h - V}{lh} = w$$

37. Solve for C.

$$P = R - C$$

$$\underline{-R \quad -R}$$

$$P - R = 0 - C$$

$$P - R = -C$$

$$-1\left(P - R\right) = -1\left(-C\right)$$

$$-P + R = C$$

$$R - P = C$$

39. Solve for r.

$$I = Prt$$

$$\frac{I}{Pt} = \frac{Prt}{Pt}$$

$$\frac{I}{Pt} = r$$

41. Solve for C.

$$P = \frac{C}{n}$$

$$n \cdot P = n \cdot \frac{C}{n}$$

$$nP = C$$

43. Solve for r.

$$A = P + Prt$$

$$\underline{-P \quad -P}$$

$$A - P = 0 + Prt$$

$$A - P = Prt$$

$$\frac{A - P}{Pt} = \frac{Prt}{Pt}$$

$$\frac{A - P}{Pt} = r$$

45. Solve for t.

$$d = rt$$

$$\frac{d}{r} = \frac{rt}{r}$$

$$\frac{d}{r} = t$$

47. Solve for d.

$$W = Fd$$

$$\frac{W}{F} = \frac{Fd}{F}$$

$$\frac{W}{F} = d$$

49. Solve for t.

$$P = \frac{W}{t}$$

$$t \cdot P = t \cdot \frac{W}{t}$$

$$tP = W$$

$$\frac{tP}{P} = \frac{W}{P}$$

$$t = \frac{W}{P}$$

51. Solve for t.

$$v = -32t + v_0$$

$$\underline{-v_0 \quad -v_0}$$

$$v - v_0 = -32t + 0$$

$$v - v_0 = -32t$$

$$\frac{v - v_0}{-32} = \frac{-32t}{-32}$$

$$\frac{v - v_0}{-32} = t$$

$$\text{or } t = \frac{v_0 - v}{32}$$

53. Solve for C.

$$F = \frac{9}{5}C + 32$$

$$\underline{-32 \quad -32}$$

$$F - 32 = \frac{9}{5}C + 0$$

$$F - 32 = \frac{9}{5}C$$

$$\frac{5}{9}\left(F - 32\right) = \frac{5}{9} \cdot \frac{9}{5}C$$

$$\frac{5}{9}\left(F - 32\right) = C$$

55. Solve for m.

$$F = G\frac{Mm}{R^2}$$

$$R^2 \cdot F = R^2 \cdot G\frac{Mm}{R^2}$$

$$R^2 F = GMm$$

$$\frac{R^2 F}{GM} = \frac{GMm}{GM}$$

$$\frac{FR^2}{GM} = m$$

57. Mistake: Subtracted the coefficient 7 instead of dividing.

Correct: $t = \dfrac{54 - 3n}{7}$

59. Mistake: Multiplied 5 by –2.
Correct: $3m - 2 = 5nk$

$$\underline{+2 \quad +2}$$
$$3m = 5nk + 2$$
$$\frac{3m}{3} = \frac{5nk + 2}{3}$$
$$m = \frac{5nk + 2}{3}$$

Review Exercises

1. $7n + 4$ 2. $3(x + 2) - 9$

3. Solve the equation for x.
$$4x - 7 = -5$$
$$\underline{+7 \quad +7}$$
$$4x + 0 = 2$$
$$4x = 2$$
$$\frac{4x}{4} = \frac{2}{4}$$
$$x = 0.5$$

4. Solve the equation for x.
$$\frac{3}{4}x = -\frac{5}{8}$$
$$\frac{\cancel{4}}{\cancel{3}} \cdot \left(\frac{\cancel{3}}{\cancel{4}} x \right) = \frac{^1\cancel{4}}{3} \cdot \left(-\frac{5}{\cancel{8}_2} \right)$$
$$x = -\frac{5}{6}$$

5. Solve the equation for n.
$$3(n - 5) = 7n - 12$$
$$3n - 15 = 7n - 12$$
$$\underline{+15 \qquad +15}$$
$$3n + 0 = 7n + 3$$
$$3n = 7n + 3$$
$$\underline{-7n \quad -7n}$$
$$-4n = 0 + 3$$
$$-4n = 3$$
$$\frac{-4n}{-4} = \frac{3}{-4}$$
$$n = -\frac{3}{4}$$

6. Solve the equation for y.
$$\frac{1}{4}y + 6 = \frac{1}{5}(y + 10)$$
$$20 \cdot \left(\frac{1}{4}y + 6 \right) = 20 \cdot \frac{1}{5}(y + 10)$$
$$\frac{^5\cancel{20}}{1} \cdot \frac{1}{\cancel{4}}y + 120 = \frac{^4\cancel{20}}{1} \cdot \frac{1}{\cancel{5}}(y + 10)$$
$$5y + 120 = 4y + 40$$
$$\underline{-120 \qquad -120}$$
$$5y + 0 = 4y - 80$$
$$5y = 4y - 80$$
$$\underline{-4y \quad -4y}$$
$$1y = 0 - 80$$
$$y = -80$$

Exercise Set 2.5

1. Translate and solve.
$$y + 3 = -8$$
$$\underline{-3 \quad -3}$$
$$y + 0 = -11$$
$$y = -11$$

3. Translate and solve.
$$6 - x = -3$$
$$\underline{-6 \qquad -6}$$
$$0 - x = -9$$
$$-x = -9$$
$$\frac{-x}{-1} = \frac{-9}{-1}$$
$$x = 9$$

5. Translate and solve.

$$11m = -99$$

$$\frac{11m}{11} = \frac{-99}{11}$$

$$m = -9$$

7. Translate and solve.

$$m \div 4 = 1.6$$

$$\frac{m}{4} = 1.6$$

$$4\left(\frac{m}{4}\right) = 4(1.6)$$

$$m = 6.4$$

9. Let x be the number.

$$\frac{4}{5}x = \frac{5}{8}$$

$$\frac{5}{4} \cdot \frac{4}{5}x = \frac{5}{4} \cdot \frac{5}{8}$$

$$x = \frac{25}{32}$$

11. Let w be the number.

$$7 + 3w = 34$$

$$\underline{-7 \qquad -7}$$

$$0 + 3w = 27$$

$$3w = 27$$

$$\frac{3w}{3} = \frac{27}{3}$$

$$w = 9$$

13. Let a be the number.

$$5a - 9 = 76$$

$$\underline{+9 \quad +9}$$

$$5a + 0 = 85$$

$$5a = 85$$

$$\frac{5a}{5} = \frac{85}{5}$$

$$a = 17$$

15. Let t be the number.

$$8(8 + t) = 160$$

$$64 + 8t = 160$$

$$\underline{-64 \qquad -64}$$

$$0 + 8t = 96$$

$$8t = 96$$

$$\frac{8t}{8} = \frac{96}{8}$$

$$t = 12$$

17. Translate and solve.

$$-3(x - 2) = 12$$

$$-3x + 6 = 12$$

$$\underline{-6 \quad -6}$$

$$-3x + 0 = 6$$

$$-3x = 6$$

$$\frac{-3x}{-3} = \frac{6}{-3}$$

$$x = -2$$

19. Let g be the number.

$$\frac{1}{3}(g + 2) = 1$$

$$3 \cdot \frac{1}{3}(g + 2) = 3 \cdot 1$$

$$g + 2 = 3$$

$$\underline{-2 \quad -2}$$

$$g + 0 = 1$$

$$g = 1$$

21. Let m be the number.

$$3m - 11 = 5 + m$$

$$\underline{+11 \quad +11}$$

$$3m + 0 = 16 + m$$

$$3m = 16 + m$$

$$\underline{-m \qquad -m}$$

$$2m = 16 + 0$$

$$\frac{2m}{2} = \frac{16}{2}$$

$$m = 8$$

23. Let r be the number.

$$10 = \frac{r}{4} - 1$$

$$\underline{+1 \qquad +1}$$

$$11 = \frac{r}{4} + 0$$

$$11 = \frac{r}{4}$$

$$4 \cdot 11 = 4 \cdot \frac{r}{4}$$

$$44 = r$$

25. Let a be the number.

$$2a - 3(a + 4) = -3$$
$$2a - 3a - 12 = -3$$
$$-a - 12 = -3$$
$$\underline{+12 \quad +12}$$
$$-a + 0 = 9$$
$$-a = 9$$
$$\frac{-a}{-1} = \frac{9}{-1}$$
$$a = -9$$

27. Let d be the number.

$$(2d - 8) + (d - 12) = 13$$
$$2d - 8 + d - 12 = 13$$
$$3d - 20 = 13$$
$$\underline{+20 \quad +20}$$
$$3d + 0 = 33$$
$$3d = 33$$
$$\frac{3d}{3} = \frac{33}{3}$$
$$d = 11$$

29. Let x be the number.

$$5\left(x + \frac{2}{3}\right) = 3x - \frac{2}{3}$$
$$5x + \frac{10}{3} = 3x - \frac{2}{3}$$
$$3 \cdot \left(5x + \frac{10}{3}\right) = 3 \cdot \left(3x - \frac{2}{3}\right)$$
$$15x + 10 = 9x - 2$$
$$\underline{-10 \qquad -10}$$
$$15x + 0 = 9x - 12$$
$$15x = 9x - 12$$
$$\underline{-9x \quad -9x}$$
$$6x = 0 - 12$$
$$6x = -12$$
$$\frac{6x}{6} = -\frac{12}{6}$$
$$x = -2$$

31. Let x be the number.

$$2x - 4 = 16 + 3x$$
$$\underline{-3x \qquad -3x}$$
$$-1x - 4 = 16 + 0$$
$$-x - 4 = 16$$
$$\underline{+4 \quad +4}$$
$$-x + 0 = 20$$
$$-x = 20$$
$$\frac{-x}{-1} = \frac{20}{-1}$$
$$x = -20$$

33. Let x be the number.

$$\frac{2}{5}x = \frac{1}{2}x - 2$$
$$10 \cdot \frac{2}{5}x = 10\left(\frac{1}{2}x - 2\right)$$
$$4x = 5x - 20$$
$$\underline{-5x \quad -5x}$$
$$-1x = 0 - 20$$
$$-x = -20$$
$$\frac{-x}{-1} = \frac{-20}{-1}$$
$$x = 20$$

35. Let n be the number.

$$\frac{n - 3}{3} = \frac{n - 1}{4}$$
$$12 \cdot \left(\frac{n - 3}{3}\right) = 12 \cdot \left(\frac{n - 1}{4}\right)$$
$$4(n - 3) = 3 \cdot (n - 1)$$
$$4n - 12 = 3n - 3$$
$$\underline{+12 \qquad +12}$$
$$4n + 0 = 3n + 9$$
$$4n = 3n + 9$$
$$\underline{-3n \quad -3n}$$
$$1n = 0 + 9$$
$$n = 9$$

37. Let x be the number.

$$-4(2-x)=2(4-3x)-6$$
$$-8+4x=8-6x-6$$
$$-8+4x=2-6x$$
$$\underline{+6x+6x}$$
$$-8+10x=2+0$$
$$-8+10x=2$$
$$\underline{+8+8}$$
$$0+10x=10$$
$$10x=10$$
$$\frac{10x}{10}=\frac{10}{10}$$
$$x=1$$

39. Three added to four times a number is seven.

41. Six times the sum of a number and four is equal to the product of negative ten and the number.

43. One-half of the difference of a number and three will result in two-thirds of the difference of the number and eight.

45. Five hundredths of a number added to six-hundredths of the difference of the number and eleven is twenty-two.

47. The sum of two-thirds, three-fourths, and one-half of the same number will equal ten.

49. Mistake: Subtraction translated in reverse order.
Correct: $n-10=40$

51. Mistake: Multiplied 5 times the unknown number instead of the difference, which requires parentheses.
Correct: $5(x-6)=-2$

53. Mistake: Subtracted the unknown number instead of the sum, which requires parentheses.
Correct: $2t-(t+3)=-6$

55. Translation: $P=2(l+w)$

a) $P=2(l+w)$
$P=2(24+18)$
$P=2(42)$
$P=84$ ft.

b) $P=2(l+w)$
$P=2(12.5+18)$
$P=2(30.5)$
$P=61$ cm

c) $P=2(l+w)$
$$P=2\left(8\frac{1}{4}+10\frac{3}{8}\right)$$
$$P=2\left(\frac{33}{4}+\frac{83}{8}\right)$$
$$P=2\left(\frac{66}{8}+\frac{83}{8}\right)$$
$$P=2\left(\frac{149}{8}\right)$$
$$P=\frac{149}{4}=37\frac{1}{4}\text{ in.}$$

57. Translation: $P=b+2s$

a) $P=b+2s$
$P=5+2(11)$
$P=5+22$
$P=27$ in.

b) $P=b+2s$
$P=1.5+2(0.6)$
$P=1.5+1.2$
$P=2.7$ m

c) $P=b+2s$
$$P=15\frac{1}{4}+2\left(9\frac{1}{2}\right)$$
$$P=15\frac{1}{4}+19$$
$$P=34\frac{1}{4}\text{ in.}$$

59. Translation: $I=Prt$

a) $I=Prt$
$I=4000(0.05)(2)$
$I=\$400$

b) $I=Prt$
$$I=500(0.03)\left(\frac{1}{2}\right)$$
$I=\$7.50$

c) $I=Prt$
$$I=2000(0.06)\left(\frac{1}{4}\right)$$
$I=\$30$

61.   Translation: $t \approx \dfrac{e-d}{153.8}$

    a) $t \approx \dfrac{e-d}{153.8}$

       $t \approx \dfrac{12{,}500 - 5000}{153.8}$

       $t \approx \dfrac{7500}{153.8}$

       $t \approx 49$ sec.

    b) $t \approx \dfrac{e-d}{153.8}$

       $t \approx \dfrac{12{,}500 - 4500}{153.8}$

       $t \approx \dfrac{8000}{153.8}$

       $t \approx 52$ sec.

    c) $t \approx \dfrac{e-d}{153.8}$

       $t \approx \dfrac{12{,}500 - 3000}{153.8}$

       $t \approx \dfrac{9500}{153.8}$

       $t \approx 62$ sec.

Review Exercises

1.   True. 19 is equal to 19.

2. $-\lvert -6 \rvert \quad ? \quad -2(3)$

      $-6 \; = \; -6$

3. $5(4) - 23 \quad ? \quad 2(3-8)$

      $20 - 23 \quad ? \quad 2(-5)$

         $-3 \; > \; -10$

4.   Solve the equation for x.

    $7x + 9 = 4x - 6$

    $\underline{\quad -9 \qquad\quad -9 \quad}$

    $7x + 0 = 4x - 15$

      $7x = 4x - 15$

    $\underline{-4x \qquad -4x \quad}$

      $3x = 0 - 15$

      $3x = -15$

      $\dfrac{3x}{3} = -\dfrac{15}{3}$

        $x = -5$

Check: $7(-5) + 9 \;\overset{?}{=}\; 4(-5) - 6$

         $-35 + 9 \;\overset{?}{=}\; -20 - 6$

           $-26 \;=\; -26$

5.   Solve the equation for x.

$$\frac{5}{6} = -\frac{3}{4}x$$

$$-\frac{\overset{2}{\cancel{4}}}{3}\left(\frac{5}{\underset{3}{\cancel{6}}}\right) = -\frac{4}{3}\left(-\frac{3}{4}x\right)$$

$$-\frac{10}{9} = x$$

Check: $\dfrac{5}{6} \;\overset{?}{=}\; -\dfrac{3}{4}\left(-\dfrac{10}{9}\right)$

          $\dfrac{5}{6} \;\overset{?}{=}\; \dfrac{30}{36}$

          $\dfrac{5}{6} \;=\; \dfrac{5}{6}$

6.   Solve the equation for y.

$$\frac{1}{3}(y-2) = \frac{3}{5}y + 6$$

$$\frac{1}{3}y - \frac{2}{3} = \frac{3}{5}y + 6$$

$$15\left(\frac{1}{3}y - \frac{2}{3}\right) = 15\left(\frac{3}{5}y + 6\right)$$

$$\overset{5}{\cancel{15}}\cdot\frac{1}{\cancel{3}}y - \overset{5}{\cancel{15}}\cdot\frac{2}{\cancel{3}} = \overset{3}{\cancel{15}}\cdot\frac{3}{\cancel{5}}y + 15\cdot 6$$

     $5y - 10 = 9y + 90$

     $\underline{\;+10 \qquad\quad +10\;}$

     $5y + 0 = 9y + 100$

       $5y = 9y + 100$

     $\underline{-9y \qquad -9y\;}$

     $-4y = 0 + 100$

     $-4y = 100$

     $\dfrac{-4y}{-4} = \dfrac{100}{-4}$

        $y = -25$

Check: $\dfrac{1}{3}(-25-2) \;\overset{?}{=}\; \dfrac{3}{5}(-25) + 6$

        $\dfrac{1}{3}(-27) \;\overset{?}{=}\; -\dfrac{75}{5} + 6$

         $\dfrac{-27}{3} \;\overset{?}{=}\; -15 + 6$

           $-9 \;=\; -9$

Exercise Set 2.6

1. a) The ratio of carbon atoms to total atoms in the molecule is $\dfrac{12}{12+22+11} = \dfrac{12}{45} = \dfrac{4}{15}$.

    b) The ratio of carbon atoms to hydrogen atoms is $\dfrac{12}{22} = \dfrac{6}{11}$.

    c) The ratio of hydrogen atoms to oxygen atoms is $\dfrac{22}{11} = \dfrac{2}{1}$.

3. The ratio of rise to horizontal length is $\dfrac{0.75}{1.5} = \dfrac{1}{2}$.

5. The ratio of rotations of the back wheel to pedal rotations is $\dfrac{1\frac{1}{3}}{2} = 1\frac{1}{3} \div 2 = \dfrac{4}{3} \cdot \dfrac{1}{2} = \dfrac{2}{3}$.

7. The ratio of miles to hours is $\dfrac{254}{4} = \dfrac{127}{2} = \dfrac{63.5}{1}$. This unit ratio means that he drives an average of 63.5 miles per hour.

9. The ratio of cost to bananas is $\dfrac{0.75}{6} = \dfrac{0.25}{2} = \dfrac{0.125}{1}$. This unit ratio means that each banana costs 12.5 cents.

11. $\dfrac{73.98}{4.73} \approx \dfrac{15.64}{1}$; The price of the stock is \$15.64 for every \$1 earned in 2012.

13. The 8-ounce unit ratio is $\dfrac{1.42}{8} = \dfrac{0.1775}{1}$ and the 24-ounce unit ratio is $\dfrac{3.59}{24} \approx \dfrac{0.1496}{1}$. The 24-ounce container is better because it costs less per ounce. (The unit ratio of price to quantity is less.)

15. a) $\dfrac{1023}{5535} \approx \dfrac{0.18}{1}$

    b) $\dfrac{2229}{4824} \approx \dfrac{0.46}{1}$

    c) The ratio of aggravated assault hate crimes to total hate crimes was greater in 2010 than in 2008.

17. a) $\dfrac{35,541}{28,224} \approx \dfrac{1.26}{1}$; people in the 25–34 age group with an associate degree earn \$1.26 for every \$1.00 earned by high school graduates.

    b) $\dfrac{48,445}{35,541} \approx \dfrac{1.36}{1}$

    c) In general, people 25–34 years of age who have a bachelor's degree make more money than high school graduates or people with associate degrees.

19. Yes, because the cross products are equal.
$$5 \cdot 15 \stackrel{?}{=} 25 \cdot 3$$
$$75 = 75$$

21. No, because the cross products are not equal.
$$30 \cdot 20 \stackrel{?}{=} 25 \cdot 25$$
$$600 \neq 625$$

23. No, because the cross products are not equal.
$$6.2 \cdot 7 \stackrel{?}{=} 4 \cdot 15$$
$$43.4 \neq 60$$

25. Yes, because the cross products are equal.
$$3.5 \cdot 10 \stackrel{?}{=} 14 \cdot 2.5$$
$$35 = 35$$

27. No, because the cross products are not equal.
$$\left(10\frac{1}{2}\right) \cdot \frac{2}{5} \stackrel{?}{=} \frac{3}{10} \cdot 15$$
$$\frac{21}{2} \cdot \frac{2}{5} \stackrel{?}{=} \frac{3}{\overset{}{\underset{2}{10}}} \cdot \frac{\overset{3}{\cancel{15}}}{1}$$
$$\frac{21}{5} \neq \frac{9}{2}$$

29. $\dfrac{2}{3} = \dfrac{7}{x}$
$$x \cdot 2 = 3 \cdot 7$$
$$2x = 21$$
$$\frac{2x}{2} = \frac{21}{2}$$
$$x = 10.5$$

31. $\dfrac{-9}{2} = \dfrac{63}{n}$
$$n(-9) = 2 \cdot 63$$
$$-9n = 126$$
$$\frac{-9n}{-9} = \frac{126}{-9}$$
$$n = -14$$

33. $\dfrac{21}{5} = \dfrac{h}{2.5}$

$2.5 \cdot 21 = 5 \cdot h$

$52.5 = 5h$

$\dfrac{52.5}{5} = \dfrac{5h}{5}$

$10.5 = h$

35. $\dfrac{-17}{b} = \dfrac{8.5}{4}$

$4 \cdot (-17) = b \cdot 8.5$

$-68 = 8.5b$

$-\dfrac{68}{8.5} = \dfrac{8.5b}{8.5}$

$-8 = b$

37. $\dfrac{-1.5}{60} = \dfrac{-2\frac{7}{8}}{t}$

$t \cdot (-1.5) = 60 \cdot \left(-2\dfrac{7}{8}\right)$

$-1.5t = -172.5$

$\dfrac{-1.5t}{-1.5} = \dfrac{-172.5}{-1.5}$

$t = 115$

39. $\dfrac{2\frac{1}{4}}{6\frac{2}{3}} = \dfrac{d}{10\frac{2}{3}}$

$\dfrac{\frac{9}{4}}{\frac{20}{3}} = \dfrac{d}{\frac{32}{3}}$

$\dfrac{\overset{8}{\cancel{32}}}{\cancel{3}} \cdot \dfrac{\overset{3}{\cancel{9}}}{\cancel{4}} = \dfrac{20}{3} \cdot d$

$24 = \dfrac{20}{3}d$

$\dfrac{3}{\underset{5}{\cancel{20}}} \cdot \overset{6}{\cancel{24}} = \dfrac{\cancel{3}}{\cancel{20}} \cdot \dfrac{\cancel{20}}{\cancel{3}}d$

$\dfrac{18}{5} = d \ \text{ or } \ d = 3\dfrac{3}{5}$

41. $\dfrac{2}{3} = \dfrac{8}{x+8}$

$(x+8) \cdot 2 = 3 \cdot 8$

$2x + 16 = 24$

$\underline{\quad -16 \quad -16}$

$2x + 0 = 8$

$2x = 8$

$\dfrac{2x}{2} = \dfrac{8}{2}$

$x = 4$

43. $\dfrac{4x-4}{8} = \dfrac{2x+1}{5}$

$5 \cdot (4x-4) = 8 \cdot (2x+1)$

$20x - 20 = 16x + 8$

$\underline{-16x \qquad -16x}$

$4x - 20 = 0 + 8$

$4x - 20 = 8$

$\underline{\quad +20 \quad +20}$

$4x + 0 = 28$

$4x = 28$

$\dfrac{4x}{4} = \dfrac{28}{4}$

$x = 7$

45. Translate to a proportion and solve.

$\dfrac{110 \text{ miles}}{3 \text{ hours}} = \dfrac{x \text{ miles}}{5 \text{ hours}}$

$5 \cdot 110 = 3 \cdot x$

$550 = 3x$

$\dfrac{550}{3} = \dfrac{3x}{3}$

$183\dfrac{1}{3} \text{ miles} = x$

47. Translate to a proportion and solve.

$\dfrac{\$1040 \text{ taxes}}{\$154,000 \text{ value}} = \dfrac{x \text{ taxes}}{\$200,000 \text{ value}}$

$200,000 \cdot 1040 = 154,000 \cdot x$

$208,000,000 = 154,000x$

$\dfrac{208,000,000}{154,000} = \dfrac{154,000x}{154,000}$

$\$1350.65 = x$

49. Translate to a proportion and solve.

$\dfrac{\$1575}{20 \text{ business days}} = \dfrac{\$x}{120 \text{ business days}}$

$120 \cdot 1575 = 20 \cdot x$

$189,000 = 20x$

$\dfrac{189,000}{20} = \dfrac{20x}{20}$

$\$9450 = x$

51. Translate to a proportion and solve.
$$\frac{1.8946 \text{ US}}{1 \text{ British}} = \frac{800 \text{ US}}{x \text{ British}}$$
$$x \cdot 1.8946 = 1 \cdot 800$$
$$1.8946x = 800$$
$$\frac{1.8946x}{1.8946} = \frac{800}{1.8946}$$
$$x \approx 422.25 \text{ pounds}$$

53. Translate to a proportion and solve.
$$\frac{1\frac{1}{2} \text{ in.}}{1 \text{ ft.}} = \frac{x \text{ in.}}{3.25 \text{ ft.}}$$
$$3.25 \cdot \left(1\frac{1}{2}\right) = 1 \cdot x$$
$$4.875 \text{ in.} = x$$

55. Translate to a proportion and solve.
$$\frac{1\frac{1}{2} \text{ teaspoons sugar}}{12 \text{ pancakes}} = \frac{x \text{ teaspoons sugar}}{20 \text{ pancakes}}$$
$$20 \cdot 1\frac{1}{2} = 12 \cdot x$$
$$30 = 12x$$
$$\frac{30}{12} = \frac{12x}{12}$$
$$2\frac{1}{2} \text{ tsp.} = x$$

57. Translate to a proportion and solve.
$$\frac{630 \text{ miles}}{17.5 \text{ gallons}} = \frac{x \text{ miles}}{12 \text{ gallons}}$$
$$12 \cdot 630 = 17.5 \cdot x$$
$$7560 = 17.5x$$
$$\frac{7560}{17.5} = \frac{17.5x}{17.5}$$
$$432 \text{ miles} = x$$

59. Translate to a proportion and solve.
$$\frac{264 \text{ characters}}{56.6 \text{ seconds}} = \frac{300 \text{ characters}}{x \text{ minutes}}$$
$$x \cdot 264 = 56.6 \cdot 300$$
$$264x = 16,980$$
$$\frac{264x}{264} = \frac{16,980}{264}$$
$$x \approx 64.3 \text{ sec.}$$

61. Translate to a proportion and solve.
$$\frac{9 \text{ pounds}}{1000 \text{ square feet}} = \frac{x \text{ pounds}}{25,000 \text{ square feet}}$$
$$25,000 \cdot 9 = 1000 \cdot x$$
$$225,000 = 1000x$$
$$\frac{225,000}{1000} = \frac{1000x}{1000}$$
$$225 \text{ lb.} = x$$

63. Begin by calculating the total square footage of each room: living room = $20 \cdot 15 = 300 \text{ ft.}^2$ and dining room = $18 \cdot 12 = 216 \text{ ft.}^2$ Now translate to a proportion and solve.
$$\frac{\$49.95}{300 \text{ ft.}^2} = \frac{\$x}{216 \text{ ft.}^2}$$
$$216 \cdot 49.95 = 300 \cdot x$$
$$10,789.20 = 300x$$
$$\frac{10,789.20}{300} = \frac{300x}{300}$$
$$\$35.96 = x$$
Cost to clean the living room + cost to clean the dining room = total cost
$$\$49.95 + \$35.96 = \$85.91$$

65. Translate to a proportion and solve.
$$\frac{17 \text{ tagged}}{45 \text{ total}} = \frac{25 \text{ tagged}}{x \text{ total}}$$
$$x \cdot 17 = 45 \cdot 25$$
$$17x = 1125$$
$$\frac{17x}{17} = \frac{1125}{17}$$
$$x \approx 66 \text{ deer}$$

67. Translate to a proportion and solve.
$$\frac{17 \text{ generators}}{750,000 \text{ people}} = \frac{14 \text{ generators}}{x \text{ people}}$$
$$x \cdot 17 = 750,000 \cdot 14$$
$$17x = 10,500,000$$
$$\frac{17x}{17} = \frac{10,500,000}{17}$$
$$x \approx 617,647 \text{ people}$$

69. Translate to a proportion and solve.

$$\frac{16}{20} = \frac{9}{x}$$

$$x \cdot 16 = 20 \cdot 9$$

$$16x = 180$$

$$\frac{16x}{16} = \frac{180}{16}$$

$$x = 11.25 \text{ cm}$$

71. Translate to proportions and solve.

$$\frac{8}{a} = \frac{9\frac{1}{2}}{5\frac{1}{4}} \qquad \frac{10\frac{2}{3}}{c} = \frac{9\frac{1}{2}}{5\frac{1}{4}}$$

$$\frac{8}{a} = \frac{\frac{19}{2}}{\frac{21}{4}} \qquad \frac{\frac{32}{3}}{c} = \frac{\frac{19}{2}}{\frac{21}{4}}$$

$$\frac{21}{4} \cdot 8 = a \cdot \frac{19}{2} \qquad \frac{\overset{7}{\cancel{21}}}{\cancel{4}} \cdot \frac{\overset{8}{\cancel{32}}}{\cancel{3}} = c \cdot \frac{19}{2}$$

$$42 = \frac{19}{2}a \qquad 56 = \frac{19}{2}c$$

$$\frac{2}{19} \cdot 42 = \frac{2}{19} \cdot \frac{19}{2}a \qquad \frac{2}{19} \cdot 56 = \frac{2}{19} \cdot \frac{19}{2}c$$

$$4\frac{8}{19} \text{ ft.} = a \qquad 5\frac{17}{19} \text{ ft.} = c$$

$$b = 5\frac{1}{4} \text{ ft.}$$

73. Translate to a proportion and solve.

$$\frac{1.8}{x} = \frac{1.2}{200}$$

$$200 \cdot 1.8 = x \cdot 1.2$$

$$360 = 1.2x$$

$$\frac{360}{1.2} = \frac{1.2x}{1.2}$$

$$300 \text{ m} = x$$

75. Translate to a proportion and solve.

$$\frac{5}{16.5} = \frac{8}{d}$$

$$d \cdot 5 = 16.5 \cdot 8$$

$$5d = 132$$

$$\frac{5d}{5} = \frac{132}{5}$$

$$d = 26.4 \text{ m}$$

Review Exercises

1. Yes, 0.58 is a rational number because it can be written as a ratio of integers, $\frac{58}{100}$.

2. $\frac{42}{350} = \frac{3 \cdot \cancel{14}}{25 \cdot \cancel{14}} = \frac{3}{25}$

3. $(0.15)(20.4) = 3.06$

4. $45.2 \div 100 = 0.452$

5. $0.75n = 240$

$$\frac{0.75n}{0.75} = \frac{240}{0.75}$$

$$n = 320$$

6. $60p = 24$

$$\frac{60p}{60} = \frac{24}{60}$$

$$p = 0.4 \text{ or } \frac{2}{5}$$

Exercise Set 2.7

1. $20\% = \frac{20}{100} = \frac{1}{5} = 0.2$

3. $15\% = \frac{15}{100} = \frac{3}{20} = 0.15$

5. $14.8\% = \frac{14.8}{100} = 0.148$

$$= \frac{148}{1000} = \frac{37}{250}$$

7. $3.75\% = \frac{3.75}{100} = 0.0375$

$$= \frac{375}{10,000} = \frac{3}{80}$$

9. $45\frac{1}{2}\% = \frac{45\frac{1}{2}}{100}$

$$= 45\frac{1}{2} \div 100$$

$$= \frac{91}{2} \cdot \frac{1}{100}$$

$$= \frac{91}{200} = 0.455$$

11. $33\dfrac{1}{3}\% = \dfrac{33\dfrac{1}{3}}{100}$

$\phantom{33\dfrac{1}{3}\%} = 33\dfrac{1}{3} \div 100$

$\phantom{33\dfrac{1}{3}\%} = \dfrac{100}{3} \cdot \dfrac{1}{100}$

$\phantom{33\dfrac{1}{3}\%} = \dfrac{1}{3} = 0.\overline{3}$

13. $\dfrac{3}{5} = \dfrac{3}{\cancel{5}} \cdot \dfrac{\cancel{100}^{20}}{1}\% = 60\%$

15. $\dfrac{3}{8} = \dfrac{3}{\cancel{8}} \cdot \dfrac{\cancel{100}^{25}}{1}\% = \dfrac{75}{2}\% = 37.5\%$

17. $\dfrac{5}{6} = \dfrac{5}{\cancel{6}} \cdot \dfrac{\cancel{100}^{50}}{1}\% = \dfrac{250}{3}\% = 83.3\%$

19. $\dfrac{2}{3} = \dfrac{2}{3} \cdot \dfrac{100}{1}\% = \dfrac{200}{3}\% = 66.7\%$

21. $0.96 = 0.96 \cdot 100\% = 96\%$

23. $0.8 = 0.8 \cdot 100\% = 80\%$

25. $0.09 = 0.09 \cdot 100\% = 9\%$

27. $1.2 = 1.2 \cdot 100\% = 120\%$

29. $0.028 = 0.028 \cdot 100\% = 2.8\%$

31. $4.051 = 4.051 \cdot 100\% = 405.1\%$

33. Translate "80% of 35 is what number?" word for word. Let n represent the unknown number.
$0.80 \cdot 35 = n$

$28 = n$

The number is 28.

35. Translate "2.5% of 124 is what number?" word for word. Let n represent the unknown number.
$0.025 \cdot 124 = n$

$3.1 = n$

The number is 3.1.

37. Translate "what number is $9\dfrac{1}{4}\%$ of 64?" word for word. Let n represent the unknown number.
$n = 0.0925 \cdot 64$

$ = 5.92$

The number is 5.92.

39. Translate "120% of 86 is what number?" word for word. Let n represent the unknown number.
$1.2 \cdot 86 = n$

$103.2 = n$

The number is 103.2.

41. Translate "30% of what number is 15?" word for word. Let n represent the unknown number.
$0.3 \cdot n = 15$

$0.3n = 15$

$\dfrac{0.3n}{0.3} = \dfrac{15}{0.3}$

$n = 50$

The number is 50.

43. Translate "16.4 is 20.5% of what number?" word for word. Let n represent the unknown number.
$16.4 = 0.205 \cdot n$

$16.4 = 0.205n$

$\dfrac{16.4}{0.205} = \dfrac{0.205n}{0.205}$

$80 = n$

The number is 80.

45. Translate "7.8 is $12\dfrac{1}{2}\%$ of what number?" word for word. Let n represent the unknown number.
$7.8 = 0.125 \cdot n$

$7.8 = 0.125n$

$\dfrac{7.8}{0.125} = \dfrac{0.125n}{0.125}$

$62.4 = n$

The number is 62.4.

47. Translate "what percent of 45 is 9?" word for word. Let p represent the unknown percent.
$p \cdot 45 = 9$

$45p = 9$

$\dfrac{45p}{45} = \dfrac{9}{45}$

$p = 0.20$

To write 0.20 as a percent, multiply by 100%. The percent is $0.20 \cdot 100\% = 20\%$.

49. Translate "what percent of 38 is 39.9?" word for word. Let p represent the unknown percent.

$$p \cdot 38 = 39.9$$
$$38p = 39.9$$
$$\frac{38p}{38} = \frac{39.9}{38}$$
$$p = 1.05$$

To write 1.05 as a percent, multiply by 100%. The percent is $1.05 \cdot 100\% = 105\%$.

51. Translate "what percent is 18 out of 27?" to a proportion. Let p represent the unknown percent.

$$\frac{p}{100} = \frac{18}{27}$$
$$27 \cdot p = 100 \cdot 18$$
$$27p = 1800$$
$$\frac{27p}{27} = \frac{1800}{27}$$
$$p = 66.\overline{6}$$
$$p\% = 66.\overline{6}\%$$

The percent is $66.\overline{6}\%$.

53. Translate "what percent is 12 of 15?" to a proportion. Let p represent the unknown percent.

$$\frac{p}{100} = \frac{12}{15}$$
$$15 \cdot p = 100 \cdot 12$$
$$15p = 1200$$
$$\frac{15p}{15} = \frac{1200}{15}$$
$$p = 80$$
$$p\% = 80\%$$

The percent is 80%.

55. This situation is the same as "86% of 50 is what number?" Translate word for word. Let n represent the unknown number.

$$0.86 \cdot 50 = n$$
$$43 = n$$

She answered 43 questions correctly.

57. This situation is the same as "4% of $120,000 is what number?" Translate word for word. Let n represent the unknown number.

$$0.04 \cdot 120,000 = n$$
$$4800 = n$$

The tax will be $4800 annually or $4800 \div 12 = \$400$ per month.

59. This situation is the same as "28% of $3210 is what number?" Translate word for word. Let n represent the unknown number.

$$0.28 \cdot 3210 = n$$
$$898.80 = n$$

The most a lender will allow for a mortgage payment is $898.80.

61. This situation is the same as "83.4% of 6934.6 is what number?" Translate word for word. Let n represent the unknown number.

$$0.834 \cdot 6934.6 = n$$
$$5783.5 \approx n$$

A total of 5783.5 tg of carbon dioxide were emitted by the United States.

63. This situation is the same as "20% of what number is $8.50?" Translate word for word. Let n represent the unknown number.

$$0.2 \cdot n = 8.50$$
$$0.2n = 8.50$$
$$\frac{0.2n}{0.2} = \frac{8.50}{0.2}$$
$$n = 42.5$$

The cost of the meal was $42.50.

65. This situation is the same as "2.5% of what number is $7.50?" Translate word for word. Let n represent the unknown number.

$$0.025 \cdot n = 7.50$$
$$0.025n = 7.50$$
$$\frac{0.025n}{0.025} = \frac{7.50}{0.025}$$
$$n = 300$$

The tickets cost $300.

67. This situation is the same as "7194 is what percent of 15,904?" Translate word for word. Let p represent the unknown percent.

$$7194 = p \cdot 15,904$$
$$7194 = 15,904p$$
$$\frac{7194}{15,904} = \frac{15,904p}{15,904}$$
$$0.452 \approx p$$

To write 0.452 as a percent, multiply by 100%. The percent is $0.452 \cdot 100\% = 45.2\%$. He scored approximately 45.2% of his field goal attempts.

69. This situation is the same as "220,958,853 is what percent of 308,745,538?" Translate word for word. Let p represent the unknown percent.

$$220,958,853 = p \cdot 308,745,538$$
$$220,958,853 = 308,745,538\,p$$
$$\frac{220,958,853}{308,745,538} = \frac{308,745,538\,p}{308,745,538}$$
$$0.716 \approx p$$

About 71.6% of the total population is 21 years of age or older.

71. This situation is the same as "185.9 is what percent of 2336.6?" Translate word for word. Let p represent the unknown percent.

$$185.9 = p \cdot 2336.6$$
$$185.9 = 2336.6\,p$$
$$\frac{185.9}{2336.6} = \frac{2336.6\,p}{2336.6}$$
$$0.08 \approx p$$

Approximately 8.0% of the energy use in the United States comes from nuclear energy.

73. United States:

$$570.1 = p \cdot 2336.6$$
$$570.1 = 2336.6\,p$$
$$\frac{570.1}{2336.6} = \frac{2336.6\,p}{2336.6}$$
$$0.244 \approx p$$

Approximately 24.4% of the United States' energy use comes from gas.
China:

$$42.3 = p \cdot 1554.0$$
$$42.3 = 1554.0\,p$$
$$\frac{42.3}{1554.0} = \frac{1554.0\,p}{1554.0}$$
$$0.027 \approx p$$

Approximately 2.7% of China's energy use comes from gas.
Russia:

$$364.6 = p \cdot 679.6$$
$$364.6 = 679.6\,p$$
$$\frac{364.6}{679.6} = \frac{679.6\,p}{679.6}$$
$$0.536 \approx p$$

Approximately 53.6% of Russia's energy use comes from gas.
Russia has the highest percentage.

75. Find the total cost for the quarter: $124.2 + 248.6 + 115.7 = \$488.5$ thousand. This situation is the same as "what percent of 488.5 is 115.7?"

Translate word for word. Let p represent the unknown percent.

$$p \cdot 488.5 = 115.7$$
$$488.5\,p = 115.7$$
$$\frac{488.5\,p}{488.5} = \frac{115.7}{488.5}$$
$$p \approx 0.237$$

The percent spent on research and development is about 23.7%.

77. Translate to an equation. Let p represent the unknown percent.

$$0.10(50 \text{ ml}) + 0.25(100 \text{ ml}) = p(150 \text{ ml})$$
$$5 + 25 = 150\,p$$
$$30 = 150\,p$$
$$\frac{30}{150} = \frac{150\,p}{150}$$
$$0.20 = p$$

Margaret will have a 20% solution of HCl.

79. Begin by finding the sales tax. Translate as "5% of $124.45 is what number?" Let x represent the unknown amount of sales tax.

$$5\% \cdot 124.45 = x$$
$$0.05 \cdot 124.45 = x$$
$$6.22 \approx x$$

The sales tax is $6.22.
Now add the grocery purchases to the sales tax to get the total cost of the purchase: $124.45 + $6.22 = $130.67. The total cost is $130.67.

81. Begin by finding the amount of the discount. Translate as "10% of 949 is what number?" Let x represent the unknown amount of the discount.

$$0.10 \cdot 949 = x$$
$$\$94.90 = x$$

The discount is $94.90. The sale price is the original price minus the discount:
$949 – $94.90 = $854.10.

83. Let x represent the price of the sofa before the sales tax was added. Then the amount of the sales tax is 7% of the original price, or $0.07x$.

$$x + 0.07x = 2675$$
$$1.07x = 2675$$
$$\frac{1.07x}{1.07} = \frac{2675}{1.07}$$
$$x = 2500$$

The price of the sofa before the sales tax was added was $2500.

85. Find the amount of the increase in sales from November to December: $4550 – $2600 = $1950. The initial amount is $2600. Let p represent the unknown percent.

$$\frac{p}{100} = \frac{\text{Amount of increase}}{\text{Initial amount}}$$

$$\frac{p}{100} = \frac{1950}{2600}$$

$$2600p = 195,000$$

$$\frac{2600p}{2600} = \frac{195,000}{2600}$$

$$p = 75$$

The sales increased 75% from November to December.

87. Find the amount of the increase in the number of bachelor's degrees awarded: 943,381 – 845,000 = 98,381. The initial amount is 845,000. Let p represent the unknown percent.

$$\frac{p}{100} = \frac{\text{Amount of increase}}{\text{Initial amount}}$$

$$\frac{p}{100} = \frac{98,381}{845,000}$$

$$845,000p = 9,838,100$$

$$\frac{845,000p}{845,000} = \frac{9,838,100}{845,000}$$

$$p \approx 11.6$$

The degrees awarded to women increased by about 11.6%.

89. Find the decrease in the cost of the calculator: $120 – $79 = $41. The initial amount is $120. Let p represent the unknown percent.

$$\frac{p}{100} = \frac{\text{Amount of decrease}}{\text{Initial amount}}$$

$$\frac{p}{100} = \frac{41}{120}$$

$$120p = 4100$$

$$\frac{120p}{120} = \frac{4100}{120}$$

$$p = 34.1\overline{6}$$

The percent of decrease is $34.1\overline{6}\%$.

91. Find the decrease in carbon monoxide concentration: 2.4 – 1.4 = 1 part per million. The initial concentration is 2.4 parts per million. Let p represent the unknown percent.

$$\frac{p}{100} = \frac{\text{Amount of decrease}}{\text{Initial amount}}$$

$$\frac{p}{100} = \frac{1}{2.4}$$

$$2.4p = 100$$

$$\frac{2.4p}{2.4} = \frac{100}{2.4}$$

$$p = 41.6\overline{6}$$

The percent of decrease in the carbon monoxide concentration from 2000 to 2010 was approximately 41.7%.

93. Find the increase in the minimum wage: $7.25 – $0.25 = $7.00. The initial minimum wage $0.25. Let p represent the unknown percent.

$$\frac{p}{100} = \frac{\text{Amount of increase}}{\text{Initial amount}}$$

$$\frac{p}{100} = \frac{7.00}{0.25}$$

$$0.25p = 700$$

$$\frac{0.25p}{0.25} = \frac{700}{0.25}$$

$$p = 2800$$

The minimum waged increased by 2800%.

95. Let x represent the unknown initial number of single fathers in 1970. The percent increase is 597.5%. We know that in 2010, the number of single fathers is 2,790,000, the ending number. The amount of the increase in single fathers is given by taking the ending number of single fathers and subtracting the initial number of single fathers. Therefore, the amount of increase is given by $2,790,000 - x$.

$$\frac{p}{100} = \frac{\text{Amount of increase}}{\text{Initial amount}}$$

$$\frac{597.5}{100} = \frac{2,790,000 - x}{x}$$

$$597.5x = 279,000,000 - 100x$$

$$\underline{+100x} \qquad\qquad \underline{+100x}$$

$$697.5x = 279,000,000 + 0$$

$$\frac{697.5x}{697.5} = \frac{279,000,000}{697.5}$$

$$x = 400,000$$

In 1970, there were 400,000 single fathers.

97. Since there was a change of –22.6%, that means that Black Monday's closing price was 100% – 22.6% = 77.4% of the previous day's closing price. Translate as "77.4% of what number is 1739?" Let x represent the previous day's closing price.

$$0.774 \cdot x = 1739$$
$$0.774x = 1739$$
$$\frac{0.774x}{0.774} = \frac{1739}{0.774}$$
$$x \approx 2246.77$$

The closing price on October 16 was approximately 2246.77.

Review Exercises

1. $-3^0 = -1$

2. $-3(x+7) = -3x - 21$

3. $7xy + 4x - 8xy + 3y$
$$= 7xy - 8xy + 4x + 3y$$
$$= -xy + 4x + 3y$$

4. Solve the equation for n.
$$3(n+9) = n - 11$$
$$3n + 27 = n - 11$$
$$\underline{-n \qquad -n}$$
$$2n + 27 = 0 - 11$$
$$2n + 27 = -11$$
$$\underline{-27 \quad -27}$$
$$2n + 0 = -38$$
$$2n = -38$$
$$\frac{2n}{2} = \frac{-38}{2}$$
$$n = -19$$

5. $\dfrac{1}{5} = \dfrac{n}{10.5}$
$$10.5 \cdot 1 = 5 \cdot n$$
$$10.5 = 5n$$
$$\frac{10.5}{5} = \frac{5n}{5}$$
$$2.1 = n$$

6. $\dfrac{4}{7} = \dfrac{6}{x}$
$$x \cdot 4 = 7 \cdot 6$$
$$4x = 42$$
$$\frac{4x}{4} = \frac{42}{4}$$
$$x = 10.5 \text{ cm}$$

Exercise Set 2.8

1. a) $\{x \mid x \geq -3\}$ b) $[-3, \infty)$

c)

3. a) $\{h \mid h < 6\}$ b) $(-\infty, 6)$

c)

5. a) $\left\{n \mid n < -\dfrac{2}{3}\right\}$ b) $\left(-\infty, -\dfrac{2}{3}\right)$

c)

7. a) $\{t \mid t \geq 2.4\}$ b) $[2.4, \infty)$

c)

9. a) $\{x \mid -3 < x < 6\}$ b) $(-3, 6)$

c)

11. a) $\{n \mid 0 \leq n \leq 5\}$ b) $[0, 5]$

c)

13. This number line shows the set of all real numbers greater than or equal to -4.

Set-builder notation: $\{x \mid x \geq -4\}$

Interval notation: $[-4, \infty)$

15. This number line shows the set of all real numbers less than 5.

Set-builder notation: $\{x \mid x < 5\}$

Interval notation: $(-\infty, 5)$

17. This number line shows the set of all real numbers greater than or equal to −4 and less than 3.

    Set-builder notation: $\{x \mid -4 \le x < 3\}$

    Interval notation: $[-4, 3)$

19. a) $n - 3 > 2$

    $\underline{+3 \quad +3}$

    $n + 0 > 5$

    $n > 5$

    b) $\{n \mid n > 5\}$ c) $(5, \infty)$

    d)
       2 3 4 5 6 7 8

21. a) $z + 2 \le -4$

    $\underline{-2 \quad -2}$

    $z + 0 \le -6$

    $z \le -6$

    b) $\{z \mid z \le -6\}$ c) $(-\infty, -6]$

    d)
       −8 −7 −6 −5 −4 −3

23. a) $16y \ge 32$

    $\dfrac{16y}{16} \ge \dfrac{32}{16}$

    $y \ge 2$

    b) $\{y \mid y \ge 2\}$ c) $[2, \infty)$

    d)
       0 1 2 3 4

25. a) $-3x \ge -12$

    $\dfrac{-3x}{-3} \le \dfrac{-12}{-3}$

    $x \le 4$

    b) $\{x \mid x \le 4\}$ c) $(-\infty, 4]$

    d)
       −1 0 1 2 3 4 5

27. a) $\dfrac{2}{3}x \ge 4$

    $\dfrac{\cancel{3}}{\cancel{2}} \cdot \dfrac{\cancel{2}}{\cancel{3}}x \ge \dfrac{3}{1\cancel{2}} \cdot \cancel{4}^{2}$

    $x \ge 6$

    b) $\{x \mid x \ge 6\}$ c) $[6, \infty)$

    d)
       3 4 5 6 7 8 9

29. a) $-\dfrac{3}{4}m < -6$

    $-\dfrac{\cancel{4}}{\cancel{3}} \cdot -\dfrac{\cancel{3}}{\cancel{4}}m > -\dfrac{4}{1\cancel{3}} \cdot -\cancel{6}^{2}$

    $m > 8$

    b) $\{m \mid m > 8\}$ c) $(8, \infty)$

    d)
       4 5 6 7 8 9 10 11

31. a) $5y + 1 > 16$

    $\underline{-1 \quad -1}$

    $5y > 15$

    $\dfrac{5y}{5} > \dfrac{15}{5}$

    $y > 3$

    b) $\{y \mid y > 3\}$ c) $(3, \infty)$

    d)
       0 1 2 3 4 5 6

33. a) $1 - 6x < 25$

    $\underline{-1 \qquad\quad -1}$

    $0 - 6x < 24$

    $-6x < 24$

    $\dfrac{-6x}{-6} > \dfrac{24}{-6}$

    $x > -4$

    b) $\{x \mid x > -4\}$ c) $(-4, \infty)$

    d)
       −5 −4 −3 −2 −1 0 1

35. a) $\dfrac{a}{2} + 1 < \dfrac{3}{2}$

    $2\left(\dfrac{a}{2} + 1\right) < 2\left(\dfrac{3}{2}\right)$

    $a + 2 < 3$

    $\underline{-2 \quad -2}$

    $a + 0 < 1$

    $a < 1$

    b) $\{a \mid a < 1\}$ c) $(-\infty, 1)$

    d)
       0 1 2 3 4

37. a) Solve for f.

$$10 + 2f \le 2 - 2f$$

$$\underline{-10 \qquad\quad -10}$$

$$0 + 2f \le -8 - 2f$$

$$2f \le -8 - 2f$$

$$\underline{+2f \qquad\quad +2f}$$

$$4f \le -8 + 0$$

$$4f \le -8$$

$$\frac{4f}{4} \le \frac{-8}{4}$$

$$f \le -2$$

b) $\{f \mid f \le -2\}$ c) $(-\infty, -2]$

d)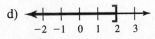
$$-4\ -3\ -2\ -1\ \ 0$$

39. a) Solve for u.

$$3 - 6u \ge -5 - 2u$$

$$\underline{+2u \qquad\quad +2u}$$

$$3 - 4u \ge -5 + 0$$

$$3 - 4u \ge -5$$

$$\underline{-3 \qquad\quad -3}$$

$$0 - 4u \ge -8$$

$$-4u \ge -8$$

$$\frac{-4u}{-4} \le \frac{-8}{-4}$$

$$u \le 2$$

b) $\{u \mid u \le 2\}$ c) $(-\infty, 2]$

d) ![number line]
$$-2\ -1\ \ 0\ \ 1\ \ 2\ \ 3$$

41. a) Solve for c.

$$2(c - 3) < 3(c + 2)$$

$$2c - 6 < 3c + 6$$

$$\underline{-3c \qquad\quad -3c}$$

$$-c - 6 < 0 + 6$$

$$-c - 6 < 6$$

$$\underline{+6 \quad +6}$$

$$-c + 0 < 12$$

$$-c < 12$$

$$\frac{-c}{-1} > \frac{12}{-1}$$

$$c > -12$$

b) $\{c \mid c > -12\}$ c) $(-12, \infty)$

d) ![number line]
$$-14\ -13\ -12\ -11\ -10\ -9$$

43. a) Solve for w.

$$10(9w + 13) - 5(3w - 1) \le 8(11w + 12)$$

$$90w + 130 - 15w + 5 \le 88w + 96$$

$$75w + 135 \le 88w + 96$$

$$\underline{-75w \qquad\qquad -75w}$$

$$0 + 135 \le 13w + 96$$

$$135 \le 13w + 96$$

$$\underline{-96 \qquad\qquad -96}$$

$$39 \le 13w + 0$$

$$39 \le 13w$$

$$\frac{39}{13} \le \frac{13w}{13}$$

$$3 \le w$$

$$w \ge 3$$

b) $\{w \mid w \ge 3\}$ c) $[3, \infty)$

d) ![number line]
$$0\ \ 1\ \ 2\ \ 3\ \ 4\ \ 5$$

45. a) Solve for x.

$$\frac{1}{2}(3x + 1) \le \frac{1}{3}(4x - 5)$$

$$6 \cdot \frac{1}{2}(3x + 1) \le 6 \cdot \frac{1}{3}(4x - 5)$$

$$3(3x + 1) \le 2(4x - 5)$$

$$9x + 3 \le 8x - 10$$

$$\underline{-3 \qquad\quad -3}$$

$$9x + 0 \le 8x - 13$$

$$9x \le 8x - 13$$

$$\underline{-8x \quad -8x}$$

$$1x \le 0 - 13$$

$$x \le -13$$

b) $\{x \mid x \le -13\}$ c) $(-\infty, -13]$

d) ![number line]
$$-15\ -14\ -13\ -12\ -11\ -10$$

47. a) Solve for n.

$$\frac{1}{6}(2n+4)-\frac{1}{3}(3n-15)>1$$

$$6\cdot\left[\frac{1}{6}(2n+4)-\frac{1}{3}(3n-15)\right]>6\cdot1$$

$$2n+4-2(3n-15)>6$$

$$2n+4-6n+30>6$$

$$-4n+34>6$$

$$\underline{\quad-34\quad-34\quad}$$

$$-4n+0>-28$$

$$-4n>-28$$

$$\frac{-4n}{-4}<\frac{-28}{-4}$$

$$n<7$$

b) $\{n\,|\,n<7\}$ c) $(-\infty,7)$

d)

$$4 \quad 5 \quad 6 \quad 7 \quad 8 \quad 9$$

49. a) Solve for l.

$$0.4l-0.37<0.3(l+2.1)$$

$$0.4l-0.37<0.3l+0.63$$

$$100(0.4l-0.37)<100(0.3l+0.63)$$

$$40l-37<30l+63$$

$$\underline{-30l\qquad\quad-30l}$$

$$10l-37<0+63$$

$$10l-37<63$$

$$\underline{\quad+37\quad+37\quad}$$

$$10l+0<100$$

$$10l<100$$

$$\frac{10l}{10}<\frac{100}{10}$$

$$l<10$$

b) $\{l\,|\,l<10\}$ c) $(-\infty,10)$

d)

$$5 \quad 6 \quad 7 \quad 8 \quad 9 \quad 10 \quad 11$$

51. a) Solve for t.

$$0.6t+0.2(t-8)\le0.5t-0.6$$

$$0.6t+0.2t-1.6\le0.5t-0.6$$

$$0.8t-1.6\le0.5t-0.6$$

$$10(0.8t-1.6)\le10(0.5t-0.6)$$

$$8t-16\le5t-6$$

$$\underline{-5t\qquad\quad-5t}$$

$$3t-16\le0-6$$

$$3t-16\le-6$$

$$\underline{\quad+16\quad+16\quad}$$

$$3t+0\le10$$

$$3t\le10$$

$$\frac{3t}{3}\le\frac{10}{3}$$

$$t\le3.\overline{3}$$

b) $\{t\,|\,t\le3.\overline{3}\}$ c) $(-\infty,3.\overline{3}\,]$

d)

$$2 \qquad 3 \quad 3.\overline{3} \;\; 3.\overline{6} \quad 4 \qquad\quad 5$$

53. Let n be the number.

$$n-4>24$$

$$\underline{\quad+4\quad+4\quad}$$

$$n+0>28$$

$$n>28$$

55. Let x be the number.

$$\frac{4}{9}x\le-8$$

$$\frac{\cancel{9}}{\cancel{4}}\cdot\frac{\cancel{4}}{\cancel{9}}x\le\frac{9}{1\cancel{4}}\cdot\left(-\cancel{8}^{2}\right)$$

$$x\le-18$$

57. Let y be the number.

$$8y-36\ge60$$

$$\underline{\quad+36\quad+36\quad}$$

$$8y+0\ge96$$

$$8y\ge96$$

$$\frac{8y}{8}\ge\frac{96}{8}$$

$$y\ge12$$

59. Let a be the number.

$$\frac{1}{2}a + 5 \le 2$$

$$\underline{-5 \quad -5}$$

$$\frac{1}{2}a + 0 \le -3$$

$$2 \cdot \frac{1}{2}a \le 2 \cdot (-3)$$

$$a \le -6$$

61. Let x be the number.

$$4x - 8 < 2x$$

$$\underline{-2x \qquad -2x}$$

$$2x - 8 < 0$$

$$\underline{+8 \quad +8}$$

$$2x + 0 < 8$$

$$2x < 8$$

$$\frac{2x}{2} < \frac{8}{2}$$

$$x < 4$$

63. Let x be the number.

$$25 \ge 6x + 7$$

$$\underline{-7 \qquad -7}$$

$$18 \ge 6x + 0$$

$$18 \ge 6x$$

$$\frac{18}{6} \ge \frac{6x}{6}$$

$$3 \ge x$$

$$x \le 3$$

65. Let l be the length of the rectangular playground and $w = 17$. We know that $lw = A$ and that the area may not exceed 170 square feet, or $A \le 170$.
Translation: $lw \le 170$

$$lw \le 170$$

$$l(17) \le 170$$

$$\frac{17l}{17} \le \frac{170}{17}$$

$$l \le 10$$

The length may not exceed 10 feet.

67. Let h be the height of the storage box, $w = 27$, and $l = 41$. The surface area, $SA = 2lw + 2lh + 2hw$, must be at least 4254 square inches, or $SA \ge 4254$.

Translation: $2lw + 2lh + 2hw \ge 4254$.

$$2lw + 2lh + 2hw \ge 4254$$

$$2(41)(27) + 2(41)h + 2h(27) \ge 4254$$

$$2214 + 82h + 54h \ge 4254$$

$$2214 + 136h \ge 4254$$

$$\underline{-2214 \qquad\qquad -2214}$$

$$0 + 136h \ge 2040$$

$$136h \ge 2040$$

$$\frac{136h}{136} \ge \frac{2040}{136}$$

$$h \ge 15$$

The height must equal or exceed 15 in.

69. Let r be the radius of the circular table. The circumference, $C = 2\pi r$, must be no more than 150 inches.
Translation: $2\pi r \le 150$

$$2\pi r \le 150$$

$$\frac{2\pi r}{2\pi} \le \frac{150}{2\pi}$$

$$r \le \frac{150}{2(3.14)}$$

$$r \le 23.89$$

The radius can be at most 23.89 inches.

71. Let t be the time Jon spent driving. Jon's rate r never exceeds 65 mph, or $r \le 65$. Because $rt = d$, we know that $r = \dfrac{d}{t}$.

Translation: $\dfrac{d}{t} \le 65$

$$\frac{d}{t} \le 65$$

$$\frac{410}{t} \le 65$$

$$t \cdot \frac{410}{t} \le t \cdot 65$$

$$410 \le 65t$$

$$\frac{410}{65} \le \frac{65t}{65}$$

$$\frac{82}{13} \le t$$

$$t \ge 6\frac{4}{13}$$

His time driving will be at least $6\dfrac{4}{13}$ hours.

73. Let R be the revenue that must be generated.

Translation: $P = R - C$ and $P \geq 850{,}000$

$$R - 625{,}000 \geq 850{,}000$$

$$R \geq \$1{,}475{,}000$$

The revenue must be at least $\$1{,}475{,}000$.

75. Let x be the fifth test score.

Translation: $\dfrac{82 + 91 + 95 + 84 + x}{5} \geq 90$

$$\dfrac{82 + 91 + 95 + 84 + x}{5} \geq 90$$

$$\dfrac{352 + x}{5} \geq 90$$

$$5\left(\dfrac{352 + x}{5}\right) \geq 5(90)$$

$$352 + x \geq 450$$

$$352 - 352 + x \geq 450 - 352$$

$$x \geq 98$$

Tina needs to score at least 98 on her fifth test to get an A in the course.

77. $V = iR$ and $V \leq 12$

$$iR \leq 12$$

$$i \cdot 8 \leq 12$$

$$8i \leq 12$$

$$\dfrac{8i}{8} \leq \dfrac{12}{8}$$

$$i \leq 1.5$$

The current must be less than or equal to 1.5 A.

Review Exercises

1. $\{1, 2, 3, 4, 5, 6, 7, 8, 9, 10\}$

2. $\dfrac{2450}{140} = \dfrac{\cancel{70} \cdot 35}{\cancel{70} \cdot 2} = \dfrac{35}{2}$

3. Let x be the number.

$2(x - 5) + 6$

4. Let $a = 1$, $b = -2$, and $c = -3$.

$$b^2 - 4ac = (-2)^2 - 4 \cdot 1 \cdot (-3)$$

$$= 4 - 4(1)(-3)$$

$$= 4 + 12$$

$$= 16$$

5. Solve for n.

$$3\frac{1}{5}n = -108$$

$$\frac{16}{5}n = -108$$

$$\frac{5}{16} \cdot \frac{16}{5}n = -\overset{27}{\cancel{108}} \cdot \frac{5}{\underset{4}{\cancel{16}}}$$

$$n = -\frac{135}{4} \text{ or } -33\frac{3}{4}$$

6. $5n = 7131.04$

$$\frac{5n}{5} = \frac{7131.04}{5}$$

$$n = 1426.208$$

Chapter 2 Review Exercises

1. For $y + 6 + 3y - 4 = 14$, let $y = 3$.

$$3 + 6 + 3(3) - 4 \overset{?}{=} 14$$

$$3 + 6 + 9 - 4 \overset{?}{=} 14$$

$$14 = 14$$

Yes, $y = 3$ is a solution.

2. For $\frac{1}{4}(n - 1) = \frac{1}{2}(n + 1)$, let $n = 5$.

$$\frac{1}{4}(5 - 1) \overset{?}{=} \frac{1}{2}(5 + 1)$$

$$\frac{1}{4} \cdot 4 \overset{?}{=} \frac{1}{2} \cdot 6$$

$$1 \neq 3$$

No, $n = 5$ is not a solution.

3. For $x^2 + 6.1 = 3x - 0.1$, let $x = -0.1$.

$$(-0.1)^2 + 6.1(-0.1) \overset{?}{=} 3(-0.1) - 0.1$$

$$0.01 + 6.1(-0.1) \overset{?}{=} -0.3 - 0.1$$

$$0.01 - 0.61 \overset{?}{=} -0.4$$

$$-0.6 \neq -0.4$$

No, $x = -0.1$ is not a solution.

4. For $\frac{5}{6}m-1=\frac{m}{3}+\frac{1}{2}$, let $m=3$.

$$\frac{5}{6}(3)-1 \overset{?}{=} \frac{3}{3}+\frac{1}{2}$$

$$\frac{5}{2}-1 \overset{?}{=} 1+\frac{1}{2}$$

$$\frac{3}{2} = \frac{3}{2}$$

Yes, $m=3$ is a solution.

5. Let $b=9$ m and $h=6$ m.

$$A = bh$$
$$= (9)(6)$$
$$= 54 \text{ m}^2$$

6. Let $r=8$ in.

$$A = \pi r^2$$
$$\approx (3.14)(8)^2$$
$$\approx 200.96 \text{ in.}^2$$

7. Let $l=9$ cm, $w=5$ cm, and $h=10$ cm.

$$V = \frac{1}{3}lwh$$
$$= \frac{1}{{}^1\cancel{3}}({}^3\cancel{9})(5)(10)$$
$$= (3)(50)$$
$$= 150 \text{ cm}^3$$

8. Let $r=2$ in. and $h=8$ in.

$$V = \pi r^2 h$$
$$\approx (3.14)(2)^2(8)$$
$$\approx (3.14)(4)(8)$$
$$\approx 100.48 \text{ in.}^3$$

9. To find the total area of the entertainment center, begin by sketching three rectangles. Find total area by adding the area of each rectangle.

$$A = lw + lw + lw$$

$$A = 2.5(2.75)+3(2)+2.5(2.75)$$

$$A = 6.875+6+6.875$$

$$A = 19.75 \text{ ft.}^2$$

The total area of the entertainment center is 19.75 ft.^{2}

10. To find the area of the composite figure, begin by sketching a rectangle and a triangle. Find the total area by adding the area of each figure.

$$A = lw+\frac{1}{2}bh$$

$$A = (15)(6)+\frac{1}{2}(6)(4.2)$$

$$A = 90+12.6$$

$$A = 102.6 \text{ in.}^2$$

The total area is 102.6 in.^{2}

11. Solve the equation for x.

$$x+7 = -5$$
$$\underline{-7 \quad -7}$$
$$x+0 = -12$$
$$x = -12$$

Check:
$$x+7 = -5$$
$$-12+7 \overset{?}{=} -5$$
$$-5 = -5$$

12. Solve the equation for x.

$$\frac{2}{3} = x+\frac{3}{4}$$

$${}^4\cancel{12}\left(\frac{2}{\cancel{3}_1}\right) = 12\left(x+\frac{3}{4}\right)$$

$$8 = 12x+9$$
$$\underline{-9 \qquad -9}$$
$$-1 = 12x+0$$

$$\frac{-1}{12} = \frac{12x}{12}$$

$$-\frac{1}{12} = x$$

Check:
$$\frac{2}{3} = x+\frac{3}{4}$$
$$\frac{2}{3} \overset{?}{=} -\frac{1}{12}+\frac{3}{4}$$
$$\frac{8}{12} \overset{?}{=} -\frac{1}{12}+\frac{9}{12}$$
$$\frac{8}{12} = \frac{8}{12}$$

13. Solve the equation for x.

$$-6x = 30$$

$$\frac{-6x}{-6} = \frac{30}{-6}$$

$$x = -5$$

Check: $-6(-5) \overset{?}{=} 30$
$$30 = 30$$

14. Solve the equation for y.
$$-\frac{2}{3}y = -12$$

$$-\frac{3}{2}\cdot\left(-\frac{2}{3}y\right) = -\frac{3}{\cancel{2}}\cdot\left(-6\right)$$

$$y = 18$$

Check: $-\frac{2}{3}\cdot 18 \overset{?}{=} -12$

$$-\frac{36}{3} \overset{?}{=} -12$$

$$-12 = -12$$

15. Solve the equation for n.
$$4n - 9 = 15$$

$$\underline{+9 \quad +9}$$

$$4n + 0 = 24$$

$$4n = 24$$

$$\frac{4n}{4} = \frac{24}{4}$$

$$n = 6$$

Check: $4(6) - 9 \overset{?}{=} 15$

$$24 - 9 \overset{?}{=} 15$$

$$15 = 15$$

16. Solve the equation for a.
$$5(2a - 3) - 6(a + 2) = -7$$

$$10a - 15 - 6a - 12 = -7$$

$$4a - 27 = -7$$

$$\underline{+27 \quad +27}$$

$$4a + 0 = 20$$

$$4a = 20$$

$$\frac{4a}{4} = \frac{20}{4}$$

$$a = 5$$

Check: $5(2(5) - 3) - 6(5 + 2) \overset{?}{=} -7$

$$5(10 - 3) - 6(7) \overset{?}{=} -7$$

$$5(7) - 6(7) \overset{?}{=} -7$$

$$35 - 42 \overset{?}{=} -7$$

$$-7 = -7$$

17. Solve the equation for t.
$$12t + 9 = 4t - 7$$

$$\underline{-4t \qquad -4t}$$

$$8t + 9 = 0 - 7$$

$$8t + 9 = -7$$

$$\underline{-9 \quad -9}$$

$$8t + 0 = -16$$

$$8t = -16$$

$$\frac{8t}{8} = \frac{-16}{8}$$

$$t = -2$$

Check: $12(-2) + 9 \overset{?}{=} 4(-2) - 7$

$$-24 + 9 \overset{?}{=} -8 - 7$$

$$-15 = -15$$

18. Solve the equation for c.
$$2(c + 4) - 5c = 17 - 8c$$

$$2c + 8 - 5c = 17 - 8c$$

$$-3c + 8 = 17 - 8c$$

$$\underline{+8c \qquad +8c}$$

$$5c + 8 = 17 + 0$$

$$5c + 8 = 17$$

$$\underline{-8 \quad -8}$$

$$5c + 0 = 9$$

$$5c = 9$$

$$\frac{5c}{5} = \frac{9}{5}$$

$$c = \frac{9}{5}$$

Check:

$$2\left(\frac{9}{5} + 4\right) - 5\left(\frac{9}{5}\right) \overset{?}{=} 17 - 8\left(\frac{9}{5}\right)$$

$$2\left(\frac{9}{5} + \frac{20}{5}\right) - \cancel{5}\left(\frac{9}{\cancel{5}}\right) \overset{?}{=} 17 - 8\left(\frac{9}{5}\right)$$

$$2\left(\frac{29}{5}\right) - 9 \overset{?}{=} 17 - 8\left(\frac{9}{5}\right)$$

$$\frac{58}{5} - 9 \overset{?}{=} 17 - \frac{72}{5}$$

$$\frac{58}{5} - \frac{45}{5} \overset{?}{=} \frac{85}{5} - \frac{72}{5}$$

$$\frac{13}{5} = \frac{13}{5}$$

19. Solve the equation for c.
$$1-2(3c-5)=10-6c$$
$$1-6c+10=10-6c$$
$$11-6c=10-6c$$
The expressions on each side of the equation have the same variable term but different constant terms, so the equation is a contradiction and has no solution.

20. Solve the equation for b.
$$5(b-1)-3b=7b-5(1+b)$$
$$5b-5-3b=7b-5-5b$$
$$2b-5=2b-5$$
Because the linear equation is an identity, every real number is a solution.

21. Solve the equation for u.
$$5(u-1)-(u-2)=10-(2u+1)$$
$$5u-5-u+2=10-2u-1$$
$$4u-3=-2u+9$$
$$\underline{+2u\qquad\quad +2u}$$
$$6u-3=0+9$$
$$6u-3=9$$
$$\underline{+3\quad +3}$$
$$6u+0=12$$
$$6u=12$$
$$\frac{6u}{6}=\frac{12}{6}$$
$$u=2$$

Check: $5(2-1)-(2-2)\overset{?}{=}10-(2(2)+1)$
$$5(1)-0\overset{?}{=}10-(4+1)$$
$$5-0\overset{?}{=}10-5$$
$$5=5$$

22. Solve the equation for m.
$$\frac{3}{4}m-\frac{1}{2}=\frac{2}{3}m+1$$
$$12\left(\frac{3}{4}m-\frac{1}{2}\right)=12\left(\frac{2}{3}m+1\right)$$
$$\overset{3}{\cancel{12}}\cdot\frac{3}{\cancel{4}}m-\overset{6}{\cancel{12}}\cdot\frac{1}{\cancel{2}}=\overset{4}{\cancel{12}}\cdot\frac{2}{\cancel{3}}m+12\cdot1$$
$$9m-6=8m+12$$
$$\underline{-8m\qquad\; -8m}$$
$$m-6=0+12$$
$$m-6=12$$
$$\underline{+6\quad +6}$$
$$m+0=18$$
$$m=18$$

Check: $\dfrac{3}{2\cancel{4}}\left(\overset{9}{\cancel{18}}\right)-\dfrac{1}{2}\overset{?}{=}\dfrac{2}{\cancel{3}}\left(\overset{6}{\cancel{18}}\right)+1$
$$\frac{27}{2}-\frac{1}{2}\overset{?}{=}12+1$$
$$\frac{26}{2}\overset{?}{=}13$$
$$13=13$$

23. Solve the equation for x.
$$\frac{1}{12}(4x+9)=\frac{2}{3}x$$
$$\frac{12}{1}\cdot\frac{1}{12}(4x+9)=\frac{\cancel{12}}{1}\cdot\overset{4}{\frac{2}{\cancel{3}}}x$$
$$4x+9=8x$$
$$\underline{-4x\qquad -4x}$$
$$0+9=4x$$
$$9=4x$$
$$\frac{9}{4}=\frac{4x}{4}$$
$$\frac{9}{4}=x$$

Check: $\dfrac{1}{12}\left(4\left(\dfrac{9}{4}\right)+9\right)\overset{?}{=}\dfrac{\cancel{2}}{\cancel{3}}\left(\dfrac{\overset{3}{\cancel{9}}}{\underset{2}{\cancel{4}}}\right)$
$$\frac{1}{12}(9+9)\overset{?}{=}\frac{3}{2}$$
$$\frac{1}{\underset{2}{\cancel{12}}}\cdot\left(\overset{3}{\cancel{18}}\right)\overset{?}{=}\frac{3}{2}$$
$$\frac{3}{2}=\frac{3}{2}$$

24. Solve the equation for x.

$$\frac{5}{3} + 7x - 2x = \frac{1}{3}(9x + 2)$$

$$\frac{5}{3} + 5x = \frac{1}{3} \cdot 9x + \frac{1}{3} \cdot 2$$

$$\frac{5}{3} + 5x = 3x + \frac{2}{3}$$

$$3\left(\frac{5}{3} + 5x\right) = 3\left(3x + \frac{2}{3}\right)$$

$$\cancel{3} \cdot \frac{5}{\cancel{3}} + 3 \cdot 5x = 3 \cdot 3x + \cancel{3} \cdot \frac{2}{\cancel{3}}$$

$$5 + 15x = 9x + 2$$

$$\underline{\quad -9x \quad\quad -9x \quad}$$

$$5 + 6x = 0 + 2$$

$$5 + 6x = 2$$

$$\underline{-5 \quad\quad\quad -5 \quad}$$

$$0 + 6x = -3$$

$$6x = -3$$

$$\frac{6x}{6} = \frac{-3}{6}$$

$$x = -\frac{1}{2}$$

Check:

$$\frac{5}{3} + 7\left(-\frac{1}{2}\right) - 2\left(-\frac{1}{2}\right) \overset{?}{=} \frac{1}{3}\left(9\left(-\frac{1}{2}\right) + 2\right)$$

$$\frac{5}{3} - \frac{7}{2} + 1 \overset{?}{=} \frac{1}{3}\left(-\frac{9}{2} + 2\right)$$

$$\frac{10}{6} - \frac{21}{6} + \frac{6}{6} \overset{?}{=} \frac{1}{3}\left(-\frac{9}{2} + \frac{4}{2}\right)$$

$$-\frac{11}{6} + \frac{6}{6} \overset{?}{=} \frac{1}{3}\left(-\frac{5}{2}\right)$$

$$-\frac{5}{6} \overset{?}{=} -\frac{5}{6}$$

25. Solve the equation for m.

$$2 - \frac{2}{3}m = 6\left(m - \frac{1}{3}\right) - 4m + 2$$

$$2 - \frac{2}{3}m = 6m - 2 - 4m + 2$$

$$2 - \frac{2}{3}m = 2m$$

$$3\left(2 - \frac{2}{3}m\right) = 3(2m)$$

$$3 \cdot 2 - 3 \cdot \frac{2}{3}m = 6m$$

$$6 - 2m = 6m$$

$$\underline{\quad +2m \quad\quad +2m \quad}$$

$$6 + 0 = 8m$$

$$6 = 8m$$

$$\frac{6}{8} = \frac{8m}{8}$$

$$\frac{3}{4} = m$$

Check:

$$2 - \frac{\cancel{2}}{\cancel{3}}\left(\frac{\cancel{3}}{\underset{2}{\cancel{4}}}\right) \overset{?}{=} 6\left(\frac{3}{4} - \frac{1}{3}\right) - 4\left(\frac{3}{4}\right) + 2$$

$$2 - \frac{1}{2} \overset{?}{=} 6\left(\frac{9}{12} - \frac{4}{12}\right) - 3 + 2$$

$$\frac{4}{2} - \frac{1}{2} \overset{?}{=} \cancel{6}\left(\frac{5}{\underset{2}{\cancel{12}}}\right) - 3 + 2$$

$$\frac{3}{2} \overset{?}{=} \frac{5}{2} - 3 + 2$$

$$\frac{3}{2} \overset{?}{=} \frac{5}{2} - \frac{6}{2} + \frac{4}{2}$$

$$\frac{3}{2} = \frac{3}{2}$$

26. Solve the equation for x.

$$1.4x - 0.5(9 - 6x) = 6 + 2.4x$$

$$10\left[1.4x - 0.5(9 - 6x)\right] = 10\left[6 + 2.4x\right]$$

$$14x - 5(9 - 6x) = 60 + 24x$$

$$14x - 45 + 30x = 60 + 24x$$

$$44x - 45 = 60 + 24x$$

$$\underline{-24x \qquad\qquad -24x}$$

$$20x - 45 = 60 + 0$$

$$20x - 45 = 60$$

$$\underline{+45 \qquad +45}$$

$$20x + 0 = 105$$

$$20x = 105$$

$$\frac{20x}{20} = \frac{105}{20}$$

$$x = 5.25$$

Check:

$$1.4(5.25) - 0.5(9 - 6(5.25)) \overset{?}{=} 6 + 2.4(5.25)$$

$$7.35 - 0.5(9 - 31.5) \overset{?}{=} 6 + 12.6$$

$$7.35 - 0.5(-22.5) \overset{?}{=} 18.6$$

$$7.35 + 11.25 \overset{?}{=} 18.6$$

$$18.6 = 18.6$$

27. Solve the equation for x.

$$3(x + 2) - 4 = x + 2x + 2$$

$$3x + 6 - 4 = 3x + 2$$

$$3x + 2 = 3x + 2$$

Because the linear equation is an identity, every real number is a solution.

28. Solve the equation for x.

$$4x - 3(x + 5) = 2(2x - 3) - 3(x + 5)$$

$$4x - 3x - 15 = 4x - 6 - 3x - 15$$

$$x - 15 = x - 6 - 15$$

$$x - 15 = x - 21$$

The expressions on each side of the equation have the same variable term but different constant terms, so the equation is a contradiction and has no solution.

29. Let w be the width of the shower curtain.

Translation: $A = lw$

$$A = lw$$

$$5760 = 72w$$

$$\frac{5760}{72} = \frac{72w}{72}$$

$$80 \text{ in.} = w$$

The shower curtain will be 80 inches wide.

30. Use the formula for the area of a trapezoid. Let $A = 58$, $h = 10$, and $a = 4.5$. Solve for b.

$$A = \frac{1}{2}h(a + b)$$

$$58 = \frac{1}{2}(10)(4.5 + b)$$

$$58 = 5(4.5 + b)$$

$$58 = 22.5 + 5b$$

$$\underline{-22.5 \qquad -22.5}$$

$$35.5 = 0 + 5b$$

$$35.5 = 5b$$

$$\frac{35.5}{5} = \frac{5b}{5}$$

$$7.1 \text{ in.} = b$$

31. Let d be the distance that Randy has already skied.

Translation: $\dfrac{1}{3}(1.8) = d$

$$\frac{1}{3}(1.8) = d$$

$$0.6 \text{ mi.} = d$$

Now subtract the distance he has skied from the total trail length to find the distance he has left to ski. $1.8 - 0.6 = 1.2$ mi.

32. Let l be the length of the trunk.

Translation: $SA = 2lw + 2lh + 2hw$

$$SA = 2lw + 2lh + 2hw$$

$$2610 = 2l(36) + 2l(15) + 2(15)(36)$$

$$2610 = 72l + 30l + 1080$$

$$2610 = 102l + 1080$$

$$\underline{-1080 \qquad\qquad -1080}$$

$$1530 = 102l + 0$$

$$1530 = 102l$$

$$\frac{1530}{102} = \frac{102l}{102}$$

$$15 \text{ in.} = l$$

The length is 15 in.

33. Solve for x.

$$x + y = 1$$

$$\underline{-y \qquad -y}$$

$$x + 0 = 1 - y$$

$$x = 1 - y$$

34. Solve for d.

$$C = \pi d$$

$$\frac{C}{\pi} = \frac{\pi d}{\pi}$$

$$\frac{C}{\pi} = d$$

35. Solve for h.

$$A = \frac{1}{2}bh$$

$$2 \cdot A = 2 \cdot \frac{1}{2}bh$$

$$2A = bh$$

$$\frac{2A}{b} = \frac{bh}{b}$$

$$\frac{2A}{b} = h$$

36. Solve for m.

$$F = \frac{mv^2}{r}$$

$$r \cdot F = r \cdot \frac{mv^2}{r}$$

$$Fr = mv^2$$

$$\frac{Fr}{v^2} = \frac{mv^2}{v^2}$$

$$\frac{Fr}{v^2} = m$$

37. Solve for h.

$$V = \frac{1}{3}\pi r^2 h$$

$$3 \cdot V = 3 \cdot \frac{1}{3}\pi r^2 h$$

$$3V = \pi r^2 h$$

$$\frac{3V}{\pi r^2} = \frac{\pi r^2 h}{\pi r^2}$$

$$\frac{3V}{\pi r^2} = h$$

38. Solve for m.

$$y = mx + b$$

$$\underline{-b} \qquad \underline{-b}$$

$$y - b = mx + 0$$

$$y - b = mx$$

$$\frac{y - b}{x} = \frac{mx}{x}$$

$$\frac{y - b}{x} = m$$

39. Solve for w.

$$P = 2l + 2w$$

$$\underline{-2l} \quad \underline{-2l}$$

$$P - 2l = 0 + 2w$$

$$P - 2l = 2w$$

$$\frac{P - 2l}{2} = \frac{2w}{2}$$

$$\frac{P - 2l}{2} = w$$

40. Solve for h.

$$A = \frac{1}{2}h(a + b)$$

$$2 \cdot A = 2 \cdot \frac{1}{2}h(a + b)$$

$$2A = h(a + b)$$

$$\frac{2A}{(a + b)} = \frac{h(a + b)}{(a + b)}$$

$$\frac{2A}{a + b} = h$$

41. Let x be the number.

$$6x = -18$$

$$\frac{6x}{6} = \frac{-18}{6}$$

$$x = -3$$

42. Let m be the number.

$$\frac{1}{2}m - 3 = \frac{1}{4}m - 9$$

$$4\left(\frac{1}{2}m - 3\right) = 4\left(\frac{1}{4}m - 9\right)$$

$$4\cdot\frac{1}{2}m - 4\cdot 3 = 4\cdot\frac{1}{4}m - 4\cdot 9$$

$$2m - 12 = m - 36$$

$$\underline{-m \qquad\qquad -m}$$

$$m - 12 = 0 - 36$$

$$m - 12 = -36$$

$$\underline{+12 \qquad +12}$$

$$m + 0 = -24$$

$$m = -24$$

43. Let v be the number.

$$2(v + 4) - 1 = 1$$

$$2v + 8 - 1 = 1$$

$$2v + 7 = 1$$

$$\underline{-7 \quad -7}$$

$$2v + 0 = -6$$

$$2v = -6$$

$$\frac{2v}{2} = \frac{-6}{2}$$

$$v = -3$$

44. Let y be the number.

$$2(y - 1) + 3y = 6y - 20$$

$$2y - 2 + 3y = 6y - 20$$

$$5y - 2 = 6y - 20$$

$$\underline{-6y \qquad\quad -6y}$$

$$-y - 2 = 0 - 20$$

$$-y - 2 = -20$$

$$\underline{+2 \quad +2}$$

$$-y + 0 = -18$$

$$-y = -18$$

$$\frac{-y}{-1} = \frac{-18}{-1}$$

$$y = 18$$

45. The ratio of sugar to flour is

$$\frac{\frac{3}{4}}{2} = \frac{3}{4} \div 2 = \frac{3}{4}\cdot\frac{1}{2} = \frac{3}{8}.$$

46. The ratio of price to weight is $\dfrac{0.89}{10.5} \approx \dfrac{0.085}{1}$.

This unit ratio means that each ounce costs 8.5 cents.

47. No, because the cross products are not equal.

$$20\cdot 2 \overset{?}{=} 5\cdot 10$$

$$40 \neq 50$$

48. Yes, because the cross products are equal.

$$4(-2) \overset{?}{=} 8(-1)$$

$$-8 = -8$$

49. $$\frac{3}{5} = \frac{15}{x}$$

$$x\cdot 3 = 5\cdot 15$$

$$3x = 75$$

$$\frac{3x}{3} = \frac{75}{3}$$

$$x = 25$$

50. $$\frac{-11}{12} = \frac{n}{-24}$$

$$-24(-11) = 12\cdot n$$

$$264 = 12n$$

$$\frac{264}{12} = \frac{12n}{12}$$

$$22 = n$$

51. $$\frac{7}{p} = \frac{\frac{2}{5}}{4\frac{1}{2}}$$

$$\frac{7}{p} = \frac{\frac{2}{5}}{4\frac{1}{2}}$$

$$4\frac{1}{2}\cdot 7 = p\cdot\frac{2}{5}$$

$$\frac{9}{2}\cdot\frac{7}{1} = \frac{2}{5}p$$

$$\frac{63}{2} = \frac{2}{5}p$$

$$\frac{63}{2}\cdot\frac{5}{2} = \frac{5}{2}\cdot\frac{2}{5}p$$

$$\frac{315}{4} = p$$

$$p = \frac{315}{4} = 78\frac{3}{4}$$

52. $$\frac{c}{12.7} = \frac{-5}{3}$$

$$3\cdot c = 12.7\cdot(-5)$$

$$3c = -63.5$$

$$\frac{3c}{3} = \frac{-63.5}{3}$$

$$c = -21.1\overline{6}$$

53. Translate to a proportion and solve.

$$\frac{\$5.85}{4.5 \text{ lb.}} = \frac{\$x}{7 \text{ lb.}}$$

$$7(5.85) = 4.5 \cdot x$$

$$40.95 = 4.5x$$

$$\frac{40.95}{4.5} = \frac{4.5x}{4.5}$$

$$\$9.10 = x$$

54. Translate to a proportion and solve.

$$\frac{209 \text{ miles}}{23.7 \text{ gallons}} = \frac{x \text{ miles}}{25 \text{ gallons}}$$

$$25 \cdot 209 = 23.7 \cdot x$$

$$5225 = 23.7x$$

$$\frac{5225}{23.7} = \frac{23.7x}{23.7}$$

$$220.5 \text{ mi.} \approx x$$

55. Translate to a proportion.

$$\frac{10}{15} = \frac{8}{y}$$

$$y \cdot 10 = 15 \cdot 8$$

$$10y = 120$$

$$\frac{10y}{10} = \frac{120}{10}$$

$$y = 12 \text{ in.}$$

56. Translate to proportions.

$$\frac{7}{12} = \frac{2\frac{3}{4}}{a} \qquad\qquad \frac{7}{12} = \frac{b}{8}$$

$$a \cdot 7 = 12 \cdot 2.75 \qquad 8 \cdot 7 = 12 \cdot b$$

$$7a = 33 \qquad\qquad 56 = 12b$$

$$\frac{7a}{7} = \frac{33}{7} \qquad\qquad \frac{\overset{14}{\cancel{56}}}{\underset{3}{\cancel{12}}} = \frac{12b}{12}$$

$$a = 4\frac{5}{7} \text{ in.} \qquad\qquad 4\frac{2}{3} \text{ in.} = b$$

$$\frac{7}{12} = \frac{7\frac{7}{8}}{c}$$

$$c \cdot 7 = \overset{3}{\cancel{12}} \cdot \frac{63}{\underset{2}{\cancel{8}}}$$

$$7c = \frac{189}{2}$$

$$\frac{1}{7} \cdot 7c = \frac{1}{\cancel{7}} \cdot \frac{\overset{27}{\cancel{189}}}{2}$$

$$c = 13\frac{1}{2} \text{ in.}$$

57. $15\% = \dfrac{15}{100} = \dfrac{3}{20} = 0.15$

58. $82.5\% = \dfrac{82.5}{100} = 0.825$

$$= \frac{825}{1000} = \frac{33}{40}$$

59. $12\dfrac{1}{2}\% = \dfrac{12\frac{1}{2}}{100}$

$$= 12\frac{1}{2} \div 100$$

$$= \frac{\overset{1}{\cancel{25}}}{2} \cdot \frac{1}{\underset{4}{\cancel{100}}}$$

$$= \frac{1}{8} = 0.125$$

60. $33\dfrac{1}{3}\% = \dfrac{33\dfrac{1}{3}}{100}$

$\quad\quad = 33\dfrac{1}{3} \div 100$

$\quad\quad = \dfrac{100}{3} \cdot \dfrac{1}{100}$

$\quad\quad = \dfrac{1}{3} = 0.\overline{3}$

61. $\dfrac{2}{5} = \dfrac{2}{\cancel{5}} \cdot \dfrac{\overset{20}{\cancel{100}}}{1}\% = \dfrac{40\%}{1} = 40\%$

62. $\dfrac{4}{11} = \dfrac{4}{11} \cdot \dfrac{100}{1}\% = \dfrac{400}{11}\% = 36\dfrac{4}{11}\%$ or $36.\overline{36}\%$

63. $0.35 = 0.35 \cdot 100\% = 35\%$

64. $2.016 = 2.016 \cdot 100\% = 201.6\%$

65. Translate "16% of 91 is what number?" word for word. Let n represent the unknown number.
$0.16 \cdot 91 = n$
$\quad\quad 14.56 = n$
The number is 14.56.

66. Translate "14.5% of what number is 42?" word for word. Let n represent the unknown number.
$0.145 \cdot n = 42$
$\quad\quad 0.145n = 42$
$\quad\quad \dfrac{0.145n}{0.145} = \dfrac{42}{0.145}$
$\quad\quad\quad n \approx 289.7$
The number is 289.7.

67. Translate "6.5 is what percent of 20?" word for word. Let p represent the unknown percent.
$\quad 6.5 = p \cdot 20$
$\quad 6.5 = 20p$
$\quad \dfrac{6.5}{20} = \dfrac{20p}{20}$
$0.325 = p$
To write 0.325 as a percent, multiply by 100%.
The percent is $0.325 \cdot 100\% = 32.5\%$.

68. Translate "what percent is 18 out of 32?" to a proportion. Let p represent the unknown percent.
$\quad \dfrac{p}{100} = \dfrac{18}{32}$
$\quad 32 \cdot p = 100 \cdot 18$
$\quad\quad 32p = 1800$
$\quad\quad \dfrac{32p}{32} = \dfrac{1800}{32}$
$\quad\quad\quad p = 56.25$
The percent is 56.25%

69. This situation is the same as "47% of 1016 is what number?" Translate word for word. Let n represent the unknown number.
$0.47 \cdot 1016 = n$
$\quad\quad 477.52 = n$
Approximately 478 knew someone who lost a job.

70. Begin by finding the sales tax. Translate as "6% of $759.99 is what number?" Let x represent the unknown amount of sales tax.
$0.06 \cdot 759.99 = x$
$\quad\quad 45.60 \approx x$
Now add the cost of the table to the sales tax to get the total cost of the purchase: $759.99 + $45.60 = $805.59. The total cost is $805.59.

71. This situation is the same as "what percent is 96 out of 162?" Let p represent the unknown percent.
$\quad \dfrac{p}{100} = \dfrac{96}{162}$
$\quad 162 \cdot p = 100 \cdot 96$
$\quad\quad 162p = 9600$
$\quad\quad \dfrac{162p}{162} = \dfrac{9600}{162}$
$\quad\quad\quad p \approx 59.3$
The Red Sox won approximately 59.3% of its games.

72. Find the decrease in the price: $75.4 - 48.2 = 27.2$. The initial amount is 75.4. Let p represent the unknown percent.

$$\frac{p}{100} = \frac{\text{Amount of decrease}}{\text{Initial amount}}$$

$$\frac{p}{100} = \frac{27.2}{75.4}$$

$$75.4p = 2720$$

$$\frac{75.4p}{75.4} = \frac{2720}{75.4}$$

$$p \approx 36.1$$

73. a) Solve for x.
$$-6x > 18$$

$$\frac{-6x}{-6} < \frac{18}{-6}$$

$$x < -3$$

b) $\{x | x < -3\}$ c) $(-\infty, -3)$

d)

74. a) Solve for x.
$$-3x + 2 \le 5$$

$$\underline{-2 \quad -2}$$

$$-3x + 0 \le 3$$

$$-3x \le 3$$

$$\frac{-3x}{-3} \ge \frac{3}{-3}$$

$$x \ge -1$$

b) $\{x | x \ge -1\}$ c) $[-1, \infty)$

d)

75. a) Solve for z.
$$1 - 2z + 4 \le -3(z+1) + 7$$

$$5 - 2z \le -3z - 3 + 7$$

$$5 - 2z \le -3z + 4$$

$$\underline{+3z \quad +3z}$$

$$5 + 1z \le 0 + 4$$

$$5 + z \le 4$$

$$\underline{-5 \quad -5}$$

$$0 + z \le -1$$

$$z \le -1$$

b) $\{z | z \le -1\}$ c) $(-\infty, -1]$

d)

76. a) Solve for v.
$$\frac{1}{4}v < 3 + v$$

$$4 \cdot \frac{1}{4}v < 4(3+v)$$

$$v < 12 + 4v$$

$$\underline{-4v \qquad -4v}$$

$$-3v < 12 + 0$$

$$-3v < 12$$

$$\frac{-3v}{-3} > \frac{12}{-3}$$

$$v > -4$$

b) $\{v | v > -4\}$ c) $(-4, \infty)$

d)

77. a) Solve for m.
$$\frac{1}{2}(m-1) > 5 - \frac{7}{2}$$

$$2 \cdot \frac{1}{2}(m-1) > 2\left(5 - \frac{7}{2}\right)$$

$$m - 1 > 2 \cdot 5 - 2 \cdot \frac{7}{2}$$

$$m - 1 > 10 - 7$$

$$m - 1 > 3$$

$$\underline{+1 \quad +1}$$

$$m + 0 > 4$$

$$m > 4$$

b) $\{m | m > 4\}$ c) $(4, \infty)$

d)

78. a) Solve for c.
$$2c - (5c - 7) + 4c \ge 8 - 10$$

$$2c - 5c + 7 + 4c \ge -2$$

$$c + 7 \ge -2$$

$$\underline{-7 \quad -7}$$

$$c + 0 \ge -9$$

$$c \ge -9$$

b) $\{c | c \ge -9\}$ c) $[-9, \infty)$

d)

79. Let p be the number.
$$13 > 3 - 10p$$
$$\underline{-3 \quad -3}$$
$$10 > 0 - 10p$$
$$10 > -10p$$
$$\frac{10}{-10} < \frac{-10p}{-10}$$
$$-1 < p$$
$$p > -1$$

80. Let z be the number.
$$3 + 2z < 3(z - 5)$$
$$3 + 2z < 3z - 15$$
$$\underline{-2z \quad -2z}$$
$$3 + 0 < z - 15$$
$$3 < z - 15$$
$$\underline{+15 \quad +15}$$
$$18 < z + 0$$
$$18 < z$$
$$z > 18$$

81. Let v be the number.
$$-\frac{1}{2}x \geq 4$$
$$-2\left(-\frac{1}{2}x\right) \leq -2(4)$$
$$x \leq -8$$

82. Let k be the number.
$$-2 \leq 1 - \frac{1}{4}k$$
$$4(-2) \leq 4\left(1 - \frac{1}{4}k\right)$$
$$-8 \leq 4 - k$$
$$\underline{-4 = -4}$$
$$-12 \leq 0 - k$$
$$-12 \leq -k$$
$$\frac{-12}{-1} \geq \frac{-k}{-1}$$
$$12 \geq k$$
$$k \leq 12$$

83. Let d be the diameter of the circular flowerbed.
Translation: $C = \pi d$ and $C \leq 10$
$$\pi d \leq 10$$
$$d \leq \frac{10}{\pi}$$
$$d \leq \frac{10}{3.14}$$
$$d \leq 3.18 \text{ ft.}$$
The diameter must not exceed 3.18 ft.

84. Let R be the revenue that the company must make.
Translation: $P = R - C$ and
$$P \geq 350,000$$
$$R - C \geq 350,000$$
$$R - 475,000 \geq 350,000$$
$$\underline{+475,000 \quad +475,000}$$
$$R + 0 \geq 825,000$$
$$R \geq \$825,000$$
The company's revenue must be at least $825,000.

Chapter 2 Practice Test

1. Equation because there is an equal sign.

2. Nonlinear because there is a variable raised to an exponent other than 1.

3. For $3x^2 + x = 10x$, let $x = 3$.
$$3(3)^2 + (3) \overset{?}{=} 10(3)$$
$$3(9) + 3 \overset{?}{=} 30$$
$$27 + 3 \overset{?}{=} 30$$
$$30 = 30$$
Yes, $x = 3$ is a solution because it makes the equation true.

4. Solve the equation for a.

$$-5.6 + 8a = 7a + 7.6$$

$$10(-5.6 + 8a) = (7a + 7.6)10$$

$$-56 + 80a = 70a + 76$$

$$\underline{\quad -70a \qquad -70a\quad}$$

$$-56 + 10a = \quad 0 + 76$$

$$-56 + 10a = 76$$

$$\underline{+56 \qquad\qquad +56}$$

$$0 + 10a = 132$$

$$\frac{10a}{10} = \frac{132}{10}$$

$$a = 13.2$$

Check: $\quad -5.6 + 8a \overset{?}{=} 7a + 7.6$

$$-5.6 + 8(13.2) \overset{?}{=} 7(13.2) + 7.6$$

$$-5.6 + 105.6 \overset{?}{=} 92.4 + 7.6$$

$$100 = 100$$

5. Solve the equation for x.

$$8 + 5(x - 3) = 9x - (6x + 1)$$

$$8 + 5x - 15 = 9x - 6x - 1$$

$$5x - 7 = 3x - 1$$

$$\underline{-3x \qquad -3x}$$

$$2x - 7 = \quad 0 - 1$$

$$2x - 7 = -1$$

$$\underline{+7 \quad +7}$$

$$2x + 0 = 6$$

$$2x = 6$$

$$\frac{2x}{2} = \frac{6}{2}$$

$$x = 3$$

Check: $8 + 5(x - 3) \overset{?}{=} 9x - (6x + 1)$

$$8 + 5(3 - 3) \overset{?}{=} 9(3) - (6(3) + 1)$$

$$8 + 5(0) \overset{?}{=} 27 - (18 + 1)$$

$$8 + 0 \overset{?}{=} 27 - 19$$

$$8 = 8$$

6. Solve the equation for c.

$$1 - 3(c + 4) + 8c = 1 - 42 + 5(c + 7)$$

$$1 - 3c - 12 + 8c = 1 - 42 + 5c + 35$$

$$-11 + 5c = -6 + 5c$$

$$\underline{\quad -5c \qquad\quad -5c\quad}$$

$$-11 + 0 = -6 + 0$$

$$-11 = -6$$

Because the equation is a contradiction, it has no solution.

7. Solve the equation for m.

$$-\frac{1}{8}(6m - 32) = \frac{3}{4}m - 2$$

$$-\frac{1}{\cancel{8}} \cdot \overset{3}{\cancel{6}}\, m - \left(-\frac{1}{\cancel{8}}\right)\overset{4}{\cancel{32}} = \frac{3}{4}m - 2$$

$$-\frac{3}{4}m + 4 = \frac{3}{4}m - 2$$

$$4\left(-\frac{3}{4}m + 4\right) = 4\left(\frac{3}{4}m - 2\right)$$

$$-3m + 16 = 3m - 8$$

$$\underline{+3m \qquad\qquad +3m}$$

$$0 + 16 = 6m - 8$$

$$16 = 6m - 8$$

$$\underline{+8 \qquad\qquad +8}$$

$$24 = 6m + 0$$

$$\frac{24}{6} = \frac{6m}{6}$$

$$4 = m$$

Check: $\quad -\frac{1}{8}(6m - 32) \overset{?}{=} \frac{3}{4}m - 2$

$$-\frac{1}{8}(6(4) - 32) \overset{?}{=} \frac{3}{4}(4) - 2$$

$$-\frac{1}{8}(24 - 32) \overset{?}{=} 3 - 2$$

$$-\frac{1}{8}(-8) \overset{?}{=} 1$$

$$1 = 1$$

8. Solve the equation for x.

$$2x + y = 7$$

$$\underline{-y \quad\quad -y}$$

$$2x + 0 = 7 - y$$

$$2x = 7 - y$$

$$\frac{2x}{2} = \frac{7 - y}{2}$$

$$x = \frac{7 - y}{2}$$

9. Solve the equation for h.

$$A = \frac{1}{2}bh$$

$$2 \cdot A = 2 \cdot \frac{1}{2}bh$$

$$2A = bh$$

$$\frac{2A}{b} = \frac{bh}{b}$$

$$\frac{2A}{b} = h$$

10.
$$\frac{m}{8} = \frac{5}{12}$$

$$12 \cdot m = 8 \cdot 5$$

$$12m = 40$$

$$\frac{12m}{12} = \frac{40}{12}$$

$$m = \frac{10}{3} = 3\frac{1}{3}$$

11. Write 22 over 100, then simplify.

$$22\% = \frac{22}{100} = \frac{11}{50} = 0.22$$

12. Write $3\frac{1}{3}$ over 100, then simplify.

$$3\frac{1}{3}\% = \frac{3\frac{1}{3}}{100}$$

$$= 3\frac{1}{3} \div 100$$

$$= \frac{\cancel{10}}{3} \cdot \frac{1}{\cancel{100}_{10}}$$

$$= \frac{1}{30} = 0.0\overline{3}$$

13. $3.2 = 3.2 \cdot 100\% = 320\%$

14. $\dfrac{2}{5} = \dfrac{2}{5} \cdot \dfrac{100}{1}\% = \dfrac{200}{5}\% = 40\%$

15. Set-builder notation: $\left\{x \mid x \geq 3\right\}$

Interval notation: $[3, \infty)$

Graph:

16. Set-builder notation: $\left\{x \mid -1 \leq x < 4\right\}$

Interval notation: $[-1, 4)$

Graph:

17. a) Solve the inequality for m.

$$-5m + 3 > -12$$

$$\underline{-3 \quad\quad -3}$$

$$-5m + 0 > -15$$

$$\frac{-5m}{-5} < \frac{-15}{-5}$$

$$m < 3$$

b) $\left\{m \mid m < 3\right\}$

c) $(-\infty, 3)$

d)

18. a) Solve the inequality for x.

$$-2(2 - x) \geq -20$$

$$-4 + 2x \geq -20$$

$$\underline{+4 \quad\quad\quad\quad +4}$$

$$0 + 2x \geq -16$$

$$\frac{2x}{2} \geq \frac{-16}{2}$$

$$x \geq -8$$

b) $\left\{x \mid x \geq -8\right\}$

c) $[-8, \infty)$

d)

19. a) Solve the inequality for p.

$$6(p-2)-5 > 3-4(p+6)$$
$$6p-12-5 > 3-4p-24$$
$$6p-17 > -21-4p$$
$$\underline{+4p +4p}$$
$$10p-17 > -21+0$$
$$10p-17 > -21$$
$$\underline{+17 +17}$$
$$10p+0 > -4$$
$$\frac{10p}{10} > -\frac{4}{10}$$
$$p > -\frac{2}{5}$$

b) $\left\{ p \mid p > -\dfrac{2}{5} \right\}$ c) $\left(-\dfrac{2}{5}, \infty \right)$

d)

20. a) Solve the inequality for l.

$$\frac{1}{5}(l+10) \ge \frac{1}{2}(l+5)$$
$$10 \cdot \frac{1}{5}(l+10) \ge 10 \cdot \frac{1}{2}(l+5)$$
$$2(l+10) \ge 5(l+5)$$
$$2l+20 \ge 5l+25$$
$$\underline{-2l -2l}$$
$$0+20 \ge 3l+25$$
$$20 \ge 3l+25$$
$$\underline{-25 -25}$$
$$-5 \ge 3l+0$$
$$\frac{-5}{3} \ge \frac{3l}{3}$$
$$-\frac{5}{3} \ge l$$
$$l \le -\frac{5}{3}$$

b) $\left\{ l \mid l \le -\dfrac{5}{3} \right\}$ c) $\left(-\infty, -\dfrac{5}{3} \right]$

d)

21. Let n represent the unknown number. "Two-thirds of a number minus one-sixth is the same as twice the number" translates to the equation $\dfrac{2}{3}n - \dfrac{1}{6} = 2n$. To solve the equation, begin by clearing the fractions, that is, by multiplying through by the LCD, 6.

$$\frac{2}{3}n - \frac{1}{6} = 2n$$
$$6 \cdot \left(\frac{2}{3}n - \frac{1}{6} \right) = 6 \cdot (2n)$$
$$4n-1 = 12n$$
$$\underline{-4n -4n}$$
$$0-1 = 8n$$
$$\frac{-1}{8} = \frac{8n}{8}$$
$$-\frac{1}{8} = n$$

22. Let n represent the unknown number. "Three subtracted from five times the difference of a number and two is equal to ten minus four times the difference of the number and one" translates to the equation $5(n-2)-3 = 10-4(n-1)$.

Solve the equation for n.

$$5(n-2)-3 = 10-4(n-1)$$
$$5n-10-3 = 10-4n+4$$
$$5n-13 = 14-4n$$
$$\underline{+4n +4n}$$
$$9n-13 = 14+0$$
$$9n-13 = 14$$
$$\underline{+13 +13}$$
$$9n+0 = 27$$
$$\frac{9n}{9} = \frac{27}{9}$$
$$n = 3$$

23. Translate "12 is what percent of 60?" word for word. Let p represent the unknown percent.

$$12 = p \cdot 60$$
$$12 = 60p$$
$$\frac{12}{60} = \frac{60p}{60}$$
$$0.20 = p$$

To write 0.20 as a percent, multiply by 100%. The percent is $0.20 \cdot 100\% = 20\%$.

24. Translate "70 is 40% of what number?" word for word. Let n represent the unknown number.

$$70 = 0.40 \cdot n$$
$$70 = 0.40n$$
$$\frac{70}{0.40} = \frac{0.40n}{0.40}$$
$$175 = n$$

The number is 175.

25. Let n represent the unknown number. "One minus a number is greater than twice the number" translates to the inequality $1-n>2n$. Solve the inequality for n.
$$1-n>2n$$
$$\underline{+n \qquad +n}$$
$$1+0>3n$$
$$\frac{1}{3}>\frac{3n}{3}$$
$$\frac{1}{3}>n$$
$$n<\frac{1}{3}$$

26. Use the formula for the perimeter of a rectangle, $P=2(l+w)$. Let l be 4 ft. and w be 6 ft.
$$P=2(l+w)$$
$$P=2(4+6)$$
$$P=2\cdot10$$
$$P=20\text{ ft.}$$
The perimeter of the window is 20 ft.

27. Find the area of the rectangle if it were whole, and then subtract the area of the triangle. For the rectangle, let $l=14$ and $w=6.5$. For the triangle, let $b=6.5$ and $h=7$.
$$A=lw-\frac{1}{2}bh$$
$$=(14)(6.5)-\frac{1}{2}(6.5)(7)$$
$$=91-22.75$$
$$=68.25$$
The area of the figure is 68.25 in^2.

28. Let w be the number of weeks. Since there are 52 weeks in one year, we will substitute 52 for w.
$$C=10+10w$$
$$C=10+10(52)$$
$$C=10+520$$
$$C=530$$
The TV will cost $530.

29. The ratio of rise to horizontal length is $\frac{8}{20}=\frac{2}{5}$.

30. $$\frac{3.77}{14.5}=\frac{0.26}{1}$$
The unit ratio $\frac{0.26}{1}$ indicates that one ounce of cereal costs 26 cents.

31. Because the figures are similar, the lengths of the corresponding sides are proportional. The corresponding sides can be shown as follows:
$$\frac{\text{Side of smaller triangle}}{\text{Side of larger triangle}}=\frac{5}{7.2}=\frac{9}{d}=\frac{12.5}{18}.$$
Write and solve a proportion to find the missing side.
$$\frac{5}{7.2}=\frac{9}{d}$$
$$5\cdot d=7.2\cdot9$$
$$5d=64.8$$
$$\frac{5d}{5}=\frac{64.8}{5}$$
$$d=12.96$$
The missing length is $d=12.96$ cm.

32. This situation is the same as asking "37.1% of 500 is what number?" Translate word for word. Let n represent the unknown number.
$$0.371\cdot500=n$$
$$185.5=n$$
If 500 Americans were surveyed, about 186 of them would say that they are in excellent health.

33. Begin by finding the increase in her bill by subtracting the original bill amount ($148.40) from the final bill amount ($166.95): Increase $=166.95-148.40=18.55$. This situation is the same as asking "The increase in her bill is what percent of the original bill?" Translate word for word. Let p represent the unknown percent.
$$18.55=p\cdot148.40$$
$$18.55=148.40p$$
$$\frac{18.55}{148.40}=\frac{148.40p}{148.40}$$
$$0.125=p$$
To write 0.125 as a percent, multiply by 100%. Her bill increased by $0.125\cdot100\%=12.5\%$.

34. Let x represent the fifth test score. "If his average on five tests is at least 92" means that his average must be greater than or equal to 92.

Translation: $\dfrac{90+84+89+96+x}{5} \geq 92$

$$\dfrac{359+x}{5} \geq 92$$

$$5 \cdot \left(\dfrac{359+x}{5}\right) \geq 5 \cdot (92)$$

$$359+x \geq 460$$

$$\underline{-359 -359}$$

$$x \geq 101$$

Dan must score at least 101 on the last test to be exempt from the final.

Chapters 1 and 2 Cumulative Review

1. True

2. False because you do multiplication and division in order from left to right when using the order of operations.

3. False because "four less than a number" is translated as $n-4$.

4. True

5. itself

6. LCD

7. $120 = 2 \cdot 60$
$$= 2 \cdot 2 \cdot 30$$
$$= 2 \cdot 2 \cdot 2 \cdot 15$$
$$= 2 \cdot 2 \cdot 2 \cdot 3 \cdot 5$$
$$= 2^3 \cdot 3 \cdot 5$$

8. The expression is defined when the denominator equals 0. Set the denominator equal to 0 and solve for x to find the values that make the expression undefined.
$$x - 3 = 0$$
$$x = 3$$

9. Distributive Property

10. Set the cross products equal to each other and solve for the variable.

11. $\left(-\dfrac{2}{3}\right)^2 = \left(-\dfrac{2}{3}\right)\left(-\dfrac{2}{3}\right) = \dfrac{4}{9}$

12. $\sqrt{16} + \sqrt{9} = 4 + 3$
$$= 7$$

13. $\dfrac{5-2(-1)^2}{3\sqrt{64+36}} = \dfrac{5-2(1)}{3\sqrt{100}} = \dfrac{5-2}{3 \cdot 10} = \dfrac{3}{30} = \dfrac{1}{10}$

14. $-\left|3 \cdot (-5)\right| + \left|(-1)^3\right| = -\left|-15\right| + \left|-1\right|$
$$= -15 + 1$$
$$= -14$$

15. $2\left(-\dfrac{1}{2}\right)^2 - 4\left(\dfrac{3}{5}\right) + 3 = \overset{1}{\cancel{2}}\left(\dfrac{1}{\underset{2}{\cancel{4}}}\right) - 4\left(\dfrac{3}{5}\right) + 3$
$$= \dfrac{1}{2} - \dfrac{12}{5} + 3$$
$$= \dfrac{5}{10} - \dfrac{24}{10} + \dfrac{30}{10}$$
$$= \dfrac{11}{10}$$

16. $-6 - 2(3+0.2)^2 = -6 - 2(3.2)^2$
$$= -6 - 2(10.24)$$
$$= -6 - 20.48$$
$$= -26.48$$

17. $-3\left(\dfrac{1}{6}x - 9\right) = -3\left(\dfrac{1}{6}x\right) - (-3)(9)$
$$= -\dfrac{1}{2}x + 27$$

18. Solve the equation for c.
$$3.23c - 8.75 = 1.41c + 7.63$$
$$\underline{-1.41c -1.41c}$$
$$1.82c - 8.75 = 0 + 7.63$$
$$1.82c - 8.75 = 7.63$$
$$\underline{+8.75 +8.75}$$
$$1.82c + 0 = 16.38$$
$$\dfrac{1.82c}{1.82} = \dfrac{16.38}{1.82}$$
$$c = 9$$

Check: $3.23(9) - 8.75 \overset{?}{=} 1.41(9) + 7.63$
$$29.07 - 8.75 \overset{?}{=} 12.69 + 7.63$$
$$20.32 = 20.32$$

19. Solve the equation for m.

$$2m+\frac{1}{2}=4m-3\frac{1}{2}+8m$$

$$2m+\frac{1}{2}=12m-3\frac{1}{2}$$

$$2\cdot\left(2m+\frac{1}{2}\right)=2\cdot\left(12m-\frac{7}{2}\right)$$

$$4m+1=24m-7$$

$$\underline{-4m\qquad\quad-4m}$$

$$0+1=20m-7$$

$$\underline{+7\qquad\quad+7}$$

$$8=20m+0$$

$$\frac{8}{20}=\frac{20m}{20}$$

$$\frac{2}{5}=m$$

Check: $2\left(\dfrac{2}{5}\right)+\dfrac{1}{2}\overset{?}{=}4\left(\dfrac{2}{5}\right)-3\dfrac{1}{2}+8\left(\dfrac{2}{5}\right)$

$$\frac{4}{5}+\frac{5}{10}\overset{?}{=}\frac{8}{5}-\frac{7}{2}+\frac{16}{5}$$

$$\frac{8}{10}+\frac{5}{10}\overset{?}{=}\frac{16}{10}-\frac{35}{10}+\frac{32}{10}$$

$$\frac{13}{10}=\frac{13}{10}$$

20. $6x-2=3\left(2x-\dfrac{2}{3}\right)$

$6x-2=6x-2$

Because the linear equation is an identity, every real number is a solution.

21. Solve the equation for p.

$$-7=-6(p+2)+5(2p-3)$$

$$-7=-6p-12+10p-15$$

$$-7=4p-27$$

$$\underline{+27\qquad\quad+27}$$

$$20=4p+0$$

$$\frac{20}{4}=\frac{4p}{4}$$

$$5=p$$

Check: $-7\overset{?}{=}-6(5+2)+5(2\cdot5-3)$

$$-7\overset{?}{=}-6(7)+5(7)$$

$$-7\overset{?}{=}-42+35$$

$$-7=-7$$

22. Solve for b.

$$A=\frac{1}{2}bh$$

$$2\cdot A=2\cdot\frac{1}{2}bh$$

$$2A=bh$$

$$\frac{2A}{h}=\frac{bh}{h}$$

$$\frac{2A}{h}=b$$

23. Solve for c.

$$P=\ a+b+c$$

$$\underline{-a-b\quad-a-b}$$

$$P-a-b=0+c$$

$$P-a-b=c$$

24. a) Solve for h.

$$-3h-8\le19$$

$$\underline{+8\quad+8}$$

$$-3h+0\le27$$

$$-3h\le27$$

$$\frac{-3h}{-3}\ge\frac{27}{-3}$$

$$h\ge-9$$

b) $\{h\,|\,h\ge-9\}$ c) $[-9,\infty)$

d)

25. a) Solve for m.

$$9m-(2m+3)<4(m-5)+2$$

$$9m-2m-3<4m-20+2$$

$$7m-3<\ 4m-18$$

$$\underline{-4m\qquad\quad-4m}$$

$$3m-3<0-18$$

$$3m-3<-18$$

$$\underline{+3\quad+3}$$

$$3m+0<-15$$

$$3m<-15$$

$$\frac{3m}{3}<-\frac{15}{3}$$

$$m<-5$$

b) $\{m\,|\,m<-5\}$ c) $(-\infty,-5)$

d)

26. Find the area of the trapezoid first using the formula $\frac{1}{2}h(a+b)$ with $h=10$, $a=7$, and $b=16$. Then subtract the area s^2 of the square in the center using $s=4$.

$$A = \frac{1}{2}h(a+b) - s^2$$

$$= \frac{1}{2}(10)(7+16) - 4^2$$

$$= \frac{1}{2}(10)(23) - 16$$

$$= 5(23) - 16$$

$$= 115 - 16$$

$$= 99$$

The area of the shaded region is 99 cm^2.

27. Use the formula $h = -16t^2 + s_0$ where $s_0 = 2400$ and $t = 8$.

$$h = -16t^2 + s_0$$

$$h = -16(8)^2 + 2400$$

$$h = -16(64) + 2400$$

$$h = -1024 + 2400$$

$$h = 1376$$

Its height after 8 seconds is 1376 ft.

28. Translate "what is 15% of 42.3?" word for word. Let n be the unknown number.

$$n = 0.15 \cdot 42.3$$

$$n = 6.345$$

The number is 6.345.

29. Let t represent the score on the third test. Translate to an inequality.

$$\frac{82 + 70 + t}{3} \geq 80$$

$$3 \cdot \left(\frac{82 + 70 + t}{3} \right) \geq 3 \cdot 80$$

$$82 + 70 + t \geq 240$$

$$152 + t \geq 240$$

$$\underline{-152 \qquad -152}$$

$$0 + t \geq 88$$

$$t \geq 88$$

He must score at least an 88 on his third test to have an average of at least 80.

30. Translate to a proportion:

$$\frac{120 \text{ miles}}{4 \text{ gallons}} = \frac{x \text{ miles}}{10 \text{ gallons}}$$

$$10 \cdot 120 = 4 \cdot x$$

$$1200 = 4x$$

$$\frac{1200}{4} = \frac{4x}{4}$$

$$300 \text{ mi.} = x$$

Chapter 3
Graphing Linear Equations and Inequalities

Exercise Set 3.1

1. $A(4, 3)$; a vertical line intersects the *x*-axis at 4, and a horizontal line intersects the *y*-axis at 3. $B(-3, 1)$; a vertical line intersects the *x*-axis at −3, and a horizontal line intersects the *y*-axis at 1. $C(0, -2)$; a vertical line intersects the *x*-axis at 0, and a horizontal line intersects the *y*-axis at −2. $D(2, -5)$; a vertical line intersects the *x*-axis at 2, and a horizontal line intersects the *y*-axis at −5.

3. $A(-4, 0)$; a vertical line intersects the *x*-axis at −4, and a horizontal line intersects the *y*-axis at 0. $B(2, 4)$; a vertical line intersects the *x*-axis at 2, and a horizontal line intersects the *y*-axis at 4. $C(-2, -3)$; a vertical line intersects the *x*-axis at −2, and a horizontal line intersects the *y*-axis at −3. $D(4, -4)$; a vertical line intersects the *x*-axis at 4, and a horizontal line intersects the *y*-axis at −4.

5.   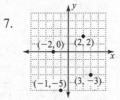   7.

9. Quadrant II because the first coordinate is negative and the second coordinate is positive.

11. Quadrant I because both coordinates are positive.

13. Quadrant IV because the first coordinate is positive and the second coordinate is negative.

15. Quadrant III because both coordinates are negative.

17. On the *y*-axis because the *x*-coordinate is 0.

19. On the *x*-axis because the *y*-coordinate is 0.

21. Begin by plotting the points.

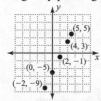

Because the points can be connected to form a straight line, the set of points is linear.

23. Begin by plotting the points.

Because the points do not form a straight line, the set of points is nonlinear.

25. Begin by plotting the points.

Because the points can be connected to form a straight line, the set of points is linear.

27. Begin by plotting the points.

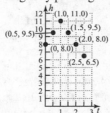

Because the points do not form a straight line, the set of points is nonlinear.

29. Begin by plotting the points.

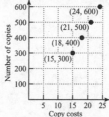

Because the points can be connected to form a straight line, the set of points is linear.

31. Answers may vary. Some possible answers are $(-1, -5)$, $(0, -3)$, and $(2, 1)$.

33. *A*: (–6, –4) because a vertical line intersects the
x-axis at –6, and a horizontal line intersects the
y-axis at –4.
B: (–3, 2) because a vertical line intersects the
x-axis at –3, and a horizontal line intersects the
y-axis at 2.
C: (5, 2) because a vertical line intersects the
x-axis at 5, and a horizontal line intersects the
y-axis at 2.

D: (8, –4) because a vertical line intersects the
x-axis at 8, and a horizontal line intersects the
y-axis at –4.

35. a) The coordinates of each vertex of the
parallelogram in its original position are
(–3, 1), (2, 1), (1, –3), and (–4, –3).
b) The coordinates of each vertex of the
parallelogram in its new position are (0, 2),
(5, 2), (4, –2), and (–1, –2).
c) The x-coordinate of each vertex of the
parallelogram in its new position is 3 units
more than the vertex x-coordinates in the
original position. The y-coordinate of each
vertex of the parallelogram in its new position
is 1 unit more than the vertex y-coordinates in
the original position. A rule that describes
how to move each point on the parallelogram
in its original position to obtain each point on
the parallelogram in its new position is
$(x + 3, y + 1)$.

37. a)

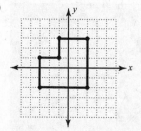

b) To find the perimeter, add the lengths of all
of the sides:
$3 + 5 + 5 + 3 + 2 + 2 = 20$ units.
c) To find the area, separate the figure into two
rectangles. Find the area of each and then
add the areas together.
Area 1 = 3(5) = 15
Area 2 = 2(3) = 6.
The sum of the two areas is $15 + 6 = 21$
square units.

Review Exercises

1.
$$\begin{array}{cccccccc} -4 & -3 & -2 & -1 & 0 & 1 & 2 \end{array}$$

2. For $2x - 3y$, let $x = 4$ and $y = 2$.
$$2x - 3y = 2(4) - 3(2)$$
$$= 8 - 6$$
$$= 2$$

3. For $-4x - 5y$, let $x = -3$ and $y = 0$.
$$-4x - 5y = -4(-3) - 5(0)$$
$$= 12 - 0$$
$$= 12$$

4. For $-\dfrac{1}{3}x + 2$, let $x = 6$.
$$-\frac{1}{3}x + 2 = -\frac{1}{3}(6) + 2$$
$$= -2 + 2$$
$$= 0$$

5. $2x - 5 = 10$
$$\underline{+5 \quad +5}$$
$$2x + 0 = 15$$
$$2x = 15$$
$$\frac{2x}{2} = \frac{15}{2}$$
$$x = \frac{15}{2}$$

6.
$$-\frac{5}{6}x = \frac{3}{4}$$
$$-\frac{6}{5} \cdot \left(-\frac{5}{6}x\right) = -\frac{6}{5} \cdot \left(\frac{3}{4}\right)$$
$$x = -\frac{18}{20}$$
$$x = -\frac{9}{10}$$

Exercise Set 3.2

1. Replace x with 2 and y with 3 and see if the
equation is true.
$$x + 2y = 8$$
$$2 + 2(3) \overset{?}{=} 8$$
$$2 + 6 \overset{?}{=} 8$$
$$8 = 8$$
Yes, (2, 3) is a solution for the equation.

3. Replace x with 4 and y with 9 and see if the equation is true.

    $$3x - 2y = -6$$
    $$3(4) - 2(9) \overset{?}{=} -6$$
    $$12 - 18 \overset{?}{=} -6$$
    $$-6 = -6$$

    Yes, (4, 9) is a solution for the equation.

5. Replace x with -5 and y with 2 and see if the equation is true.

    $$y - 4x = 3$$
    $$2 - 4(-5) \overset{?}{=} 3$$
    $$2 + 20 \overset{?}{=} 3$$
    $$22 \neq 3$$

    No, $(-5, 2)$ is not a solution for the equation.

7. Replace x with 9 and y with 0 and see if the equation is true.

    $$y = -2x + 18$$
    $$0 \overset{?}{=} -2(9) + 18$$
    $$0 \overset{?}{=} -18 + 18$$
    $$0 = 0$$

    Yes, (9, 0) is a solution for the equation.

9. Replace x with 6 and y with -2 and see if the equation is true.

    $$y = -\frac{2}{3}x$$
    $$-2 \overset{?}{=} -\frac{2}{3}(6)$$
    $$-2 \overset{?}{=} -\frac{12}{3}$$
    $$-2 \neq -4$$

    No, $(6, -2)$ is not a solution for the equation.

11. Replace x with -8 and y with -8 and see if the equation is true.

    $$y = \frac{3}{4}x - 2$$
    $$-8 \overset{?}{=} \frac{3}{4}(-8) - 2$$
    $$-8 \overset{?}{=} -\frac{24}{4} - 2$$
    $$-8 \overset{?}{=} -6 - 2$$
    $$-8 = -8$$

    Yes, $(-8, -8)$ is a solution for the equation.

13. Replace x with $-1\frac{2}{5}$ and y with 0 and see if the equation is true.

    $$y - 3x = 5$$
    $$0 - 3\left(-1\frac{2}{5}\right) \overset{?}{=} 5$$
    $$0 - 3\left(-\frac{7}{5}\right) \overset{?}{=} 5$$
    $$\frac{21}{5} \neq 5$$

    No, $\left(-1\frac{2}{5}, 0\right)$ is not a solution for the equation.

15. Replace x with 2.2 and y with -11.2 and see if the equation is true.

    $$y + 6x = -2$$
    $$-11.2 + 6(2.2) \overset{?}{=} -2$$
    $$-11.2 + 13.2 \overset{?}{=} -2$$
    $$2 \neq -2$$

    No, $(2.2, -11.2)$ is not a solution for the equation.

17. $(-1, -1)$, $(0, 0)$, $(1, 1)$

    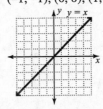

19. $(-1, -2)$, $(0, 0)$, $(1, 2)$

    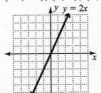

21. (−1, 5), (0, 0), (1, −5)

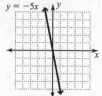

23. (0, −3), (2, −1), (4, 1)

25. (0, −2), (4, −6), (−4, 2)

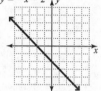

27. (0, −5), (1, −3), (2, −1)

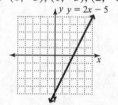

29. (−1, 6), (0, 4), (1, 2)

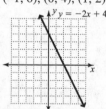

31. (−2, −1), (0, 0), (2, 1)

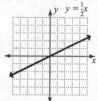

33. (0, 0), (−3, 2), (3, −2)

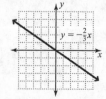

35. (−3, 6), (0, 4), (3, 2)

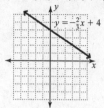

37. (3, −5), (5, −3), (6, −2)

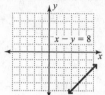

39. (1, 4), (3, 0), (5, −4)

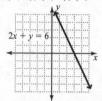

41. (0, 4), (6, 0), (3, 2)

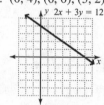

43. (0, 4), (−3, 0), (3, 8)

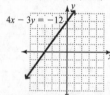

45. (1, −4), (2, 0), (3, 4)

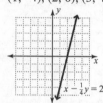

47. (−1, −5), (0, −5), (1, −5)

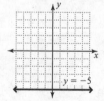

49. $(7, -1), (7, 0), (7, 1)$

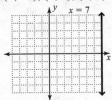

51. $(-1, -2.9), (0, -2.5), (1, -2.1)$

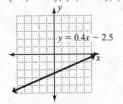

53. As a gets larger, the graph gets steeper.

55. The graph will shift up b units on the y-axis.

57. a) Replace the n with 2 and solve.

$$c = 40n + 80$$

$$c = 40(2) + 80$$

$$c = 80 + 80$$

$$c = \$160$$

If the labor is 2 hours, the total cost is $160.

b) Replace the c with 240 and solve for n.

$$c = 40n + 80$$

$$240 = 40n + 80$$

$$160 = 40n$$

$$4 = n$$

If the charges are $240, the client was charged for 4 hours.

c)

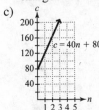

d) It represents the initial charge, which is $80.

59. a) Find the number of copies above 300
$400 - 300 = 100$. Replace the n with 100 and calculate.

$$c = 0.03n + 15$$

$$c = 0.03(100) + 15$$

$$c = 3 + 15$$

$$c = 18$$

The total cost for 400 copies is $18.

b) Replace the c with 21 and solve for n.

$$c = 0.03n + 15$$

$$21 = 0.03n + 15$$

$$6 = 0.03n$$

$$200 = n$$

There were 200 copies made above the initial 300. So $300 + 200 = 500$ total copies were made.

c)

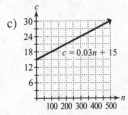

61. a) Replace the m with 25 and calculate.

$$g = -3.5m + 500$$

$$g = -3.5(25) + 500$$

$$g = -87.5 + 500$$

$$g = 412.5$$

After 25 minutes, 412.5 gallons of water remain in the tub.

b) Pumping out half the water in the tub means that 250 gallons remain in the tub. Replace the g with 250 and solve for m.

$$g = -3.5m + 500$$

$$250 = -3.5m + 500$$

$$-250 = -3.5m$$

$$71.4 \approx m$$

It takes approximately 71.4 minutes to pump out half of the water.

c) Pumping out all of the water means that 0 gallons remain in the hot tub. Replace the g with 0 and solve for m.

$$g = -3.5m + 500$$

$$0 = -3.5m + 500$$

$$-500 = -3.5m$$

$$143 \approx m$$

It takes approximately 143 minutes to pump out all of the water.

d)
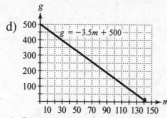

e) It represents the original amount of water in the hot tub, which is 500 gallons.

f) It represents the amount of time it takes for the hot tub to empty (0 gallons of water remaining).

63. a) Replace n with 4 and calculate.

$$c = -300n + 2400$$
$$c = -300(4) + 2400$$
$$c = -1200 + 2400$$
$$c = 1200$$

After 4 years, the computer is worth $1200.

b) Half of the computer's original value is $1200. Replace c with 1200 and solve for n.

$$c = -300n + 2400$$
$$1200 = -300n + 2400$$
$$-1200 = -300n$$
$$4 = n$$

The computer is worth half its initial value after 4 years.

c) Replace c with 0 and solve.

$$c = -300n + 2400$$
$$0 = -300n + 2400$$
$$-2400 = -300n$$
$$8 = n$$

The computer will have a value of $0 after 8 years.

d)
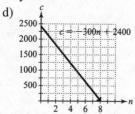

Review Exercises

1. For $\frac{2}{3}x + 2$, let $x = 3$.

$$\frac{2}{3}(3) + 2 = \frac{6}{3} + 2 = 2 + 2 = 4$$

2. Use the formula $P = 2l + 2w$ where $l = 8.5$ and $w = 7.25$.

$$P = 2l + 2w$$
$$P = 2(8.5) + 2(7.25)$$
$$P = 17 + 14.5$$
$$P = 31.5$$

The perimeter is 31.5 inches.

3. Use the formula $A = \frac{1}{2}bh$ where $b = 4.2$ and $h = 2.7$.

$$A = \frac{1}{2}bh$$
$$A = \frac{1}{2}(4.2)(2.7)$$
$$A = 5.67$$

The area is 5.67 square inches.

4.
$$0 = 3x - 5$$
$$5 = 3x$$
$$\frac{5}{3} = x$$

5.
$$v = mx + k$$
$$v - k = mx$$
$$\frac{v - k}{m} = \frac{mx}{m}$$
$$\frac{v - k}{m} = x$$

6. Plot the points and draw a line through them.

Exercise Set 3.3

1. The line intersects the x-axis at $(1, 0)$ and the y-axis at $(0, 3)$. Therefore, the x-intercept is $(1, 0)$ and the y-intercept is $(0, 3)$.

3. The line intersects the x-axis at $(-3, 0)$ and the y-axis at $(0, 4)$. Therefore, the x-intercept is $(-3, 0)$ and the y-intercept is $(0, 4)$.

5. The line intersects the x-axis at $(3, 0)$ but does not intersect the y-axis at all. Therefore, the x-intercept is $(3, 0)$ and there is no y-intercept.

7. The line intersects the y-axis at $(0, -2)$ but does not intersect the x-axis at all. Therefore, the y-intercept is $(0, -2)$ and there is no x-intercept.

9. $x - y = 4$

x-intercept	y-intercept
$y = 0: x - 0 = 4$	$x = 0: 0 - y = 4$
$x = 4$	$-y = 4$
$(4, 0)$	$\dfrac{-y}{-1} = \dfrac{4}{-1}$
	$y = -4$
	$(0, -4)$

11. $2x + 3y = 6$

x-intercept	y-intercept
$y = 0: 2x + 3(0) = 6$	$x = 0: 2(0) + 3y = 6$
$2x = 6$	$3y = 6$
$\dfrac{2x}{2} = \dfrac{6}{2}$	$\dfrac{3y}{3} = \dfrac{6}{3}$
$x = 3$	$y = 2$
$(3, 0)$	$(0, 2)$

13. $3x - 4y = -10$

x-intercept

$y = 0: 3x - 4(0) = -10$

$3x = -10$

$\dfrac{3x}{3} = \dfrac{-10}{3}$

$x = -\dfrac{10}{3}$

$\left(-\dfrac{10}{3}, 0\right)$

y-intercept

$x = 0: 3(0) - 4y = -10$

$-4y = -10$

$\dfrac{-4y}{-4} = \dfrac{-10}{-4}$

$y = \dfrac{5}{2}$

$\left(0, \dfrac{5}{2}\right)$

15. $y = 3x - 6$

x-intercept	y-intercept
$y = 0: 0 = 3x - 6$	$x = 0: y = 3(0) - 6$
$6 = 3x$	$y = 0 - 6$
$2 = x$	$y = -6$
$(2, 0)$	$(0, -6)$

17. $y = 2x + 5$

x-intercept	y-intercept
$y = 0: 0 = 2x + 5$	$x = 0: y = 2(0) + 5$
$-5 = 2x$	$y = 0 + 5$
$-\dfrac{5}{2} = x$	$y = 5$
$\left(-\dfrac{5}{2}, 0\right)$	$(0, 5)$

19. $y = 2x$

x-intercept	y-intercept
$y = 0: 0 = 2x$	$x = 0: y = 2(0)$
$\dfrac{0}{2} = \dfrac{2x}{2}$	$y = 0$
$0 = x$	$(0, 0)$
$(0, 0)$	

21. $2x + 3y = 0$

x-intercept	y-intercept
$y = 0: 2x + 3(0) = 0$	$x = 0: 2(0) + 3y = 0$
$2x = 0$	$3y = 0$
$x = 0$	$y = 0$
$(0, 0)$	$(0, 0)$

23. $x = 5$ is a vertical line parallel to the y-axis that passes through the x-axis at the point $(5, 0)$. This point is the x-intercept. There is no y-intercept.

25. $y = -2$ is a horizontal line parallel to the x-axis that passes through the y-axis at the point $(0, -2)$. This point is the y-intercept. There is no x-intercept.

27. $x + y = 7$

x-intercept	y-intercept
Let $y = 0$:	Let $x = 0$:
$x + y = 7$	$x + y = 7$
$x + 0 = 7$	$0 + y = 7$
$x = 7$	$y = 7$
$(7, 0)$	$(0, 7)$

29.　$3x - y = 6$

　　x-intercept

　　Let $y = 0$:

　　　$3x - y = 6$

　　　$3x - 0 = 6$

　　　　$3x = 6$

　　　　　$x = 2$

　　　　$(2, 0)$

　　y-intercept

　　Let $x = 0$:

　　　$3x - y = 6$

　　　$3(0) - y = 6$

　　　　　$-y = 6$

　　　　　　$y = -6$

　　　　$(0, -6)$

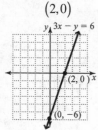

31.　$2x + 3y = 6$

　　x-intercept

　　Let $y = 0$:

　　　$2x + 3y = 6$

　　　$2x + 3(0) = 6$

　　　　$2x = 6$

　　　　　$x = 3$

　　　　$(3, 0)$

　　y-intercept

　　Let $x = 0$:

　　　$2x + 3y = 6$

　　　$2(0) + 3y = 6$

　　　　$3y = 6$

　　　　　$y = 2$

　　　　$(0, 2)$

33.　$5x + 2y = 10$

　　x-intercept

　　Let $y = 0$:

　　　$5x + 2y = 10$

　　　$5x + 2(0) = 10$

　　　　$5x = 10$

　　　　　$x = 2$

　　　　$(2, 0)$

　　y-intercept

　　Let $x = 0$:

　　　$5x + 2y = 10$

　　　$5(0) + 2y = 10$

　　　　$2y = 10$

　　　　　$y = 5$

　　　　$(0, 5)$

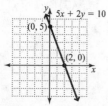

35.　$4x + 3y = -12$

　　x-intercept

　　Let $y = 0$:

　　　$4x + 3y = -12$

　　　$4x + 3(0) = -12$

　　　　$4x = -12$

　　　　　$x = -3$

　　　　$(-3, 0)$

　　y-intercept

　　Let $x = 0$:

　　　$4x + 3y = -12$

　　　$4(0) + 3y = -12$

　　　　$3y = -12$

　　　　　$y = -4$

　　　　$(0, -4)$

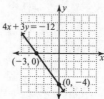

37.　$5x - 3y = -15$

　　x-intercept

　　Let $y = 0$:

　　　$5x - 3y = -15$

　　　$5x - 3(0) = -15$

　　　　$5x = -15$

　　　　　$x = -3$

　　　　$(-3, 0)$

　　y-intercept

　　Let $x = 0$:

　　　$5x - 3y = -15$

　　　$5(0) - 3y = -15$

　　　　$-3y = -15$

　　　　　$y = 5$

　　　　$(0, 5)$

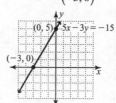

39.　$y = 2x$

　　x-intercept

　　Let $y = 0$:

　　　$0 = 2x$

　　　$0 = x$

　　　$(0, 0)$

　　Extra points:

　　Let $x = 1$:

　　　$y = 2x$

　　　$y = 2(1)$

　　　$y = 2$

　　　$(1, 2)$

　　y-intercept

　　Let $x = 0$:

　　　$y = 2(0)$

　　　$y = 0$

　　　$(0, 0)$

　　Let $x = 2$:

　　　$y = 2x$

　　　$y = 2(2)$

　　　$y = 4$

　　　$(2, 4)$

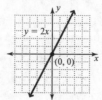

41. $y = \dfrac{2}{5}x$

x-intercept
Let $y = 0$:

$y = \dfrac{2}{5}x$

$0 = \dfrac{2}{5}x$

$0 = x$

$(0,0)$

y-intercept
Let $x = 0$:

$y = \dfrac{2}{5}x$

$y = \dfrac{2}{5}(0)$

$y = 0$

$(0,0)$

Extra points:
Let $x = 5$:

$y = \dfrac{2}{5}x$

$y = \dfrac{2}{5}(5)$

$y = 2$

$(5,2)$

Let $x = -5$

$y = \dfrac{2}{5}x$

$y = \dfrac{2}{5}(-5)$

$y = -2$

$(-5,-2)$

43. $y = x - 4$

x-intercept
Let $y = 0$:

$y = x - 4$

$0 = x - 4$

$4 = x$

$(4,0)$

y-intercept
Let $x = 0$:

$y = x - 4$

$y = 0 - 4$

$y = -4$

$(0,-4)$

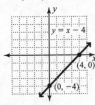

45. $y = 2x - 6$

x-intercept
Let $y = 0$:

$y = 2x - 6$

$0 = 2x - 6$

$6 = 2x$

$3 = x$

$(3,0)$

y-intercept
Let $x = 0$:

$y = 2x - 6$

$y = 2(0) - 6$

$y = -6$

$(0,-6)$

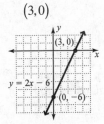

47. $6x + 3y = 9$

x-intercept
Let $y = 0$:

$6x + 3y = 9$

$6x + 3(0) = 9$

$6x = 9$

$x = \dfrac{3}{2}$

$\left(\dfrac{3}{2}, 0\right)$

y-intercept
Let $x = 0$:

$6x + 3y = 9$

$6(0) + 3y = 9$

$3y = 9$

$y = 3$

$(0,3)$

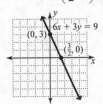

49. $4x + 2y = 5$

x-intercept
Let $y = 0$:

$4x + 2y = 5$

$4x + 2(0) = 5$

$4x = 5$

$x = \dfrac{5}{4}$

$\left(\dfrac{5}{4}, 0\right)$

y-intercept
Let $x = 0$:

$4x + 2y = 5$

$4(0) + 2y = 5$

$2y = 5$

$y = \dfrac{5}{2}$

$\left(0, \dfrac{5}{2}\right)$

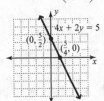

51. $x = -4$

The x-intercept is $(-4, 0)$ and there is no y-intercept. The graph is a vertical line through $(-4, 0)$.

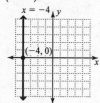

53. $y = -2$

The y-intercept is $(0, -2)$ and there is no x-intercept. The graph is a horizontal line through $(0, -2)$.

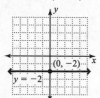

55. $y - 2 = 0$

$\quad\quad y = 2$

The y-intercept is $(0, 2)$ and there is no x-intercept. The graph is a horizontal line through $(0, 2)$.

57. $x - 2 = 0$

$\quad\quad x = 2$

The x-intercept is $(2, 0)$ and there is no y-intercept. The graph is a vertical line through $(2, 0)$.

59. The answer is d because the x-coordinate of the x-intercept is positive and the y-coordinate of the y-intercept is negative.

61. The answer is c because the x-coordinate of the x-intercept is positive and the y-coordinate of the y-intercept is negative.

63. Since all of the coordinates would be 2.7 units from the origin (0, 0), the coordinates would be (–2.7, 0), (2.7, 0), (0, –2.7), (0, 2.7).

65. a) Total sales are described by the expression $15x + 20y$.

b) The sales goal is described by the equation $15x + 20y = 2000$.

c) $y = 0$: $15x + 20(0) = 2000$

$$15x = 2000$$

$$\frac{15x}{15} = \frac{2000}{15}$$

$$x = 133\frac{1}{3}$$

$$\left(133\frac{1}{3}, 0\right)$$

$x = 0$: $15(0) + 20y = 2000$

$$20y = 2000$$

$$\frac{20y}{20} = \frac{2000}{20}$$

$$y = 100$$

$$(0, 100)$$

d) The number of units required to meet the sales goal if only one size is sold.

e) Answers may vary. Some possibilities are (20, 85), (40, 70), or (60, 55).

f)

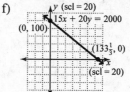

Review Exercises

1. $\dfrac{4 - (-2)}{-2 - 1} = \dfrac{4 + 2}{-3} = \dfrac{6}{-3} = -2$

2. $\dfrac{-3 - 4}{-2 - (-2)} = \dfrac{-7}{-2 + 2} = \dfrac{-7}{0}$, which is undefined because division by 0 is undefined.

3. $\quad\quad P = 2l + 2w$

$\quad\quad P - 2l = 2w$

$\quad\quad \dfrac{P - 2l}{2} = w$

4. $3x + 4 = 5(x + 6)$

    $3x + 4 = 5x + 30$

    $-26 = 2x$

    $\dfrac{-26}{2} = \dfrac{2x}{2}$

    $-13 = x$

5. The point $(0, 4)$ is located on the y-axis.

6. For $x + 3y = -5$, let $x = -8$ and $y = -2$.

    $x + 3y = -5$

    $-8 + 3(-2) \overset{?}{=} -5$

    $-8 - 6 \overset{?}{=} -5$

    $-14 \neq -5$

    No, $(-8, -2)$ is not a solution for the equation.

Exercise Set 3.4

1. Create a table of ordered pairs, then plot those ordered pairs and connect the points to graph the lines.

x	$y = \dfrac{1}{2}x$	$y = x$	$y = 2x$
0	0	0	0
1	$\dfrac{1}{2}$	1	2
2	1	2	4

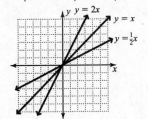

3. Create a table of ordered pairs, then plot those ordered pairs and connect the points to graph the lines.

x	$y = x$	$y = \dfrac{1}{2}x$	$y = \dfrac{1}{5}x$
0	0	0	0
2	2	1	$\dfrac{2}{5}$
5	5	$\dfrac{5}{2}$	1

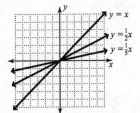

5. Create a table of ordered pairs, then plot those ordered pairs and connect the points to graph the lines.

x	$y = -\dfrac{1}{2}x$	$y = -x$	$y = -2x$
0	0	0	0
2	-1	-2	-4
-2	1	2	4

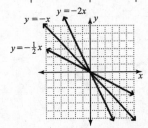

7. Create a table of ordered pairs, then plot those ordered pairs and connect the points to graph the lines.

x	$y = x$	$y = x + 2$	$y = x - 2$
0	0	2	-2
1	1	3	-1
2	2	4	0

9. $y = 2x + 3$

$m = 2$

$(0, 3)$

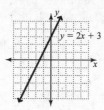

11. $y = -2x - 1$

$m = -2$

$(0, -1)$

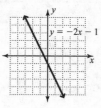

13. $y = \dfrac{1}{3}x + 2$

$m = \dfrac{1}{3}$

$(0, 2)$

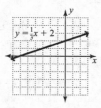

15. $y = \dfrac{3}{4}x - 2$

$m = \dfrac{3}{4}$

$(0, -2)$

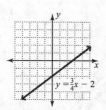

17. $y = -\dfrac{2}{3}x + 8$

$m = -\dfrac{2}{3}$

$(0, 8)$

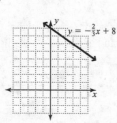

19. $2x + y = 4$ $m = -2$

$y = -2x + 4$ $(0, 4)$

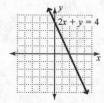

21. $3x - y = -1$ $m = 3$

$-y = -3x - 1$ $(0, 1)$

$y = 3x + 1$

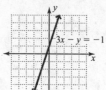

23. $2x + 3y = 6$ $m = -\dfrac{2}{3}$

$3y = -2x + 6$

$y = -\dfrac{2}{3}x + 2$ $(0, 2)$

25. $x + 2y = -4$ $m = -\dfrac{1}{2}$

$2y = -x - 4$

$y = -\dfrac{1}{2}x - 2$ $(0, -2)$

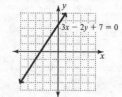

27. $3x - 2y + 7 = 0$ $m = \dfrac{3}{2}$

$3x - 2y = -7$

$-2y = -3x - 7$ $\left(0, \dfrac{7}{2}\right)$

$y = \dfrac{3}{2}x + \dfrac{7}{2}$

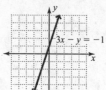

29. $2y = -3x + 5$ $\qquad m = -\dfrac{3}{2}$

$\qquad y = -\dfrac{3}{2}x + \dfrac{5}{2}$ $\qquad \left(0, \dfrac{5}{2}\right)$

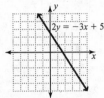

31. $0.6x - 0.2y = 1$ $\qquad m = 3$

$\qquad -0.2y = -0.6x + 1$ $\qquad (0, -5)$

$\qquad y = 3x - 5$

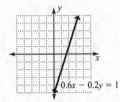

33. d

35. a

37. f

39. g

41. h

43. Let $(x_1, y_1) = (4, 1)$ and $(x_2, y_2) = (8, 11)$.

$\qquad m = \dfrac{y_2 - y_1}{x_2 - x_1} = \dfrac{11 - 1}{8 - 4} = \dfrac{10}{4} = \dfrac{5}{2}$

45. Let $(x_1, y_1) = (2, 4)$ and $(x_2, y_2) = (5, 2)$.

$\qquad m = \dfrac{y_2 - y_1}{x_2 - x_1} = \dfrac{2 - 4}{5 - 2} = -\dfrac{2}{3}$

47. Let $(x_1, y_1) = (-3, 5)$ and $(x_2, y_2) = (4, 7)$.

$\qquad m = \dfrac{y_2 - y_1}{x_2 - x_1} = \dfrac{7 - 5}{4 - (-3)} = \dfrac{2}{7}$

49. Let $(x_1, y_1) = (10, -12)$ and $(x_2, y_2) = (4, -4)$.

$\qquad m = \dfrac{y_2 - y_1}{x_2 - x_1} = \dfrac{-4 - (-12)}{4 - 10} = \dfrac{8}{-6} = -\dfrac{4}{3}$

51. Let $(x_1, y_1) = (3, -3)$ and $(x_2, y_2) = (-15, -15)$.

$\qquad m = \dfrac{y_2 - y_1}{x_2 - x_1} = \dfrac{-15 - (-3)}{-15 - 3} = \dfrac{-12}{-18} = \dfrac{2}{3}$

53. Let $(x_1, y_1) = (0, 5)$ and $(x_2, y_2) = (4, 0)$.

$\qquad m = \dfrac{y_2 - y_1}{x_2 - x_1} = \dfrac{0 - 5}{4 - 0} = -\dfrac{5}{4}$

55. Let $(x_1, y_1) = (-3, 5)$ and $(x_2, y_2) = (4, 5)$.

$\qquad m = \dfrac{y_2 - y_1}{x_2 - x_1} = \dfrac{5 - 5}{4 - (-3)} = \dfrac{0}{7} = 0$

57. Let $(x_1, y_1) = (6, 1)$ and $(x_2, y_2) = (6, -8)$.

$\qquad m = \dfrac{y_2 - y_1}{x_2 - x_1} = \dfrac{-8 - 1}{6 - 6} = \dfrac{-9}{0}$; undefined

59. $m = 2$; $\quad (0, b) = (0, 3)$

$\qquad y = mx + b$

$\qquad y = 2x + 3$

61. $m = -3$; $\quad (0, b) = (0, -9)$

$\qquad y = mx + b$

$\qquad y = -3x - 9$

63. $m = \dfrac{3}{4}$; $\quad (0, b) = (0, 5)$

$\qquad y = mx + b$

$\qquad y = \dfrac{3}{4}x + 5$

65. $m = -\dfrac{2}{5}$; $\quad (0, b) = \left(0, \dfrac{7}{8}\right)$

$\qquad y = mx + b$

$\qquad y = -\dfrac{2}{5}x + \dfrac{7}{8}$

67. $m = 0.8$; $\quad (0, b) = (0, -5.1)$

$\qquad y = mx + b$

$\qquad y = 0.8x - 5.1$

69. $m = -3$; $\quad (0, b) = (0, 0)$

$\qquad y = mx + b$

$\qquad y = -3x$

71. $m = 2$; $(0, b) = (0, 3)$

$\qquad y = mx + b$

$\qquad y = 2x + 3$

73. $m = -1$; $(0, b) = (0, 2)$

$\qquad y = mx + b$

$\qquad y = -x + 2$

75. $m = \dfrac{2}{3}; (0, b) = (0, -1)$

$y = mx + b$

$y = \dfrac{2}{3}x - 1$

77. $m = -\dfrac{1}{4}; (0, b) = (0, 3)$

$y = mx + b$

$y = -\dfrac{1}{4}x + 3$

79. The line is a vertical line, so m is undefined and there is no y-intercept. The equation of the line is $x = 2$.

81. The line is horizontal, so m is 0 and the y-intercept is (0, 3). The equation of the line is $y = 3$.

83. The equation of the horizontal line is $y = -2$; the equation of the vertical line is $x = 3$.

85. $y = 0$

87. a)

b) Slope of side connecting points (3, 4) and (1, –2) is $m = \dfrac{-2-4}{1-3} = \dfrac{-6}{-2} = 3$. Slope of side connecting points (–1, 4) and (–3, –2) is $m = \dfrac{-2-4}{-3-(-1)} = \dfrac{-6}{-2} = 3$. Slope of side connecting points (3, 4) and (–1, 4) is $m = \dfrac{4-4}{-1-3} = \dfrac{0}{-4} = 0$. Slope of side connecting points (1, –2) and (–3, –2) is $m = \dfrac{-2-(-2)}{-3-1} = \dfrac{0}{-4} = 0$.

c) They are the same.

89. $m = \dfrac{3.5}{6} = 0.58\overline{3}$ or $m = \dfrac{3.5 \cdot 2}{6 \cdot 2} = \dfrac{7}{12}$

91. $m = \dfrac{29.3}{23} \approx 1.27$ or $m = \dfrac{29.3 \cdot 10}{23 \cdot 10} = \dfrac{293}{230} \approx 1.27$

Review Exercises

1. $-\dfrac{3}{2}$

2. $12x - 9y + 7 - x + 9y$

$= 12x - x - 9y + 9y + 7$

$= 11x + 7$

3. $-2(x + 7) = -2 \cdot x - 2 \cdot 7$

$\qquad\qquad = -2x - 14$

4. $Ax + By = C$

$\quad Ax = C - By$

$\quad\; x = \dfrac{C - By}{A}$

5.

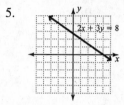

6. $y = 0: 3x - 5 \cdot 0 = 7 \qquad x = 0: 3 \cdot 0 - 5y = 7$

$\qquad\quad 3x = 7 \qquad\qquad\qquad -5y = 7$

$\qquad\quad \dfrac{3x}{3} = \dfrac{7}{3} \qquad\qquad\quad \dfrac{-5y}{-5} = \dfrac{7}{-5}$

$\qquad\quad x = \dfrac{7}{3} \qquad\qquad\qquad y = -\dfrac{7}{5}$

$\qquad\quad \left(\dfrac{7}{3}, 0\right) \qquad\qquad\qquad \left(0, -\dfrac{7}{5}\right)$

Exercise Set 3.5

1. $m = 3; (5, 2)$

$y - y_1 = m(x - x_1)$

$y - 2 = 3(x - 5)$

$y - 2 = 3x - 15$

$\quad y = 3x - 13$

3. $m = -1; (-1, -1)$

$y - y_1 = m(x - x_1)$

$y - (-1) = -1(x - (-1))$

$\quad y + 1 = -1(x + 1)$

$\quad y + 1 = -x - 1$

$\qquad y = -x - 2$

5. $m = \dfrac{2}{3}; (6,3)$

$y - y_1 = m(x - x_1)$

$y - 3 = \dfrac{2}{3}(x - 6)$

$y - 3 = \dfrac{2}{3}x - 4$

$y = \dfrac{2}{3}x - 1$

7. $m = -\dfrac{3}{4}; (-1,-5)$

$y - y_1 = m(x - x_1)$

$y - (-5) = -\dfrac{3}{4}(x - (-1))$

$y + 5 = -\dfrac{3}{4}(x + 1)$

$y + 5 = -\dfrac{3}{4}x - \dfrac{3}{4}$

$y = -\dfrac{3}{4}x - \dfrac{23}{4}$

9. $m = -\dfrac{4}{3}; (-2,4)$

$y - y_1 = m(x - x_1)$

$y - 4 = -\dfrac{4}{3}(x - (-2))$

$y - 4 = -\dfrac{4}{3}(x + 2)$

$y - 4 = -\dfrac{4}{3}x - \dfrac{8}{3}$

$y = -\dfrac{4}{3}x + \dfrac{4}{3}$

11. $m = 2; (0,0)$

$y - y_1 = m(x - x_1)$

$y - 0 = 2(x - 0)$

$y = 2x$

13. From the graph we can see that the slope of the line is $m = \dfrac{2}{3}$ and a point on the line is $(-1,0)$.

$y - y_1 = m(x - x_1)$

$y - 0 = \dfrac{2}{3}(x - (-1))$

$y = \dfrac{2}{3}(x + 1)$

$(3)y = (3)\left(\dfrac{2}{3}(x + 1)\right)$

$3y = 2(x + 1)$

$3y = 2x + 2$

$-2x + 3y = 2$

$(-1)(-2x + 3y) = (-1)(2)$

$2x - 3y = -2$

15. From the graph we can see that the slope of the line is $m = -\dfrac{1}{2}$ and a point on the line is $(3,0)$.

$y - y_1 = m(x - x_1)$

$y - 0 = -\dfrac{1}{2}(x - 3)$

$(2)(y) = (2)\left(-\dfrac{1}{2}(x - 3)\right)$

$2y = -1(x - 3)$

$2y = -x + 3$

$x + 2y = 3$

17. $(4,-3), (-1,7)$

$m = \dfrac{7 - (-3)}{-1 - 4}$

$\quad = \dfrac{10}{-5}$

$\quad = -2$

$y - y_1 = m(x - x_1)$

$y - 7 = -2(x - (-1))$

$y - 7 = -2(x + 1)$

$y - 7 = -2x - 2$

$y = -2x + 5$

19. $(3,1), (5,6)$

$m = \dfrac{6 - 1}{5 - 3} = \dfrac{5}{2}$

$y - y_1 = m(x - x_1)$

$y - 1 = \dfrac{5}{2}(x - 3)$

$y - 1 = \dfrac{5}{2}x - \dfrac{15}{2}$

$y = \dfrac{5}{2}x - \dfrac{13}{2}$

21. $(0,0),(-1,-7)$

$m = \dfrac{-7-0}{-1-0}$ $y - y_1 = m(x - x_1)$

$\quad = \dfrac{-7}{-1}$ $y - 0 = 7(x - 0)$

$\quad = 7$ $y = 7x$

23. $(8,-1),(8,-10)$

$m = \dfrac{-10-(-1)}{8-8}$ $x = 8$

$\quad = \dfrac{-9}{0}$

undefined

25. $(-9,-1),(2,-1)$

$m = \dfrac{-1-(-1)}{2-(-9)}$ $y - y_1 = m(x - x_1)$

$\quad = \dfrac{0}{11}$ $y - (-1) = 0(x - 2)$

$\quad = 0$ $y + 1 = 0$

         $y = -1$

27. $(1.4,5),(3,9.8)$

$m = \dfrac{9.8-5}{3-1.4}$ $y - y_1 = m(x - x_1)$

$\quad = \dfrac{4.8}{1.6}$ $y - 9.8 = 3(x - 3)$

$\quad = 3$ $y - 9.8 = 3x - 9$

         $y = 3x + 0.8$

29. $(3,4);(0,b)=(0,-2)$

$m = \dfrac{-2-4}{0-3}$ $y = mx + b$

$\quad = \dfrac{-6}{-3}$ $y = 2x - 2$

$\quad = 2$

31. $(0,b)=(0,3);(5,-3)$

$m = \dfrac{-3-3}{5-0}$ $y = mx + b$

$\quad = -\dfrac{6}{5}$ $y = -\dfrac{6}{5}x + 3$

33. $(2,3),(4,8)$

$y - y_1 = m(x - x_1)$

$y - 3 = \dfrac{5}{2}(x - 2)$

$2(y - 3) = 2\left(\dfrac{5}{2}(x - 2)\right)$

$m = \dfrac{8-3}{4-2}$ $2y - 6 = 5(x - 2)$

$\quad = \dfrac{5}{2}$ $2y - 6 = 5x - 10$

$-5x + 2y - 6 = -10$

$-5x + 2y = -4$

$(-1)(-5x + 2y) = (-1)(-4)$

$5x - 2y = 4$

35. $(2,0),(0,-3)$

$y - y_1 = m(x - x_1)$

$y - 0 = \dfrac{3}{2}(x - 2)$

$m = \dfrac{-3-0}{0-2}$ $y = \dfrac{3}{2}x - 3$

$\quad = \dfrac{-3}{-2}$ $2 \cdot y = 2\left(\dfrac{3}{2}x - 3\right)$

$\quad = \dfrac{3}{2}$ $2y = 3x - 6$

$-3x + 2y = -6$

$(-1)(-3x + 2y) = (-1)(-6)$

$3x - 2y = 6$

37. $(-4,-1),(7,-5)$

$m = \dfrac{-5-(-1)}{7-(-4)}$ $y - y_1 = m(x - x_1)$

$\quad = \dfrac{-4}{11}$ $y - (-5) = -\dfrac{4}{11}(x - 7)$

$11(y - (-5)) = 11\left(-\dfrac{4}{11}(x - 7)\right)$

$11(y + 5) = -4(x - 7)$

$11y + 55 = -4x + 28$

$4x + 11y + 55 = 28$

$4x + 11y = -27$

39. $(-3,2), (-6,4)$

$$y - y_1 = m(x - x_1)$$
$$y - 4 = -\frac{2}{3}(x - (-6))$$
$$y - 4 = -\frac{2}{3}(x + 6)$$

$$m = \frac{4-2}{-6-(-3)}$$
$$3(y-4) = 3\left(-\frac{2}{3}(x+6)\right)$$

$$= \frac{2}{-3}$$
$$3y - 12 = -2(x+6)$$
$$3y - 12 = -2x - 12$$

$$= -\frac{2}{3}$$
$$2x + 3y - 12 = -12$$
$$2x + 3y = 0$$

41. $(-1,0), (-1,8)$

$$m = \frac{8-0}{-1-(-1)} \qquad x = -1$$

$$= \frac{8}{0}$$
$$\text{undefined}$$

43. $(3,-1), (4,-1)$

$$m = \frac{-1-(-1)}{4-3}$$
$$y - y_1 = m(x - x_1)$$

$$= \frac{0}{1}$$
$$y - (-1) = 0(x - 4)$$
$$y + 1 = 0$$

$$= 0$$
$$y = -1$$

45. $(4,2); y = 4x + 1$

a) Use $m = 4$

$$y - y_1 = m(x - x_1)$$
$$y - 2 = 4(x - 4)$$
$$y - 2 = 4x - 16$$
$$y = 4x - 14$$

b) $\qquad y = 4x - 14$
$$-4x + y = -14$$
$$(-1)(-4x + y) = (-1)(-14)$$
$$4x - y = 14$$

47. $(-2,1); y = -3x + 1$

a) Use $m = -3$

$$y - y_1 = m(x - x_1)$$
$$y - 1 = -3(x - (-2))$$
$$y - 1 = -3(x + 2)$$
$$y - 1 = -3x - 6$$
$$y = -3x - 5$$

b) $\qquad y = -3x - 5$
$$3x + y = -5$$

49. $(5,2); y = \frac{1}{3}x + 4$

a) Use $m = \frac{1}{3}$

$$y - y_1 = m(x - x_1)$$
$$y - 2 = \frac{1}{3}(x - 5)$$
$$y - 2 = \frac{1}{3}x - \frac{5}{3}$$
$$y = \frac{1}{3}x + \frac{1}{3}$$

b) $\qquad 3 \cdot y = 3\left(\frac{1}{3}x + \frac{1}{3}\right)$
$$3y = x + 1$$
$$-x + 3y = 1$$
$$(-1)(-x + 3y) = (-1)(1)$$
$$x - 3y = -1$$

51. $(-1,-3); y = -\frac{3}{4}x+1$

   a) Use $m = -\frac{3}{4}$

$$y - y_1 = m(x - x_1)$$
$$y - (-3) = -\frac{3}{4}(x - (-1))$$
$$y + 3 = -\frac{3}{4}(x + 1)$$
$$y + 3 = -\frac{3}{4}x - \frac{3}{4}$$
$$y = -\frac{3}{4}x - \frac{15}{4}$$

   b) $4 \cdot y = 4 \cdot \left(-\frac{3}{4}x - \frac{15}{4}\right)$

$$4y = -3x - 15$$
$$3x + 4y = -15$$

53. $(2,-3); 2x + y = 5$

$$y = -2x + 5$$

   a) Use $m = -2$

$$y - y_1 = m(x - x_1)$$
$$y - (-3) = -2(x - 2)$$
$$y + 3 = -2x + 4$$
$$y = -2x + 1$$

   b) $y = -2x + 1$

$$2x + y = 1$$

55. $(-4,2); 3x - 4y = 7$

$$-4y = -3x + 7$$
$$y = \frac{3}{4}x - \frac{7}{4}$$

   a) Use $m = \frac{3}{4}$

$$y - y_1 = m(x - x_1)$$
$$y - 2 = \frac{3}{4}(x - (-4))$$
$$y - 2 = \frac{3}{4}(x + 4)$$
$$y - 2 = \frac{3}{4}x + 3$$
$$y = \frac{3}{4}x + 5$$

   b) $y = \frac{3}{4}x + 5$

$$4(y) = 4\left(\frac{3}{4}x + 5\right)$$
$$4y = 3x + 20$$
$$-3x + 4y = 20$$
$$(-1)(-3x + 4y) = (-1)(20)$$
$$3x - 4y = -20$$

57. $(-4,-1); y = 2x + 3$

   a) Use $m = -\frac{1}{2}$

$$y - y_1 = m(x - x_1)$$
$$y - (-1) = -\frac{1}{2}(x - (-4))$$
$$y + 1 = -\frac{1}{2}(x + 4)$$
$$y + 1 = -\frac{1}{2}x - 2$$
$$y = -\frac{1}{2}x - 3$$

   b) $y = -\frac{1}{2}x - 3$

$$2 \cdot y = 2\left(-\frac{1}{2}x - 3\right)$$
$$2y = -x - 6$$
$$x + 2y = -6$$

59. $(3,-4);\ y=\dfrac{1}{4}x+2$

    a)   Use $m=-4$

$$y-y_1=m(x-x_1)$$
$$y-(-4)=-4(x-3)$$
$$y+4=-4x+12$$
$$y=-4x+8$$

    b) $y=-4x+8$

$$4x+y=8$$

61. $(2,-4);\ y=-2x+7$

    a)   Use $m=\dfrac{1}{2}$

$$y-y_1=m(x-x_1)$$
$$y-(-4)=\dfrac{1}{2}(x-2)$$
$$y+4=\dfrac{1}{2}x-1$$
$$y=\dfrac{1}{2}x-5$$

    b) $y=\dfrac{1}{2}x-5$

$$2\cdot y=2\left(\dfrac{1}{2}x-5\right)$$
$$2y=x-10$$
$$-x+2y=-10$$
$$(-1)(-x+2y)=(-1)(-10)$$
$$x-2y=10$$

63. $(-2,3);\ y=\dfrac{3}{2}x-\dfrac{5}{2}$

    a)   Use $m=-\dfrac{2}{3}$

$$y-y_1=m(x-x_1)$$
$$y-3=-\dfrac{2}{3}(x-(-2))$$
$$y-3=-\dfrac{2}{3}(x+2)$$
$$y-3=-\dfrac{2}{3}x-\dfrac{4}{3}$$
$$y=-\dfrac{2}{3}x+\dfrac{5}{3}$$

    b) $y=-\dfrac{2}{3}x+\dfrac{5}{3}$

$$3\cdot y=3\left(-\dfrac{2}{3}x+\dfrac{5}{3}\right)$$
$$3y=-2x+5$$
$$2x+3y=5$$

65. $(-3,-1);\ 2x-5y=10$

$$-5y=-2x+10$$
$$y=\dfrac{2}{5}x-2$$

    a)   Use $m=-\dfrac{5}{2}$

$$y-y_1=m(x-x_1)$$
$$y-(-1)=-\dfrac{5}{2}(x-(-3))$$
$$y+1=-\dfrac{5}{2}(x+3)$$
$$y+1=-\dfrac{5}{2}x-\dfrac{15}{2}$$
$$y=-\dfrac{5}{2}x-\dfrac{17}{2}$$

    b) $y=-\dfrac{5}{2}x-\dfrac{17}{2}$

$$2\cdot y=2\cdot\left(-\dfrac{5}{2}x-\dfrac{17}{2}\right)$$
$$2y=-5x-17$$
$$5x+2y=-17$$

67. $(3, -4)$; $2x + 3y = 12$

$$3y = -2x + 12$$

$$y = -\frac{2}{3}x + 4$$

a) Use $m = \frac{3}{2}$

$$y - y_1 = m(x - x_1)$$

$$y - (-4) = \frac{3}{2}(x - 3)$$

$$y + 4 = \frac{3}{2}x - \frac{9}{2}$$

$$y = \frac{3}{2}x - \frac{17}{2}$$

b) $y = \frac{3}{2}x - \frac{17}{2}$

$$2 \cdot y = 2\left(\frac{3}{2}x - \frac{17}{2}\right)$$

$$2y = 3x - 17$$

$$-3x + 2y = -17$$

$$(-1)(-3x + 2y) = (-1)(-17)$$

$$3x - 2y = 17$$

69. The slope of the first line is 4 and the slope of the second line is 4. Because the slopes are the same and the y-intercepts are different, the lines are parallel.

71. The slope of the first line is $\frac{1}{2}$ and the slope of the second line is -2. Because the slopes are negative reciprocals, the lines are perpendicular.

73. The slope of the first line is $\frac{1}{2}$ and the slope of the second line is 2. Because the slopes are neither equal nor negative reciprocals, the lines are neither parallel nor perpendicular.

75. $4x + 6y = 5$ $8x + 12y = 9$

$\quad\ 6y = -4x + 5$ $\quad 12y = -8x + 9$

$$y = -\frac{2}{3}x + \frac{5}{6} \qquad y = -\frac{2}{3}x + \frac{3}{4}$$

$$m = -\frac{2}{3} \qquad\qquad m = -\frac{2}{3}$$

Because the slopes are the same and the y-intercepts are different, the lines are parallel.

77. $5x + 10y = 12$ $4x - 2y = 5$

$\quad\ 10y = -5x + 12$ $\quad -2y = -4x + 5$

$$y = -\frac{1}{2}x + \frac{6}{5} \qquad y = 2x - \frac{5}{2}$$

$$m = -\frac{1}{2} \qquad\qquad m = 2$$

Because the slopes are negative reciprocals, the lines are perpendicular.

79. $2x + 3y = -6$ $3x + 2y = 4$

$\quad\ 3y = -2x - 6$ $\quad 2y = -3x + 4$

$$y = -\frac{2}{3}x - 2 \qquad y = -\frac{3}{2}x + 2$$

$$m = -\frac{2}{3} \qquad\qquad m = -\frac{3}{2}$$

Because the slopes are neither equal nor negative reciprocals, the lines are neither parallel nor perpendicular.

81. The graph of the first line is horizontal, and the graph of the second line is vertical. Because the lines intersect at a 90° angle, the lines are perpendicular.

83. The graph of the first line is vertical, and the graph of the second line is vertical. Because both lines are vertical, they must be parallel.

85. a) $c = 0.3n + 7.75$

b) There are 40 eighths of a mile in 5 miles. Substitute 40 for n and calculate.

$$c = 0.3(40) + 7.75$$

$$c = 12 + 7.75$$

$$c = \$19.75$$

c)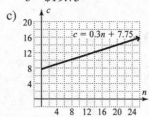

87. a)

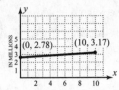

b) $(0, 2.78), (10, 3.17)$

$$m = \frac{3.17 - 2.78}{10 - 0} = \frac{0.39}{10} = 0.039$$

c) $p = 0.039n + 2.78$

d) Substitute 5 for n.

$p = 0.039n + 2.78$

$p = 0.039(5) + 2.78$

$p = 2.975$ million

According to the equation, 2.975 million females participated in high school athletics in the 2005-2006 school year.

e) Substitute 18 for n.

$p = 0.039n + 2.78$

$p = 0.039(18) + 2.78$

$p = 3.482$

According to the equation, 3.482 million females will participate in high school athletics in the 2018–2019 school year.

89. a)

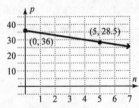

b) $(0, 36), (5, 28.5)$

$$m = \frac{28.5 - 36}{5 - 0} = \frac{-7.5}{5} = -1.5$$

c) $(0, b) = (0, 36)$

$p = -1.5n + 36$

d) (0, 36); The p-intercept indicates the initial price.

e) $p = -1.5n + 36$

$0 = -1.5n + 36$

$-36 = -1.5n$

$24 = n$

(24, 0); The n-intercept indicates that on the 24th day, the price will be $0.

f) Substitute 8 for n.

$p = -1.5(8) + 36$

$p = -12 + 36$

$p = \$24$

Review Exercises

1. $0.6 < \frac{3}{4}$

2. $\frac{3}{4}x - \frac{4}{5} = \frac{1}{2}x + 1$

$20\left(\frac{3}{4}x - \frac{4}{5}\right) = 20\left(\frac{1}{2}x + 1\right)$

$15x - 16 = 10x + 20$

$5x - 16 = 20$

$5x = 36$

$x = \frac{36}{5}$

3. $4x - 9 \geq 6x + 5$

$-2x - 9 \geq 5$

$-2x \geq 14$

$\frac{-2x}{-2} \leq \frac{14}{-2}$

$x \leq -7$

$\longleftarrow\!\!+\!\!+\!\!+\!\!+\!\!+\!\!+\!\!\longrightarrow \ x \leq -7$
$\quad -10\,\text{-}9\,\text{-}8\,\text{-}7\,\text{-}6\ \text{-}5\,\text{-}4$

4. $2(x + 1) - 5x \geq 3x - 6$

$2x + 2 - 5x \geq 3x - 6$

$-3x + 2 \geq 3x - 6$

$-6x + 2 \geq -6$

$-6x \geq -8$

$\frac{-6x}{-6} \leq \frac{-8}{-6}$

$x \leq \frac{4}{3}$

$\longleftarrow\!\!+\!\!\longrightarrow \ x \leq \frac{4}{3}$
$\quad 0 \quad 1\frac{4}{3} \ 2 \quad 3$

5. $y = -\frac{3}{4}x + 2$

The slope is $m = -\frac{3}{4}$ and the y-intercept is

$(0, 2)$.

6. $y = 0 : 2x - 7(0) = 14$ $x = 0 : 2(0) - 7y = 14$

$$2x = 14 \qquad\qquad -7y = 14$$

$$\frac{2x}{2} = \frac{14}{2} \qquad\qquad \frac{-7y}{-7} = \frac{14}{-7}$$

$$x = 7 \qquad\qquad\quad y = -2$$

$$(7,0) \qquad\qquad\quad (0,-2)$$

Exercise Set 3.6

1.   Test $(5,5)$ in $y > -x + 2$.

$$y \;>\; -x + 2$$

$$5 \;\overset{?}{>}\; -(5) + 2$$

$$5 \;>\; -3$$

This statement is true. Yes, $(5,5)$ is a solution.

3.   Test $(2,0)$ in $y \ge \dfrac{1}{2}x + 1$.

$$y \;\ge\; \frac{1}{2}x + 1$$

$$0 \;\overset{?}{\ge}\; \frac{1}{2}(2) + 1$$

$$0 \;\overset{?}{\ge}\; 1 + 1$$

$$0 \;\ge\; 2$$

This statement is false. No, $(2,0)$ is not a solution.

5.   Test $(-1,2)$ in $3x - y \le -8$.

$$3x - y \;\le\; -8$$

$$3(-1) - 2 \;\overset{?}{\le}\; -8$$

$$-3 - 2 \;\overset{?}{\le}\; -8$$

$$-5 \;\le\; -8$$

This statement is false. No, $(-1,2)$ is not a solution.

7.   Test $(-1,-2)$ in $x - 2y > -7$.

$$x - 2y \;>\; -7$$

$$-1 - 2(-2) \;\overset{?}{>}\; -7$$

$$-1 + 4 \;\overset{?}{>}\; -7$$

$$3 \;>\; -7$$

This statement is true. Yes, $(-1,-2)$ is a solution.

9. $y \le -x + 1$

Begin by graphing the related equation $y = -x + 1$ with a solid line. Now choose $(0,0)$ as a test point.

$$y \le -x + 1$$

$$0 \overset{?}{\le} -(0) + 1$$

$$0 \le 1$$

Because $(0,0)$ satisfies the inequality, shade the region which includes $(0,0)$.

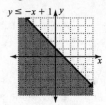

11. $y > -2x + 5$

Begin by graphing the related equation $y = -2x + 5$ with a dashed line. Now choose $(0,0)$ as a test point.

$$y > -2x + 5$$

$$0 \overset{?}{>} -2(0) + 5$$

$$0 > 5$$

Because $(0,0)$ does not satisfy the inequality, shade the side of the line on the opposite side from $(0,0)$.

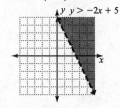

13. $y > 2x$

Begin by graphing the related equation $y = 2x$ with a dashed line. Now choose $(1,0)$ as a test point.

$y > 2x$

$0 \overset{?}{>} 2(1)$

$0 > 2$

Because $(1,0)$ does not satisfy the inequality, shade the side of the line on the opposite side from $(1,0)$.

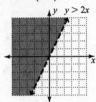

15. $y > \dfrac{1}{3}x$

Begin by graphing the related equation $y = \dfrac{1}{3}x$ with a dashed line. Now choose $(1,0)$ as a test point.

$y > \dfrac{1}{3}x$

$0 \overset{?}{>} \dfrac{1}{3}(1)$

$0 > \dfrac{1}{3}$

Because $(1,0)$ does not satisfy the inequality, shade the side of the line on the opposite side from $(1,0)$.

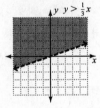

17. $y \geq \dfrac{3}{4}x + 2$

Begin by graphing the related equation $y = \dfrac{3}{4}x + 2$ with a solid line. Now choose $(0,0)$ as a test point.

$y \geq \dfrac{3}{4}x + 2$

$0 \overset{?}{\geq} \dfrac{3}{4}(0) + 2$

$0 \geq 2$

Because $(0,0)$ does not satisfy the inequality, shade the side of the line on the opposite side from $(0,0)$.

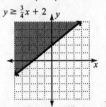

19. $3x + y \leq 9$

Begin by graphing the related equation $3x + y = 9$ with a solid line. Now choose $(0,0)$ as a test point.

$3x + y \leq 9$

$3(0) + 0 \overset{?}{\leq} 9$

$0 \leq 9$

Because $(0,0)$ satisfies the inequality, shade the region which includes $(0,0)$.

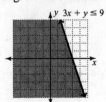

21. $5x - 2y < 10$

Begin by graphing the related equation $5x - 2y = 10$ with a dashed line. Now choose $(0,0)$ as a test point.

$$5x - 2y < 10$$

$$5(0) - 2(0) \overset{?}{<} 10$$

$$0 - 0 \overset{?}{<} 10$$

$$0 < 10$$

Because $(0,0)$ satisfies the inequality, shade the region which includes $(0,0)$.

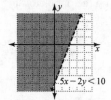

23. $3x + y \geq 8$

Begin by graphing the related equation $3x + y = 8$ with a solid line. Now choose $(0,0)$ as a test point.

$$3x + y \geq 8$$

$$3(0) + (0) \overset{?}{\geq} 8$$

$$0 + 0 \overset{?}{\geq} 8$$

$$0 \geq 8$$

Because $(0,0)$ does not satisfy the inequality, shade the side of the line on the opposite side from $(0,0)$.

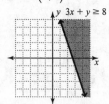

25. $2x + 3y < 9$

Begin by graphing the related equation $2x + 3y = 9$ with a dashed line. Now choose $(0,0)$ as a test point.

$$2x + 3y < 9$$

$$2(0) + 3(0) \overset{?}{<} 9$$

$$0 + 0 \overset{?}{<} 9$$

$$0 < 9$$

Because $(0,0)$ satisfies the inequality, shade the region which includes $(0,0)$.

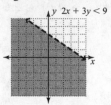

27. $3x + y \geq 0$

Begin by graphing the related equation $3x + y = 0$ with a solid line. Now choose $(1,0)$ as a test point.

$$3x + y \geq 0$$

$$3(1) + 0 \overset{?}{\geq} 0$$

$$3 + 0 \overset{?}{\geq} 0$$

$$3 \geq 0$$

Because $(1,0)$ satisfies the inequality, shade the region which includes $(1,0)$.

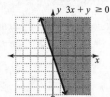

29. $x > -4$

Begin by graphing the related equation $x = -4$ with a dashed line. Now choose $(0,0)$ as a test point.

$$x > -4$$

$$0 > -4$$

Because $(0,0)$ satisfies the inequality, shade the region which includes $(0,0)$.

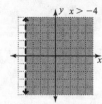

31. $y \leq 2$

Begin by graphing the related equation $y = 2$ with a solid line. Now choose $(0,0)$ as a test point.

$y \leq 2$

$0 \leq 2$

Because $(0,0)$ satisfies the inequality, shade the region which includes $(0,0)$.

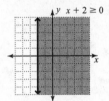

33. $x + 2 \geq 0$

Begin by graphing the related equation $x + 2 = 0$ with a solid line. Now choose $(0,0)$ as a test point.

$x + 2 \geq 0$

$0 + 2 \overset{?}{\geq} 0$

$2 \geq 0$

Because $(0,0)$ satisfies the inequality, shade the region which includes $(0,0)$.

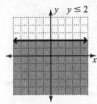

35. $y - 3 < 0$

Begin by graphing the related equation $y - 3 = 0$ with a dashed line. Now choose $(0,0)$ as a test point.

$y - 3 < 0$

$0 - 3 \overset{?}{<} 0$

$-3 < 0$

Because $(0,0)$ satisfies the inequality, shade the region which includes $(0,0)$.

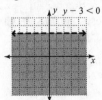

37. a) x represents the number of large bottles sold; y represents the number of small bottles sold.

b)

c) The boundary line represents combinations of bottle sizes sold to produce a revenue of exactly $280,000 (breakeven). The shaded region represents combinations that produce a revenue greater than $280,000 (profit).

d) Answers will vary; three possible combinations are (80,000, 80,000), (20,000, 160,000), and (50,000, 120,000).

e) Answers will vary; three possible combinations are (150,000, 0), (90,000, 100,000), and (80,000, 120,000).

f) No, only combinations of whole numbers are possible because we cannot sell fractions of a bottle of hand lotion.

39. a) Let x represent the number of 1-gallon containers and let y represent the number of 5-gallon containers. An inequality that has Andrea's total cost within the amount of the gift certificate is $5.5x + 12.5y \leq 100$

b)
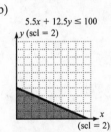

c) The combinations of plant purchases that would cost exactly $100.

d) The combinations of plants purchased that would cost less than $100.

e) Answers will vary; some combinations are (2, 4), (6, 2), and (8, 4).

f) Yes, (0, 8). Because she cannot purchase fractional numbers of plants, the combination must be whole numbers and (0, 8) is the only combination of whole numbers that costs exactly $100.

41. a) Let l represent length and let w represent width. Then an inequality in which the maximum perimeter is 180 feet is $2l + 2w \le 180$.

    b)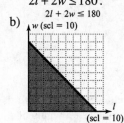

    c) The combinations of lengths and widths that yield a perimeter of exactly 180 feet.

    d) The combinations of lengths and widths that yield a perimeter that is less than 180 feet.

    e) Answers will vary; some combinations are (10, 80), (20, 70), and (40, 50).

    f) Answers will vary; some combinations are (10, 10), (20, 20), and (30, 30).

Review Exercises

1. $\{a, e, i, o, u\}$

2. Let $x = 8$.
    $$\frac{3}{4}x - 5 = \frac{3}{4}(8) - 5$$
    $$= \frac{24}{4} - 5$$
    $$= 6 - 5$$
    $$= 1$$

3. $3x - 4(x + 7) = -15$
    $$3x - 4x - 28 = -15$$
    $$-x - 28 = -15$$
    $$-x = 13$$
    $$-1 \cdot (-x) = -1 \cdot (13)$$
    $$x = -13$$
    Check: $3(-13) - 4(-13 + 7) = -15$
    $$-39 - 4(-6) \overset{?}{=} -15$$
    $$-39 + 24 \overset{?}{=} -15$$
    $$-15 = -15$$

4. Let l be the length of the rectangle and let w be the width. The perimeter of a rectangle is given by the formula $P = 2l + 2w$. The length is four more than the width, so $l = w + 4$. Substitute the expression for l and the value of P into the formula and solve for w.

$$P = 2l + 2w$$
$$88 = 2(w + 4) + 2w$$
$$88 = 2w + 8 + 2w$$
$$88 = 4w + 8$$
$$80 = 4w$$
$$20 = w$$

Substitute the value of w into the expression for l to find the length.

$$l = 20 + 4 = 24$$

So, the dimensions of the rectangle are 20 ft by 24 ft.

5. $3x - 2y = 8$

x-intercept	*y*-intercept

    $y = 0$: $3x - 2(0) = 8$ $x = 0$: $3(0) - 2y = 8$
    $$3x = 8 \qquad\qquad\qquad -2y = 8$$
    $$\frac{3x}{3} = \frac{8}{3} \qquad\qquad \frac{-2y}{-2} = \frac{8}{-2}$$
    $$x = \frac{8}{3} \qquad\qquad\qquad y = -4$$
    $$\left(\frac{8}{3}, 0\right) \qquad\qquad\qquad (0, -4)$$

6.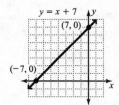

Exercice Set 3.7

1. Domain: $\{2, 3, 1, -1\}$; Range: $\{1, -2, 4, -1\}$

3. Domain: $\{4, 2, 3, -4\}$; Range: $\{1, -5, 0\}$

5. Domain: $\{0, -3, 2, 8\}$; Range: $\{0, -3, 9, -16\}$

7. Yes, this relation is a function because each sector in the domain has exactly one cost value in the range.

9. No, this relation is not a function because several elements in the domain (positions) are assigned to multiple people in the range.

11. Yes, this relation is a function because each title in the domain is assigned exactly one units sold in the range.

13. No, this relation is not a function because several of the average high temperatures are assigned to more than one month in the range.

15. Yes, every element in the domain is assigned to exactly one element in the range.

17. No, an element in the domain is assigned to more than one element in the range. (8 is assigned to both 2 and –1).

19. Yes, every element in the domain is assigned to exactly one element in the range.

21. No, elements in the domain are assigned to more than one element in the range.

23. No, it fails the vertical line test.

25. Yes, it passes the vertical line test.

27. No, it fails the vertical line test.

29. Yes, it passes the vertical line test.

31. Domain: $\{x|1980 \le x \le 2010\}$ or $[1980, 2010]$

    Range: $\{y|17.7 \text{ million} \le y \le 104.1 \text{ million}\}$ or

    $[17.7 \text{ million}, 104.1 \text{ million}]$

    This is a function because every element in the domain is assigned to exactly one element in the range.

33. Domain: {1900, 1910, 1920, 1930, 1940, 1950, 1960, 1970, 1980, 1990, 2000, 2010}
    Range: {4.7, 5.4, 6.2, 6.9, 8.0, 8.8, 10.4, 11.6, 12.9, 13.2, 13.6, 14.7}
    This is a function because every element in the domain is assigned to exactly one element in the range.

35. Domain: $\{x|-4 \le x \le 4\}$ or $[-4, 4]$

    Range: $\{y|-3 \le y \le 2\}$ or $[-3, 2]$

    This is a function because the graph passes the Vertical Line Test.

37. Domain: $\{x|x \le 0\}$ or $(-\infty, 0]$

    Range: all real numbers or $(-\infty, \infty)$

    This is not a function because the Vertical Line Test fails.

39. Domain: all real numbers or $(-\infty, \infty)$

    Range: $\{y|y \ge -1\}$ or $[-1, \infty)$

    This is a function because the graph passes the Vertical Line Test.

41. $f(x) = 2x - 5$

    a) $f(0) = 2(0) - 5$
    $= 0 - 5$
    $= -5$

    b) $f(1) = 2(1) - 5$
    $= 2 - 5$
    $= -3$

    c) $f(-2) = 2(-2) - 5$
    $= -4 - 5$
    $= -9$

    d) $f\left(\frac{1}{2}\right) = 2\left(\frac{1}{2}\right) - 5$
    $= 1 - 5$
    $= -4$

43. $f(x) = x^2 - 3x + 7$

    a) $f(0) = 0^2 - 3(0) + 7$
    $= 0 - 0 + 7$
    $= 7$

    b) $f(1) = 1^2 - 3(1) + 7$
    $= 1 - 3 + 7$
    $= 5$

    c) $f(-1) = (-1)^2 - 3(-1) + 7$
    $= 1 + 3 + 7$
    $= 11$

    d) $f(-2) = (-2)^2 - 3(-2) + 7$
    $= 4 + 6 + 7$
    $= 17$

45. $f(x) = \sqrt{x+3}$

    a) $f(0) = \sqrt{0+3}$
    $= \sqrt{3}$

    b) $f(-4) = \sqrt{-4+3}$
    $= \sqrt{-1}$
    not a real number

    c) $f(-1) = \sqrt{-1+3}$
    $= \sqrt{2}$

    d) $f(a) = \sqrt{a+3}$

47. $f(x) = \sqrt{x^2 - 4}$

   a) $f(2) = \sqrt{2^2 - 4}$
      $= \sqrt{4 - 4}$
      $= \sqrt{0}$
      $= 0$

   b) $f(1) = \sqrt{1^2 - 4}$
      $= \sqrt{1 - 4}$
      $= \sqrt{-3}$
      not a real number

   c) $f(-2) = \sqrt{(-2)^2 - 4}$
      $= \sqrt{4 - 4}$
      $= \sqrt{0}$
      $= 0$

   d) $f(3) = \sqrt{3^2 - 4}$
      $= \sqrt{9 - 4}$
      $= \sqrt{5}$

49. $f(x) = \dfrac{1}{x - 1}$

   a) $f(0) = \dfrac{1}{0 - 1}$
      $= \dfrac{1}{-1}$
      $= -1$

   b) $f(1) = \dfrac{1}{1 - 1}$
      $= \dfrac{1}{0}$
      undefined

   c) $f(-1) = \dfrac{1}{-1 - 1}$
      $= -\dfrac{1}{2}$

   d) $f(a) = \dfrac{1}{a - 1}$

51. $f(x) = x^2 - 2x - 3$

   a) $f(0) = 0^2 - 2(0) - 3$
      $= -3$

   b) $f(0.1) = (0.1)^2 - 2(0.1) - 3$
      $= 0.01 - 0.2 - 3$
      $= -3.19$

   c) $f(a) = a^2 - 2a - 3$

   d) $f(-a) = (-a)^2 - 2(-a) - 3$
      $= a^2 + 2a - 3$

53. $f(x) = \dfrac{1}{3}x - 2$

   a) $f(0) = \dfrac{1}{3}(0) - 2$
      $= 0 - 2$
      $= -2$

   b) $f(1) = \dfrac{1}{3}(1) - 2$
      $= \dfrac{1}{3} - 2$
      $= \dfrac{1}{3} - \dfrac{6}{3}$
      $= -\dfrac{5}{3}$

   c) $f(-1) = \dfrac{1}{3}(-1) - 2$
      $= -\dfrac{1}{3} - 2$
      $= -\dfrac{1}{3} - \dfrac{6}{3}$
      $= -\dfrac{7}{3}$

   d) $f(a) = \dfrac{1}{3}(a) - 2$
      $= \dfrac{1}{3}a - 2$

55. $f(x) = \dfrac{x+3}{x-4}$

   a) $f(0) = \dfrac{0+3}{0-4}$

   $= -\dfrac{3}{4}$

   b) $f(4) = \dfrac{4+3}{4-4}$

   $= \dfrac{7}{0}$

   undefined

   c) $f(-2) = \dfrac{-2+3}{-2-4}$

   $= -\dfrac{1}{6}$

   d) $f(3) = \dfrac{3+3}{3-4}$

   $= \dfrac{6}{-1}$

   $= -6$

57. $f(x) = \dfrac{1}{\sqrt{x+2}}$

   a) $f(-2) = \dfrac{1}{\sqrt{-2+2}}$

   $= \dfrac{1}{\sqrt{0}}$

   $= \dfrac{1}{0}$

   undefined

   b) $f(1) = \dfrac{1}{\sqrt{1+2}}$

   $= \dfrac{1}{\sqrt{3}}$

   c) $f(-3) = \dfrac{1}{\sqrt{-3+2}}$

   $= \dfrac{1}{\sqrt{-1}}$

   not a real number

   d) $f(a) = \dfrac{1}{\sqrt{a+2}}$

59. $f(x) = |4x-2|$

   a) $f(0) = |4(0)-2|$

   $= |0-2|$

   $= |-2|$

   $= 2$

   b) $f(1.5) = |4(1.5)-2|$

   $= |6-2|$

   $= |4|$

   $= 4$

   c) $f(-1) = |4(-1)-2|$

   $= |-4-2|$

   $= |-6|$

   $= 6$

   d) $f(-2) = |4(-2)-2|$

   $= |-8-2|$

   $= |-10|$

   $= 10$

61. $f(x) = x^2 - 3x + 2$, $g(x) = \dfrac{x-4}{x+2}$

   a) $f(2) = (2)^2 - 3(2) + 2$

   $= 4 - 6 + 2$

   $= 0$

   b) $g(2) = \dfrac{2-4}{2+2}$

   $= \dfrac{-2}{4}$

   $= -\dfrac{1}{2}$

   c) $f(-3) = (-3)^2 - 3(-3) + 2$

   $= 9 + 9 + 2$

   $= 20$

   d) $g(-3) = \dfrac{-3-4}{-3+2}$

   $= \dfrac{-7}{-1}$

   $= 7$

63. $f(x) = 2x^2 + 3x + 1$

 a) $f(2a) = 2(2a)^2 + 3(2a) + 1$

   $= 2(4a^2) + 6a + 1$

   $= 8a^2 + 6a + 1$

 b) $f(-3a) = 2(-3a)^2 + 3(-3a) + 1$

   $= 18a^2 - 9a + 1$

 c) $f\left(\dfrac{1}{2}a\right) = 2\left(\dfrac{1}{2}a\right)^2 + 3\left(\dfrac{1}{2}a\right) + 1$

   $= 2\left(\dfrac{1}{4}a^2\right) + \dfrac{3}{2}a + 1$

   $= \dfrac{1}{2}a^2 + \dfrac{3}{2}a + 1$

 d) $f(0.1a) = 2(0.1a)^2 + 3(0.1a) + 1$

   $= 2(0.01a^2) + 0.3a + 1$

   $= 0.02a^2 + 0.3a + 1$

65. a) $f(1) = 3$

 b) $f(2) = 4$

 c) $f(0) = 2$

67. a) $f(0) = 0$

 b) $f(2) = -2$

 c) $f(3)$ is undefined

69.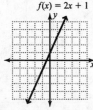
$f(x) = 2x + 1$

71.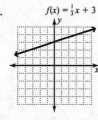
$f(x) = \frac{1}{3}x + 3$

73.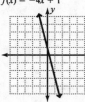
$f(x) = -4x + 1$

75.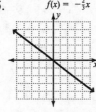
$f(x) = -\frac{2}{3}x$

77.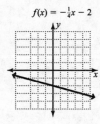
$f(x) = -\frac{1}{4}x - 2$

79. $C(x) = 20x + 150$

 a) Substitute 10 for x and calculate.

   $C(10) = 20(10) + 150 = 200 + 150 = \$350$

 b) Substitute 15 for x and calculate.

   $C(15) = 20(15) + 150 = 300 + 150 = \$450$

 c) The cost of producing 30 items is \$750.

81. $V(x) = -1500x + 18,000$

 a) Substitute 3 for x and calculate.

   $V(3) = -1500(3) + 18,000$

   $= -4500 + 18,000$

   $= \$13,500$

 b) Substitute 6 for x and calculate.

   $V(6) = -1500(6) + 18,000$

   $= -9000 + 18,000$

   $= \$9000$

 c) The value of the car after ten years is \$3000.

Review Exercises

1. $\dfrac{2}{5} - \dfrac{5}{8} \div \dfrac{3}{4} = \dfrac{2}{5} - \dfrac{5}{8} \cdot \dfrac{4}{3}$

   $= \dfrac{2}{5} - \dfrac{5}{6}$

   $= \dfrac{12}{30} - \dfrac{25}{30}$

   $= -\dfrac{13}{30}$

2. $x + 3y - 5x + 7 - 3y - 9 = -4x - 2$

3.   Commutative Property of Addition.

4. $2x - y = 4$

   $\underline{x\text{-intercept}}$ $\underline{y\text{-intercept}}$
   Let $y = 0$:   Let $x = 0$:
   $2x - 0 = 4$ $2(0) - y = 4$
   $2x = 4$ $-y = 4$
   $x = 2$ $\dfrac{-y}{-1} = \dfrac{4}{-1}$
   $(2, 0)$ $y = -4$
                $(0, -4)$

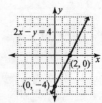

5. $3x - 2y = 6$

    $-2y = -3x + 6$

    $y = \dfrac{3}{2}x - 3$

6. $2x + 3y = 9 \qquad\qquad 4x + 6y = 12$

    $3y = -2x + 9 \qquad\quad 6y = -4x + 12$

    $y = -\dfrac{2}{3}x + 3 \qquad\quad y = -\dfrac{2}{3}x + 2$

    Both lines have slopes of $m = -\dfrac{2}{3}$ and different

    y-intercepts. Therefore, the lines are parallel.

Chapter 3 Review Exercises

1. $A(4, 3)$; a vertical line intersects the x-axis at 4, and a horizontal line intersects the y-axis at 3.
    $B(-2, 4)$; a vertical line intersects the x-axis at -2, and a horizontal line intersects the y-axis at 4.
    $C(-4, -2)$; a vertical line intersects the x-axis at -4, and a horizontal line intersects the y-axis at -2.
    $D(2, -1)$; a vertical line intersects the x-axis at 2, and a horizontal line intersects the y-axis at -1.

2. $A(3, 1)$; a vertical line intersects the x-axis at 3, and a horizontal line intersects the y-axis at 1.
    $B(0, 4)$; a vertical line intersects the x-axis at 0, and a horizontal line intersects the y-axis at 4.
    $C(-3, 0)$; a vertical line intersects the x-axis at -3, and a horizontal line intersects the y-axis at 0.
    $D(0, -2)$; a vertical line intersects the x-axis at 0, and a horizontal line intersects the y-axis at -2.

3. 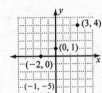   4.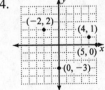

5. Quadrant II because the first coordinate is negative and the second coordinate is positive.

6. Quadrant III because both coordinates are negative.

7. Quadrant I because both coordinates are positive.

8. Quadrant IV because the first coordinate is positive and the second coordinate is negative.

9. Replace x with -1 and y with 3 and see if the equation is true.

    $$x + 3y = 8$$
    $$-1 + 3(3) \stackrel{?}{=} 8$$
    $$-1 + 9 \stackrel{?}{=} 8$$
    $$8 = 8$$

    Yes, $(-1, 3)$ is a solution for the equation.

10. Replace x with 2 and y with 1 and see if the equation is true.

    $$2x - y = 3$$
    $$2(2) - 1 \stackrel{?}{=} 3$$
    $$4 - 1 \stackrel{?}{=} 3$$
    $$3 = 3$$

    Yes, $(2, 1)$ is a solution for the equation.

11. Replace x with 0 and y with 0 and see if the equation is true.

    $$6x = 2y + 1$$
    $$6(0) \stackrel{?}{=} 2(0) + 1$$
    $$0 \stackrel{?}{=} 0 + 1$$
    $$0 \neq 1$$

    No, $(0, 0)$ is not a solution for the equation.

12. Replace x with -2 and y with -2 and see if the equation is true.

    $$y = \dfrac{2}{3}x$$
    $$-2 \stackrel{?}{=} \dfrac{2}{3}(-2)$$
    $$-2 \neq -\dfrac{4}{3}$$

    No, $(-2, -2)$ is not a solution for the equation.

13. $(0, 0), (1, -2), (-1, 2)$

    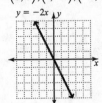

14. $(0,7),(1,6),(3,4)$

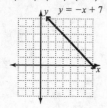

15. $(0,2),(3,0),(-3,4)$

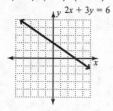

16. $(0,3),(3,2),(-3,4)$

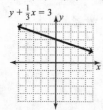

17. a) Substitute 24,000 for s and calculate.

$$p = 0.05(24,000) + 1000$$

$$p = 1200 + 1000$$

$$p = \$2200$$

b) Substitute 0 for s and calculate.

$$p = 0.05(0) + 1000 = 0 + 1000 = 1000$$

(0, 1000); The p-intercept indicates the person's gross pay if he has \$0 in sales during the month.

c)

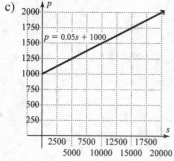

18.

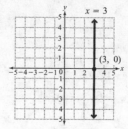

19.

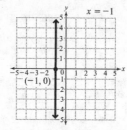

20.

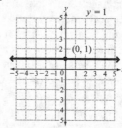

21.

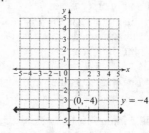

22. $3x + 2y = 6$

<u>x-intercept</u>	<u>y-intercept</u>
Let $y = 0$:	Let $x = 0$:
$3x + 2(0) = 6$	$3(0) + 2y = 6$
$3x = 6$	$2y = 6$
$x = 2$	$y = 3$
$(2,0)$	$(0,3)$

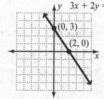

23. $y = -2x$

_x_-intercept	_y_-intercept
Let $y = 0$:	Let $x = 0$:
$0 = -2x$	$y = -2(0)$
$x = 0$	$y = 0$
$(0,0)$	$(0,0)$

The line goes through $(0,0), (-1,2), (1,-2)$.

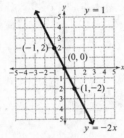

24. $y = -\dfrac{1}{5}x + 3$

_x_-intercept	_y_-intercept
Let $y = 0$:	Let $x = 0$:
$0 = -\dfrac{1}{5}x + 3$	$y = -\dfrac{1}{5}(0) + 3$
$\dfrac{1}{5}x = 3$	$y = 0 + 3$
$5 \cdot \dfrac{1}{5}x = 5 \cdot 3$	$y = 3$
$x = 15$	$(0,3)$
$(15,0)$	

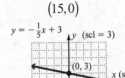

25. $2x = 5y + 10$

_x_-intercept	_y_-intercept
Let $y = 0$:	Let $x = 0$:
$2x = 5(0) + 10$	$2(0) = 5y + 10$
$2x = 10$	$0 = 5y + 10$
$x = 5$	$-10 = 5y$
$(5,0)$	$-2 = y$
	$(0,-2)$

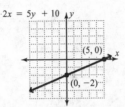

26. $7 = y$

The _y_-intercept is (0, 7) and there is no _x_-intercept. The graph is a horizontal line through (0, 7).

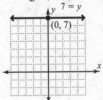

27. $x = -1$

The _x_-intercept is (−1, 0) and there is no _y_-intercept. The graph is a vertical line through (−1, 0).

28. $y = -2x + 3$

$m = -2; (0,3)$

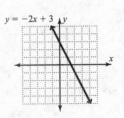

29. $y = -\dfrac{1}{4}x + 2$

$m = -\dfrac{1}{4}; (0, 2)$

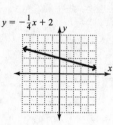

30. $y = 3x$

$m = 3; (0, 0)$

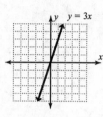

31. $x + y = 5$

$y = -x + 5$

$m = -1; (0, 5)$

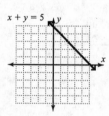

32. $x - 3y = 7$

$-3y = -x + 7$

$y = \dfrac{1}{3}x - \dfrac{7}{3}$

$m = \dfrac{1}{3}; \left(0, -\dfrac{7}{3}\right)$

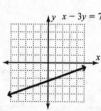

33. $2x - 3y = 8$

$-3y = -2x + 8$

$y = \dfrac{2}{3}x - \dfrac{8}{3}$

$m = \dfrac{2}{3}; \left(0, -\dfrac{8}{3}\right)$

34. Let $(x_1, y_1) = (2, 7)$ and $(x_2, y_2) = (-1, -2)$.

$m = \dfrac{y_2 - y_1}{x_2 - x_1} = \dfrac{-2 - 7}{-1 - 2} = \dfrac{-9}{-3} = 3$

35. Let $(x_1, y_1) = (-6, 2)$ and $(x_2, y_2) = (-1, -1)$.

$m = \dfrac{y_2 - y_1}{x_2 - x_1} = \dfrac{-1 - 2}{-1 - (-6)} = \dfrac{-3}{5}$

36. Let $(x_1, y_1) = (2, 8)$ and $(x_2, y_2) = (-1, 8)$.

$m = \dfrac{y_2 - y_1}{x_2 - x_1} = \dfrac{8 - 8}{-1 - 2} = \dfrac{0}{-3} = 0$

37. Let $(x_1, y_1) = (-7, 3)$ and $(x_2, y_2) = (-7, -5)$.

$m = \dfrac{y_2 - y_1}{x_2 - x_1} = \dfrac{-5 - 3}{-7 - (-7)} = \dfrac{-8}{0};$ undefined

38. $m = -1; (0, 7)$

$y = mx + b$

$y = -x + 7$

39. $m = -\dfrac{1}{5}; (0, -8)$

$y = mx + b$

$y = -\dfrac{1}{5}x - 8$

40. $m = 0.2; (0, 6)$

$y = mx + b$

$y = 0.2x + 6$

41. $m = 1; (0, 0)$

$y = mx + b$

$y = x$

42. $m = -2; (1, 7)$

$y - y_1 = m(x - x_1)$

$y - 7 = -2(x - 1)$

$y - 7 = -2x + 2$

$y = -2x + 9$

43. $m = -\dfrac{2}{5}; (3,3)$

$$y - y_1 = m(x - x_1)$$

$$y - 3 = -\dfrac{2}{5}(x - 3)$$

$$y - 3 = -\dfrac{2}{5}x + \dfrac{6}{5}$$

$$y = -\dfrac{2}{5}x + \dfrac{21}{5}$$

44. $(7,-3), (2,2)$

$m = \dfrac{2 - (-3)}{2 - 7}$　　　$y - y_1 = m(x - x_1)$

$ = \dfrac{5}{-5}$　　　　　$y - 2 = -1(x - 2)$

$ = -1$　　　　　$y - 2 = -x + 2$

　　　$y = -x + 4$

　　　$x + y = 4$

45. $(5,-3), (9,2)$

$$y - y_1 = m(x - x_1)$$

$$y - 2 = \dfrac{5}{4}(x - 9)$$

$$y - 2 = \dfrac{5}{4}x - \dfrac{45}{4}$$

$$y = \dfrac{5}{4}x - \dfrac{37}{4}$$

$m = \dfrac{2 - (-3)}{9 - 5}$　　$4 \cdot y = 4\left(\dfrac{5}{4}x - \dfrac{37}{4}\right)$

$ = \dfrac{5}{4}$　　　　　$4y = 5x - 37$

　　$-5x + 4y = -37$

　$-1(-5x + 4y) = -1(-37)$

　　$5x - 4y = 37$

46. a)

b) $m = \dfrac{12,000 - 18,000}{4 - 0} = \dfrac{-6000}{4} = -1500$

c) $v = -1500n + 18,000$

d) Substitute 9,000 (half of 18,000) for v and solve for n.

$$9000 = -1500n + 18,000$$

$$-9000 = -1500n$$

$$6 = n$$

In the year $2009 + 6 = 2015$, the copier is worth half its original value.

e) Substitute 0 for v and solve for n.

$$0 = -1500n + 18,000$$

$$-18000 = -1500n$$

$$12 = n$$

In the year $2009 + 12 = 2021$, the copier will be worth \$0.

47. $(0, b) = (0, -2)$

Use $m = 2$

$$y = 2x - 2$$

48. $(1, 5);\ 2x + 3y = 6$

$$3y = -2x + 6$$

$$y = -\dfrac{2}{3}x + 2$$

Use $m = -\dfrac{2}{3}$

$$y - y_1 = m(x - x_1)$$

$$y - 5 = -\dfrac{2}{3}(x - 1)$$

$$y - 5 = -\dfrac{2}{3}x + \dfrac{2}{3}$$

$$y = -\dfrac{2}{3}x + \dfrac{17}{3}$$

49. $(-1, -2);\ y = \dfrac{3}{5}x - 1$

Use $m = -\dfrac{5}{3}$

$$y - y_1 = m(x - x_1)$$

$$y - (-2) = -\dfrac{5}{3}(x - (-1))$$

$$y + 2 = -\dfrac{5}{3}(x + 1)$$

$$y + 2 = -\dfrac{5}{3}x - \dfrac{5}{3}$$

$$y = -\dfrac{5}{3}x - \dfrac{11}{3}$$

50. $(-2,4);\ y=-\dfrac{2}{5}x-6$

    Use $m=\dfrac{5}{2}$

    $y-y_1=m\left(x-x_1\right)$

    $y-4=\dfrac{5}{2}\left(x-(-2)\right)$

    $y-4=\dfrac{5}{2}\left(x+2\right)$

    $y-4=\dfrac{5}{2}x+5$

    $y=\dfrac{5}{2}x+9$

51. Test $(3,2)$ in $x+2y>5$.

    $\begin{aligned}x+2y &>5\\ 3+2(2) &\overset{?}{>} 5\\ 3+4 &\overset{?}{>} 5\\ 7 &> 5\end{aligned}$

    This statement is true. Yes, $(3,2)$ is a solution.

52. Test $(-1,0)$ in $y<x+2$.

    $\begin{aligned}y &< x+2\\ 0 &\overset{?}{<} -1+2\\ 0 &< 1\end{aligned}$

    This statement is true. Yes, $(-1,0)$ is a solution.

53. Test $(-5,8)$ in $2x-6y\le17$.

    $\begin{aligned}2x-6y &\le 17\\ 2(-5)-6(8) &\overset{?}{\le} 17\\ -10-48 &\overset{?}{\le} 17\\ -58 &\le 17\end{aligned}$

    This statement is true. Yes, $(-5,8)$ is a solution.

54. Test $(0,0)$ in $x+y\ge5$.

    $\begin{aligned}x+y &\ge 5\\ 0+0 &\overset{?}{\ge} 5\\ 0 &\ge 5\end{aligned}$

    This statement is false. No, $(0,0)$ is not a solution.

55. $y<3x-5$

    Begin by graphing the related equation $y=3x-5$ with a dashed line. Now choose $(0,0)$ as a test point.

    $y<3x-5$

    $0\overset{?}{<}3(0)-8$

    $0\overset{?}{<}0-8$

    $0<-8$

    Because $(0,0)$ does not satisfy the inequality, shade the side of the line on the opposite side from $(0,0)$.

    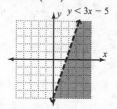

56. $y\ge-\dfrac{2}{3}x$

    Begin by graphing the related equation $y=-\dfrac{2}{3}x$ with a solid line. Now choose $(1,0)$ as a test point.

    $y\ge-\dfrac{2}{3}x$

    $0\overset{?}{\ge}-\dfrac{2}{3}(1)$

    $0\ge-\dfrac{2}{3}$

    Because $(1,0)$ satisfies the inequality, shade the region which includes $(1,0)$.

    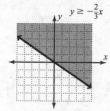

57. $x - y \geq 3$

Begin by graphing the related equation $x - y = 3$ with a solid line. Now choose $(0,0)$ as a test point.

$$x - y \geq 3$$
$$0 - 0 \overset{?}{\geq} 3$$
$$0 \geq 3$$

Because $(0,0)$ does not satisfy the inequality, shade the side of the line on the opposite side from $(0,0)$.

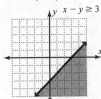

58. $-3x - 5y < -15$

Begin by graphing the related equation $-3x - 5y = -15$ with a dashed line. Now choose $(0,0)$ as a test point.

$$-3x - 5y < -15$$
$$-3(0) - 5(0) \overset{?}{<} -15$$
$$0 - 0 \overset{?}{<} -15$$
$$0 < -15$$

Because $(0,0)$ does not satisfy the inequality, shade the side of the line on the opposite side from $(0,0)$.

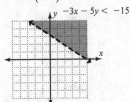

59. $x \geq -1$

Begin by graphing the related equation $x = -1$ with a solid line. Now choose $(0,0)$ as a test point.

$$x \geq -1$$
$$0 \geq -1$$

Because $(0,0)$ satisfies the inequality, shade the region which includes $(0,0)$.

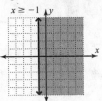

60. a) $6a + 8b \leq 12,000$

b)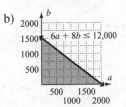

c) Answers may vary; some combinations are (400, 1200), (800, 900), (1200, 600).

d) Answers may vary; some combinations are (500, 100), (1000, 200), (200, 1100).

61. Domain: {2, –1, 3, –3}; Range {3, 5, 6, –4}

62. Domain: {Volkswagen, Honda, Toyota, Subaru, Nissan}
Range: {52.2, 49.7, 49.0, 47.8, 45.8}

63. No, this relation is not a function because instructors Carson and Webb in the domain are assigned to more than one course in the range.

64. Yes, this relation is a function because each size in the domain has exactly one price in the range.

65. Domain: all real numbers or $(-\infty, \infty)$
Range: all real numbers
This is a function because the graph passes the Vertical Line Test.

66. Domain: $\{x \mid x \leq 0\}$ or $(-\infty, 0]$
Range: all real numbers or $(-\infty, \infty)$
This is not a function because the Vertical Line Test fails.

67. Domain: $\{x \mid -4 \leq x \leq 5\}$ or $[-4, 5]$
Range: {1, 2, 3}
This is a function because the graph passes the Vertical Line Test.

68. a) $f(-3)=1$

b) $f(-1)=3$

c) $f(0)=3$

d) $f(3)=2$

69. $f(x)=x^3-5x+2$

a) $f(2)=(2)^3-5(2)+2$
$=8-10+2$
$=0$

b) $f(0)=(0)^3-5(0)+2$
$=0-0+2$
$=2$

c) $f(-3)=(-3)^3-5(-3)+2$
$=-27+15+2$
$=-10$

70. $f(x)=\dfrac{x}{x-3}$

a) $f(6)=\dfrac{6}{6-3}=\dfrac{6}{3}=2$

b) $f(3)=\dfrac{3}{3-3}=\dfrac{3}{0}$; undefined

c) $f(-5)=\dfrac{-5}{-5-3}=\dfrac{-5}{-8}=\dfrac{5}{8}$

Chapter 3 Practice Test

1. $A(2, 4)$; a vertical line intersects the x-axis at 2, and a horizontal line intersects the y-axis at 4. $B(-3, 2)$; a vertical line intersects the x-axis at -3, and a horizontal line intersects the y-axis at 2. $C(0, -5)$; a vertical line intersects the x-axis at 0, and a horizontal line intersects the y-axis at -5. $D(4, -2)$; a vertical line intersects the x-axis at 4, and a horizontal line intersects the y-axis at -2.

2.
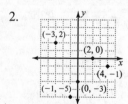

3. Quadrant I because both coordinates are positive.

4. Replace x with 1 and y with 7 and see if the equation is true.

$$y = -\frac{2}{5}x-8$$

$$7 \overset{?}{=} -\frac{2}{5}(1)-8$$

$$7 \overset{?}{=} -\frac{2}{5}-8$$

$$7 \overset{?}{=} -\frac{2}{5}-\frac{40}{5}$$

$$7 \neq -\frac{42}{5}$$

No, $(1, 7)$ is not a solution for the equation.

5. $x-2y=4$

x-intercept	y-intercept
Let $y=0$:	Let $x=0$:
$x-2(0)=4$	$x-2y=4$
$x-0=4$	$0-2y=4$
$x=4$	$-2y=4$
$(4,0)$	$\dfrac{-2y}{-2}=\dfrac{4}{-2}$
	$y=-2$
	$(0,-2)$

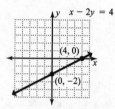

6. $2x+5y=10$

x-intercept	y-intercept
Let $y=0$:	Let $x=0$:
$2x+5y=10$	$2x+5y=10$
$2x+5(0)=10$	$2(0)+5y=10$
$2x+0=10$	$0+5y=10$
$2x=10$	$5y=10$
$\dfrac{2x}{2}=\dfrac{10}{2}$	$\dfrac{5y}{5}=\dfrac{10}{5}$
$x=5$	$y=2$
$(5,0)$	$(0,2)$

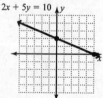

7. $y = -2x$

    x-intercept   y-intercept
    Let $y = 0$:   Let $x = 0$:
    $y = -2x$ $y = -2x$
    $0 = -2x$ $y = -2(0)$
    $\dfrac{0}{-2} = \dfrac{-2x}{-2}$ $y = 0$
    $0 = x$ $(0,0)$
    $(0,0)$

    Second point
    Choose x to be 1.
    $y = -2x$
    $y = -2(1)$
    $y = -2$
    $(1,-2)$

    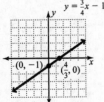

8. $y = \dfrac{3}{4}x - 1$

    x-intercept   y-intercept
    Let $y = 0$:   Let $x = 0$:
    $y = \dfrac{3}{4}x - 1$ $y = \dfrac{3}{4}x - 1$
    $0 = \dfrac{3}{4}x - 1$ $y = \dfrac{3}{4}(0) - 1$
    $1 = \dfrac{3}{4}x$ $y = 0 - 1$
    $\left(\dfrac{4}{3}\right) \cdot 1 = \left(\dfrac{4}{3}\right) \cdot \dfrac{3}{4}x$ $y = -1$
    $\dfrac{4}{3} = x$ $(0,-1)$
    $\left(\dfrac{4}{3},0\right)$

9. $y = \dfrac{3}{4}x + 11$

    Because the equation is in the form $y = mx + b$,

    the slope is $m = \dfrac{3}{4}$ and the y-intercept is $(0,11)$.

10. $5x - 3y = 8$

    Write the equation in slope-intercept form by isolating y.
    $5x - 3y = 8$
    $-3y = -5x + 8$
    $-3y = -5x + 8$
    $\dfrac{-3y}{-3} = \dfrac{-5x + 8}{-3}$
    $y = \dfrac{5}{3}x - \dfrac{8}{3}$

    Because the equation is in the form $y = mx + b$,

    the slope is $m = \dfrac{5}{3}$ and the y-intercept is

    $\left(0, -\dfrac{8}{3}\right)$.

11. Let $(x_1, y_1) = (-5,6)$ and $(x_2, y_2) = (-2,-4)$.
    $m = \dfrac{y_2 - y_1}{x_2 - x_1} = \dfrac{-4 - 6}{-2 - (-5)} = -\dfrac{10}{3}$

12. Let $(x_1, y_1) = (6,2)$ and $(x_2, y_2) = (3,2)$.
    $m = \dfrac{y_2 - y_1}{x_2 - x_1} = \dfrac{2 - 2}{3 - 6} = \dfrac{0}{-3} = 0$

13. We use $y = mx + b$, the slope-intercept form of

    the equation, replacing m with the slope, $\dfrac{3}{5}$, and

    b with the y-coordinate of the y-intercept, 4.

    $y = \dfrac{3}{5}x + 4$

14. Let $(x_1, y_1) = (8,-1)$ and $(x_2, y_2) = (-7,5)$.

    $m = \dfrac{y_2 - y_1}{x_2 - x_1}$ $y - y_1 = m(x - x_1)$

    $= \dfrac{5 - (-1)}{-7 - 8}$ $y - (-1) = -\dfrac{2}{5}(x - 8)$

    $= \dfrac{6}{-15}$ $y + 1 = -\dfrac{2}{5}x + \dfrac{16}{5}$

    $= -\dfrac{2}{5}$ $y = -\dfrac{2}{5}x + \dfrac{11}{5}$

15. Let $(x_1, y_1) = (2,1)$ and $(x_2, y_2) = (5,3)$.

$$y - y_1 = m(x - x_1)$$

$$y - 1 = \frac{2}{3}(x - 2)$$

$$y - 1 = \frac{2}{3}x - \frac{4}{3}$$

$$m = \frac{y_2 - y_1}{x_2 - x_1}$$

$$3(y - 1) = 3\left(\frac{2}{3}x - \frac{4}{3}\right)$$

$$= \frac{3 - 1}{5 - 2}$$

$$3y - 3 = 2x - 4$$

$$= \frac{2}{3}$$

$$-2x + 3y - 3 = -4$$

$$-2x + 3y = -1$$

$$(-1)(-2x + 3y) = (-1)(-1)$$

$$2x - 3y = 1$$

16. The slope of the given line $y = -3x + 4$ is -3, so the slope of the perpendicular line will be $m = \frac{1}{3}$. Let $(x_1, y_1) = (4, -2)$.

$$y - y_1 = m(x - x_1)$$

$$y - (-2) = \frac{1}{3}(x - 4)$$

$$y + 2 = \frac{1}{3}x - \frac{4}{3}$$

$$(3)(y + 2) = (3)\left(\frac{1}{3}x - \frac{4}{3}\right)$$

$$3y + 6 = x - 4$$

$$-x + 3y + 6 = -4$$

$$-x + 3y = -10$$

$$(-1)(-x + 3y) = (-1)(-10)$$

$$x - 3y = 10$$

17. a) $(0, 79.9)$ and $(7, 86.5)$

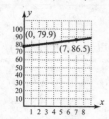

b) $m = \frac{y_2 - y_1}{x_2 - x_1}$

$$= \frac{86.5 - 79.9}{7 - 0}$$

$$= 0.943$$

c) The y-intercept is the y coordinate where $x = 0$, which is 79.9.

$$y = mx + b$$

$$y = 0.943x + 79.9$$

d) The x-coordinate corresponding to the year 2015 is 11. Use this value of x in the equation of the line.

$$y = 0.943x + 79.9$$

$$= 0.943(11) + 79.9$$

$$= 90.3 \text{ million}$$

18. Replace x with 2 and y with 7 and see if the inequality is true.

$$y \geq 2x - 9$$

$$7 \overset{?}{\geq} 2(2) - 9$$

$$7 \overset{?}{\geq} 4 - 9$$

$$7 \geq -5$$

Yes, this statement is true, so $(2, 7)$ is a solution.

19. $y \geq \frac{2}{5}x - 1$

Begin by graphing the related equation $y = \frac{2}{5}x - 1$ with a solid line. Now choose $(0, 0)$ as a test point.

$$y \geq \frac{2}{5}x - 1$$

$$0 \overset{?}{\geq} \frac{2}{5}(0) - 1$$

$$0 \overset{?}{\geq} 0 - 1$$

$$0 \geq -1$$

Because $(0, 0)$ satisfies the inequality, shade the region which includes $(0, 0)$.

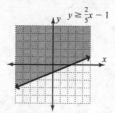

20. $x - 3y < 8$

Begin by graphing the related equation $x - 3y = 8$ with a dashed line. Now choose $(0,0)$ as a test point.

$$x - 3y < 8$$
$$0 - 3(0) \overset{?}{<} 8$$
$$0 - 0 \overset{?}{<} 8$$
$$0 < 8$$

Because $(0,0)$ satisfies the inequality, shade the region which includes $(0,0)$.

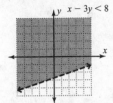

21. a) $10x + 8y \geq 360$

b)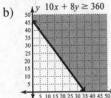

c) Answers may vary. One possible answer is $(0, 45)$.

d) Answers may vary. One possible answer is $(10, 35)$.

22. Yes, this relation is a function because each year in the domain has exactly one income value in the range.

23. No, this graph is not a function because a vertical line can be drawn that intersects the graph at two or more different points.

24. a) $f(x) = \dfrac{x^2}{x-4}$

$$f(2) = \frac{(2)^2}{2-4}$$
$$= \frac{4}{-2}$$
$$= -2$$

b) $f(x) = \dfrac{x^2}{x-4}$

$$f(4) = \frac{4^2}{4-4}$$
$$= \frac{16}{0}$$
undefined

c) $f(x) = \dfrac{x^2}{x-4}$

$$f(-3) = \frac{(-3)^2}{-3-4}$$
$$= \frac{9}{-7}$$
$$= -\frac{9}{7}$$

25. a) Domain: $\{x \mid -3 \leq x \leq 3\}$ or $[-3,3]$;
Range: $\{-2, 1, 2\}$

b) From the graph, we find that $f(1) = 2$.

Chapters 1–3 Cumulative Review

1. False, because π is an irrational number.

2. True

3. False, because $6^0 = 1$.

4. False, because $\sqrt{-36}$ is not a real number.

5. identity

6. 100

7. $16.5\% = \dfrac{16.5}{100} = \dfrac{165}{1000} = \dfrac{33}{200}$

8. Multiply all of the terms by the power of 10 that will clear the decimal from the number with the most decimal places.

9. $(-2)^3 - 3\sqrt{9+16} = (-2)^3 - 3\sqrt{25}$
$$= -8 - 3(5)$$
$$= -8 - 15$$
$$= -23$$

10. $\{4 - 2[6 + (-10)]\} + 3 = [4 - 2(-4)] + 3$
$$= (4 + 8) + 3$$
$$= 12 + 3$$
$$= 15$$

11. $-\left|-12-(-2)(-3)\right| = -\left|-12-(6)\right|$
$$= -\left|-18\right|$$
$$= -18$$

12. $-3^4 = -3\cdot3\cdot3\cdot3 = -81$

13. $-5\sqrt{25} + 20 \div (-4)\cdot2 - 6$
$$= -5(5) + 20 \div (-4)\cdot2 - 6$$
$$= -25 + (-5)\cdot2 - 6$$
$$= -25 + (-10) - 6$$
$$= -41$$

14. $4^2 - 7\cdot3 + \sqrt{144} - 25 \div (-5)$
$$= 16 - 7\cdot3 + 12 - 25 \div (-5)$$
$$= 16 - 21 + 12 - (-5)$$
$$= 16 - 21 + 12 + 5$$
$$= -5 + 12 + 5$$
$$= 7 + 5$$
$$= 12$$

15. $2x - 3y = 6$
    <u>x-intercept</u>
    $y = 0$: $2x - 3(0) = 6$
    $$2x = 6$$
    $$x = 3$$
    $$(3, 0)$$

    <u>y-intercept</u>
    $x = 0$: $2(0) - 3y = 6$
    $$-3y = 6$$
    $$y = -2$$
    $$(0, -2)$$

16. $y = -3x + 4$

x	$y = -3x + 4$	(x, y)
1	$-3(1) + 4 = 1$	$(1, 1)$
0	$-3(0) + 4 = 4$	$(0, 4)$
-1	$-3(-1) + 4 = 7$	$(-1, 7)$

17. $2y - 3x = 9$
$$2y = 3x + 9$$
$$y = \frac{3}{2}x + \frac{9}{2}$$

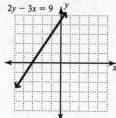

18. $x + 3y = 8$
$$3y = -x + 8$$
$$y = -\frac{1}{3}x + \frac{8}{3}$$

The slope is $m = -\dfrac{1}{3}$

19. $(5, 3), (-3, -1)$
$$m = \frac{-1-3}{-3-5} = \frac{-4}{-8} = \frac{1}{2}$$
$$y - y_1 = m(x - x_1)$$
$$y - 3 = \frac{1}{2}(x - 5)$$
$$y - 3 = \frac{1}{2}x - \frac{5}{2}$$
$$y = \frac{1}{2}x + \frac{1}{2}$$

20. $2x + 5y > 10$
Begin by graphing the related equation $2x + 5y = 10$ with a dashed line. Now choose $(0, 0)$ as a test point.
$$2x + 5y > 10$$
$$2(0) + 5(0) \overset{?}{>} 10$$
$$0 - 0 \overset{?}{>} 10$$
$$0 > 10$$

Because $(0,0)$ does not satisfy the inequality, shade the side of the line on the opposite side from $(0,0)$.

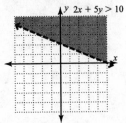

21. $\dfrac{3}{5}a - 8 = \dfrac{1}{2}$

$10\left(\dfrac{3}{5}a - 8\right) = 10\left(\dfrac{1}{2}\right)$

$6a - 80 = 5$

$6a = 85$

$a = \dfrac{85}{6}$

Check:

$\dfrac{3}{5}\left(\dfrac{85}{6}\right) - 8 \overset{?}{=} \dfrac{1}{2}$

$\dfrac{17}{2} - 8 \overset{?}{=} \dfrac{1}{2}$

$\dfrac{17}{2} - \dfrac{16}{2} \overset{?}{=} \dfrac{1}{2}$

$\dfrac{1}{2} = \dfrac{1}{2}$

22. $\dfrac{u}{14} = -\dfrac{6}{7}$

$7 \cdot u = 14 \cdot 6$

$7u = -84$

$u = -12$

Check:

$\dfrac{-12}{14} \overset{?}{=} -\dfrac{6}{7}$

$-\dfrac{6}{7} = -\dfrac{6}{7}$

23. $1.6x - 14 = 8 + 2.4(x - 5)$

$10(1.6x - 14) = 10(8 + 2.4(x - 5))$

$16x - 140 = 80 + 24(x - 5)$

$16x - 140 = 80 + 24x - 120$

$16x - 140 = 24x - 40$

$16x - 100 = 24x$

$-100 = 8x$

$-12.5 = x$

Check:

$1.6(-12.5) - 14 \overset{?}{=} 8 + 2.4(-12.5 - 5)$

$-20 - 14 \overset{?}{=} 8 + 2.4(-17.5)$

$-34 \overset{?}{=} 8 - 42$

$-34 = -34$

24. $C = 2\pi r$

$\dfrac{C}{2\pi} = \dfrac{2\pi r}{2\pi}$

$\dfrac{C}{2\pi} = r$

25. According to the graph, 23% of Internet users check their e-mail weekly. If 1000 Internet users were polled, this would be 23% of 1000, or $0.23(1000) = 230$ Internet users.

26. A married couple spends, on average, \$924 on gifts each year.

$\dfrac{p}{100} = \dfrac{924}{48,241}$

$48,241p = 92,400$

$p \approx 1.9$

A couple who makes \$48,241 per year spends approximately 1.9% of their income on gifts.

27. Write a proportion to find d, how far an automobile travels on 15 gallons of gas.

$\dfrac{d}{15} = \dfrac{224}{8}$

$d = 420$ miles

28. Find the increase in the cost from an end zone ticket to a lower-level 30–50 yard line ticket: $6917 – $2187 = $4730. The initial amount is $2187. Let p represent the unknown percent.

$$\frac{p}{100} = \frac{\text{Amount of increase}}{\text{Initial amount}}$$

$$\frac{p}{100} = \frac{4730}{2187}$$

$$2187p = 473,000$$

$$\frac{2187p}{2187} = \frac{473,000}{2187}$$

$$p \approx 216$$

The percent of increase is approximately 216%.

29. Write a proportion to find p, how many phones out of 2000 she should expect to be defective.

$$\frac{p}{2000} = \frac{15}{400}$$

$$p = 75 \text{ phones}$$

30. Write a proportion to find d, the distance between two points that are 4 inches apart on the map.

$$\frac{d}{4} = \frac{450}{2.25}$$

$$d = 800 \text{ miles}$$

Chapter 4

Systems of Linear Equations and Inequalities

Exercise Set 4.1

1. Is $(1,1)$ a solution of $\begin{cases} x = y \\ 9y - 13 = -4x \end{cases}$?

Equation 1	Equation 2
$x = y$	$9y - 13 = -4x$
$1 = 1$	$9(1) - 13 \stackrel{?}{=} -4(1)$
	$9 - 13 \stackrel{?}{=} -4$
	$-4 = -4$

   Yes, $(1,1)$ is a solution for the system because it satisfies both equations.

3. Is $(2,-3)$ a solution of $\begin{cases} 4x + 3y = -1 \\ 2x - 5y = -11 \end{cases}$?

Equation 1	Equation 2
$4(2) + 3(-3) \stackrel{?}{=} -1$	$2(2) - 5(-3) \stackrel{?}{=} -11$
$8 - 9 \stackrel{?}{=} -1$	$4 + 15 \stackrel{?}{=} -11$
$-1 = -1$	$19 \neq -11$

   No, $(2,-3)$ is not a solution for the system because it does not satisfy both equations.

5. Is $\left(\dfrac{2}{3}, \dfrac{4}{3}\right)$ a solution of $\begin{cases} x + y = 2 \\ 2x - y = 0 \end{cases}$?

Equation 1	Equation 2
$\dfrac{2}{3} + \dfrac{4}{3} \stackrel{?}{=} 2$	$2\left(\dfrac{2}{3}\right) - \dfrac{4}{3} \stackrel{?}{=} 0$
$\dfrac{6}{3} \stackrel{?}{=} 2$	$\dfrac{4}{3} - \dfrac{4}{3} \stackrel{?}{=} 0$
$2 = 2$	$0 = 0$

   Yes, $\left(\dfrac{2}{3}, \dfrac{4}{3}\right)$ is a solution for the system because it satisfies both equations.

7. Is $(2,-4)$ a solution of $\begin{cases} 0.5x + 1.25y = 4 \\ x + y = -2 \end{cases}$?

Equation 1	Equation 2
$0.5(2) + 1.25(-4) \stackrel{?}{=} 4$	$2 + (-4) \stackrel{?}{=} -2$
$1 - 5 \stackrel{?}{=} 4$	$-2 = -2$
$-4 \neq 4$	

   No, $(2,-4)$ is not a solution for the system because it does not satisfy both equations.

9. Is $(5,-2)$ a solution of $\begin{cases} x + 0.5y = 4 \\ \dfrac{1}{5}x - \dfrac{1}{2}y = 2 \end{cases}$?

Equation 1	Equation 2
$5 + 0.5(-2) \stackrel{?}{=} 4$	$\dfrac{1}{5} \cdot 5 - \dfrac{1}{2}(-2) \stackrel{?}{=} 2$
$5 - 1 \stackrel{?}{=} 4$	$1 + 1 \stackrel{?}{=} 2$
$4 = 4$	$2 = 2$

   Yes, $(5,-2)$ is a solution for the system because it satisfies both equations.

11. $\begin{cases} x + y = 5 \\ 2x - y = 7 \end{cases}$ Graph each equation.

    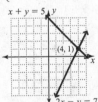

    The lines intersect at a single point, which appears to be $(4, 1)$.

13. $\begin{cases} x + y = 1 \\ y = 7 + x \end{cases}$ Graph each equation.

    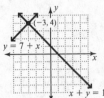

    The lines intersect at a single point, which appears to be $(-3, 4)$.

15. $\begin{cases} y = x - 5 \\ 2x - y = 8 \end{cases}$ Graph each equation.

The lines intersect at a single point, which appears to be (3, –2).

17. $\begin{cases} x = y - 4 \\ x + y = 2 \end{cases}$ Graph each equation.

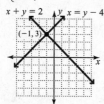

The lines intersect at a single point, which appears to be (–1, 3).

19. $\begin{cases} 4x + y = 8 \\ 2x - 3y = 18 \end{cases}$ Graph each equation.

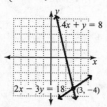

The lines intersect at a single point, which appears to be (3, –4).

21. $\begin{cases} 2x + y = 5 \\ x + 2y = -2 \end{cases}$ Graph each equation.

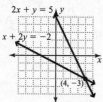

The lines intersect at a single point, which appears to be (4, –3).

23. $\begin{cases} 2x + 2y = -2 \\ 3x - 2y = 12 \end{cases}$ Graph each equation.

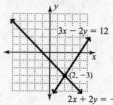

The lines intersect at a single point, which appears to be (2, –3).

25. $\begin{cases} 3y = -2x + 6 \\ 4x + 6y = 18 \end{cases}$ Graph each equation.

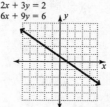

The lines appear to be parallel, so the system has no solution.

27. $\begin{cases} 2x + 3y = 2 \\ 6x + 9y = 6 \end{cases}$ Graph each equation.

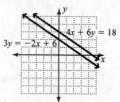

The lines appear to be identical, so the solution is the set of all ordered pairs along $2x + 3y = 2$.

29. $\begin{cases} 3x - 2y = -6 \\ x = 2 \end{cases}$ Graph each equation.

The lines intersect at a single point, which appears to be (2, 6).

31. $\begin{cases} x - 2y = -6 \\ y = 2 \end{cases}$ Graph each equation.

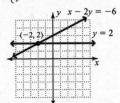

The lines intersect at a single point, which appears to be (–2, 2).

33. $\begin{cases} y = 3 \\ x = -2 \end{cases}$ Graph each equation.

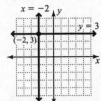

The lines intersect at a single point, which appears to be (–2, 3).

35. $\begin{cases} y = -\dfrac{2}{5}x \\ x - y = 7 \end{cases}$ Graph each equation.

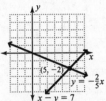

The lines intersect at a single point, which appears to be (5, – 2).

37. a) Because the lines are parallel, this system is inconsistent.
    b) This system has no solution.

39. a) Because these lines intersect in a single point, this system is consistent with independent equations.
    b) This system has one solution.

41. a) Because these lines coincide, this system is consistent with dependent equations.
    b) This system has an infinite number of solutions.

43. $\begin{cases} x - 4y = 9 \\ 2x - 3y = 8 \end{cases}$

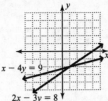

a) Because these lines intersect in a single point, this system is consistent with independent equations.
b) This system has one solution.

45. $\begin{cases} 3x + 2y = 5 \\ -6x - 4y = 1 \end{cases}$

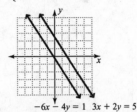

a) Because the lines are parallel, this system is inconsistent.
b) This system has no solution.

47. $\begin{cases} x - y = 1 \\ 2x - 2y = 2 \end{cases}$

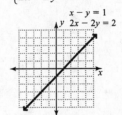

a) Because these lines coincide, this system is consistent with dependent equations.
b) This system has an infinite number of solutions.

49. $\begin{cases} 2x - y = 1 \\ x + y = 4 \end{cases}$

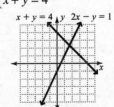

a) Because these lines intersect in a single point, this system is consistent with independent equations.
b) This system has one solution.

51. a) 5000 units
    b) $4000
    c) $u > 5000$

53. a) the One-Rate plan
    b) 375 min.
    c) more than 375 min.

Review Exercises

1. $2y - 5(y + 7) = 2y - 5y - 35$
   $$= -3y - 35$$

2. $x + 3y = 10$
   $$3y = 10 - x$$
   $$\frac{3y}{3} = \frac{10 - x}{3}$$
   $$y = \frac{10 - x}{3}$$

3. $\frac{1}{3}x - 2y = 10$
   $$3\left(\frac{1}{3}x - 2y\right) = 3 \cdot 10$$
   $$x - 6y = 30$$
   $$x = 30 + 6y$$

4. $V = \frac{1}{3}\pi r^2 h$
   $$= \frac{1}{3}\pi \cdot 2^2 \cdot 10$$
   $$= \frac{1}{3}\pi \cdot 4 \cdot 10$$
   $$= \frac{40\pi}{3} \text{ in.}^3 \text{ or } \approx 41.89 \text{ in.}^3$$

5. Let the number be x
   $$4x + 5 = x - 7$$
   $$3x = -12$$
   $$x = -4$$
   The number is –4.

6. $f(x) = 2x + 9$
   $$f(x + 4) = 2(x + 4) + 9$$
   $$= 2x + 8 + 9$$
   $$= 2x + 17$$

Exercise Set 4.2

1. $\begin{cases} x + y = 6 \\ x = 2y \end{cases}$

   $\quad x + y = 6 \qquad x = 2y$
   $\quad (2y) + y = 6 \qquad x = 2(2)$
   $\qquad 3y = 6 \qquad x = 4$
   $\qquad y = 2$
   Solution: $(4, 2)$

3. $\begin{cases} x = -3y \\ 2x - 5y = 44 \end{cases}$

   $\quad 2x - 5y = 44 \qquad x = -3y$
   $\quad 2(-3y) - 5y = 44 \qquad x = -3(-4)$
   $\quad -6y - 5y = 44 \qquad x = 12$
   $\qquad -11y = 44$
   $\qquad y = -4$
   Solution: $(12, -4)$

5. $\begin{cases} 2x - y = 4 \\ y = 3x + 6 \end{cases}$

   $\quad 2x - y = 4 \qquad y = 3x + 6$
   $\quad 2x - (3x + 6) = 4 \qquad y = 3(-10) + 6$
   $\quad 2x - 3x - 6 = 4 \qquad y = -30 + 6$
   $\quad -x - 6 = 4 \qquad y = -24$
   $\qquad -x = 10$
   $\qquad x = -10$
   Solution: $(-10, -24)$

7. $\begin{cases} y = -2x \\ 4x + 2y = 0 \end{cases}$

   $\quad 4x + 2y = 0$
   $\quad 4x + 2(-2x) = 0$
   $\quad 4x - 4x = 0$
   $\qquad 0 = 0$
   Notice that $0 = 0$ no longer has a variable and is true. The equations in the system are dependent; there are an infinite number of solutions that are all of the ordered pairs along $y = -2x$.

9. $\begin{cases} 2x+y=-4 \\ 3x+2y=-5 \end{cases}$

Solve the first equation for y. $y=-2x-4$

$$3x+2y=-5 \qquad y=-2(-3)-4$$
$$3x+2(-2x-4)=-5 \qquad y=6-4$$
$$3x-4x-8=-5 \qquad y=2$$
$$-x-8=-5$$
$$-x=3$$
$$x=-3$$

Solution: $(-3,2)$

11. $\begin{cases} 5x-y=1 \\ 3x+2y=24 \end{cases}$

Solve the first equation for y. $y=5x-1$

$$3x+2y=24 \qquad y=5x-1$$
$$3x+2(5x-1)=24 \qquad y=5(2)-1$$
$$3x+10x-2=24 \qquad y=10-1$$
$$13x-2=24 \qquad y=9$$
$$13x=26$$
$$x=2$$

Solution: $(2,9)$

13. $\begin{cases} 5x-3y=4 \\ y-x=-2 \end{cases}$

Solve the second equation for y. $y=x-2$

$$5x-3y=4 \qquad y=x-2$$
$$5x-3(x-2)=4 \qquad y=-1-2$$
$$5x-3x+6=4 \qquad y=-3$$
$$2x+6=4$$
$$2x=-2$$
$$x=-1$$

Solution: $(-1,-3)$

15. $\begin{cases} 2x=10-y \\ 3x-y=5 \end{cases}$

Solve the first equation for y. $y=10-2x$

$$3x-y=5 \qquad y=10-2x$$
$$3x-(10-2x)=5 \qquad y=10-2(3)$$
$$3x-10+2x=5 \qquad y=10-6$$
$$5x-10=5 \qquad y=4$$
$$5x=15$$
$$x=3$$

Solution: $(3,4)$

17. $\begin{cases} -x+2y=1 \\ 3x-2y=1 \end{cases}$

Solve the first equation for x. $x=2y-1$

$$3x-2y=1 \qquad x=2y-1$$
$$3(2y-1)-2y=1 \qquad x=2(1)-1$$
$$6y-3-2y=1 \qquad x=2-1$$
$$4y=4 \qquad x=1$$
$$y=1$$

Solution: $(1,1)$

19. $\begin{cases} 3x+2y=8 \\ x+4y=1 \end{cases}$

Solve the second equation for x. $x=1-4y$

$$3x+2y=8 \qquad x=1-4y$$
$$3(1-4y)+2y=8 \qquad x=1-4\left(-\frac{1}{2}\right)$$
$$3-12y+2y=8 \qquad x=1+2$$
$$3-10y=8 \qquad x=3$$
$$-10y=5$$
$$y=-\frac{1}{2}$$

Solution: $\left(3,-\frac{1}{2}\right)$

21. $\begin{cases} x-3y=-4 \\ -5x+15y=6 \end{cases}$

Solve the first equation for x. $x=3y-4$

$$-5x+15y=6$$
$$-5(3y-4)+15y=6$$
$$-15y+20+15y=6$$
$$20=6$$

Because this last statement is false, the system is inconsistent and has no solution.

23. $\begin{cases} 4x+3y=-2 \\ 8x-2y=12 \end{cases}$

Solve the second equation for y. $y=4x-6$

$$4x+3y=-2 \qquad y=4x-6$$
$$4x+3(4x-6)=-2 \qquad y=4(1)-6$$
$$4x+12x-18=-2 \qquad y=4-6$$
$$16x-18=-2 \qquad y=-2$$
$$16x=16$$
$$x=1$$

Solution: $(1,-2)$

25. $\begin{cases} 5x + 4y = 12 \\ 7x - 6y = 40 \end{cases}$

Solve the first equation for y. $y = \dfrac{12 - 5x}{4}$

$7x - 6y = 40$ $\qquad$ $y = \dfrac{12 - 5(4)}{4}$

$7x - 6\left(\dfrac{12 - 5x}{4}\right) = 40$ $\qquad$ $y = \dfrac{12 - 20}{4}$

$7x - 18 + 7.5x = 40$

$14.5x - 18 = 40$ $\qquad$ $y = \dfrac{-8}{4}$

$14.5x = 58$ $\qquad$ $y = -2$

$x = 4$

Solution: $(4, -2)$

27. $\begin{cases} 5x + 6y = 2 \\ 10x + 3y = -2 \end{cases}$

Solve the first equation for x. $\quad x = \dfrac{2 - 6y}{5}$

$10x + 3y = -2$ $\qquad$ $x = \dfrac{2 - 6\left(\dfrac{2}{3}\right)}{5}$

$10\left(\dfrac{2 - 6y}{5}\right) + 3y = -2$ $\qquad$ $x = \dfrac{2 - 4}{5}$

$2(2 - 6y) + 3y = -2$

$4 - 12y + 3y = -2$ $\qquad$ $x = -\dfrac{2}{5}$

$-9y = -6$

$y = \dfrac{2}{3}$

Solution: $\left(-\dfrac{2}{5}, \dfrac{2}{3}\right)$

29. $\begin{cases} y = 2x + 1 \\ 3x + y = 11 \end{cases}$

$3x + y = 11$ $\qquad$ $y = 2x + 1$

$3x + 2x + 1 = 11$ $\qquad$ $y = 2(2) + 1$

$5x + 1 = 11$ $\qquad$ $y = 4 + 1$

$5x = 10$ $\qquad$ $y = 5$

$x = 2$

Solution: $(2, 5)$

31. $\begin{cases} 2x + y = 4 \\ x + 2y = 5 \end{cases}$

Solve the second equation for x. $x = 5 - 2y$

$2x + y = 4$ $\qquad$ $x = 5 - 2y$

$2(5 - 2y) + y = 4$ $\qquad$ $x = 5 - 2(2)$

$10 - 4y + y = 4$ $\qquad$ $x = 5 - 4$

$10 - 3y = 4$ $\qquad$ $x = 1$

$-3y = -6$

$y = 2$

Solution: $(1, 2)$

33. Mistake: 3 was not distributed to $3y$.

Correct: $3(6 + 3y) + 2y = 5$

$18 + 9y + 2y = 5$

$18 + 11y = 5$

$11y = -13$

$y = -\dfrac{13}{11}$

$x = 6 + 3\left(-\dfrac{13}{11}\right)$ $\qquad$ $\left(\dfrac{27}{11}, -\dfrac{13}{11}\right)$

$x = 6 - \dfrac{39}{11}$

$x = \dfrac{27}{11}$

35. Let x be the smaller integer and y be the larger integer.

$\begin{cases} x + y = -8 \\ y = 10 + 2x \end{cases}$

$x + y = -8$ $\qquad$ $x + y = -8$

$x + (10 + 2x) = -8$ $\qquad$ $-6 + y = -8$

$3x + 10 = -8$ $\qquad$ $y = -2$

$3x = -18$

$x = -6$

The integers are -6 and -2.

37. Let Jon's age be J and Tony's age be T.

$\begin{cases} J = 13 + T \\ J + T = 27 \end{cases}$

$J + T = 27$ $\qquad$ $J = 13 + T$

$(13 + T) + T = 27$ $\qquad$ $J = 13 + 7$

$13 + 2T = 27$ $\qquad$ $J = 20$

$2T = 14$

$T = 7$

Tony is 7 and Jon is 20.

39. Let w be the width and l be the length. Translate to a system of equations.

$$\begin{cases} 2l + 2w = 288 \\ l = 44 + w \end{cases}$$

$$\begin{array}{ll} 2l + 2w = 288 & l = 44 + w \\ 2(44 + w) + 2w = 288 & l = 44 + 50 \\ 88 + 2w + 2w = 288 & l = 94 \\ 4w = 200 & \\ w = 50 & \end{array}$$

Length: 94 ft. by width: 50 ft.

41. Let w be the width of the rectangle and the length be l.

$$\begin{cases} l = 800 + 10w \\ 2l + 2w = 4900 \end{cases}$$

$$\begin{array}{ll} 2l + 2w = 4900 & l = 800 + 10w \\ 2(800 + 10w) + 2w = 4900 & l = 800 + 10(150) \\ 1600 + 20w + 2w = 4900 & l = 800 + 1500 \\ 22w + 1600 = 4900 & l = 2300 \\ 22w = 3300 & \\ w = 150 & \end{array}$$

The width is 150 ft. and the length is 2300 ft.

43. Let x be the measure of the angle made with the truss on one side of the support beam, and let y be the measure of the angle on the other side of the support beam.

$$\begin{cases} x + y = 180 \\ x = 3y \end{cases}$$

$$\begin{array}{ll} x + y = 180 & x = 3y \\ (3y) + y = 180 & x = 3(45) \\ 4y = 180 & x = 135° \\ y = 45° & \end{array}$$

The angle made with the truss on one side of the support beam measures 135° and the other angle measures 45°.

45. Let x be checking balance and y be savings balance. Translate to a system of equations.

$$\begin{cases} x + y = 4200 \\ y = 4x \end{cases}$$

$$\begin{array}{ll} x + y = 4200 & y = 4x \\ x + (4x) = 4200 & y = 4(840) \\ 5x = 4200 & y = 3360 \\ x = 840 & \end{array}$$

Troy has $840 in his checking account and $3360 in savings.

47. Let U be U2's earnings and B be Bon Jovi's earnings. Translate the given information to a system of equations.

$$\begin{cases} U = B + 101 \\ U + B = 483 \end{cases}$$

$$\begin{array}{ll} U + B = 483 & U = B + 101 \\ (B + 101) + B = 483 & U = 191 + 101 \\ 2B + 101 = 483 & U = 292 \\ 2B = 382 & \\ B = 191 & \end{array}$$

U2 earned $292 million and Bon Jovi earned $191 million.

49. Complete a table.

Categories	Price	Number	Revenue
3-hour	135	x	$135x$
5-hour	225	y	$225y$

Translate to a system of equations and solve.

$$\begin{cases} x + y = 42 \\ 135x + 225y = 7020 \end{cases}$$

Solve the first equation for x: $x = 42 - y$.

$$\begin{array}{ll} 135x + 225y = 7020 & x = 42 - y \\ 135(42 - y) + 225y = 7020 & x = 42 - 15 \\ 5670 - 135y + 225y = 7020 & x = 27 \\ 5670 + 90y = 7020 & \\ 90y = 1350 & \\ y = 15 & \end{array}$$

27 3-hr. courses and 15 5-hr. courses.

51. Complete a table.

Categories	Rate	Time	Distance
John	70	x	$70x$
Karen	65	y	$65y$

Translate to a system of equations.

$$\begin{cases} 70x = 65y \\ y = x + \dfrac{1}{4} \end{cases}$$

$$70x = 65y$$

$$70x = 65\left(x + \frac{1}{4}\right)$$

$$70x = 65x + \frac{65}{4}$$

$$5x = \frac{65}{4}$$

$$x = 3.25$$

John will catch up with Karen 3 hours and 15 minutes after 1:30, at 4:45 pm.

53. Complete a table.

Categories	Rate	Time	Distance
Sharon	8	x	$8x$
Rae	10	y	$10y$

Translate to a system of equations.

$$\begin{cases} 8x = 10y \\ x = y + \dfrac{1}{4} \end{cases}$$

$$8x = 10y$$

$$8\left(y + \frac{1}{4}\right) = 10y$$

$$8y + 2 = 10y$$

$$2 = 2y$$

$$1 \text{ hr.} = y$$

Rae will catch up with Sharon 1 hour after 12:33, at 1:33 pm.

Review Exercises

1. $\dfrac{1}{4}(20x - 16y) = \dfrac{20}{4}x - \dfrac{16}{4}y$
$$= 5x - 4y$$

2. $-3(x - 7y) = -3x + 21y$

3. $6x + 2y + 3x - 2y = 6x + 3x + 2y - 2y$
$$= 9x$$

4. $5x + 3y - 5x + 10y = 5x - 5x + 3y + 10y$
$$= 13y$$

5. $\quad \dfrac{1}{4}x = 20$

$$4 \cdot \frac{1}{4}x = 4 \cdot 20$$

$$x = 80$$

6. $-8y = 12$

$$\frac{-8y}{-8} = \frac{12}{-8}$$

$$y = -\frac{3}{2}$$

Exercise Set 4.3

1. $\begin{cases} x - y = 14 \\ x + y = -2 \end{cases}$

$x - y = 14$	$x - y = 14$
$x + y = -2$	$6 - y = 14$
$2x = 12$	$-y = 8$
$x = 6$	$y = -8$

Solution: $(6, -8)$

3. $\begin{cases} 3x + y = 9 \\ 2x - y = 1 \end{cases}$

$3x + y = 9$	$3x + y = 9$
$2x - y = 1$	$3(2) + y = 9$
$5x = 10$	$6 + y = 9$
$x = 2$	$y = 3$

Solution: $(2, 3)$

5. $\begin{cases} 3x + 2y = -7 \\ 5x - 2y = -1 \end{cases}$

$3x + 2y = -7$	$3x + 2y = -7$
$5x - 2y = -1$	$3(-1) + 2y = -7$
$8x = -8$	$-3 + 2y = -7$
$x = -1$	$2y = -4$
	$y = -2$

Solution: $(-1, -2)$

7. $\begin{cases} 4x + 3y = 17 \\ 2x + 3y = 13 \end{cases}$

$-1(4x + 3y = 17)$

$2x + 3y = 13$	$4x + 3y = 17$
$-4x - 3y = -17$	$4(2) + 3y = 17$
$2x + 3y = 13$	$8 + 3y = 17$
$-2x = -4$	$3y = 9$
$x = 2$	$y = 3$

Solution: $(2, 3)$

9. $\begin{cases} 5x + y = 14 \\ 2x + y = 5 \end{cases}$

$-1(5x + y = 14)$

$2x + y = 5$	$2x + y = 5$
$-5x - y = -14$	$2(3) + y = 5$
$2x + y = 5$	$6 + y = 5$
$-3x = -9$	$y = -1$
$x = 3$	

Solution: $(3, -1)$

11. $\begin{cases} 3x - 5y = -17 \\ 4x + y = -15 \end{cases}$

$3x - 5y = -17$ $4x - y = -15$

$\underline{5(4x + y = -15)}$ $4(-4) + y = -15$

$3x - 5y = -17$ $-16 + y = -15$

$\underline{20x + 5y = -75}$ $y = 1$

$23x = -92$

$x = -4$

Solution: $(-4, 1)$

13. $\begin{cases} 12x - 2y = -54 \\ 13x + 4y = -40 \end{cases}$

$2(12x - 2y = -54)$ $13x - 4y = -40$

$\underline{13x + 4y = -40}$ $13(-4) + 4y = -40$

$24x - 4y = -108$ $-52 + 4y = -40$

$\underline{13x + 4y = -40}$ $4y = 12$

$37x = -148$ $y = 3$

$x = -4$

Solution: $(-4, 3)$

15. $\begin{cases} 2x + y = -2 \\ 8x + 4y = -8 \end{cases}$

$-4(2x + y = -2)$

$\underline{8x + 4y = -8}$

$-8x - 4y = 8$

$\underline{8x + 4y = -8}$

$0 = 0$

Both variables have been eliminated, and the resulting equation, $0 = 0$, is true. This means that the equations are dependent. There are an infinite number of solutions, which are all of the ordered pairs along the line $2x + y = -2$.

17. $\begin{cases} 2x - 5y = 4 \\ -8x + 20y = -20 \end{cases}$

$4(2x - 5y = 4)$

$\underline{-8x + 20y = -20}$

$8x - 20y = 16$

$\underline{-8x + 20y = -20}$

$0 = -4$

Both variables have been eliminated, and the resulting equation, $0 = -4$, is false. Therefore, there is no solution. This system of equations is inconsistent.

19. $\begin{cases} 5x + 6y = 11 \\ 2x - 4y = -2 \end{cases}$

$2(5x + 6y = 11)$ $5x + 6y = 11$

$\underline{3(2x - 4y = -2)}$ $5(1) + 6y = 11$

$10x + 12y = 22$ $5 + 6y = 11$

$\underline{6x - 12y = -6}$ $6y = 6$

$16x = 16$ $y = 1$

$x = 1$

Solution: $(1, 1)$

21. $\begin{cases} 4x + 5y = -4 \\ 3x + 8y = -20 \end{cases}$

$-3(4x + 5y = -4)$ $4x + 5y = -4$

$\underline{4(3x + 8y = -20)}$ $4x + 5(-4) = -4$

$-12x - 15y = 12$ $4x - 20 = -4$

$\underline{12x + 32y = -80}$ $4x = 16$

$17y = -68$ $x = 4$

$y = -4$

Solution: $(4, -4)$

23. $\begin{cases} 5x + 6y = 2 \\ 10x + 3y = -2 \end{cases}$

$-2(5x + 6y = 2)$ $10x + 3y = -2$

$\underline{10x + 3y = -2}$ $10x + 3\left(\dfrac{2}{3}\right) = -2$

$-10x - 12y = -4$ $10x + 2 = -2$

$\underline{10x + 3y = -2}$ $10x = -4$

$-9y = -6$ $x = -\dfrac{2}{5}$

$y = \dfrac{2}{3}$

Solution: $\left(-\dfrac{2}{5}, \dfrac{2}{3}\right)$

25. $\begin{cases} 10x + 5y = 3.5 \\ 3x - 4y = -0.6 \end{cases}$

$4(10x + 5y = 3.5)$ $3x - 4y = -0.6$

$\underline{5(3x - 4y = -0.6)}$ $3(0.2) - 4y = -0.6$

$40x + 20y = 14$ $0.6 - 4y = -0.6$

$\underline{15x - 20y = -3}$ $-4y = -1.2$

$55x = 11$ $y = 0.3$

$x = 0.2$

Solution: $(0.2, 0.3)$

27. $\begin{cases} \dfrac{3}{2}x - \dfrac{4}{3}y = \dfrac{11}{3} \\ \dfrac{1}{4}x - \dfrac{2}{3}y = -\dfrac{7}{6} \end{cases}$

$$\dfrac{3}{2}x - \dfrac{4}{3}y = \dfrac{11}{3} \qquad \dfrac{3}{2}x - \dfrac{4}{3}y = \dfrac{11}{3}$$

$$\underline{-2\left(\dfrac{1}{4}x - \dfrac{2}{3}y = -\dfrac{7}{6}\right)} \qquad \dfrac{3}{2}(6) - \dfrac{4}{3}y = \dfrac{11}{3}$$

$$\dfrac{3}{2}x - \dfrac{4}{3}y = \dfrac{11}{3} \qquad 9 - \dfrac{4}{3}y = \dfrac{11}{3}$$

$$\underline{-\dfrac{1}{2}x + \dfrac{4}{3}y = \dfrac{7}{3}} \qquad -\dfrac{4}{3}y = -\dfrac{16}{3}$$

$$x \qquad = 6 \qquad\qquad y = 4$$

Solution: $(6, 4)$

29. $\begin{cases} 0.7x - \dfrac{1}{2}y = 9.4 \\ 0.9x + \dfrac{7}{10}y = 0 \end{cases}$

$$0.7x - 0.5y = 9.4 \qquad 0.7x - 0.5y = 9.4$$

$$\underline{0.9x + 0.7y = 0} \qquad 0.7x - 0.5(-9) = 9.4$$

$$90(0.7x - 0.5y = 9.4) \qquad 0.7x + 4.5 = 9.4$$

$$\underline{-70(0.9x + 0.7y = 0)} \qquad 0.7x = 4.9$$

$$63x - 45y = 846 \qquad\qquad x = 7$$

$$\underline{-63x - 49y = 0}$$

$$-94y = 846$$

$$y = -9$$

Solution: $(7, -9)$

31. $\begin{cases} y = 2x - 5 \\ x - y = 9 \end{cases}$

Write first equation as $-2x + y = -5$.

$$-2x + y = -5 \qquad\qquad -2x + y = -5$$

$$\underline{x - y = 9} \qquad\qquad -2(-4) + y = -5$$

$$-x = 4 \qquad\qquad 8 + y = -5$$

$$x = -4 \qquad\qquad y = -13$$

Solution: $(-4, -13)$

33. $\begin{cases} y = \dfrac{3}{4}x + 1 \\ 4x = 2y + 3 \end{cases}$

Write first equation as $-\dfrac{3}{4}x + y = 1$.

Write the second equation as $4x - 2y = 3$.

$$4\left(-\dfrac{3}{4}x + y = 1\right) \qquad 4x - 2y = 3$$

$$\underline{2(4x - 2y = 3)} \qquad 4(2) - 2y = 3$$

$$-3x + 4y = 4 \qquad\qquad 8 - 2y = 3$$

$$\underline{8x - 4y = 6} \qquad\qquad -2y = -5$$

$$5x = 10 \qquad\qquad y = \dfrac{5}{2}$$

$$x = 2$$

Solution: $\left(2, \dfrac{5}{2}\right)$

35. $\begin{cases} 0.2x - y = -3.2 \\ y = \dfrac{3}{4}x + 1 \end{cases}$

Write second equation as $-0.75x + y = 1$.

$$0.2x - y = -3.2 \qquad 0.2x - y = -3.2$$

$$\underline{-0.75x + y = 1} \qquad 0.2(4) - y = -3.2$$

$$-0.55x = -2.2 \qquad 0.8 - y = -3.2$$

$$x = 4 \qquad\qquad -y = -4$$

$$y = 4$$

Solution: $(4, 4)$

37. Mistake: Did not multiply the right side of $x - y = 1$ by -9.

Correct: $-9(x - y = 1) \qquad x - y = 1$

$$9x + 8y = 77 \qquad\qquad x - 4 = 1$$

$$-9x + 9y = -9 \qquad\qquad x = 5$$

$$\underline{9x + 8y = 77}$$

$$17y = 68$$

$$y = 4$$

Solution: $(5, 4)$

39. Let l be the length of the rectangle and w be the width of the rectangle.

$$\begin{cases} l - w = 14 \\ 2l + 2w = 60 \end{cases}$$

$$2(l - w = 14) \qquad l - w = 14$$

$$\underline{2l + 2w = 60} \qquad 22 - w = 14$$

$$2l - 2w = 28 \qquad\qquad 8 = w$$

$$\underline{2l + 2w = 60}$$

$$4l = 88$$

$$l = 22$$

The length is 22 in. and the width is 8 in.

41. Let x be the speed of the boat in still water and y be the speed of the current. Complete a table.

Categories	Rate	Time	Distance
upstream	$x-y$	3	$3(x-y)$
downstream	$x+y$	3	$3(x+y)$

Translate to a system of equations.

$$\begin{cases} 3(x-y) = 24 \\ 3(x+y) = 36 \end{cases}$$

$$\begin{array}{ll} x - y = 8 & x - y = 8 \\ x + y = 12 & 10 - y = 8 \\ \hline 2x = 20 & -y = -2 \\ x = 10 & y = 2 \end{array}$$

Still water speed is 10 mph and current is 2 mph.

43. Let x be the speed of the plane in still air and y be the speed of the wind. Translate to a system of equations.

$$\begin{cases} x+y = 500 \\ x-y = 440 \end{cases}$$

$$\begin{array}{ll} x + y = 500 & x + y = 500 \\ x - y = 440 & 470 + y = 500 \\ \hline 2x = 940 & y = 30 \\ x = 470 & \end{array}$$

The speed in still air is 470 mph and the wind speed is 30 mph.

45. Create a table.

Categories	APR	Principal	Interest
acct 1	0.04	x	$0.04x$
acct 2	0.03	y	$0.03y$

Translate to a system of equations and solve.

$$\begin{cases} x+y = 15,500 \\ 0.04x + 0.03y = 585 \end{cases}$$

$$\begin{array}{l} x \quad + \quad y = 15,500 \\ 0.04x + 0.03y = 585 \\ \hline -0.04(x+y \quad = 15,500) \\ 0.04x + 0.03y \quad = 585 \\ \hline -0.04x - 0.04y = -620 \\ 0.04x + 0.03y = 585 \\ \hline -0.01y \quad = -35 \\ y \quad = 3500 \end{array}$$

$$\begin{array}{l} x + y = 15,500 \\ x + 3500 = 15,500 \\ x = 12,000 \end{array}$$

$3500 was invested at 3% and $12,000 was invested at 4%.

47. Create a table.

Categories	APR	Principal	Interest
loan 1	0.05	x	$0.05x$
loan 2	0.08	y	$0.08y$

Translate to a system of equations and solve.

$$\begin{cases} x+y = 13,250 \\ 0.05x + 0.08y = 727 \end{cases}$$

$$\begin{array}{l} x \quad + \quad y = 13,250 \\ 0.05x + 0.08y = 727 \\ \hline -0.05(x+y \quad = 13,250) \\ 0.05x + 0.08y \quad = 727 \\ \hline -0.05x - 0.05y = -662.50 \\ 0.05x + 0.08y = 727 \\ \hline 0.03y = 64.50 \\ y = 2150 \end{array}$$

$$\begin{array}{l} x + y = 13,250 \\ x + 2150 = 13,250 \\ x = 11,100 \end{array}$$

$2150 is in the loan at 8% and $11,100 is in the loan at 5%.

49. Create a table.

Categories	APR	Principal	Interest
loan 1	0.029	x	$0.029x$
loan 2	0.06	y	$0.06y$

Translate to a system of equations and solve.

$$\begin{cases} x+y = 16,250 \\ 0.029x + 0.06y = 657.16 \end{cases}$$

$$\begin{array}{l} x \quad + \quad y = 16,250 \\ 0.029x + 0.06y = 657.16 \\ \hline -0.06(x+y \quad = 16,250) \\ 0.029x + 0.06y = 657.16 \\ \hline -0.06x - 0.06y = -975 \\ 0.029x + 0.06y = 657.16 \\ \hline -0.031x \quad = -317.84 \\ x \approx 10,250 \end{array}$$

$$\begin{array}{l} x + y = 16,250 \\ 10,250 + y = 16,250 \\ y = 6000 \end{array}$$

$6000 is owed on the loan at 6% and $10,250 is owed on the loan at 2.9%.

51. Create a table.

Solutions	Concentration	Volume	Amt of Acid
20% sol'n	0.20	x	$0.20x$
15% sol'n	0.15	80	$0.15(80)$
18% sol'n	0.18	y	$0.18y$

Translate to a system of equations and solve.

$$\begin{cases} x + 80 = y \\ 0.20x + 0.15(80) = 0.18y \end{cases}$$

$$0.20x + 0.15(80) = 0.18y$$
$$0.20x + 12 = 0.18y$$
$$0.20x + 12 = 0.18(x + 80)$$
$$0.20x + 12 = 0.18x + 14.4$$
$$0.02x + 12 = 14.4$$
$$0.02x = 2.4$$
$$x = 120$$

120 ml of the 20% solution is needed.

53. Create a table.

Solutions	Concentration	Volume	Amt of Alcohol
15% sol'n	0.15	x	$0.15x$
45% sol'n	0.45	y	$0.45y$
25% sol'n	0.25	300	$0.25(300)$

Translate to a system of equations and solve.

$$\begin{cases} x + y = 300 \\ 0.15x + 0.45y = 0.25(300) \end{cases}$$

$$\begin{array}{ll} x + y = 300 & x + y = 300 \\ \underline{0.15x + 0.45y = 75} & x + 100 = 300 \\ -0.15(x + y = 300) & x = 200 \\ \underline{0.15x + 0.45y = 75} & \\ -0.15x - 0.15y = -45 & \\ \underline{0.15x + 0.45y = 75} & \\ 0.30y = 30 & \\ y = 100 & \end{array}$$

100 ml of the 45% solution and 200 ml of the 15% solution

55. Create a table.

Categories	Value	Amount	Cost
Italian Roast	9.80	x	$9.80x$
Gold Coast	11.20	y	$11.20y$
Mix	10.43	20	$10.43(20)$

Translate to a system of equations and solve.

$$\begin{cases} x + y = 20 \\ 9.80x + 11.20y = 10.43(20) \end{cases}$$

$$\begin{array}{ll} x + \quad y = 20 & x + y = 20 \\ \underline{9.80x + 11.20y = 208.6} & x + 9 = 20 \\ -9.80(x + y = 20) & x = 11 \\ \underline{9.80x + 11.20y = 208.6} & \\ -9.80x - 9.80y = -196 & \\ \underline{9.80x + 11.20y = 208.6} & \\ 1.4y = 12.6 & \\ y = 9 & \end{array}$$

9 lbs. of Gold Coast Blend is mixed with 11 lbs. of Italian Roast.

Review Exercises

1. $4(5x + 7y - z) = 20x + 28y - 4z$

2. $3x - 4y + z - 2x + y - z = x - 3y$

3. $x + 3y = 6$
$$3y = -x + 6$$
$$y = -\frac{1}{3}x + 2$$

4. $\frac{1}{3}x - 2y = 10$
$$x - 6y = 30$$
$$x = 6y + 30$$

5. $x + y + 2z = 7$
$$x + (-1) + 2(3) = 7$$
$$x - 1 + 6 = 7$$
$$x = 2$$

6. $x - 3y + 2z = 6$
$$-2 - 3y + 2(4) = 6$$
$$-2 - 3y + 8 = 6$$
$$6 - 3y = 6$$
$$-3y = 0$$
$$y = 0$$

Exercise Set 4.4

1. Is $(3,-1,1)$ a solution of $\begin{cases} x+y+z=3 \\ 2x-2y-z=7 \\ 2x+y-2z=3 \end{cases}$?

Equation 1: $\quad x+y+z = 3$

$$3+(-1)+1 \overset{?}{=} 3$$

$$3 = 3$$

Equation 2: $\quad 2x-2y-z = 7$

$$2(3)-2(-1)-1 \overset{?}{=} 7$$

$$6+2-1 \overset{?}{=} 7$$

$$7 = 7$$

Equation 3: $\quad 2x+y-2z = 3$

$$2(3)+(-1)-2(1) \overset{?}{=} 3$$

$$6-1-2 \overset{?}{=} 3$$

$$3 = 3$$

Yes, $(3,-1,1)$ is a solution for the system because it satisfies all of the equations.

3. Is $(1,0,2)$ a solution of $\begin{cases} 2x+3y-3z=-4 \\ -2x+4y-z=-4 \\ 3x-4y+2z=5 \end{cases}$?

Equation 1: $\quad 2x+3y-3z = -4$

$$2(1)+3(0)-3(2) \overset{?}{=} -4$$

$$2+0-6 \overset{?}{=} -4$$

$$-4 = -4$$

Equation 2: $\quad -2x+4y-z = -4$

$$-2(1)+4(0)-(2) \overset{?}{=} -4$$

$$-2+0-2 \overset{?}{=} -4$$

$$-4 = -4$$

Equation 3: $\quad 3x-4y+2z = 5$

$$3(1)-4(0)+2(2) \overset{?}{=} 5$$

$$3-0+4 \overset{?}{=} 5$$

$$7 \neq 5$$

No, $(1,0,2)$ is not a solution for the system because it does not satisfy all three equations.

5. Is $(2,-2,4)$ a solution of $\begin{cases} x+2y-z=-6 \\ 2x-3y+4z=26 \\ -x+2y-3z=-18 \end{cases}$?

Equation 1: $\quad x+2y-z = -6$

$$2+2(-2)-4 \overset{?}{=} -6$$

$$2-4-4 \overset{?}{=} -6$$

$$-6 = -6$$

Equation 2: $\quad 2x-3y+4z = 26$

$$2(2)-3(-2)+4(4) \overset{?}{=} 26$$

$$4+6+16 \overset{?}{=} 26$$

$$26 = 26$$

Equation 3: $\quad -x+2y-3z = -18$

$$-(2)+2(-2)-3(4) \overset{?}{=} -18$$

$$-2-4-12 \overset{?}{=} -18$$

$$-18 = -18$$

Yes, $(2,-2,4)$ is a solution for the system because it satisfies all of the equations.

7. $\begin{cases} x+y+z=5 & \text{Eqtn. 1} \\ 2x+y-2z=-5 & \text{Eqtn. 2} \\ x-2y+z=8 & \text{Eqtn. 3} \end{cases}$

Multiply equation 1 by -1 and add to equation 3. This makes equation 4.

$$\begin{aligned} -x-y-z &= -5 \\ \underline{x-2y+z} &= \underline{8} \\ -3y &= 3 \\ y &= -1 \qquad \text{Eqtn. 4} \end{aligned}$$

Multiply equation 1 by -2 and add to equation 2. This makes equation 5.

$$-2x - 2y - 2z = -10$$
$$\underline{2x + y - 2z = -5}$$
$$-y - 4z = -15 \quad \text{Eqtn. 5}$$

Substitute equation 4 into equation 5 and solve for z.
$$-(-1) - 4z = -15$$
$$1 - 4z = -15$$
$$-4z = -16$$
$$z = 4$$

Substitute the values for y and z into equation 1 to solve for x.
$$x - 1 + 4 = 5$$
$$x + 3 = 5$$
$$x = 2$$

Solution: $(2, -1, 4)$

9. $\begin{cases} x + y + z = 2 & \text{Eqtn. 1} \\ 4x - 3y + 2z = 2 & \text{Eqtn. 2} \\ 2x + 3y - 2z = -8 & \text{Eqtn. 3} \end{cases}$

Multiply equation 1 by -2 and add to equation 3. This makes equation 4.
$$-2x - 2y - 2z = -4$$
$$\underline{2x + 3y - 2z = -8}$$
$$y - 4z = -12 \quad \text{Eqtn. 4}$$

Multiply equation 1 by -4 and add to equation 2. This makes equation 5.
$$-4x - 4y - 4z = -8$$
$$\underline{4x - 3y + 2z = 2}$$
$$-7y - 2z = -6 \quad \text{Eqtn. 5}$$

Use equations 4 and 5 to make a system of equations in two variables. Solve for y.
$$y - 4z = -12$$
$$-2(-7y - 2z = -6)$$

$$y - 4z = -12$$
$$\underline{14y + 4z = 12}$$
$$15y = 0$$
$$y = 0$$

Substitute the value for y into equation 4 and solve for z.
$$0 - 4z = -12$$
$$-4z = -12$$
$$z = 3$$

Substitute the values for y and z into equation 1 to solve for x.

$$x + 0 + 3 = 2$$
$$x + 3 = 2$$
$$x = -1$$

Solution: $(-1, 0, 3)$

11. $\begin{cases} 2x + y + 2z = 5 & \text{Eqtn. 1} \\ 3x - 2y + 3z = 4 & \text{Eqtn. 2} \\ -2x + 3y + z = 8 & \text{Eqtn. 3} \end{cases}$

Multiply equation 1 by 2 and add to equation 2. This makes equation 4.
$$4x + 2y + 4z = 10$$
$$\underline{3x - 2y + 3z = 4}$$
$$7x + 7z = 14 \quad \text{Eqtn. 4}$$

Multiply equation 1 by -3 and add to equation 3. This makes equation 5.
$$-6x - 3y - 6z = -15$$
$$\underline{-2x + 3y + z = 8}$$
$$-8x - 5z = -7 \quad \text{Eqtn. 5}$$

Use equations 4 and 5 to make a system of equations in two variables. Solve.
$$8(7x + 7z = 14)$$
$$7(-8x - 5z = -7)$$

$$56x + 56z = 112$$
$$\underline{-56x - 35z = -49}$$
$$21z = 63$$
$$z = 3$$

Substitute the value for z into equation 4 and solve for x.
$$7x + 7 \cdot 3 = 14$$
$$7x + 21 = 14$$
$$7x = -7$$
$$x = -1$$

Substitute the values for x and z into equation 1 to solve for y.
$$2(-1) + y + 2(3) = 5$$
$$-2 + y + 6 = 5$$
$$y + 4 = 5$$
$$y = 1$$

Solution: $(-1, 1, 3)$

13. $\begin{cases} x + 2y - z = 1 & \text{Eqtn. 1} \\ 2x + 4y - 2z = -8 & \text{Eqtn. 2} \\ 3x + y - 4z = 6 & \text{Eqtn. 3} \end{cases}$

Multiply equation 1 by -2 and add to equation 2. This makes equation 4.

$-2x - 4y + 2z = -2$

$\underline{2x + 4y - 2z = -8}$

$0 = -10$

Because this is a false statement, the system is inconsistent and there is no solution.

15. $\begin{cases} 4x - 2y + 3z = 6 & \text{Eqtn. 1} \\ 6x - 3y + 4.5z = 9 & \text{Eqtn. 2} \\ 12x - 6y + 9z = 18 & \text{Eqtn. 3} \end{cases}$

Multiply equation 2 by -2 and add to equation 3. This makes equation 4.

$-12x + 6y - 9z = -18$

$\underline{12x - 6y + 9z = 18}$

$0 = 0 \quad \text{Eqtn. 4}$

Multiply equation 1 by -3 and add to equation 3. This makes equation 5.

$-12x + 6y - 9z = -18$

$\underline{12x - 6y + 9z = 18}$

$0 = 0 \quad \text{Eqtn. 5}$

Because equation 4 and equation 5 are both true statements, the three equations are dependent. There are an infinite number of solutions.

17. $\begin{cases} x = 2y + z + 7 & \text{Eqtn. 1} \\ y = -3x + 2z + 1 & \text{Eqtn. 2} \\ 2x + y - z = 0 & \text{Eqtn. 3} \end{cases}$

Multiply equation 2 by -1 and add to equation 3. This makes equation 4.

$-3x - y + 2z = -1$

$\underline{2x + y - z = 0}$

$-x + z = -1 \quad \text{Eqtn. 4}$

Multiply equation 2 by 2 and add to equation 1. This makes equation 5.

$x - 2y - z = 7$

$\underline{6x + 2y - 4z = 2}$

$7x - 5z = 9 \qquad \text{Eqtn. 5}$

Use equations 4 and 5 to make a system of equations in two variables. Solve for x.

$-x + z = -1 \qquad \text{Multiply by 5.}$

$\underline{7x - 5z = 9}$

$-5x + 5z = -5$

$\underline{7x - 5z = 9}$

$2x = 4$

$x = 2$

Substitute the value for x into equation 4 and solve for z.

$-x + z = -1$

$-2 + z = -1$

$z = 1$

Substitute the values for x and z into equation 2 to solve for y.

$y = -3x + 2z + 1$

$y = -3(2) + 2(1) + 1$

$y = -6 + 2 + 1$

$y = -3$

Solution: $(2, -3, 1)$

19. $\begin{cases} x - 4y + z = 1 & \text{Eqtn. 1} \\ 3x + 2y - z = -8 & \text{Eqtn. 2} \\ x + 6y + 2z = -3 & \text{Eqtn. 3} \end{cases}$

Add equations 1 and 2. This makes equation 4.

$x - 4y + z = 1$

$\underline{3x + 2y - z = -8}$

$4x - 2y = -7 \qquad \text{Eqtn. 4}$

Multiply equation 2 by 2 and add to equation 3. This makes equation 5.

$x + 6y + 2z = -3$

$\underline{6x + 4y - 2z = -16}$

$7x + 10y = -19 \quad \text{Eqtn. 5}$

Use equations 4 and 5 to make a system of equations in two variables. Solve for x.

$4x - 2y = -7 \qquad \text{Multiply by 5}$

$\underline{7x + 10y = -19}$

$20x - 10y = -35$

$\underline{7x + 10y = -19}$

$27x = -54$

$x = -2$

Substitute the value for x into equation 4 and solve for y.

$4(-2) - 2y = -7$

$-8 - 2y = -7$

$-2y = 1$

$y = -\dfrac{1}{2}$

Substitute the values for x and y into equation 1 to solve for z.

$-2 - 4\left(-\dfrac{1}{2}\right) + z = 1$

$-2 + 2 + z = 1$

$z = 1$

Solution: $\left(-2, -\dfrac{1}{2}, 1\right)$

21. $\begin{cases} 4x+2y+3z=9 & \text{Eqtn. 1} \\ 2x-4y-z=7 & \text{Eqtn. 2} \\ 3x-2z=4 & \text{Eqtn. 3} \end{cases}$

Multiply equation 1 by 2 and add to equation 2. This makes equation 4.

$8x+4y+6z=18$
$\underline{2x-4y-z=7}$
$10x+5z=25 \qquad \text{Eqtn. 4}$

Use equations 3 and 4 to make a system of equations in two variables. Solve for z.

$-10(3x-2z=4\ )$
$\underline{3(10x+5z=25)}$
$-30x+20z=-40$
$\underline{30x+15z=75}$
$35z=35$
$z=1$

Substitute the value of z into equation 3 and solve for x.

$3x-2\cdot 1=4$
$3x-2=4$
$3x=6$
$x=2$

Substitute the values for x and z into equation 1 to solve for y.

$4\cdot 2+2y+3\cdot 1=9$
$8+2y+3=9$
$2y=-2$
$y=-1$

Solution: $(2,-1,1)$

23. $\begin{cases} 3x-2z=-1 & \text{Eqtn. 1} \\ 4x+5y=23 & \text{Eqtn. 2} \\ y+2z=-1 & \text{Eqtn. 3} \end{cases}$

Add equation 1 and equation 3. This makes equation 4.

$3x-2z=-1$
$\underline{y+2z=-1}$
$3x+y=-2 \qquad \text{Eqtn. 4}$

Use equation 2 and equation 4 to form a system. Multiply equation 4 by –5 and solve for x.

$4x+5y=23$
$3x+y=-2$
$\overline{}$
$4x+5y=23$
$\underline{-15x-5y=10}$
$-11x=33$
$x=-3$

Substitute the value for x in equation 1 and solve for z.

$3x-2z=-1$
$3(-3)-2z=-1$
$-9-2z=-1$
$-2z=8$
$z=-4$

Substitute the value for z in equation 3 and solve for y.

$y+2(-4)=-1$
$y-8=-1$
$y=7$

Solution: $(-3,\ 7,\ -4)$

25. $\begin{cases} 3x+2y=-2 & \text{Eqtn. 1} \\ 2x-3z=1 & \text{Eqtn. 2} \\ 0.4y-0.5z=-2.1 & \text{Eqtn. 3} \end{cases}$

Multiply equation 1 by 2 and multiply equation 2 by -3. Add the equations to make equation 4.

$6x+4y=-4$
$\underline{-6x+9z=-3}$
$4y+9z=-7 \qquad \text{Eqtn. 4}$

Use equations 3 and 4 to make a system of equations in two variables. Solve for z.

$0.4y-0.5z=-2.1 \qquad$ Multiply by -10
$\underline{4y+9z=-7}$
$-4y+5z=21$
$\underline{4y+9z=-7}$
$14z=14$
$z=1$

Substitute the value of z into equation 4 and solve for y.

$4y+9\cdot 1=-7$
$4y+9=-7$
$4y=-16$
$y=-4$

Substitute the value for y into equation 1 to solve for x.

$3x+2(-4)=-2$
$3x-8=-2$
$3x=6$
$x=2$

Solution: $(2,-4,1)$

27. $\begin{cases} \dfrac{3}{2}x + y - z = 0 & \text{Eqtn. 1} \\ 4y - 3z = -22 & \text{Eqtn. 2} \\ -0.2x + 0.3y = -2 & \text{Eqtn. 3} \end{cases}$

Multiply equation 1 by -3 and add to equation 2 to make equation 4.

$-\dfrac{9}{2}x - 3y + 3z = 0$

$\underline{\qquad 4y - 3z = -22}$

$-\dfrac{9}{2}x + y = -22 \qquad$ Eqtn. 4

Use equations 3 and 4 to make a system of equations in two variables. Solve for x.

$-\dfrac{9}{2}x + y \quad = -22 \qquad$ Multiply by -0.3

$\underline{-0.2x + 0.3y = -2}$

$1.35x - 0.3y = 6.6$

$\underline{-0.2x + 0.3y = -2}$

$1.15x \quad = 4.6$

$x = 4$

Substitute the value of x into equation 4 and solve for y.

$-\dfrac{9}{2} \cdot 4 + y = -22$

$-18 + y = -22$

$y = -4$

Substitute the value for y into equation 2 to solve for z.

$4(-4) - 3z = -22$

$-16 - 3z = -22$

$-3z = -6$

$z = 2$

Solution: $(4, -4, 2)$

29. Let x, y, and z represent the measures of the three angles of the triangle.

$\begin{cases} x + y + z = 180 & \text{Eqtn. 1} \\ x = 3y & \text{Eqtn. 2} \\ y = z - 5 & \text{Eqtn. 3} \end{cases}$

Substitute equation 2 into equation 1. This becomes equation 4.

$3y + y + z = 180$

$4y + z = 180 \quad$ Eqtn. 4

Use equations 3 and 4 to make a system of equations in two variables. Solve.

$y - z = -5 \qquad$ Eqtn. 3

$\underline{4y + z = 180} \qquad$ Eqtn. 4

$5y = 175$

$y = 35$

Substitute the value of y into equation 2 and solve for x.

$x = 3 \cdot 35$

$x = 105$

Substitute the value for y into equation 3 and solve for z.

$35 = z - 5$

$40 = z$

The angles are 105°, 35°, and 40°.

31. Let $b =$ the cost of a burger, $f =$ the cost of an order of fries, and $d =$ the cost of a drink.

$\begin{cases} b + f + d = 5 & \text{Eqtn. 1} \\ 3b + 2f + 2d = 12.50 & \text{Eqtn. 2} \\ 2b + 4f + 3d = 14 & \text{Eqtn. 3} \end{cases}$

Multiply equation 1 by -2 and add to equation 3 to make equation 4.

$-2b - 2f - 2d = -10$

$\underline{2b + 4f + 3d = 14}$

$2f + d = 4 \qquad$ Eqtn. 4

Multiply equation 1 by -3 and add to equation 2 to make equation 5.

$-3b - 3f - 3d = -15$

$\underline{3b + 2f + 2d = 12.50}$

$-f - d = -2.5 \qquad$ Eqtn. 5

Use equations 4 and 5 to make a system of equations in two variables. Solve.

$2f + d = 4$

$\underline{-f - d = -2.5}$

$f = 1.5$

Substitute into equation 5 to solve for d.

$-1.5 - d = -2.5$

$-d = -1$

$d = 1$

Substitute into equation 1 to solve for b.

$b + 1.5 + 1 = 5$

$b + 2.5 = 5$

$b = 2.5$

A burger costs $2.50, fries cost $1.50, and a drink costs $1.00.

33. Let a = then number of 2-point shots, b = the number of 3-point shots, and let c = the number of 1-point shots.

$$\begin{cases} 2a + 3b + c = 14 & \text{Eqtn. 1} \\ a + b + c = 7 & \text{Eqtn. 2} \\ c = a + 2 & \text{Eqtn. 3} \end{cases}$$

Substitute equation 3 into equation 1. This makes equation 4.

$$2a + 3b + c = 14$$
$$2a + 3b + (a + 2) = 14$$
$$3a + 3b = 12 \qquad \text{Eqtn. 4}$$

Substitute equation 3 into equation 2. This makes equation 5.

$$a + b + c = 7$$
$$a + b + a + 2 = 7$$
$$2a + b = 5 \qquad \text{Eqtn. 5}$$

Set up a system using equation 4 and equation 5.

$$3a + 3b = 12$$
$$-3\,(2a + b = 5)$$

$$3a + 3b = 12$$
$$\underline{-6a - 3b = -15}$$
$$-3a = -3$$
$$a = 1$$

Substitute 1 for a in equation 5 and solve for b.

$$2a + b = 5$$
$$2(1) + b = 5$$
$$2 + b = 5$$
$$b = 3$$

Substitute 1 for a in equation 3 and solve for c.

$$c = a + 2$$
$$c = 1 + 2$$
$$c = 3$$

John made $a = 1$ two-point shots, $b = 3$ three-point shots, and $c = 3$ one-point free-throw shots.

35. Let h = the number of pounds of Black Forest ham, t = the number of pounds of turkey breast, and r = the number of pounds of roast beef.

$$\begin{cases} h + t + r = 10 & \text{Eqtn. 1} \\ 11.96h + 8.76t + 9.16r = 10(9.80) & \text{Eqtn. 2} \\ t = h + r & \text{Eqtn. 3} \end{cases}$$

Substitute equation 3 into equations 1 and 2 to make a system of linear equations in two variables.

$$h + h + r + r = 10$$
$$11.96h + 8.76(h + r) + 9.16r = 98$$
$$2h + 2r = 10 \qquad \text{Multiply by } -8.96$$
$$\underline{20.72h + 17.92r = 98}$$
$$-17.92h - 17.92r = -89.6$$
$$\underline{20.72h + 17.92r = 98}$$
$$2.8h = 8.4$$
$$h = 3$$

$$2 \cdot 3 + 2r = 10 \qquad\qquad 3 + t + 2 = 10$$
$$2r = 4 \qquad\qquad\qquad t = 5$$
$$r = 2$$

3 lbs. of ham, 5 lbs. of turkey, and 2 lbs. of roast beef were used for the party tray.

37. Let x = the amount invested at 4%, y = the amount invested at 6%, and z = the amount invested at 7%.

$$\begin{cases} x + y + z = 8000 & \text{Eqtn. 1} \\ 0.04x + 0.06y + 0.07z = 475 & \text{Eqtn. 2} \\ z - x = 1500 & \text{Eqtn. 3} \end{cases}$$

Multiply equation 1 by –0.06 and add to equation 2. This makes equation 4.

$$-0.06x - 0.06y - 0.06z = -480$$
$$\underline{0.04x + 0.06y + 0.07z = 475}$$
$$-0.02x + 0.01z = -5 \quad \text{Eqtn. 4}$$

Multiply equation 3 by –0.01 and add to equation 4.

$$0.01x - 0.01z = -15$$
$$\underline{-0.02x + 0.01z = -5}$$
$$-0.01x = -20$$
$$x = 2000$$

$$z - x = 1500 \qquad\qquad x + y + z = 8000$$
$$z - 2000 = 1500 \qquad 2000 + y + 3500 = 8000$$
$$z = 3500 \qquad\qquad\qquad y = 2500$$

$2000 is invested at 4%, $2500 is invested at 6%, and $3500 is invested at 7%.

39. Let b = the number of calories burned while bicycling, w = the number of calories burned while walking, and c = the number of calories burned while climbing stairs.

$$\begin{cases} b = 120 + w & \text{Eqtn. 1} \\ c = 180 + b & \text{Eqtn. 2} \\ c = 2w & \text{Eqtn. 3} \end{cases}$$

Substitute equation 3 into equation 2. Use this new equation with equation 1 to form a system of equations in two variables.

$$\begin{array}{ll} b - w = 120 & c = 2 \cdot 300 \\ \underline{-b + 2w = 180} & c = 600 \\ w = 300 \\ b = 120 + 300 \\ b = 420 \end{array}$$

The number of calories burned while bicycling is 420, while walking is 300, and while stair climbing is 600.

Review Exercises

1. $-3(2x - 4y - z + 9) = -6x + 12y + 3z - 27$

2. $x - 0.5y = 8$
$6 - 0.5y = 8$
$-0.5y = 2$
$y = -4$

3. $f(x) = -2x + 1$
$f(1) = -2(1) + 1$
$= -2 + 1$
$= -1$

4. The y-intercept is $(0, 1)$.

5. The slope is $m = -2$.

6.

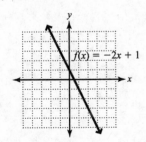

Exercise Set 4.5

1. $\begin{bmatrix} 14 & 7 & \vdots & 6 \\ 7 & 6 & \vdots & 8 \end{bmatrix}$

3. $\begin{bmatrix} 7 & -6 & \vdots & 1 \\ 0 & -2 & \vdots & 5 \end{bmatrix}$

5. $\begin{bmatrix} 1 & -3 & 1 & \vdots & 4 \\ 2 & -4 & 2 & \vdots & -4 \\ 6 & -2 & 5 & \vdots & -4 \end{bmatrix}$

7. $\begin{bmatrix} 4 & 6 & -2 & \vdots & -1 \\ 8 & 3 & 0 & \vdots & -12 \\ 0 & -1 & 2 & \vdots & 4 \end{bmatrix}$

9. $\begin{bmatrix} 1 & -3 & \vdots & -2 \\ 0 & 1 & \vdots & 2 \end{bmatrix}$ represents the system

$$\begin{cases} x - 3y = -2 \\ y = 2 \end{cases}$$

$$\begin{array}{l} x - 3y = -2 \\ x - 3(2) = -2 \\ x - 6 = -2 \\ x = 4 \end{array}$$

Solution: $(4, 2)$

11. $\begin{bmatrix} 1 & -4 & -8 & \vdots & 6 \\ 0 & 1 & -2 & \vdots & -7 \\ 0 & 0 & 1 & \vdots & 1 \end{bmatrix}$ represents the system

$$\begin{cases} x - 4y - 8z = 6 \\ y - 2z = -7 \\ z = 1 \end{cases}$$

$$\begin{array}{ll} y - 2z = -7 & x - 4y - 8z = 6 \\ y - 2(1) = -7 & x - 4(-5) - 8(1) = 6 \\ y - 2 = -7 & x + 20 - 8 = 6 \\ y = -5 & x = -6 \end{array}$$

Solution: $(-6, -5, 1)$

13. $\begin{bmatrix} 1 & 3 & \vdots & -1 \\ -2 & 5 & \vdots & 6 \end{bmatrix}$

$2R_1 + R_2 \rightarrow \begin{bmatrix} 1 & 3 & \vdots & -1 \\ 0 & 11 & \vdots & 4 \end{bmatrix}$

15. $\begin{bmatrix} 1 & -2 & 4 & \vdots & 6 \\ 0 & 2 & -1 & \vdots & -5 \\ 0 & 8 & -6 & \vdots & -3 \end{bmatrix}$

$-4R_2 + R_3 \rightarrow \begin{bmatrix} 1 & -2 & 4 & \vdots & 6 \\ 0 & 2 & -1 & \vdots & -5 \\ 0 & 0 & -2 & \vdots & 17 \end{bmatrix}$

17. $\begin{bmatrix} 4 & 8 & | & -10 \\ -1 & 3 & | & 2 \end{bmatrix}$

$\frac{1}{4}R_1 \rightarrow \begin{bmatrix} 1 & 2 & | & -2.5 \\ -1 & 3 & | & 2 \end{bmatrix}$

19. Replace R_2 with $3R_1 + R_2$.

21. Replace R_3 with $-2R_2 + R_3$.

23. $\begin{cases} x - y = 5 \\ x + y = -1 \end{cases}$ $\begin{bmatrix} 1 & -1 & | & 5 \\ 1 & 1 & | & -1 \end{bmatrix}$

$-1R_1 + R_2 \rightarrow \begin{bmatrix} 1 & -1 & | & 5 \\ 0 & 2 & | & -6 \end{bmatrix}$

$\frac{1}{2}R_2 \rightarrow \begin{bmatrix} 1 & -1 & | & 5 \\ 0 & 1 & | & -3 \end{bmatrix}$

$\begin{cases} x - y = 5 \\ y = -3 \end{cases}$

$x - y = 5$ Solution: $(2, -3)$

$x - (-3) = 5$

$x + 3 = 5$

$x = 2$

25. $\begin{cases} x + y = 3 \\ 3x - y = 1 \end{cases}$

$\begin{bmatrix} 1 & 1 & | & 3 \\ 3 & -1 & | & 1 \end{bmatrix}$

$-3R_1 + R_2 \rightarrow \begin{bmatrix} 1 & 1 & | & 3 \\ 0 & -4 & | & -8 \end{bmatrix}$

$-\frac{1}{4}R_2 \rightarrow \begin{bmatrix} 1 & 1 & | & 3 \\ 0 & 1 & | & 2 \end{bmatrix}$

$\begin{cases} x + y = 3 \\ y = 2 \end{cases}$

$x + y = 3$ Solution: $(1, 2)$

$x + 2 = 3$

$x = 1$

27. $\begin{cases} x + 2y = -7 \\ 2x - 4y = 2 \end{cases}$

$\begin{bmatrix} 1 & 2 & | & -7 \\ 2 & -4 & | & 2 \end{bmatrix}$

$-2R_1 + R_2 \rightarrow \begin{bmatrix} 1 & 2 & | & -7 \\ 0 & -8 & | & 16 \end{bmatrix}$

$-\frac{1}{8}R_2 \rightarrow \begin{bmatrix} 1 & 2 & | & -7 \\ 0 & 1 & | & -2 \end{bmatrix}$

$\begin{cases} x + 2y = -7 \\ y = -2 \end{cases}$

$x + 2y = -7$ Solution: $(-3, -2)$

$x + 2(-2) = -7$

$x - 4 = -7$

$x = -3$

29. $\begin{cases} -2x + 5y = 4 \\ x - 2y = -2 \end{cases}$

$\begin{bmatrix} -2 & 5 & | & 4 \\ 1 & -2 & | & -2 \end{bmatrix}$

$R_1 + 2R_2 \rightarrow \begin{bmatrix} -2 & 5 & | & 4 \\ 0 & 1 & | & 0 \end{bmatrix}$

$-\frac{1}{2}R_1 \rightarrow \begin{bmatrix} 1 & -\frac{5}{2} & | & -2 \\ 0 & 1 & | & 0 \end{bmatrix}$

$\begin{cases} x - \frac{5}{2}y = -2 \\ y = 0 \end{cases}$

$x - \frac{5}{2}y = -2$ Solution: $(-2, 0)$

$x - \frac{5}{2}(0) = -2$

$x - 0 = -2$

$x = -2$

31. $\begin{cases} 4x - 3y = -2 \\ 2x - 3y = -10 \end{cases}$

$$\begin{bmatrix} 4 & -3 & \vdots & -2 \\ 2 & -3 & \vdots & -10 \end{bmatrix}$$

$R_1 - 2R_2 \rightarrow \begin{bmatrix} 4 & -3 & \vdots & -2 \\ 0 & 3 & \vdots & 18 \end{bmatrix}$

$\dfrac{1}{4}R_1 \rightarrow \begin{bmatrix} 1 & -\dfrac{3}{4} & \vdots & -\dfrac{1}{2} \\ 0 & 3 & \vdots & 18 \end{bmatrix}$

$\dfrac{1}{3}R_2 \rightarrow \begin{bmatrix} 1 & -\dfrac{3}{4} & \vdots & -\dfrac{1}{2} \\ 0 & 1 & \vdots & 6 \end{bmatrix}$

$\begin{cases} x - \dfrac{3}{4}y = -\dfrac{1}{2} \\ \quad\quad y = 6 \end{cases}$

$x - \dfrac{3}{4}y = -\dfrac{1}{2}$ Solution: $(4,6)$

$x - \dfrac{3}{4}(6) = -\dfrac{1}{2}$

$x - \dfrac{9}{2} = -\dfrac{1}{2}$

$x = 4$

33. $\begin{cases} 5x + 2y = 12 \\ 2x + 3y = -4 \end{cases}$

$$\begin{bmatrix} 5 & 2 & \vdots & 12 \\ 2 & 3 & \vdots & -4 \end{bmatrix}$$

$-2R_1 + 5R_2 \rightarrow \begin{bmatrix} 5 & 2 & \vdots & 12 \\ 0 & 11 & \vdots & -44 \end{bmatrix}$

$\dfrac{1}{5}R_1 \rightarrow \begin{bmatrix} 1 & \dfrac{2}{5} & \vdots & \dfrac{12}{5} \\ 0 & 11 & \vdots & -44 \end{bmatrix}$

$\dfrac{1}{11}R_2 \rightarrow \begin{bmatrix} 1 & \dfrac{2}{5} & \vdots & \dfrac{12}{5} \\ 0 & 1 & \vdots & -4 \end{bmatrix}$

$\begin{cases} x + \dfrac{2}{5}y = \dfrac{12}{5} \\ \quad\quad y = -4 \end{cases}$

$x + \dfrac{2}{5}y = \dfrac{12}{5}$ Solution: $(4,-4)$

$x + \dfrac{2}{5}(-4) = \dfrac{12}{5}$

$x - \dfrac{8}{5} = \dfrac{12}{5}$

$x = 4$

35. $\begin{cases} x + y + z = 3 \\ 2x + y - 3z = -10 \\ 2x + 2y + z = 3 \end{cases}$

$$\begin{bmatrix} 1 & 1 & 1 & \vdots & 3 \\ 2 & 1 & -3 & \vdots & -10 \\ 2 & 2 & 1 & \vdots & 3 \end{bmatrix}$$

$-2R_1 + R_2 \rightarrow \begin{bmatrix} 1 & 1 & 1 & \vdots & 3 \\ 0 & -1 & -5 & \vdots & -16 \\ 2 & 2 & 1 & \vdots & 3 \end{bmatrix}$

$-2R_1 + R_3 \rightarrow \begin{bmatrix} 1 & 1 & 1 & \vdots & 3 \\ 0 & -1 & -5 & \vdots & -16 \\ 0 & 0 & -1 & \vdots & -3 \end{bmatrix}$

$\begin{matrix} -1R_2 \rightarrow \\ -1R_3 \rightarrow \end{matrix} \begin{bmatrix} 1 & 1 & 1 & \vdots & 3 \\ 0 & 1 & 5 & \vdots & 16 \\ 0 & 0 & 1 & \vdots & 3 \end{bmatrix}$

$\begin{cases} x + y + z = 3 \\ \quad y + 5z = 16 \\ \quad\quad\quad z = 3 \end{cases}$

$y + 5z = 16 \quad\quad x + y + z = 3$

$y + 5(3) = 16 \quad\quad x + 1 + 3 = 3$

$y + 15 = 16 \quad\quad x + 4 = 3$

$y = 1 \quad\quad\quad x = -1$

Solution: $(-1,1,3)$

37. $\begin{cases} x - y + z = -1 \\ 2x - 2y + z = 0 \\ x + 3y + 2z = 1 \end{cases}$

$$\begin{bmatrix} 1 & -1 & 1 & | & -1 \\ 2 & -2 & 1 & | & 0 \\ 1 & 3 & 2 & | & 1 \end{bmatrix}$$

$-2R_1 + R_2 \rightarrow \begin{bmatrix} 1 & -1 & 1 & | & -1 \\ 0 & 0 & -1 & | & 2 \\ 1 & 3 & 2 & | & 1 \end{bmatrix}$

$\begin{matrix} R_3 \rightarrow \\ R_2 \rightarrow \end{matrix} \begin{bmatrix} 1 & -1 & 1 & | & -1 \\ 1 & 3 & 2 & | & 1 \\ 0 & 0 & -1 & | & 2 \end{bmatrix}$

$-R_1 + R_2 \rightarrow \begin{bmatrix} 1 & -1 & 1 & | & -1 \\ 0 & 4 & 1 & | & 2 \\ 0 & 0 & -1 & | & 2 \end{bmatrix}$

$\begin{matrix} \frac{1}{4}R_2 \rightarrow \\ -1R_3 \rightarrow \end{matrix} \begin{bmatrix} 1 & -1 & 1 & | & -1 \\ 0 & 1 & \frac{1}{4} & | & \frac{1}{2} \\ 0 & 0 & 1 & | & -2 \end{bmatrix}$

$\begin{cases} x - y + z = -1 \\ \quad\quad y + \dfrac{1}{4}z = \dfrac{1}{2} \\ \quad\quad\quad\quad z = -2 \end{cases}$

$y + \dfrac{1}{4}z = \dfrac{1}{2} \quad\quad\quad x - y + z = -1$

$y + \dfrac{1}{4}(-2) = \dfrac{1}{2} \quad\quad x - 1 - 2 = -1$

$y - \dfrac{1}{2} = \dfrac{1}{2} \quad\quad\quad\quad x - 3 = -1$

$y = 1 \quad\quad\quad\quad\quad\quad x = 2$

Solution: $(2, 1, -2)$

39. $\begin{cases} 2x - y + z = 8 \\ x - 2y + 3z = 11 \\ 2x + 3y - z = -6 \end{cases}$

$$\begin{bmatrix} 2 & -1 & 1 & | & 8 \\ 1 & -2 & 3 & | & 11 \\ 2 & 3 & -1 & | & -6 \end{bmatrix}$$

$\begin{matrix} R_2 \rightarrow \\ R_1 \rightarrow \end{matrix} \begin{bmatrix} 1 & -2 & 3 & | & 11 \\ 2 & -1 & 1 & | & 8 \\ 2 & 3 & -1 & | & -6 \end{bmatrix}$

$-2R_1 + R_2 \rightarrow \begin{bmatrix} 1 & -2 & 3 & | & 11 \\ 0 & 3 & -5 & | & -14 \\ 2 & 3 & -1 & | & -6 \end{bmatrix}$

$-2R_1 + R_3 \rightarrow \begin{bmatrix} 1 & -2 & 3 & | & 11 \\ 0 & 3 & -5 & | & -14 \\ 0 & 7 & -7 & | & -28 \end{bmatrix}$

$\frac{1}{7}R_3 \rightarrow \begin{bmatrix} 1 & -2 & 3 & | & 11 \\ 0 & 3 & -5 & | & -14 \\ 0 & 1 & -1 & | & -4 \end{bmatrix}$

$\begin{matrix} R_3 \rightarrow \\ R_2 \rightarrow \end{matrix} \begin{bmatrix} 1 & -2 & 3 & | & 11 \\ 0 & 1 & -1 & | & -4 \\ 0 & 3 & -5 & | & -14 \end{bmatrix}$

$-3R_2 + R_3 \rightarrow \begin{bmatrix} 1 & -2 & 3 & | & 11 \\ 0 & 1 & -1 & | & -4 \\ 0 & 0 & -2 & | & -2 \end{bmatrix}$

$-\frac{1}{2}R_3 \rightarrow \begin{bmatrix} 1 & -2 & 3 & | & 11 \\ 0 & 1 & -1 & | & -4 \\ 0 & 0 & 1 & | & 1 \end{bmatrix}$

$\begin{cases} x - 2y + 3z = 11 \\ \quad\quad y - z = -4 \\ \quad\quad\quad\quad z = 1 \end{cases}$

$y - z = -4 \quad\quad\quad x - 2y + 3z = 11$

$y - 1 = -4 \quad\quad x - 2(-3) + 3(1) = 11$

$y = -3 \quad\quad\quad\quad x + 6 + 3 = 11$

$x = 2$

Solution: $(2, -3, 1)$

41. $\begin{cases} 3x+2y-3z=1 \\ -2x+3y-4z=7 \\ 5x-2y+z=-5 \end{cases}$

$$\begin{bmatrix} 3 & 2 & -3 & | & 1 \\ -2 & 3 & -4 & | & 7 \\ 5 & -2 & 1 & | & -5 \end{bmatrix}$$

$R_2+R_1 \rightarrow \begin{bmatrix} 1 & 5 & -7 & | & 8 \\ -2 & 3 & -4 & | & 7 \\ 5 & -2 & 1 & | & -5 \end{bmatrix}$

$2R_1+R_2 \rightarrow \begin{bmatrix} 1 & 5 & -7 & | & 8 \\ 0 & 13 & -18 & | & 23 \\ 5 & -2 & 1 & | & -5 \end{bmatrix}$

$-5R_1+R_3 \rightarrow \begin{bmatrix} 1 & 5 & -7 & | & 8 \\ 0 & 13 & -18 & | & 23 \\ 0 & -27 & 36 & | & -45 \end{bmatrix}$

$\begin{matrix} -R_3-2R_2 \rightarrow \\ \frac{1}{9}R_3 \rightarrow \end{matrix} \begin{bmatrix} 1 & 5 & -7 & | & 8 \\ 0 & 1 & 0 & | & -1 \\ 0 & -3 & 4 & | & -5 \end{bmatrix}$

$3R_2+R_3 \rightarrow \begin{bmatrix} 1 & 5 & -7 & | & 8 \\ 0 & 1 & 0 & | & -1 \\ 0 & 0 & 4 & | & -8 \end{bmatrix}$

$\frac{1}{4}R_3 \rightarrow \begin{bmatrix} 1 & 5 & -7 & | & 8 \\ 0 & 1 & 0 & | & -1 \\ 0 & 0 & 1 & | & -2 \end{bmatrix}$

$$\begin{cases} x+5y-7z=8 \\ y=-1 \\ z=-2 \end{cases}$$

$$x+5y-7z=8$$
$$x+5(-1)-7(-2)=8$$
$$x-5+14=8$$
$$x=-1$$

Solution: $(-1,-1,-2)$

43. $\begin{cases} 3x-6y+z=-10 \\ 7y-z=2 \\ 2x+4z=14 \end{cases}$

$$\begin{bmatrix} 3 & -6 & 1 & | & -10 \\ 0 & 7 & -1 & | & 2 \\ 2 & 0 & 4 & | & 14 \end{bmatrix}$$

$\begin{matrix} R_3 \rightarrow \\ \\ R_1 \rightarrow \end{matrix} \begin{bmatrix} 2 & 0 & 4 & | & 14 \\ 0 & 7 & -1 & | & 2 \\ 3 & -6 & 1 & | & -10 \end{bmatrix}$

$\frac{1}{2}R_1 \rightarrow \begin{bmatrix} 1 & 0 & 2 & | & 7 \\ 0 & 7 & -1 & | & 2 \\ 3 & -6 & 1 & | & -10 \end{bmatrix}$

$-3R_1+R_3 \rightarrow \begin{bmatrix} 1 & 0 & 2 & | & 7 \\ 0 & 7 & -1 & | & 2 \\ 0 & -6 & -5 & | & -31 \end{bmatrix}$

$6R_2+7R_3 \rightarrow \begin{bmatrix} 1 & 0 & 2 & | & 7 \\ 0 & 7 & -1 & | & 2 \\ 0 & 0 & -41 & | & -205 \end{bmatrix}$

$\begin{matrix} \frac{1}{7}R_2 \rightarrow \\ -\frac{1}{41}R_3 \rightarrow \end{matrix} \begin{bmatrix} 1 & 0 & 2 & | & 7 \\ 0 & 1 & -\frac{1}{7} & | & \frac{2}{7} \\ 0 & 0 & 1 & | & 5 \end{bmatrix}$

$$\begin{cases} x+2z=7 \\ y-\frac{1}{7}z=\frac{2}{7} \\ z=5 \end{cases}$$

$y-\frac{1}{7}z=\frac{2}{7}$

$y-\frac{1}{7}(5)=\frac{2}{7}$

$y-\frac{5}{7}=\frac{2}{7}$

$y=1$

$x+2z=7$

$x+2(5)=7$

$x+10=7$

$x=-3$

Solution: $(-3,1,5)$

45. $\begin{cases} 2x+5y-z=10 \\ x-y-z=14 \\ x-6y=20 \end{cases}$

$$\begin{bmatrix} 2 & 5 & -1 & | & 10 \\ 1 & -1 & -1 & | & 14 \\ 1 & -6 & 0 & | & 20 \end{bmatrix}$$

$\begin{matrix} R_2 \to \\ R_1 \to \\ \\ \end{matrix} \begin{bmatrix} 1 & -1 & -1 & | & 14 \\ 2 & 5 & -1 & | & 10 \\ 1 & -6 & 0 & | & 20 \end{bmatrix}$

$\begin{matrix} \\ -2R_1+R_2 \to \\ -R_1+R_3 \to \end{matrix} \begin{bmatrix} 1 & -1 & -1 & | & 14 \\ 0 & 7 & 1 & | & -18 \\ 0 & -5 & 1 & | & 6 \end{bmatrix}$

$\frac{1}{7}R_2 \to \begin{bmatrix} 1 & -1 & -1 & | & 14 \\ 0 & 1 & \frac{1}{7} & | & -\frac{18}{7} \\ 0 & -5 & 1 & | & 6 \end{bmatrix}$

$5R_2+R_3 \to \begin{bmatrix} 1 & -1 & -1 & | & 14 \\ 0 & 1 & \frac{1}{7} & | & -\frac{18}{7} \\ 0 & 0 & \frac{12}{7} & | & -\frac{48}{7} \end{bmatrix}$

$\frac{7}{12}R_3 \to \begin{bmatrix} 1 & -1 & -1 & | & 14 \\ 0 & 1 & \frac{1}{7} & | & -\frac{18}{7} \\ 0 & 0 & 1 & | & -4 \end{bmatrix}$

$\begin{cases} x-y-z=14 \\ y+\frac{1}{7}z=-\frac{18}{7} \\ z=-4 \end{cases}$

$y+\frac{1}{7}z=-\frac{18}{7}$ $x-y-z=14$

$y+\frac{1}{7}(-4)=-\frac{18}{7}$ $x-(-2)-(-4)=14$

$y-\frac{4}{7}=-\frac{18}{7}$ $x+2+4=14$

$y=-2$ $x=8$

Solution: $(8,-2,-4)$

47. $(3,-4)$ 49. $(-6,8)$

51. $(4,-3,5)$ 53. $(7,-6,5)$

55. Mistake: In the second step, $4R_1+R_2$ is calculated incorrectly.
Correct: The correct calculation is
$\begin{bmatrix} 1 & 3 & | & 13 \\ 0 & 11 & | & 26 \end{bmatrix}$. The solution is $\left(\frac{65}{11}, \frac{26}{11}\right)$.

57. Let x = the cost of a grilled chicken sandwich and y = the cost of a drink.
$\begin{cases} 3x+2y=12.90 \\ 7x+4y=29.30 \end{cases}$

$$\begin{bmatrix} 3 & 2 & | & 12.9 \\ 7 & 4 & | & 29.3 \end{bmatrix}$$

$-7R_1+3R_2 \to \begin{bmatrix} 3 & 2 & | & 12.9 \\ 0 & -2 & | & -2.4 \end{bmatrix}$

$\begin{matrix} \frac{1}{3}R_1 \to \\ -\frac{1}{2}R_2 \to \end{matrix} \begin{bmatrix} 1 & \frac{2}{3} & | & 4.3 \\ 0 & 1 & | & 1.2 \end{bmatrix}$

$\begin{cases} x+\frac{2}{3}y=4.3 \\ y=1.2 \end{cases}$

$x+\frac{2}{3}y=4.3$

$x+\frac{2}{3}(1.2)=4.3$

$x+0.8=4.3$

$x=3.5$

The chicken sandwich costs \$3.50 and the drink costs \$1.20.

59. Let n = the length of the Nile and a = the length of the Amazon.

$$\begin{cases} n+a=8050 \\ n=250+a \end{cases}$$

$$\begin{bmatrix} 1 & 1 & \vdots & 8050 \\ 1 & -1 & \vdots & 250 \end{bmatrix}$$

$$-R_1+R_2 \to \begin{bmatrix} 1 & 1 & \vdots & 8050 \\ 0 & -2 & \vdots & -7800 \end{bmatrix}$$

$$-\tfrac{1}{2}R_2 \to \begin{bmatrix} 1 & 1 & \vdots & 8050 \\ 0 & 1 & \vdots & 3900 \end{bmatrix}$$

$$\begin{cases} n+a=8050 \\ a=3900 \end{cases}$$

$$n+a=8050$$
$$n+3900=8050$$
$$n=4150$$

The Amazon River is 3900 mi. and the Nile River is 4150 mi.

61. Let c = the amount invested in the CD at 5% interest and m = the amount invested in the money market account at 6% interest.

$$\begin{cases} c+m=10,000 \\ 0.05c+0.06m=536 \end{cases}$$

$$\begin{bmatrix} 1 & 1 & \vdots & 10,000 \\ 0.05 & 0.06 & \vdots & 536 \end{bmatrix}$$

$$-0.05R_1+R_2 \to \begin{bmatrix} 1 & 1 & \vdots & 10,000 \\ 0 & 0.01 & \vdots & 36 \end{bmatrix}$$

$$100R_2 \to \begin{bmatrix} 1 & 1 & \vdots & 10,000 \\ 0 & 1 & \vdots & 3600 \end{bmatrix}$$

$$\begin{cases} c+m=10,000 \\ m=3600 \end{cases}$$

$$c+m=10,000$$
$$c+3600=10,000$$
$$c=6400$$

$6400 was invested in the CD and $3600 was invested in the money market account.

63. Let c = the cost of a CD, b = the cost of a book, and d = the cost of a DVD.

$$\begin{cases} 2c+4b+3d=164 \\ 5c+2b+2d=160 \\ c+b=d \end{cases}$$

$$\begin{bmatrix} 2 & 4 & 3 & \vdots & 164 \\ 5 & 2 & 2 & \vdots & 160 \\ 1 & 1 & -1 & \vdots & 0 \end{bmatrix}$$

$$\begin{matrix} R_3 \to \\ \\ R_1 \to \end{matrix} \begin{bmatrix} 1 & 1 & -1 & \vdots & 0 \\ 5 & 2 & 2 & \vdots & 160 \\ 2 & 4 & 3 & \vdots & 164 \end{bmatrix}$$

$$\begin{matrix} -5R_1+R_2 \to \\ -2R_1+R_3 \to \end{matrix} \begin{bmatrix} 1 & 1 & -1 & \vdots & 0 \\ 0 & -3 & 7 & \vdots & 160 \\ 0 & 2 & 5 & \vdots & 164 \end{bmatrix}$$

$$\begin{matrix} -\tfrac{1}{3}R_2 \to \\ 2R_2+3R_3 \to \end{matrix} \begin{bmatrix} 1 & 1 & -1 & \vdots & 0 \\ 0 & 1 & -\dfrac{7}{3} & \vdots & -\dfrac{160}{3} \\ 0 & 0 & 29 & \vdots & 812 \end{bmatrix}$$

$$\tfrac{1}{29}R_3 \to \begin{bmatrix} 1 & 1 & -1 & \vdots & 0 \\ 0 & 1 & -\dfrac{7}{3} & \vdots & -\dfrac{160}{3} \\ 0 & 0 & 1 & \vdots & 28 \end{bmatrix}$$

$$\begin{cases} c+b-d=0 \\ b-\dfrac{7}{3}d=-\dfrac{160}{3} \\ d=28 \end{cases}$$

$$b-\frac{7}{3}d=-\frac{160}{3} \qquad c+b-d=0$$
$$b-\frac{7}{3}(28)=-\frac{160}{3} \qquad c+12-28=0$$
$$b-\frac{196}{3}=-\frac{160}{3} \qquad c-16=0$$
$$b=12 \qquad c=16$$

The CDs cost $16 each, the books cost $12 each, and the DVDs cost $28 each.

65. Let t = the number of touchdowns, e = the number of extra points, and f = the number of field goals.

$$\begin{cases} 6t + e + 3f = 31 \\ t = e \\ t = f + 3 \end{cases} = \begin{cases} 6t + e + 3f = 31 \\ t - e = 0 \\ t - f = 3 \end{cases}$$

$$\begin{bmatrix} 6 & 1 & 3 & | & 31 \\ 1 & -1 & 0 & | & 0 \\ 1 & 0 & -1 & | & 3 \end{bmatrix}$$

$$\begin{matrix} R_3 \to \\ \\ R_1 \to \end{matrix} \begin{bmatrix} 1 & 0 & -1 & | & 3 \\ 1 & -1 & 0 & | & 0 \\ 6 & 1 & 3 & | & 31 \end{bmatrix}$$

$$\begin{matrix} \\ -R_1 + R_2 \to \\ -6R_1 + R_3 \to \end{matrix} \begin{bmatrix} 1 & 0 & -1 & | & 3 \\ 0 & -1 & 1 & | & -3 \\ 0 & 1 & 9 & | & 13 \end{bmatrix}$$

$$\begin{matrix} \\ -1R_2 \to \\ R_2 + R_3 \to \end{matrix} \begin{bmatrix} 1 & 0 & -1 & | & 3 \\ 0 & 1 & -1 & | & 3 \\ 0 & 0 & 10 & | & 10 \end{bmatrix}$$

$$\begin{matrix} \\ \\ \tfrac{1}{10}R_3 \to \end{matrix} \begin{bmatrix} 1 & 0 & -1 & | & 3 \\ 0 & 1 & -1 & | & 3 \\ 0 & 0 & 1 & | & 1 \end{bmatrix}$$

$$\begin{cases} t - f = 3 \\ e - f = 3 \\ f = 1 \end{cases}$$

$$\begin{array}{ll} e - f = 3 & t - f = 3 \\ e - 1 = 3 & t - 1 = 3 \\ e = 4 & t = 4 \end{array}$$

New York scored 4 touchdowns, 4 extra points, and 1 field goal.

67.

$$\begin{cases} 4v_1 - v_2 = 30 \\ -2v_1 + 5v_2 - v_3 = 10 \\ -v_2 + 5v_3 = 4 \end{cases}$$

$$\begin{bmatrix} 4 & -1 & 0 & | & 30 \\ -2 & 5 & -1 & | & 10 \\ 0 & -1 & 5 & | & 4 \end{bmatrix}$$

$$\begin{matrix} \\ R_3 \to \\ R_2 \to \end{matrix} \begin{bmatrix} 4 & -1 & 0 & | & 30 \\ 0 & -1 & 5 & | & 4 \\ -2 & 5 & -1 & | & 10 \end{bmatrix}$$

$$\begin{matrix} \\ -1R_2 \to \\ R_1 + 2R_3 \to \end{matrix} \begin{bmatrix} 4 & -1 & 0 & | & 30 \\ 0 & 1 & -5 & | & -4 \\ 0 & 9 & -2 & | & 50 \end{bmatrix}$$

$$\begin{matrix} \\ \\ -9R_2 + R_3 \to \end{matrix} \begin{bmatrix} 4 & -1 & 0 & | & 30 \\ 0 & 1 & -5 & | & -4 \\ 0 & 0 & 43 & | & 86 \end{bmatrix}$$

$$\begin{matrix} \tfrac{1}{4}R_1 \to \\ \\ \tfrac{1}{43}R_3 \to \end{matrix} \begin{bmatrix} 1 & -\tfrac{1}{4} & 0 & | & \tfrac{15}{2} \\ 0 & 1 & -5 & | & -4 \\ 0 & 0 & 1 & | & 2 \end{bmatrix}$$

$$\begin{cases} v_1 - \tfrac{1}{4}v_2 = \tfrac{15}{2} \\ v_2 - 5v_3 = -4 \\ v_3 = 2 \end{cases}$$

$$\begin{array}{ll} v_2 - 5v_3 = -4 & v_1 - \tfrac{1}{4}v_2 = \tfrac{15}{2} \\ v_2 - 5(2) = -4 & v_1 - \tfrac{1}{4}(6) = \tfrac{15}{2} \\ v_2 - 10 = -4 & v_1 - \tfrac{3}{2} = \tfrac{15}{2} \\ v_2 = 6 & v_1 = 9 \end{array}$$

$$v_1 = 9, \quad v_2 = 6, \quad v_3 = 2$$

Review Exercises

1. $(-3)(4) - (-4)(5) = -12 + 20 = 8$

2. $2(2 - 8) + 3\left[-1 - (-6)\right] - 4(4 - 6)$
$= 2(-6) + 3(5) - 4(-2)$
$= -12 + 15 + 8$
$= 11$

3. $\dfrac{(-5)(-1) - (9)(3)}{(2)(-1) - (3)(3)} = \dfrac{5 - 27}{-2 - 9} = \dfrac{-22}{-11} = 2$

4. $\dfrac{2\left(\dfrac{5}{4}\right)-8\left(\dfrac{1}{4}\right)}{\left(\dfrac{1}{2}\right)\left(\dfrac{5}{4}\right)-\left(\dfrac{3}{2}\right)\left(\dfrac{1}{4}\right)}=\dfrac{\dfrac{5}{2}-2}{\dfrac{5}{8}-\dfrac{3}{8}}=\dfrac{\dfrac{1}{2}}{\dfrac{2}{8}}=2$

5. $\dfrac{(1.1)(-0.2)-(1.7)(-0.5)}{(0.2)(-0.2)-(0.5)(-0.5)}=\dfrac{-0.22+0.85}{-0.04+0.25}$

$$=\dfrac{0.63}{0.21}$$

$$=3$$

6. $\dfrac{1\left[-45-(-9)\right]+7(-9-2)-2(-27-30)}{1(6-3)-2(-9-2)-2\left[9-(-4)\right]}$

$$=\dfrac{1\left[-45+9\right]+7(-11)-2(-57)}{1(3)-2(-11)-2\left[9+4\right]}$$

$$=\dfrac{1\left[-36\right]-77+114}{3+22-2\cdot13}$$

$$=\dfrac{-36-77+114}{3+22-26}$$

$$=\dfrac{1}{-1}$$

$$=-1$$

Exercise Set 4.6

1. $\begin{cases}x+y>4\\x-y<6\end{cases}$

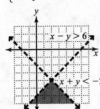

3. $\begin{cases}x+y<-2\\x-y>6\end{cases}$

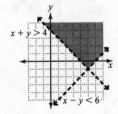

5. $\begin{cases}2x+y\ge-1\\x-y\le5\end{cases}$

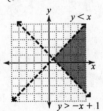

7. $\begin{cases}x+y<3\\x-2y\ge2\end{cases}$

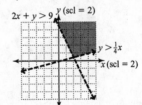

9. $\begin{cases}2x+y>9\\y>\dfrac{1}{4}x\end{cases}$

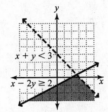

11. $\begin{cases}y<x\\y>-x+1\end{cases}$

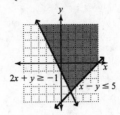

13. $\begin{cases}3x>-4y\\3x+4y\le-8\end{cases}$

15. $\begin{cases} y < 3x+1 \\ 3x-y \le 4 \end{cases}$

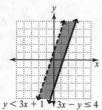

$y < 3x+1$ $3x-y \le 4$

17. $\begin{cases} x > 2 \\ 3x-2y > 6 \end{cases}$

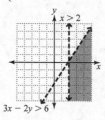

$3x-2y > 6$

19. $\begin{cases} x \ge -3y \\ y \ge 2x \end{cases}$

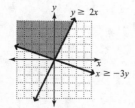

$y \ge 2x$ $x \ge -3y$

21. $\begin{cases} y > 2 \\ x < -1 \end{cases}$

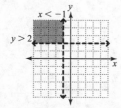

$x < -1$ $y > 2$

23. $\begin{cases} 2x+y \ge 1 \\ x-y > -1 \\ x > 2 \end{cases}$

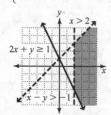

$x > 2$ $2x+y \ge 1$ $x-y > -1$

25. $x + y = 3$ should be a dashed line.

27. The wrong area is shaded. It should be the region containing the point (1, 5).

29. a) $V + M \ge 1100$

    $M \ge 500$

    b) $V \le 800$

    $M \le 800$

    c)

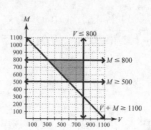

    d) Answers will vary. Some possible (V, M) pairs are (500, 700), (600, 500), or (700, 600).

31. a) $\begin{cases} R+D \ge 100 \\ 15.95R + 20.95D \ge 1700 \end{cases}$

    b)

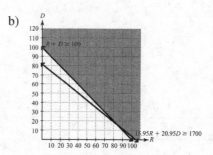

    c) Answers will vary. Some possible (R, D) pairs are (30, 90), (40, 80), and (50, 60).

Review Exercises

1. $3^2 = 9$

2. $-10^2 = -10 \cdot 10 = -100$

3. $\sqrt{25} = 5$

4. $\sqrt{\dfrac{16}{49}} = \dfrac{\sqrt{16}}{\sqrt{49}} = \dfrac{4}{7}$

5. $f(x) = 3x - 5$

    $f(3) = 3(3) - 5$

    $\quad\quad = 9 - 5$

    $\quad\quad = 4$

6. $f(x) = 3x^2 - 2x + 5$

   $f(-2) = 3(-2)^2 - 2(-2) + 5$

   $\quad\quad\ = 3(4) + 4 + 5$

   $\quad\quad\ = 12 + 4 + 5$

   $\quad\quad\ = 21$

Chapter 4 Review Exercises

1.   Is $(4,3)$ a solution of $\begin{cases} x - y = 1 \\ -x + y = -1 \end{cases}$?

   Equation 1   Equation 2

   $x - y = 1$ $-x + y = -1$

   $\overset{?}{4 - 3 = 1}$ $\overset{?}{-4 + 3 = -1}$

   $1 = 1$ $-1 = -1$

   Yes, $(4,3)$ is a solution for the system because it satisfies both equations.

2.   Is $(1,1)$ a solution of $\begin{cases} 2x - y = 7 \\ x + y = 8 \end{cases}$?

   Equation 1   Equation 2

   $2x - y = 7$ $x + y = 8$

   $\overset{?}{2(1) - 1 = 7}$ $\overset{?}{1 + 1 = 8}$

   $\overset{?}{2 - 1 = 7}$ $2 \neq 8$

   $1 \neq 7$

   No, $(1,1)$ is not a solution for the system because it does not satisfy both equations.

3. $\begin{cases} 4x = y \\ 3x + y = -7 \end{cases}$ Graph each equation.

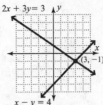

   The lines intersect at a single point, which appears to be $(-1,-4)$.

4. $\begin{cases} x - y = 4 \\ 2x + 3y = 3 \end{cases}$ Graph each equation.

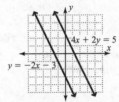

   The lines intersect at a single point, which appears to be $(3,-1)$.

5. $\begin{cases} y = -2x - 3 \\ 4x + 2y = 5 \end{cases}$ Graph each equation.

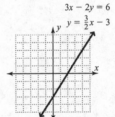

   The lines appear to be parallel, so the system has no solution.

6. $\begin{cases} 3x - 2y = 6 \\ y = \dfrac{3}{2}x - 3 \end{cases}$ Graph each equation.

   The lines appear to be identical, so the solution is the set of all ordered pairs that solve $3x - 2y = 6$.

7.   a) Because these lines intersect in a single point, this system is consistent with independent equations.
   b) This system has one solution.

8.   a) Because the lines are parallel, this system is inconsistent.
   b) This system has no solution.

9. $\begin{cases} x - 3y = 10 \\ 2x + 3y = 5 \end{cases} \rightarrow \begin{cases} y = \dfrac{1}{3}x - \dfrac{10}{3} \\ y = -\dfrac{2}{3}x + \dfrac{5}{3} \end{cases}$

a) Because these lines have different slopes, the graphs are different and the lines intersect in a single point. This system is consistent with independent equations.

b) This system has one solution.

10. $\begin{cases} x - 5y = 10 \\ 2x - 10y = 20 \end{cases} \rightarrow \begin{cases} y = \dfrac{1}{5}x - 2 \\ y = \dfrac{1}{5}x - 2 \end{cases}$

a) The lines have the same slope and the same y-intercept, so the graphs are identical. Because these lines coincide, this system is consistent with dependent equations.

b) This system has an infinite number of solutions.

11. $\begin{cases} 3x + 10y = 2 \\ x - 2y = 6 \end{cases}$

Solve the second equation for x: $x = 6 + 2y$

$\begin{aligned} 3x + 10y &= 2 \\ 3(6 + 2y) + 10y &= 2 \\ 18 + 6y + 10y &= 2 \\ 16y &= -16 \\ y &= -1 \end{aligned}$ $\qquad \begin{aligned} x &= 6 + 2(-1) \\ x &= 6 - 2 \\ x &= 4 \end{aligned}$

Solution: $(4, -1)$

12. $\begin{cases} 2x + 5y = 8 \\ x - 10y = 9 \end{cases}$

Solve the second equation for x: $x = 9 + 10y$

$\begin{aligned} 2x + 5y &= 8 \\ 2(9 + 10y) + 5y &= 8 \\ 18 + 20y + 5y &= 8 \\ 25y &= -10 \\ y &= -\dfrac{2}{5} \end{aligned}$ $\qquad \begin{aligned} x &= 9 + 10\left(-\dfrac{2}{5}\right) \\ x &= 9 - 4 \\ x &= 5 \end{aligned}$

Solution: $\left(5, -\dfrac{2}{5}\right)$

13. $\begin{cases} 3y - 4x = 6 \\ y = \dfrac{4}{3}x + 2 \end{cases}$

The second equation is solved for y.

$\begin{aligned} 3y - 4x &= 6 \\ 3\left(\dfrac{4}{3}x + 2\right) - 4x &= 6 \\ 4x + 6 - 4x &= 6 \\ 6 &= 6 \end{aligned}$

Notice that $6 = 6$ no longer has a variable and is true. The equations in the system are dependent; there are an infinite number of solutions that are all of the ordered pairs that solve $3y - 4x = 6$.

14. $\begin{cases} 8x - 2y = 12 \\ y - 4x = 3 \end{cases}$

Solve the second equation for y: $y = 4x + 3$

$\begin{aligned} 8x - 2y &= 12 \\ 8x - 2(4x + 3) &= 12 \\ 8x - 8x - 6 &= 12 \\ -6 &\neq 12 \end{aligned}$

Because this last statement is false, the system is inconsistent and has no solution.

15. Let c = the number of visitors from Canada (in thousands) and m = the number of visitors from Mexico (in thousands).

$\begin{cases} c + m = 34{,}828 \\ c - m = 7846 \end{cases}$

$\begin{aligned} c + m &= 34{,}828 \\ \underline{c - m} &= \underline{7846} \\ 2c &= 42{,}674 \\ c &= 21{,}337 \end{aligned}$ $\qquad \begin{aligned} c + m &= 34{,}828 \\ 21{,}337 + m &= 34{,}828 \\ m &= 13{,}491 \end{aligned}$

The United States was visited by 21,337 thousand visitors from Canada and 13,491 thousand visitors from Mexico.

16. Let w be the percent of carbon dioxide emissions contributed by Western Europe and n be the percent of carbon dioxide emissions contributed by North America. Translate to a system of equations and solve.

$\begin{cases} w + n = 43 \\ n = 2w - 4 \end{cases}$

$\begin{aligned} w + n &= 43 \\ w + (2w - 4) &= 43 \\ 3w &= 47 \\ w &= \dfrac{47}{3} = 15\dfrac{2}{3} \end{aligned}$ $\qquad \begin{aligned} w + n &= 43 \\ 15\dfrac{2}{3} + n &= 43 \\ n &= 27\dfrac{1}{3} \end{aligned}$

Western Europe: $15\dfrac{2}{3}\%$

North America: $27\dfrac{1}{3}\%$

17. Let x = the measure of the smaller angle and y = the measure of the larger angle. If two angles are complementary, the sum of their measures is 90°. Translate to a system of equations and solve.

$$\begin{cases} x + y = 90 \\ x = \dfrac{1}{3}y - 10 \end{cases}$$

$$x + y = 90 \qquad\qquad x + y = 90$$
$$\left(\dfrac{1}{3}y - 10\right) + y = 90 \qquad x + 75 = 90$$
$$\qquad\qquad\qquad x = 15$$
$$\dfrac{4}{3}y = 100$$
$$y = 75$$

The angles are 75° and 15°.

18. Complete a table.

Categories	Rate	Time	Distance
Yolanda	60	x	$60x$
Dee	70	y	$70y$

Translate to a system of equations.

$$\begin{cases} 60x = 70y \\ x = y + \dfrac{1}{4} \end{cases}$$

$$60x = 70y$$
$$60\left(y + \dfrac{1}{4}\right) = 70y$$
$$60y + 15 = 70y$$
$$15 = 10y$$
$$1.5 = y$$

Yolanda will catch up with Dee 1.5 hours after 4:15, at 5:45 pm.

19. $$\begin{cases} x + y = 4 \\ x - y = -2 \end{cases}$$

$$\begin{array}{ll} x + y = 4 & x + y = 4 \\ \underline{x - y = -2} & 1 + y = 4 \\ 2x \phantom{{}= 2} = 2 & y = 3 \\ x \phantom{{}=} = 1 & \end{array}$$

Solution: $(1, 3)$

20. $$\begin{cases} 3x + 2y = 4 \\ 2x - 3y = 7 \end{cases}$$

$$\begin{array}{ll} 3(3x + 2y = 4) & 3x + 2y = 4 \\ \underline{2(2x - 3y = 7)} & 3 \cdot 2 + 2y = 4 \\ 9x + 6y = 12 & 6 + 2y = 4 \\ \underline{4x - 6y = 14} & 2y = -2 \\ 13x \phantom{{}= 26} = 26 & y = -1 \\ x \phantom{{}=2} = 2 & \end{array}$$

Solution: $(2, -1)$

21. $$\begin{cases} 0.25x + 0.75y = 4 \\ x - y = -4 \end{cases}$$

$$\begin{array}{ll} -4(0.25x + 0.75y = 4) & x - y = -4 \\ \underline{x - y = -4} & x - 5 = -4 \\ -x - 3y = -16 & x = 1 \\ \underline{x - y = -4} & \\ -4y = -20 & \\ y = 5 & \end{array}$$

Solution: $(1, 5)$

22. $$\begin{cases} \dfrac{1}{5}x - \dfrac{1}{3}y = 2 \\ x + y = 2 \end{cases}$$

$$\begin{array}{ll} 3\left(\dfrac{1}{5}x - \dfrac{1}{3}y = 2\right) & x + y = 2 \\ \underline{x + y = 2} & 5 + y = 2 \\ \dfrac{3}{5}x - y = 6 & y = -3 \\ \underline{x + y = 2} & \\ \dfrac{8}{5}x = 8 & \\ x = 5 & \end{array}$$

Solution: $(5, -3)$

23. Complete a table.

Categories	Price	Number	Revenue
adults	7	x	$7x$
children	5.50	y	$5.50y$

Translate to a system of equations.

$$\begin{cases} x + y = 139 \\ 7x + 5.50y = 835 \end{cases}$$

Solve the first equation for y. $y = 139 - x$

$$\begin{array}{ll} 7x + 5.50y = 835 & y = 139 - x \\ 7x + 5.50(139 - x) = 835 & y = 139 - 47 \\ 7x + 764.5 - 5.50x = 835 & y = 92 \\ \qquad\qquad 1.5x = 70.5 & \\ \qquad\qquad x = 47 & \end{array}$$

47 adult tickets and 92 children's tickets were sold.

24. Create a table.

Categories	APR	Principal	Interest
fund 1	0.07	x	$0.07x$
fund 2	0.09	y	$0.09y$

Translate to a system of equations.

$$\begin{cases} x + y = 5000 \\ 0.07x + 0.09y = 380 \end{cases}$$

Solve the first equation for x. $x = 5000 - y$

$$0.07x + 0.09y = 380$$
$$0.07(5000 - y) + 0.09y = 380$$
$$350 - 0.07y + 0.09y = 380$$
$$0.02y = 30$$
$$y = 1500$$

$$x = 5000 - y$$
$$x = 5000 - 1500$$
$$x = 3500$$

He invested $1500 at 9% and $3500 at 7%.

25. Let x be the speed of the plane in still air and y be the speed of the jet stream.
Complete a table.

Categories	Rate	Time	Distance
against	$x - y$	5	$5(x - y)$
with	$x + y$	3	$3(x + y)$

Translate to a system of equations.

$$\begin{cases} 3(x + y) = 450 \\ 5(x - y) = 450 \end{cases} \rightarrow \begin{cases} x + y = 150 \\ x - y = 90 \end{cases}$$

$$\begin{array}{l} x + y = 150 \\ \underline{x - y = 90} \\ 2x = 240 \\ x = 120 \end{array}$$

The speed of the plane in still air is 120 mph.

26. Create a table.

Solutions	Concentration	Volume	Amt of Alcohol
10% sol'n	0.10	x	$0.10x$
60% sol'n	0.60	y	$0.60y$
30% sol'n	0.30	50	$0.30(50)$

Translate to a system of equations.

$$\begin{cases} x + y = 50 \\ 0.10x + 0.60y = 0.30(50) \end{cases}$$

Solve the first equation for y: $y = 50 - x$

$$0.10x + 0.60y = 0.30(50)$$
$$0.10x + 0.60(50 - x) = 15$$
$$0.10x + 30 - 0.60x = 15$$
$$-0.50x = -15$$
$$x = 30$$

30 ml of the 10% solution will be needed.

27. $\begin{cases} x+y+z=-2 & \text{Eqtn. 1} \\ 2x+3y+4z=-10 & \text{Eqtn. 2} \\ 3x-2y-3z=12 & \text{Eqtn. 3} \end{cases}$

Multiply equation 1 by -2 and add to equation 2. This makes equation 4.

$-2x-2y-2z=4$

$\underline{2x+3y+4z=-10}$

$\quad\quad\quad y+2z=-6 \quad$ Eqtn. 4

Multiply equation 1 by -3 and add to equation 3. This makes equation 5.

$-3x-3y-3z=6$

$\underline{3x-2y-3z=12}$

$\quad\quad -5y-6z=18 \quad$ Eqtn. 5

Use equations 4 and 5 to make a system of equations in two variables. Solve.

$y+2z=-6 \quad\quad$ Multiply by 5.

$\underline{-5y-6z=18}$

$5y+10z=-30$

$\underline{-5y-6z=18}$

$\quad\quad 4z=-12$

$\quad\quad\quad z=-3$

$y+2(-3)=-6 \quad\quad x+0-3=-2$

$\quad\quad y-6=-6 \quad\quad\quad x-3=-2$

$\quad\quad\quad y=0 \quad\quad\quad\quad\quad x=1$

Solution: $(1,0,-3)$

28. $\begin{cases} x+y+z=0 & \text{Eqtn. 1} \\ 2x-4y+3z=-12 & \text{Eqtn. 2} \\ 3x-3y+4z=2 & \text{Eqtn. 3} \end{cases}$

Multiply equation 1 by -2 and add to equation 2. This makes equation 4.

$-2x-2y-2z=0$

$\underline{2x-4y+3z=-12}$

$\quad\quad -6y+z=-12 \quad$ Eqtn 4

Multiply equation 1 by -3 and add to equation 3. This makes equation 5

$-3x-3y-3z=0$

$\underline{3x-3y+4z=2}$

$\quad\quad -6y+z=2 \quad$ Eqtn. 5

Use equations 4 and 5 to form a system of equations in two variables. Solve.

$-6y+z=-12 \quad\quad$ Multiply by -1

$\underline{-6y+z=2}$

$6y-z=12$

$\underline{-6y+z=2}$

$\quad\quad 0\ne 14$

Because this last statement is false, the system is inconsistent and has no solution.

29. $\begin{cases} 3x+4y=-3 & \text{Eqtn. 1} \\ -2y+3z=12 & \text{Eqtn. 2} \\ 4x-3z=6 & \text{Eqtn. 3} \end{cases}$

Multiply equation 2 by 2 and add to equation 1. This makes equation 4.

$3x+4y=-3$

$\underline{-4y+6z=24}$

$\quad 3x+6z=21 \quad$ Eqtn. 4

Use equations 3 and 4 to make a system of equations in two variables. Solve.

$4x-3z=6 \quad\quad$ Multiply by 2.

$\underline{3x+6z=21}$

$8x-6z=12$

$\underline{3x+6z=21}$

$\quad 11x=33$

$\quad\quad x=3$

$4\cdot 3-3z=6 \quad\quad 3\cdot 3+4y=-3$

$\quad 12-3z=6 \quad\quad 9+4y=-3$

$\quad\quad -3z=-6 \quad\quad\quad 4y=-12$

$\quad\quad\quad z=2 \quad\quad\quad\quad y=-3$

Solution: $(3,-3,2)$

30. $\begin{cases} 2x+2y-3z=5 & \text{Eqtn. 1} \\ x+3y-4z=0 & \text{Eqtn. 2} \\ x+2y+z=5 & \text{Eqtn. 3} \end{cases}$

Multiply equation 2 by -2 and add to equation 1. This makes equation 4.

$2x+2y-3z=5$

$\underline{-2x-6y+8z=0}$

$\quad -4y+5z=5 \quad$ Eqtn. 4

Multiply equation 2 by -1 and add to equation 3. This makes equation 5.

$-x-3y+4z=0$

$\underline{x+2y+z=5}$

$\quad -y+5z=5 \quad$ Eqtn. 5

Use equations 4 and 5 to make a system of equations in two variables. Solve.

$-4y + 5z = 5$ Multiply by -1.

$$-y + 5z = 5$$
$$\underline{4y - 5z = -5}$$
$$-y + 5z = 5$$
$$\overline{3y = 0}$$
$$y = 0$$

$$\begin{array}{ll} -y + 5z = 5 & x + 2y + z = 5 \\ 0 + 5z = 5 & x + 2(0) + 1 = 5 \\ 5z = 5 & x + 1 = 5 \\ z = 1 & x = 4 \end{array}$$

Solution: $(4, 0, 1)$

31. Let v = the number of vitamin supplements, f = the number of rolls of film, and c = the number of bags of candy.

$$\begin{cases} 8v + 4f + 3c = 32 & \text{Eqtn. 1} \\ v + f + c = 8 & \text{Eqtn. 2} \\ f = c - 1 & \text{Eqtn. 3} \end{cases}$$

Multiply equation 2 by -8 and add it to equation 1. This will make equation 4.

$$8v + 4f + 3c = 32$$
$$\underline{-8v - 8f - 8c = -64}$$
$$-4f - 5c = -32 \qquad \text{Eqtn. 4}$$

Use equations 3 and 4 to make a system of linear equations in two variables. Solve.

$$4(f - c = -1)$$
$$\underline{-4f - 5c = -32}$$
$$4f - 4c = -4$$
$$\underline{-4f - 5c = -32}$$
$$-9c = -36$$
$$c = 4$$

$$\begin{array}{ll} f - 4 = -1 & v + 3 + 4 = 8 \\ f = 3 & v + 7 = 8 \\ & v = 1 \end{array}$$

He purchased 1 vitamin supplement, 3 rolls of film, and 4 bags of candy.

32. Let f = the number of first-place finishes, s = the number of second-place finishes, and t = the number of third-place finishes.

$$\begin{cases} 5f + 3s + t = 38 & \text{Eqtn. 1} \\ f = 1 + s & \text{Eqtn. 2} \\ t = 3s & \text{Eqtn. 3} \end{cases}$$

Substitute equations 2 and 3 into equation 1 and solve for s.

$$\begin{array}{lll} 5f + 3s + t = 38 & t = 3s & f = 1 + s \\ 5(1 + s) + 3s + 3s = 38 & t = 3(3) & f = 1 + 3 \\ 5 + 5s + 3s + 3s = 38 & t = 9 & f = 4 \\ 11s = 33 & & \\ s = 3 & & \end{array}$$

There were 4 first-place finishes, 3 second-place finishes, and 9 third-place finishes.

33. $\begin{cases} x - 4y = 8 \\ x + 2y = 2 \end{cases}$

$$\begin{bmatrix} 1 & -4 & | & 8 \\ 1 & 2 & | & 2 \end{bmatrix}$$

$$-1R_1 + R_2 \rightarrow \begin{bmatrix} 1 & -4 & | & 8 \\ 0 & 6 & | & -6 \end{bmatrix}$$

$$\tfrac{1}{6}R_2 \rightarrow \begin{bmatrix} 1 & -4 & | & 8 \\ 0 & 1 & | & -1 \end{bmatrix}$$

$$\begin{cases} x - 4y = 8 \\ y = -1 \end{cases}$$

$$\begin{array}{ll} x - 4y = 8 & \text{Solution: } (4, -1) \\ x - 4(-1) = 8 & \\ x + 4 = 8 & \\ x = 4 & \end{array}$$

34. $\begin{cases} x+y+z=2 \\ 2x+y+2z=1 \\ 3x+2y+z=1 \end{cases}$

$$\begin{bmatrix} 1 & 1 & 1 & | & 2 \\ 2 & 1 & 2 & | & 1 \\ 3 & 2 & 1 & | & 1 \end{bmatrix}$$

$$\begin{matrix} -2R_1+R_2 \rightarrow \\ -3R_1+R_3 \rightarrow \end{matrix} \begin{bmatrix} 1 & 1 & 1 & | & 2 \\ 0 & -1 & 0 & | & -3 \\ 0 & -1 & -2 & | & -5 \end{bmatrix}$$

$$-1R_2 \rightarrow \begin{bmatrix} 1 & 1 & 1 & | & 2 \\ 0 & 1 & 0 & | & 3 \\ 0 & -1 & -2 & | & -5 \end{bmatrix}$$

$$R_2+R_3 \rightarrow \begin{bmatrix} 1 & 1 & 1 & | & 2 \\ 0 & 1 & 0 & | & 3 \\ 0 & 0 & -2 & | & -2 \end{bmatrix}$$

$$-\tfrac{1}{2}R_3 \rightarrow \begin{bmatrix} 1 & 1 & 1 & | & 2 \\ 0 & 1 & 0 & | & 3 \\ 0 & 0 & 1 & | & 1 \end{bmatrix}$$

$$\begin{cases} x+y+z=2 \\ \quad\quad y=3 \\ \quad\quad z=1 \end{cases}$$

$x+y+z=2$ Solution: $(-2,3,1)$

$x+3+1=2$

$x+4=2$

$x=-2$

35. $\begin{cases} x+y>-5 \\ x-y\ge-1 \end{cases}$

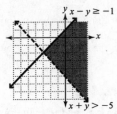

36. $\begin{cases} 2x+y>1 \\ 3x-y\le-1 \end{cases}$

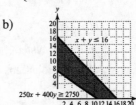

37. a) $\begin{cases} x+y\le16 \\ 250x+400y\ge2750 \\ x\ge0 \\ y\ge0 \end{cases}$

b)

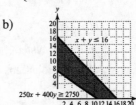

c) Answers may vary. One example is (8, 4).

Chapter 4 Practice Test

1. Is $(-1,3)$ a solution of $\begin{cases} 3x+2y=3 \\ 4x-y=-7 \end{cases}$?

Equation 1	Equation 2

$\quad\quad 3x+2y=3 \quad\quad\quad 4x-y=-7$

$3(-1)+2(3)\overset{?}{=}3 \quad\quad 4(-1)-3\overset{?}{=}-7$

$\quad\quad -3+6\overset{?}{=}3 \quad\quad\quad -4-3\overset{?}{=}-7$

$\quad\quad\quad 3=3 \quad\quad\quad\quad\quad -7=-7$

Yes, $(-1,3)$ is a solution for the system because it satisfies both equations.

2. Is $(2,2,3)$ a solution of $\begin{cases} 2x-3y+z=1 \\ -2x+y-3z=11 \\ 3x+y+3z=14 \end{cases}$?

Equation 1: $2x-3y+z=1$

$$2(2)-3(2)+(3)\overset{?}{=}1$$

$$4-6+3\overset{?}{=}1$$

$$1=1$$

Equation 2: $-2x+y-3z=11$

$$-2(2)+(2)-3(3)\overset{?}{=}11$$

$$-4+2-9\overset{?}{=}11$$

$$-11\ne11$$

Equation 3: $3x + y + 3z = 14$

$$3(2) + (2) + 3(3) \overset{?}{=} 14$$

$$6 + 2 + 9 \overset{?}{=} 14$$

$$17 \neq 14$$

No, $(2, 2, 3)$ is not a solution for the system because it does not satisfy all three equations.

3. $\begin{cases} x + 3y = 1 \\ -2x + y = 5 \end{cases}$ Graph each equation.

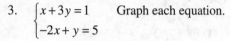
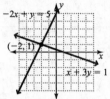

The lines intersect at a single point, which appears to be $(-2, 1)$.

4. $\begin{cases} 2x + y = 15 \\ y = 7 - x \end{cases}$

$\begin{aligned} 2x + y &= 15 \\ 2x + (7 - x) &= 15 \\ x + 7 &= 15 \\ x &= 8 \end{aligned}$ $\begin{aligned} y &= 7 - x \\ y &= 7 - 8 \\ y &= -1 \end{aligned}$

Solution: $(8, -1)$

5. $\begin{cases} x - 2y = 1 \\ 3x - 5y = 4 \end{cases}$

Solve the first equation for x. $x = 2y + 1$

$\begin{aligned} 3x - 5y &= 4 \\ 3(2y + 1) - 5y &= 4 \\ 6y + 3 - 5y &= 4 \\ y &= 1 \end{aligned}$ $\begin{aligned} x &= 2y + 1 \\ x &= 2 \cdot 1 + 1 \\ x &= 3 \end{aligned}$

Solution: $(3, 1)$

6. $\begin{cases} 3x - 2y = -8 \\ 2x + 3y = -14 \end{cases}$

$3x - 2y = -8$ Muliply by 3.

$\underline{2x + 3y = -14}$ Multiply by 2.

$9x - 6y = -24$

$\underline{4x + 6y = -28}$

$13x = -52$

$x = -4$

$3x - 2y = -8$

$3(-4) - 2y = -8$

$-12 - 2y = -8$

$-2y = 4$

$y = -2$

Solution: $(-4, -2)$

7. $\begin{cases} 4x + 6y = 2 \\ 6x + 9y = 3 \end{cases}$

$4x + 6y = 2$ Muliply by -3

$\underline{6x + 9y = 3}$ Multiply by 2

$-12x - 18y = -6$

$\underline{12x + 18y = 6}$

$0 = 0$

Notice that $0 = 0$ no longer has a variable and is true. The equations in the system are dependent; there are an infinite number of solutions that are all ordered pairs that solve $4x + 6y = 2$.

8. $\begin{cases} x + 2y + z = 2 & \text{Eqtn. 1} \\ x + 4y - z = 12 & \text{Eqtn. 2} \\ 3x - 3y - 2z = -11 & \text{Eqtn. 3} \end{cases}$

Multiply equation 1 by -3 and add to equation 3. This makes equation 4.

$-3x - 6y - 3z = -6$

$\underline{3x - 3y - 2z = -11}$

$-9y - 5z = -17$ Eqtn. 4

Multiply equation 1 by -1 and add to equation 2. This makes equation 5.

$-x - 2y - z = -2$

$\underline{x + 4y - z = 12}$

$2y - 2z = 10$ Eqtn. 5

Use equations 4 and 5 to make a system of equations in two variables. Solve.

$-9y-5z=-17$　　　Multiply by 2.

$\underline{2y-2z=10}$　　　Multipy by 9.

$-18y-10z=-34$

$\underline{18y-18z=90}$

$\quad -28z=56$

$\quad\quad z=-2$

$2y-2(-2)=10$　　Eqtn. 5

$\quad 2y+4=10$

$\quad\quad 2y=6$

$\quad\quad y=3$

$x+2\cdot3-2=2$　　Eqtn. 1

$\quad x+6-2=2$

$\quad\quad x+4=2$

$\quad\quad x=-2$

Solution: $(-2,3,-2)$

9. $\begin{cases} x+2y-3z=-9 & \text{Eqtn. 1} \\ 3x-y+2z=-8 & \text{Eqtn. 2} \\ 4x-3y+3z=-13 & \text{Eqtn. 3} \end{cases}$

Multiply equation 1 by -3 and add to equation 2. This makes equation 4.

$-3x-6y+9z=27$

$\underline{\quad 3x-y+2z=-8}$

$\quad -7y+11z=19$　　Eqtn. 4

Multiply equation 1 by -4 and add to equation 3. This makes equation 5.

$-4x-8y+12z=36$

$\underline{\quad 4x-3y+3z=-13}$

$\quad -11y+15z=23$　　Eqtn. 5

Use equations 4 and 5 to make a system of equations in two variables. Solve.

$-7y+11z=19$　　　Multiply by -11.

$\underline{-11y+15z=23}$　　Multipy by 7.

$77y-121z=-209$

$\underline{-77y+105z=161}$

$\quad -16z=-48$

$\quad\quad z=3$

$-7y+11\cdot3=19$　　Eqtn. 4

$\quad -7y+33=19$

$\quad\quad -7y=-14$

$\quad\quad y=2$

$x+2\cdot2-3\cdot3=-9$　　Eqtn. 1

$x+4-9=-9$

$x-5=-9$

$x=-4$

Solution: $(-4,2,3)$

10. $\begin{cases} x+2y=-6 \\ 3x+4y=-10 \end{cases}$

$\begin{bmatrix} 1 & 2 & | & -6 \\ 3 & 4 & | & -10 \end{bmatrix}$

$-3R_1+R_2 \rightarrow \begin{bmatrix} 1 & 2 & | & -6 \\ 0 & -2 & | & 8 \end{bmatrix}$

$-\dfrac{1}{2}R_2 \rightarrow \begin{bmatrix} 1 & 2 & | & -6 \\ 0 & 1 & | & -4 \end{bmatrix}$

$\begin{cases} x+2y=-6 \\ \quad\quad y=-4 \end{cases}$

$x+2(-4)=-6$

$\quad x-8=-6$

$\quad\quad x=2$

Solution: $(2,-4)$

11. $\begin{cases} x+y+z=6 & \text{Eqtn. 1} \\ 3x-2y+3z=-7 & \text{Eqtn. 2} \\ 4x-2y+z=-12 & \text{Eqtn. 3} \end{cases}$

$\begin{bmatrix} 1 & 1 & 1 & | & 6 \\ 3 & -2 & 3 & | & -7 \\ 4 & -2 & 1 & | & -12 \end{bmatrix}$

$\begin{matrix} -3R_1+R_2 \rightarrow \\ -4R_1+R_2 \rightarrow \end{matrix} \begin{bmatrix} 1 & 1 & 1 & | & 6 \\ 0 & -5 & 0 & | & -25 \\ 0 & -6 & -3 & | & -36 \end{bmatrix}$

$-\dfrac{1}{5}R_2 \rightarrow \begin{bmatrix} 1 & 1 & 1 & | & 6 \\ 0 & 1 & 0 & | & 5 \\ 0 & -6 & -3 & | & -36 \end{bmatrix}$

$6R_2+R_3 \rightarrow \begin{bmatrix} 1 & 1 & 1 & | & 6 \\ 0 & 1 & 0 & | & 5 \\ 0 & 0 & -3 & | & -6 \end{bmatrix}$

$-\dfrac{1}{3}R_3 \rightarrow \begin{bmatrix} 1 & 1 & 1 & | & 6 \\ 0 & 1 & 0 & | & 5 \\ 0 & 0 & 1 & | & 2 \end{bmatrix}$

$$\begin{cases} x+y+z=6 \\ \quad\quad y=5 \\ \quad\quad\quad z=2 \end{cases}$$

$$x+5+2=6$$
$$x=-1$$

Solution: $(-1,5,2)$

12. $\begin{cases} 3x+4y=14 \\ 2x-3y=-19 \end{cases}$

$3x+4y=14$ Multiply by 3.

$\underline{2x-3y=-19}$ Multiply by 4.

$9x+12y=42$

$\underline{8x-12y=-76}$

$17x \quad\quad =-34$

$x \quad\quad =-2$

$3x+4y=14$

$3(-2)+4y=14$

$-6+4y=14$

$4y=20$

$y=5$

Solution: $(-2,5)$

13. $\begin{cases} x+y+z=-1 & \text{Eqtn. 1} \\ 3x-2y+4z=0 & \text{Eqtn. 2} \\ 2x+5y-z=-11 & \text{Eqtn. 3} \end{cases}$

$$\begin{bmatrix} 1 & 1 & 1 & | & -1 \\ 3 & -2 & 4 & | & 0 \\ 2 & 5 & -1 & | & -11 \end{bmatrix}$$

$$\begin{matrix} \\ -3R_1+R_2 \to \\ -2R_1+R_2 \to \end{matrix} \begin{bmatrix} 1 & 1 & 1 & | & -1 \\ 0 & -5 & 1 & | & 3 \\ 0 & 3 & -3 & | & -9 \end{bmatrix}$$

$$-\tfrac{1}{5}R_2 \to \begin{bmatrix} 1 & 1 & 1 & | & -1 \\ 0 & 1 & -\tfrac{1}{5} & | & -\tfrac{3}{5} \\ 0 & 3 & -3 & | & -9 \end{bmatrix}$$

$$-3R_2+R_3 \to \begin{bmatrix} 1 & 1 & 1 & | & -1 \\ 0 & 1 & -\tfrac{1}{5} & | & -\tfrac{3}{5} \\ 0 & 0 & -\tfrac{12}{5} & | & -\tfrac{36}{5} \end{bmatrix}$$

$$-\tfrac{5}{12}R_3 \to \begin{bmatrix} 1 & 1 & 1 & | & -1 \\ 0 & 1 & -\tfrac{1}{5} & | & -\tfrac{3}{5} \\ 0 & 0 & 1 & | & 3 \end{bmatrix}$$

$$\begin{cases} x+y+z=-1 \\ y-\tfrac{1}{5}z=-\tfrac{3}{5} \\ \quad\quad z=3 \end{cases} \quad\quad y-\tfrac{1}{5}(3)=-\tfrac{3}{5}$$
$$\quad\quad\quad\quad\quad\quad\quad\quad\quad y=0$$

$$x+0+3=-1$$
$$x=-4 \quad\quad\quad\quad \text{Solution: } (-4,0,3)$$

14. $\begin{cases} 2x-3y<1 \\ x+2y\le-2 \end{cases}$

15. Let a be the number of accountants and let w be the number of waiters/waitresses. We are given that the combined number of accountants and waiters/waitresses who got headaches is 163. We are also given that 9 more accountants than waiters/waitresses got headaches. Translate to a system of equations.

$$\begin{cases} a+w=163 \\ 9+w=a \end{cases}$$

Substitute $9+w$ for a in the first equation.

$a+w=163 \quad\quad\quad\quad 9+w=a$

$9+w+w=163 \quad\quad\quad 9+77=a$

$2w=154 \quad\quad\quad\quad\quad 86=a$

$w=77$

Therefore, 77 waiters/waitresses and 86 accountants got headaches.

16. Let l be the number of people considering the internet to be a library and h be the number of people considering the internet to be a highway. We are given that a total of 240 people were polled, and that 3 times as many people responded "library" as opposed to "highway," Translate to a system of equations.

$$\begin{cases} l+h=240 \\ 3h=l \end{cases}$$

Substitute $3h$ for l in the first equation.

$l+h=240 \quad\quad\quad\quad 3h=l$

$3h+h=240 \quad\quad\quad 3(60)=l$

$4h=240 \quad\quad\quad\quad 180=l$

$h=60$

So, 180 people considered the internet to be a library.

17. Let x represent the rate of the boat in still-water and y represent the amount that the current increases or decreases the boat's rate. We are told that with the current, the boat took 3 hours to go 30 miles. We are also given that against the current, the boat took 3 hours to go 12 miles. Create a table.

Categories	Rate	Time	Distance
With	$x+y$	3	30
Against	$x-y$	3	12

Translate to a system of equations.
$$\begin{cases} 3(x+y) = 30 \\ 3(x-y) = 12 \end{cases}$$

Divide each equation by 3 to get the following system.
$$\begin{cases} x+y = 10 \\ x-y = 4 \end{cases}$$

Using elimination, solve the system.

$$\begin{array}{ll} x+y = 10 & x+y = 10 \\ \underline{x-y = 4} & 7+y = 10 \\ 2x = 14 & y = 3 \\ x = 7 & \end{array}$$

The speed of the boat in still water is 7 mph.

18. Let x represent the amount invested at 6% interest (fund 1) and let y represent the amount invested at 8% interest (fund 2). We are given that a total of $12,000 was invested in the two funds. We are also given that the interest earned after one year was $880. Create a table.

Categories	APR	Principal	Interest
fund 1	0.06	x	$0.06x$
fund 2	0.08	y	$0.08y$

Translate to a system of equations.
$$\begin{cases} x+y = 12,000 \\ 0.06x + 0.08y = 880 \end{cases}$$

Solve the first equation for x. $x = 12,000 - y$, and then substitute into the second equation.
$$0.06x + 0.08y = 880$$
$$0.06(12,000 - y) + 0.08y = 880$$
$$720 - 0.06y + 0.08y = 880$$
$$0.02y = 160$$
$$y = 8000$$

$$x = 12,000 - y$$
$$x = 12,000 - 8,000$$
$$x = 4000$$

So, $4000 was invested in fund 1 at 6% and $8000 was invested in fund 2 at 8%.

19. Let a represent the number of adult tickets, c the number of child tickets, and s the number of student tickets. Translate to a system of equations.
$$\begin{cases} c+s+a = 800 & \text{Eqtn. 1} \\ 3c+5s+8a = 4750 & \text{Eqtn. 2} \\ a = 50+2c & \text{Eqtn. 3} \end{cases}$$

Multiply equation 1 by -5 and add to equation 2. This makes equation 4.

$$\begin{array}{l} -5c - 5s - 5a = -4000 \\ \underline{3c+5s+8a = 4750} \\ -2c+3a = 750 \quad \text{Eqtn. 4} \end{array}$$

Use equations 3 and 4 to form a system of equations in two variables. Solve.

$$-1(-2c+a = 50)$$
$$\underline{-2c+3a = 750}$$

$$\begin{array}{l} 2c - a = -50 \\ \underline{-2c+3a = 750} \\ 2a = 700 \\ a = 350 \end{array}$$

$$350 = 50 + 2c \quad \text{Eqtn. 3}$$
$$300 = 2c$$
$$150 = c$$
$$150 + s + 350 = 800 \quad \text{Eqtn. 1}$$
$$s + 500 = 800$$
$$s = 300$$

Therefore, there were 150 child tickets, 300 student tickets, and 350 adult tickets sold.

20. a) Let l represent the length of the garden and let w represent the width. Translate to a system of equations.
$$\begin{cases} l > 0 \\ w > 0 \\ 2l + 2w \le 200 \\ l \ge w + 10 \end{cases}$$

b)

c) Answers may vary. One example is (60, 20).

Chapters 1–4 Cumulative Review

1. False; the slope of this line is -3.

2. False; $-(-4)^3 = -(-64) = 64$.

3. True

4. $8x+12$

5. $\dfrac{c}{d}$

6. perpendicular

7. $-\left(6-2^2\right)^2 +15 = -\left(6-4\right)^2 +15$
$\qquad\qquad = -(2)^2 +15$
$\qquad\qquad = -4+15$
$\qquad\qquad = 11$

8. $-\left|2-8\right|+3(2) = -\left|-6\right|+3(2)$
$\qquad\qquad\qquad = -6+6$
$\qquad\qquad\qquad = 0$

9. $\dfrac{3}{8}x+4y+12-\dfrac{1}{2}x-\dfrac{7}{10}y-7$
$\quad = \dfrac{3}{8}x+\dfrac{40}{10}y+12-\dfrac{4}{8}x-\dfrac{7}{10}y-7$
$\quad = -\dfrac{1}{8}x+\dfrac{33}{10}y+5$

10. $3.6-4(0.7p-15) = 2.2p-1.4$
$\quad 3.6-2.8p+60 = 2.2p-1.4$
$\qquad\quad 63.6 = 5p-1.4$
$\qquad\qquad 65 = 5p$
$\qquad\qquad 13 = p$

11. $2(x-3)+4x = 6(x-2)$
$\quad 2x-6+4x = 6x-12$
$\qquad 6x-6 = 6x-12$
$\qquad\quad -6 = -12$
Because the last statement is false, this equation is a contradiction. Thus, it has no solution.

12. $4z-32-5z < 6z+13+2z$
$\quad -1z-32 < 8z+13$
$\qquad -9z < 45$
$\qquad\quad z > -5$

13. $\dfrac{4}{3x-3} = \dfrac{7}{4x+1}$
$\quad 4(4x+1) = 7(3x-3)$
$\quad 16x+4 = 21x-21$
$\qquad -5x = -25$
$\qquad\quad x = 5$

14. $A = P+Prt$
$A-P = Prt$
$\dfrac{A-P}{Pr} = t$

15. $x+2 = 0$
$\quad x = -2$
So, $\dfrac{x}{x+2}$ is undefined when $x = -2$.

16. Find the area of the trapezoid first, and then subtract the area of the rectangle. Let $h = 130$ ft, $a = 140$ ft, and $b = 100$ ft.
$A = \dfrac{1}{2}h(a+b)$
$\quad = \dfrac{1}{2}(130)(140+100)$
$\quad = 15,600 \text{ ft}^2$
Now let $l = 50$ ft and $w = 40$ ft.
$A = lw$
$\quad = (40)(50)$
$\quad = 2000 \text{ ft}^2$
So the area to be sodded is
$15,600 - 2000 = 13,600 \text{ ft}^2$.

17. $y = 3x+1$

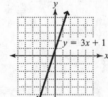

18. $4x+5y = 10$
$\quad 5y = -4x+10$
$\qquad y = -\dfrac{4}{5}x+2$

19. $x = -4$

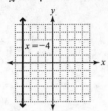

20. $2x - 7y < 14$

$$-7y < -2x + 14$$

$$y > \frac{2}{7}x - 2$$

21. Begin with point slope form and replace the values of m, x_1, and y_1.

$$y - (-1) = -\frac{1}{5}(x - (-2))$$

$$y + 1 = -\frac{1}{5}(x + 2)$$

$$y + 1 = -\frac{1}{5}x - \frac{2}{5}$$

$$y = -\frac{1}{5}x - \frac{7}{5}$$

22. Use the slope formula to find m. Let $(x_1, y_1) = (4, -7)$ and $(x_2, y_2) = (-2, -1)$.

$$m = \frac{y_2 - y_1}{x_2 - x_1}$$

$$= \frac{-1 - (-7)}{-2 - 4}$$

$$= \frac{6}{-6}$$

$$= -1$$

Now use point slope form and the point $(4, -7)$.

$$y - (-7) = -1(x - 4)$$

$$y + 7 = -x + 4$$

$$y = -x - 3$$

23. First find m.

$$3x + 4y = 4$$

$$4y = -3x + 4$$

$$y = -\frac{3}{4} + 1$$

So, $m = -\frac{3}{4}$. Now use point slope form and the point $(-4, 2)$.

$$y - 2 = -\frac{3}{4}(x - (-4))$$

$$y - 2 = -\frac{3}{4}(x + 4)$$

$$y - 2 = -\frac{3}{4}x - 3$$

$$y = -\frac{3}{4}x - 1$$

$$\frac{3}{4}x + y = -1$$

$$3x + 4y = -4$$

24. $\begin{cases} 4x - 3y = -2 \\ 6x - 7y = 7 \end{cases}$

$4x - 3y = -2$ Multiply by 7.

$6x - 7y = 7$ Multiply by 3.

$\underline{28x - 21y = -14}$

$\underline{18x - 21y = 21}$ Subtract

$10x \quad\quad = -35$

$x \quad\quad = -3.5$

$4(-3.5) - 3y = -2$

$-14 - 3y = -2$ Solution: $(-3.5, -4)$

$-3y = 12$

$y = -4$

25. $\begin{cases} x+y+z=5 & \text{Eqtn. 1} \\ 2x+y-2z=-5 & \text{Eqtn. 2} \\ x-2y+z=8 & \text{Eqtn. 3} \end{cases}$

Multiply equation 1 by -1 and add to equation 3. This makes equation 4.

$$\begin{aligned} -x-y-z &= -5 \\ \underline{x-2y+z} &= 8 \\ -3y &= 3 \\ y &= -1 \qquad \text{Eqtn. 4} \end{aligned}$$

Multiply equation 1 by -2 and add to equation 2. This makes equation 5.

$$\begin{aligned} -2x-2y-2z &= -10 \\ \underline{2x+y-2z} &= -5 \\ -y-4z &= -15 \quad \text{Eqtn. 5} \end{aligned}$$

Substitute equation 4 into equation 5 and solve for z.

$$\begin{aligned} -(-1)-4z &= -15 \\ 1-4z &= -15 \\ -4z &= -16 \\ z &= 4 \end{aligned}$$

Substitute the values for y and z into equation 1 to solve for x.

$$\begin{aligned} x-1+4 &= 5 \\ x+3 &= 5 \\ x &= 2 \end{aligned}$$

Solution: $(2,-1,4)$

26. $\begin{cases} 4x-3y<12 \\ 3x+y>6 \end{cases}$

$$\begin{array}{ll} 4x-3y<12 & 3x+y>6 \\ -3y<-4x+12 & y>-3x+6 \\ y>\dfrac{4}{3}x-4 & \end{array}$$

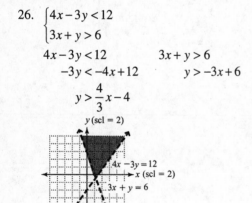

27. Let l be the length and w be the width.

$$\begin{cases} 2l+2w=420 \\ l=w+50 \end{cases}$$

The second equation is solved for l, so substitute the value of l into the first equation.

$$\begin{aligned} 2(w+50)+2w &= 420 \\ 2w+100+2w &= 420 \\ 4w+100 &= 420 \\ w &= 80 \end{aligned}$$

Now, solve for l.

$$\begin{aligned} l &= 80+50 \\ l &= 130 \end{aligned}$$

The length is 130 ft and the width is 80 ft.

28. Let g be the greater angle and s be the smaller angle.

$$\begin{cases} g+s=180 \\ g=2s-15 \end{cases}$$

The second equation is solved for g, so substitute the value of g into the first equation.

$$\begin{aligned} 2s-15+s &= 180 \\ 3s-15 &= 180 \\ 3s &= 195 \\ s &= 65 \end{aligned}$$

Now, solve for g.

$$\begin{aligned} g &= 2(65)-15 \\ &= 130-15 \\ &= 115 \end{aligned}$$

The measure of the greater angle is $115°$ and the measure of the smaller angle is $65°$.

29. Make a table.

Category	APR	Principal	Interest
Account 1	0.03	x	$0.03x$
Account 2	0.025	y	$0.025y$

We know the total principal is $50,000, and the total combined interest is $1425. Translate to a system of equations and solve.

$$\begin{cases} x+y=50,000 \\ 0.03x+0.025y=1425 \end{cases}$$

$$\begin{array}{l} x+y=50,000 \qquad \text{Multiply by } -0.03 \\ \underline{0.03x+0.025y=1425} \\ -0.03x-0.03y=-1500 \\ \underline{0.03x+0.025y=1425} \\ \qquad -0.005y=-75 \\ \qquad y=15,000 \end{array}$$

Substitute this value into the first equation.

$$\begin{aligned} x+15,000 &= 50,000 \\ x &= 35,000 \end{aligned}$$

So, she invested $35,000 at 3% and $15,000 at 2.5%.

30. Let x be the number of gallons of inside paint, and y be the number of gallons of outside paint.

$$\begin{cases} x + y = 24 \\ 22x + 28y = 588 \end{cases}$$

Solve the first equation for y.

$$x + y = 24$$
$$y = 24 - x$$

Now substitute this value into the second equation.

$$22x + 28(24 - x) = 588$$
$$22x + 672 - 28x = 588$$
$$-6x = -84$$
$$x = 14$$

Now, solve for y.

$$14 + y = 24$$
$$y = 10$$

So, the contractor bought 14 gallons of inside paint and 10 gallons of outside paint.

Chapter 5

Polynomials

Exercise Set 5.1

1. $4^0 = 1$

3. $-2^3 = -[2 \cdot 2 \cdot 2] = -8$

5. $(-5)^3 = (-5)(-5)(-5) = -125$

7. $-4^2 = -[4 \cdot 4] = -16$

9. $\left(\dfrac{3}{4}\right)^4 = \dfrac{3}{4} \cdot \dfrac{3}{4} \cdot \dfrac{3}{4} \cdot \dfrac{3}{4} = \dfrac{81}{256}$

11. $(-0.2)^3 = (-0.2)(-0.2)(-0.2) = -0.008$

13. $(-2.3)^2 = (-2.3)(-2.3) = 5.29$

15. $a^{-5} = \dfrac{1}{a^5}$

17. $2^{-3} = \dfrac{1}{2^3} = \dfrac{1}{8}$

19. $(-4)^{-3} = \dfrac{1}{(-4)^3} = -\dfrac{1}{64}$

21. $-2^{-4} = -\dfrac{1}{2^4} = -\dfrac{1}{16}$

23. $(-0.2)^{-3} = \dfrac{1}{(-0.2)^3} = \dfrac{1}{-0.008} = -125$

25. $5m^{-2} = \dfrac{5}{m^2}$

27. $-6x^{-7} = -\dfrac{6}{x^7}$

29. $\left(\dfrac{m}{n}\right)^{-5} = \left(\dfrac{n}{m}\right)^5 = \dfrac{n^5}{m^5}$

31. $\left(\dfrac{5}{7}\right)^{-2} = \left(\dfrac{7}{5}\right)^2 = \dfrac{49}{25}$

33. $\dfrac{1}{b^{-6}} = b^6$

35. $\dfrac{1}{4^{-3}} = 4^3 = 64$

37. $(-24)^{-5} = \dfrac{1}{(-24)^5} = -\dfrac{1}{7,962,624}$

39. $\left(-\dfrac{5}{8}\right)^{-3} = \left(-\dfrac{8}{5}\right)^3 = -\dfrac{8^3}{5^3} = -\dfrac{512}{125}$

41. $-0.05^{-4} = -\dfrac{1}{0.05^4} = -\dfrac{1}{0.00000625} = -160,000$

43. $(3.7)^{-6} \approx 0.00039$

45. Mistake: The expression was interpreted to mean $(-3)^4$, but it means the additive inverse of 3^4.
    Correct: -81

47. Mistake: The negative sign in the exponent was passed to the product instead of the original expression being inverted.
    Correct: $\dfrac{1}{25}$

49. $1.65 \times 10^{10} = 16,500,000,000$

51. $2.998 \times 10^8 = 299,800,000$

53. $6.378 \times 10^6 = 6,378,000$

55. $9.292 \times 10^7 = 92,920,000$

57. $2.5 \times 10^{-6} = 0.0000025$

59. 1.67×10^{-27}
    $= 0.00000000000000000000000000167$

61. $2.95 \times 10^{-9} = 0.00000000295$

63. 6.645×10^{-27}
    $= 0.000000000000000000000000006645$

65. $25,000,000,000,000 = 2.5 \times 10^{13}$

67. $5,300,000,000 = 5.3 \times 10^9$

69. $23,420,000 = 2.342 \times 10^7$

71. $1,300,000,000 = 1.3 \times 10^9$

73. $0.000000586 = 5.86 \times 10^{-7}$

75. $0.00000000000000000000000000000091094 = 9.1094 \times 10^{-31}$

77. $0.0000625 = 6.25 \times 10^{-5}$

79. $0.00000005 = 5 \times 10^{-8}$

81. $8.95 \times 10^5, 7.2 \times 10^6, 7.5 \times 10^6, 9 \times 10^7, 1.3 \times 10^8$

83. $\dfrac{2.95 \times 10^{-9}}{7.6 \times 10^{-11}} \approx 38.8$

The wavelength of ultraviolet light is about 38.8 times the wavelength of X-rays.

Review Exercises

1. $-13 + 2 = -11$

2. $-4(y - 3) = -4 \cdot y - 4 \cdot (-3)$
$= -4y + 12$

3. $\dfrac{15}{100} = \dfrac{40}{x}$
$15x = 4000$
$\dfrac{15x}{15} = \dfrac{4000}{15}$
$x = 266\dfrac{2}{3}$

4. $m = \dfrac{y_2 - y_1}{x_2 - x_1}$
$= \dfrac{-2 - (-1)}{-3 - 2}$
$= \dfrac{-1}{-5}$
$= \dfrac{1}{5}$

5. $2x - 3y = 6$
$-3y = -2x + 6$
$\dfrac{-3y}{-3} = \dfrac{-2x}{-3} + \dfrac{6}{-3}$
$y = \dfrac{2}{3}x - 2$
slope $= \dfrac{2}{3}$, y-intercept: $(0, -2)$

6. $f(x) = x^3 - 5x^2 + 7$
$f(-3) = (-3)^3 - 5(-3)^2 + 7$
$= -27 - 5(9) + 7$
$= -27 - 45 + 7$
$= -65$

Exercise Set 5.2

1. $-3x^2$ is a monomial because it is the product of a constant and a variable having a whole-number exponent of 2.

3. $\dfrac{1}{5}$ is a monomial because it is a constant.

5. $\dfrac{6m}{4y^3}$ is not a monomial because if it is rewritten as a product, one of the variables would have a negative exponent.

7. $4x^2 - 3x + 7$ is not a monomial because it is not a product of a constant and variables.

9. y is a monomial because it is a variable with a whole-number exponent.

11. $3m^4n^2$ is a monomial because it is the product of a constant and two variables with whole-number exponents.

13. $-5m^3$
coefficient: -5
degree: 3

15. $-xy^4 = -1x^1y^4$
coefficient: -1
degree: 5

17. $-9 = -9x^0$
coefficient: -9
degree: 0

19. $4.2n^3p = 4.2n^3p^1$
coefficient: 4.2
degree: 4

21. $16abc = 16a^1b^1c^1$
coefficient: 16
degree: 3

23. $y = y^1$
coefficient: 1
degree: 1

25. $6x + 8y + 3z$ is a trinomial because it has three terms; the degree is 1 because it is the greatest degree of all the terms.

27. $-7m^2n$ is a monomial because it has one term; the degree is 3.

29. $5.2x^3 - 3x^2 + 4.1x - 11$ has no special polynomial name because it has more than three terms; the degree is 3 because it is the greatest degree of all the terms.

31. $5u^3 - 16u$ is a binomial because it has two terms; the degree is 3 because it is the greatest degree of all the terms.

33. -21 is a monomial because it is a constant; the degree is 0.

35. $\dfrac{25}{x} - x^2$ is not a polynomial because $\dfrac{25}{x}$ is not a monomial.

37. $19 - 7y^4 + 3y - 2y^3 - 7$ has degree 4 because it is the greatest degree of all the terms.

39. $16z^4 + 7z^2 - z^5 - 4z^3 + z$ has degree 5 because it is the greatest degree of all the terms.

41. $2u^2 + 5u^3 - u^7 + 13u^4 - 3u$ has degree 7 because it is the greatest degree of all the terms.

43. $3ab^2 - 6a^2b^2 + 4a - 7b^3$ has degree 4 because it is the greatest degree of all the terms.

45. $-8x^4y + 5x^3y^3 - 5x^2y^6 + 3xy^7 + 2x^6$ has degree 8 because it is the greatest degree of all the terms.

47. $-3xy^2 = -3(-5)(2)^2$
$ = -3(-5)(4)$
$ = 60$

49. $x^2 - 6x + 1 = 3^2 - 6(3) + 1$
$ = 9 - 6(3) + 1$
$ = 9 - 18 + 1$
$ = -8$

51. $a^3 + 0.5ab + 2.4b$
$= (-1)^3 + 0.5(-1)(-2) + 2.4(-2)$
$= -1 + 0.5(-1)(-2) + 2.4(-2)$
$= -1 + 1 - 4.8$
$= -4.8$

53. $\dfrac{1}{2}x^2 + \dfrac{2}{3}x + 3 = \dfrac{1}{2}(6)^2 + \dfrac{2}{3}(6) + 3$
$\phantom{\dfrac{1}{2}x^2 + \dfrac{2}{3}x + 3} = \dfrac{1}{2}(36) + \dfrac{12}{3} + 3$
$\phantom{\dfrac{1}{2}x^2 + \dfrac{2}{3}x + 3} = 18 + 4 + 3$
$\phantom{\dfrac{1}{2}x^2 + \dfrac{2}{3}x + 3} = 25$

55. a) $-16t^2 + h_0 = -16(0.5)^2 + 50$
$ = -16(0.25) + 50$
$ = -4 + 50$
$ = 46$

After 0.5 second, the height of the object is 46 feet.

b) $-16t^2 + h_0 = -16(1.2)^2 + 50$
$ = -16(1.44) + 50$
$ = -23.04 + 50$
$ = 26.96$

After 1.2 seconds, the height of the object is 26.96 feet.

57. a) $5p^2 - 6p + 3 = 5(3)^2 - 6(3) + 3$
$ = 5(9) - 6(3) + 3$
$ = 45 - 18 + 3$
$ = 30$

At a price of \$3, 30 units are sold for every 1000 people.

b) $5p^2 - 6p + 3 = 5(3.5)^2 - 6(3.5) + 3$
$ = 5(12.25) - 6(3.5) + 3$
$ = 61.25 - 21 + 3$
$ = 43.25$

At a price of \$3.50, about 43 units are sold for every 1000 people.

59. a) $\dfrac{4}{3}\pi r^3 + \pi r^2 h = \dfrac{4}{3}\pi(4)^3 + \pi(4)^2(20)$
$\phantom{\dfrac{4}{3}\pi r^3 + \pi r^2 h} = \dfrac{4}{3}\pi(64) + \pi(16)(20)$
$\phantom{\dfrac{4}{3}\pi r^3 + \pi r^2 h} \approx 268.1 + 1005.3$
$\phantom{\dfrac{4}{3}\pi r^3 + \pi r^2 h} \approx 1273.4$

The volume of the tank is about 1273.4 cubic feet.

b) $\dfrac{4}{3}\pi r^3 + \pi r^2 h = \dfrac{4}{3}\pi(3)^3 + \pi(3)^2(15)$
$\phantom{\dfrac{4}{3}\pi r^3 + \pi r^2 h} = \dfrac{4}{3}\pi(27) + \pi(9)(15)$
$\phantom{\dfrac{4}{3}\pi r^3 + \pi r^2 h} \approx 113.1 + 424.1$
$\phantom{\dfrac{4}{3}\pi r^3 + \pi r^2 h} \approx 537.2$

The volume of the tank is about 537.2 cubic feet.

61. a) $-0.2x^2 + 4.25x + 26$

    $= -0.2(0)^2 + 4.25(0) + 26$

    $= 26$

    In 1986, 26 million international visitors came to the United States.

  b) $-0.2x^2 + 4.25x + 26$

    $= -0.2(7)^2 + 4.25(7) + 26$

    $= -9.8 + 29.75 + 26$

    $= 45.95$

    In 1993, 45.95 million international visitors came to the United States.

  c) $-0.2x^2 + 4.25x + 26$

    $= -0.2(17)^2 + 4.25(17) + 26$

    $= -57.8 + 72.25 + 26$

    $= 40.45$

    In 2003, 40.45 million international visitors came to the United States.

  d) $-0.2x^2 + 4.25x + 26$

    $= -0.2(26)^2 + 4.25(26) + 26$

    $= -135.2 + 110.5 + 26$

    $= 1.3$

    The model predicts that in 2012, 1.3 million international visitors will come to the United States.

    Answers will vary, but the prediction is most likely not reasonable because of the increase in world population and unforeseen social factors that might affect travel. It is not likely that the number will decrease this much.

63. $5x^7 - 3x^5 + 7x^4 - 8x + 14$

65. $-8r^6 + 18r^5 + 7r^3 - 3r^2 + 4r$

67. $-12w^4 - w^3 + 11w^2 + 5w + 20$

69. $3x + 4 + 2x - 7 = 3x + 2x + 4 - 7$

                           $= 5x - 3$

71. $7 - 6a + 2a - 2 = -6a + 2a + 7 - 2$

                               $= -4a + 5$

73. $7x^2 + 2x + 4x - 9x^2 = 7x^2 - 9x^2 + 2x + 4x$

                                  $= -2x^2 + 6x$

75. $\dfrac{1}{2}x^2 + \dfrac{1}{3}x + \dfrac{4}{3}x + \dfrac{3}{2}x^2 = \dfrac{1}{2}x^2 + \dfrac{3}{2}x^2 + \dfrac{1}{3}x + \dfrac{4}{3}x$

                        $= \dfrac{4}{2}x^2 + \dfrac{5}{3}x$

                        $= 2x^2 + \dfrac{5}{3}x$

77. $\dfrac{5}{3}a^2 + \dfrac{2}{5}a - \dfrac{1}{2}a^2 - \dfrac{5}{3}a$

    $= \dfrac{5}{3}a^2 - \dfrac{1}{2}a^2 + \dfrac{2}{5}a - \dfrac{5}{3}a$

    $= \dfrac{10}{6}a^2 - \dfrac{3}{6}a^2 + \dfrac{6}{15}a - \dfrac{25}{15}a$

    $= \dfrac{7}{6}a^2 - \dfrac{19}{15}a$

79. $2x^2 + 5x - 7x^2 + 8 - 5x + 6$

    $= 2x^2 - 7x^2 + 5x - 5x + 8 + 6$

    $= -5x^2 + 14$

81. $15 - 4y + 8y^2 - 4y - 3y^2 + 7 - y$

    $= 8y^2 - 3y^2 - 4y - 4y - y + 15 + 7$

    $= 5y^2 - 9y + 22$

83. $9k - 5k^2 + 6k^3 + 2k^2 + k^4 - 3k - 3k^4$

    $= k^4 - 3k^4 + 6k^3 - 5k^2 + 2k^2 + 9k - 3k$

    $= -2k^4 + 6k^3 - 3k^2 + 6k$

85. $7a^2 - 5a^4 + 6a^3 - 5a^4 + 12 - 6a^3 + a + 4$

    $= -5a^4 - 5a^4 + 6a^3 - 6a^3 + 7a^2 + a + 12 + 4$

    $= -10a^4 + 7a^2 + a + 16$

87. $12v^3 + 19v^2 - 20 - v^5 - 5v^3 + 16v^2 - 20v^5 + v^3$

    $= -v^5 - 20v^5 + 12v^3 - 5v^3 + v^3$

       $+ 19v^2 + 16v^2 - 20$

    $= -21v^5 + 8v^3 + 35v^2 - 20$

89. $6a - 3b - 2a + b = 6a - 2a - 3b + b$

                      $= 4a - 2b$

91. $3y^2 + 5y - 2y - 7y^2 = 3y^2 - 7y^2 + 5y - 2y$

                           $= -4y^2 + 3y$

93. $2x^2 + 3y - 4x^2 + 6y - 5x^2 - 2x^2$

    $= 2x^2 - 4x^2 - 5x^2 - 2x^2 + 3y + 6y$

    $= -9x^2 + 9y$

95. $\dfrac{1}{3}a^2 + \dfrac{3}{5}a - \dfrac{3}{2}a + \dfrac{5}{4}a^2 + \dfrac{1}{3}a$

$= \dfrac{1}{3}a^2 + \dfrac{5}{4}a^2 + \dfrac{3}{5}a - \dfrac{3}{2}a + \dfrac{1}{3}a$

$= \dfrac{4}{12}a^2 + \dfrac{15}{12}a^2 + \dfrac{18}{30}a - \dfrac{45}{30}a + \dfrac{10}{30}a$

$= \dfrac{19}{12}a^2 - \dfrac{17}{30}a$

97. $y^6 + 2yz^4 - 10yz + 3yz^4 - 3z^5 - 7z^3 + 4yz - 10$

$= y^6 + 2yz^4 + 3yz^4 - 10yz + 4yz - 3z^5 - 7z^3 - 10$

$= y^6 + 5yz^4 - 6yz - 3z^5 - 7z^3 - 10$

99. $-3w^2z - 9 + 6wz^2 + 3w^2 - w^2z + 8 + 9w^2 - 7wz^2$

$= -3w^2z - w^2z + 6wz^2 - 7wz^2 + 3w^2$

$\quad + 9w^2 - 9 + 8$

$= -4w^2z - wz^2 + 12w^2 - 1$

Review Exercises

1. $-\dfrac{3}{4} + \dfrac{1}{6} = -\dfrac{9}{12} + \dfrac{2}{12} = -\dfrac{7}{12}$

2. $-14.6 - (-10.3) = -14.6 + 10.3 = -4.3$

3. $-7^0 = -1$

4. $200,000,000,000 = 2 \times 10^{11}$

5. $\dfrac{1}{4}x = 3$

$4 \cdot \dfrac{1}{4}x = 4 \cdot 3$

$x = 12$

6. Let x represent the number of liters of the 40% solution. Using the relationship Concentration × Whole solution volume = volume of the particular chemical, complete a table.

Solutions	Concen- tration	Vol of sol.	Vol of chemical
40%	0.40	x	$0.40x$
20%	0.20	200	$0.20(200)$
35%	0.35	$x+200$	$0.35(x+200)$

Translate to an equation.

$$0.40x + 0.20(200) = 0.35(x+200)$$
$$100\big(0.40x + 0.20(200)\big) = 100\big(0.35(x+200)\big)$$
$$40x + 20(200) = 35(x+200)$$
$$40x + 4000 = 35x + 7000$$
$$5x = 3000$$
$$x = 600$$

600 L of 40% solution should be added.

Exercise Set 5.3

1. $(3x+2) + (5x-1) = 3x + 5x + 2 - 1$

$\qquad\qquad\qquad = 8x + 1$

3. $(8y+7) + (2y+5) = 8y + 2y + 7 + 5$

$\qquad\qquad\qquad\qquad = 10y + 12$

5. $(2x+3y) + (5x-3y) = 2x + 5x + 3y - 3y$

$\qquad\qquad\qquad\qquad = 7x$

7. $(2x+5) + (3x^2 - 4x + 7) = 3x^2 + 2x - 4x + 5 + 7$

$\qquad\qquad\qquad\qquad\qquad = 3x^2 - 2x + 12$

9. $(z^2 - 3z + 7) + (5z^2 - 8z - 9)$

$= z^2 + 5z^2 - 3z - 8z + 7 - 9$

$= 6z^2 - 11z - 2$

11. $(3r^2 - 2r + 10) + (2r^2 - 5r - 11)$

$= 3r^2 + 2r^2 - 2r - 5r + 10 - 11$

$= 5r^2 - 7r - 1$

13. $(4y^2 - 8y + 1) + (5y^2 + 8y + 2)$

$= 4y^2 + 5y^2 - 8y + 8y + 1 + 2$

$= 9y^2 + 3$

15. $\left(x^2 - \dfrac{2}{3}x + 3\right) + \left(2x^2 + \dfrac{3}{4}x - 2\right)$

$= x^2 + 2x^2 - \dfrac{2}{3}x + \dfrac{3}{4}x + 3 - 2$

$= x^2 + 2x^2 - \dfrac{8}{12}x + \dfrac{9}{12}x + 3 - 2$

$= 3x^2 + \dfrac{1}{12}x + 1$

17. $\left(4r^2 + 2.7r - 3.6\right) + \left(-2r^2 - 1.3r - 4.2\right)$

$= 4r^2 - 2r^2 + 2.7r - 1.3r - 3.6 - 4.2$

$= 2r^2 + 1.4r - 7.8$

19. $\left(9a^3 + 5a^2 - 3a - 1\right) + \left(-4a^3 - 2a^2 + 6a - 2\right)$

$= 9a^3 - 4a^3 + 5a^2 - 2a^2 - 3a + 6a - 1 - 2$

$= 5a^3 + 3a^2 + 3a - 3$

21. $\left(7p^3 - 9p^2 + 5p - 1\right) + \left(-4p^3 + 8p^2 + 2p + 10\right)$

$= 7p^3 - 4p^3 - 9p^2 + 8p^2 + 5p + 2p - 1 + 10$

$= 3p^3 - p^2 + 7p + 9$

23. $\left(-5w^4 - 3w^3 - 8w^2 + w - 14\right)$

$\qquad + \left(-3w^4 + 6w^3 + w^2 + 12w + 5\right)$

$= -5w^4 - 3w^4 - 3w^3 + 6w^3 - 8w^2$

$\qquad + w^2 + w + 12w - 14 + 5$

$= -8w^4 + 3w^3 - 7w^2 + 13w - 9$

25. $\left(a^3b^2 - 6ab^2 - 6a^2b + 5ab + 5b^2 - 6\right)$

$\qquad + \left(-4a^3b^2 + 4ab^2 - 2a^2b - ab - 2b^2 - 2\right)$

$= a^3b^2 - 4a^3b^2 - 6ab^2 + 4ab^2 - 6a^2b - 2a^2b$

$\qquad + 5ab - ab + 5b^2 - 2b^2 - 6 - 2$

$= -3a^3b^2 - 2ab^2 - 8a^2b + 4ab + 3b^2 - 8$

27. $\left(-2u^4 + 6uv^3 - 8u^2v^3 + 7u^3v - u + v^2 + 9\right)$

$\qquad + \left(-5u^4 - 6uv^3 + 9u^2v^3 + 8u - 14v^2 - 12\right)$

$= -2u^4 - 5u^4 + 6uv^3 - 6uv^3 - 8u^2v^3 + 9u^2v^3$

$\qquad + 7u^3v - u + 8u + v^2 - 14v^2 + 9 - 12$

$= -7u^4 + u^2v^3 + 7u^3v + 7u - 13v^2 - 3$

29. $\left(-8mnp - 6m^2n^2 - 13mn^2p + 14m^2n - 12n + 6\right)$

$\qquad + \left(-4nmp - 10m^2n^2 + mn^2p - 19n + 3\right)$

$= -8mnp - 4nmp - 6m^2n^2 - 10m^2n^2 - 13mn^2p$

$\qquad + mn^2p + 14m^2n - 12n - 19n + 6 + 3$

$= -12mnp - 16m^2n^2 - 12mn^2p + 14m^2n - 31n + 9$

31. $\left(4a^4 - \dfrac{2}{3}a^3b + \dfrac{3}{5}ab^3 - 3b^4\right)$

$\qquad + \left(-2a^4 + \dfrac{1}{4}a^3b + \dfrac{1}{3}ab^3 + 2b^4\right)$

$= 4a^4 - 2a^4 - \dfrac{2}{3}a^3b + \dfrac{1}{4}a^3b$

$\qquad + \dfrac{3}{5}ab^3 + \dfrac{1}{3}ab^3 - 3b^4 + 2b^4$

$= 4a^4 - 2a^4 - \dfrac{8}{12}a^3b + \dfrac{3}{12}a^3b$

$\qquad + \dfrac{9}{15}ab^3 + \dfrac{5}{15}ab^3 - 3b^4 + 2b^4$

$= 2a^4 - \dfrac{5}{12}a^3b + \dfrac{14}{15}ab^3 - b^4$

33. $\left(3.6a^2bc + 2.3ab^2c^2 - 5.7a^2b^2c^2\right)$

$\qquad + \left(-1.8a^2bc - 4.1ab^2c^2 + 3.3a^2b^2c^2\right)$

$= 3.6a^2bc - 1.8a^2bc + 2.3ab^2c^2$

$\qquad - 4.1ab^2c^2 - 5.7a^2b^2c^2 + 3.3a^2b^2c^2$

$= 1.8a^2bc - 1.8ab^2c^2 - 2.4a^2b^2c^2$

35. $(x + 7) + (x - 5) + (3x - 4) = 5x - 2$

37. $(6a + 2) + (a - 8) + (6a + 2) + (a - 8) = 14a - 12$

39. $(6x + 3) - (2x + 1) = (6x + 3) + (-2x - 1)$

$\qquad\qquad = 4x + 2$

41. $\left(18a^2 + 3\right) - \left(5a^2 + 4\right) = \left(18a^2 + 3\right) + \left(-5a^2 - 4\right)$

$\qquad\qquad\qquad = 13a^2 - 1$

43. $\left(2x^2 - 3x + 4\right) - (6x + 2)$

$= \left(2x^2 - 3x + 4\right) + (-6x - 2)$

$= 2x^2 - 9x + 2$

45. $\left(6z^2 - 3z + 1\right) - \left(4z^2 + 4z - 8\right)$

$= \left(6z^2 - 3z + 1\right) + \left(-4z^2 - 4z + 8\right)$

$= 2z^2 - 7z + 9$

47. $\left(8a^2 + 11a - 12\right) - \left(-4a^2 + a + 3\right)$

$= \left(8a^2 + 11a - 12\right) + \left(4a^2 - a - 3\right)$

$= 12a^2 + 10a - 15$

49. $\left(6u^3+3u^2-8u+5\right)-\left(7u^3+5u^2-2u-3\right)$

$=\left(6u^3+3u^2-8u+5\right)+\left(-7u^3-5u^2+2u+3\right)$

$=-u^3-2u^2-6u+8$

51. $\left(-w^5+3w^4+8w^2-w-1\right)$

$\quad-\left(10w^5+3w^2-w-3\right)$

$=\left(-w^5+3w^4+8w^2-w-1\right)$

$\quad+\left(-10w^5-3w^2+w+3\right)$

$=-11w^5+3w^4+5w^2+2$

53. $\left(-5p^4+8p^3-9p^2+3p+1\right)$

$\quad-\left(6p^4+5p^3+4p^2-2p+7\right)$

$=\left(-5p^4+8p^3-9p^2+3p+1\right)$

$\quad+\left(-6p^4-5p^3-4p^2+2p-7\right)$

$=-11p^4+3p^3-13p^2+5p-6$

55. $\left(16v^4+21v^2+4v-8\right)$

$\quad-\left(5v^4-3v^3-2v^2+9v+1\right)$

$=\left(16v^4+21v^2+4v-8\right)$

$\quad+\left(-5v^4+3v^3+2v^2-9v-1\right)$

$=11v^4+3v^3+23v^2-5v-9$

57. $\left(x^2+xy+y^2\right)-\left(x^2-xy+y^2\right)$

$=\left(x^2+xy+y^2\right)+\left(-x^2+xy-y^2\right)$

$=2xy$

59. $\left(4xy+5xz-6yz\right)-\left(10xy-xz-8yz\right)$

$=\left(4xy+5xz-6yz\right)+\left(-10xy+xz+8yz\right)$

$=-6xy+6xz+2yz$

61. $\left(18p^3q^2-p^3q^3+6pq^2-4pq+16q^2-4\right)$

$\quad-\left(-p^3q^2-p^3q^3+4pq^2-3pq-6\right)$

$=\left(18p^3q^2-p^3q^3+6pq^2-4pq+16q^2-4\right)$

$\quad+\left(p^3q^2+p^3q^3-4pq^2+3pq+6\right)$

$=19p^3q^2+2pq^2-pq+16q^2+2$

63. $\left(5.6a^2b-2.3ab^2-4.3a^2b^2\right)$

$\quad-\left(3.7a^2b-3.4ab^2+2.2a^2b^2\right)$

$=\left(5.6a^2b-2.3ab^2-4.3a^2b^2\right)$

$\quad+\left(-3.7a^2b+3.4ab^2-2.2a^2b^2\right)$

$=1.9a^2b+1.1ab^2-6.5a^2b^2$

65. $\left(54.95x+92.20y+25.35z\right)$

$\quad-\left(22.85x+56.75y+19.34z\right)$

$=\left(54.95x+92.20y+25.35z\right)$

$\quad+\left(-22.85x-56.75y-19.34z\right)$

$=32.1x+35.45y+6.01z$

67. $\left(6.50a+3.38b+25.00c\right)$

$\quad-\left(2.28a+1.75b+12.87c\right)$

$=\left(6.50a+3.38b+25.00c\right)$

$\quad+\left(-2.28a-1.75b-12.87c\right)$

$=4.22a+1.63b+12.13c$

69. Mistake: Only the sign of the first term in the subtrahend was changed instead of all three terms.

Correct: $\left(3x^2-5x+10\right)+\left(-7x^2+3x-5\right)$

$\qquad\qquad=-4x^2-2x+5$

Review Exercises

1. $(2.3)(-7.5)=-17.25$

2. $(-5)(-2)(1.5)=15$

3. $10^{-4}=0.0001$

4. $10^3\cdot10^4=10^7$

5. $5.89\times10^7=58,900,000$

6. $(20\text{ m})\times(15\text{ m})=300\text{ m}^2$

Exercise Set 5.4

1. $a^3\cdot a^4=a^{3+4}=a^7$

3. $2^3\cdot2^{10}=2^{3+10}=2^{13}$

5. $a^3\cdot a^2b=a^{3+2}b=a^5b$

7. $2x\cdot3x^2=2\cdot3\cdot x^{1+2}=6x^3$

9. $(-2mn)(4m^2n) = -2 \cdot 4 \cdot m^{1+2} \cdot n^{1+1} = -8m^3n^2$

11. $\left(\frac{5}{8}st\right)\left(-\frac{2}{7}s^2t^7\right) = \frac{5}{\overset{\cancel{8}}{4}}\left(-\frac{\overset{1}{\cancel{2}}}{7}\right)s^{1+2}t^{1+7}$

$= -\frac{5}{28}s^3t^8$

13. $(2.3ab^2c)(1.2a^2b^3c^5) = (2.3)(1.2)a^{1+2}b^{2+3}c^{1+5}$

$= 2.76a^3b^5c^6$

15. $(r^2)(2r^2)(-3r) = 2(-3)r^{2+2+1} = -6r^5$

17. $3xz^2(-4x^3z^2)(z^5) = 3(-4)x^{1+3}z^{2+2+5}$

$= -12x^4z^9$

19. $(5xyz)(9x^2y^4)(2x^3z^2) = 5 \cdot 9 \cdot 2 \cdot x^{1+2+3}y^{1+4}z^{1+2}$

$= 90x^6y^5z^3$

21. $(5q^2r^4s^9)(-q^3rs^2)(-r^5s^{10})$

$= 5(-1)(-1)q^{2+3}r^{4+1+5}s^{9+2+10}$

$= 5q^5r^{10}s^{21}$

23. $-5a^2\left(-\frac{1}{15}abc\right)\left(-\frac{3}{4}a^3b^2c\right)$

$= -\frac{\overset{1}{\cancel{5}}}{1}\left(-\frac{1}{\overset{\cancel{15}}{3}}\right)\left(-\frac{\overset{1}{\cancel{3}}}{4}\right)a^{2+1+3}b^{1+2}c^{1+1}$

$= -\frac{1}{4}a^6b^3c^2$

25. $(0.4wxy)(w^2y^2)(2.5x^2y^2)$

$= 0.4 \cdot 2.5 \cdot w^{1+2} \cdot x^{1+2} \cdot y^{1+2+2}$

$= w^3x^3y^5$

27. **Mistake:** The exponents of x were multiplied instead of added.

Correct: $7 \cdot 5 \cdot x^{3+4} \cdot y = 35x^7y$

29. **Mistake:** The exponents of x were not added.

Correct: $9 \cdot 6 \cdot x^{3+1} \cdot y^{2+4} \cdot z = 54x^4y^6z$

31. $w \cdot 5w = 5w^2$

33. $0.5w(9w)(w) = 4.5w^3$

35. a) height $= \frac{1}{2}b$, top side $= \frac{1}{4}b$

b) $A = \frac{1}{2} \cdot \frac{1}{2}b \cdot \left(b + \frac{1}{4}b\right)$

$= \frac{1}{4}b\left(b + \frac{1}{4}b\right)$

$= \frac{1}{4}b\left(\frac{4}{4}b + \frac{1}{4}b\right)$

$= \frac{1}{4}b\left(\frac{5}{4}b\right)$

$= \frac{5}{16}b^2$

37. $(2 \times 10^5)(6 \times 10^3) = 2 \times 6 \times 10^{5+3}$

$= 12 \times 10^8$

$= 1.2 \times 10^9$

39. $(3.2 \times 10^5)(7.5 \times 10^8) = 3.2 \times 7.5 \times 10^{5+8}$

$= 24 \times 10^{13}$

$= 2.4 \times 10^{14}$

41. $(2.9 \times 10^8)(6.3 \times 10^{-4}) = 2.9 \times 6.3 \times 10^{8+(-4)}$

$= 18.27 \times 10^4$

$= 1.827 \times 10^5$

43. $(4.23 \times 10^{-8})(2.7 \times 10^3) = 4.23 \times 2.7 \times 10^{-8+3}$

$= 11.421 \times 10^{-5}$

$= 1.1421 \times 10^{-4}$

45. $(9.2 \times 10^{-3})(6.1 \times 10^{-8}) = 9.2 \times 6.1 \times 10^{-3+(-8)}$

$= 56.12 \times 10^{-11}$

$= 5.612 \times 10^{-10}$

47. $A = lw = 350(120) = 42,000 = 4.2 \times 10^4$

The area is 4.2×10^4 sq. mi.

49. $C = 2\pi r$

$\approx 2 \times 3.14 \times 9.3 \times 10^7$

$\approx 58.404 \times 10^7$

$\approx 5.8404 \times 10^8$ mi.

51. $V = \dfrac{4}{3}\pi r^3$

$\approx \dfrac{4}{3} \times 3.14 \times \left(6.96 \times 10^8\right)^3$

$\approx \dfrac{4}{3} \times 3.14 \times 6.96^3 \times 10^{8 \cdot 3}$

$\approx \dfrac{4}{3} \times 3.14 \times 337.153 \times 10^{24}$

$\approx 1412 \times 10^{24}$

$\approx 1.412 \times 10^{27} \ \text{m}^3$

53. $E = hv$

$= \left(6.626 \times 10^{-34}\right)\left(4.5 \times 10^{14}\right)$

$= 6.626\left(4.5\right) \times 10^{-34+14}$

$= 29.817 \times 10^{-20}$

$= 2.9817 \times 10^{-19} \ \text{joule}$

55. $E = mc^2$

$= \left(2.4 \times 10^{-10}\right)\left(3 \times 10^8\right)^2$

$= \left(2.4 \times 10^{-10}\right)\left(3^2 \times 10^{8 \cdot 2}\right)$

$= \left(2.4 \times 10^{-10}\right)\left(9 \times 10^{16}\right)$

$= 2.4 \cdot 9 \times 10^{-10+16}$

$= 21.6 \times 10^6$

$= 2.16 \times 10^7 \ \text{joules}$

57. $\left(x^2\right)^3 = x^{2 \cdot 3} = x^6$

59. $\left(3^2\right)^4 = 3^{2 \cdot 4} = 3^8$

61. $\left(-h^2\right)^4 = (-1)^4 h^{2 \cdot 4} = h^8$

63. $\left(-y^2\right)^3 = (-1)^3 y^{2 \cdot 3} = -y^6$

65. $(xy)^3 = x^3 y^3$

67. $\left(3x^2\right)^3 = 3^3 x^{2 \cdot 3} = 27x^6$

69. $\left(-2x^3 y^2\right)^3 = (-2)^3 x^{3 \cdot 3} y^{2 \cdot 3} = -8x^9 y^6$

71. $\left(\dfrac{1}{4} m^2 n\right)^3 = \left(\dfrac{1}{4}\right)^3 m^{2 \cdot 3} n^3$

$= \dfrac{1}{64} m^6 n^3$

73. $\left(-2p^4 q^4 r^2\right)^6 = (-2)^6 p^{4 \cdot 6} q^{4 \cdot 6} r^{2 \cdot 6}$

$= 64 p^{24} q^{24} r^{12}$

75. $\left(-0.2x^2 y^5 z\right)^3 = (-0.2)^3 x^{2 \cdot 3} y^{5 \cdot 3} z^3$

$= -0.008 x^6 y^{15} z^3$

77. $\left(2x^3\right)\left(3x^2\right)^2 = 2x^3 \cdot 3^2 x^{2 \cdot 2}$

$= 2 \cdot 9 \cdot x^{3+4}$

$= 18x^7$

79. $\left(rs^2\right)\left(r^2 s\right)^2 = rs^2 \cdot r^{2 \cdot 2} s^2$

$= r^{1+4} s^{2+2}$

$= r^5 s^4$

81. $(2ab)^2 \left(5a^2 b\right)^2 = 2^2 a^2 b^2 \cdot 5^2 a^{2 \cdot 2} b^2$

$= 4 \cdot 25 \cdot a^{2+4} b^{2+2}$

$= 100 a^6 b^4$

83. $\left(\dfrac{1}{4} a^2 b^3 c\right)^2 \left(\dfrac{1}{4} a^2 b^3 c\right)^3 = \left(\dfrac{1}{4} a^2 b^3 c\right)^{2+3}$

$= \left(\dfrac{1}{4} a^2 b^3 c\right)^5$

$= \left(\dfrac{1}{4}\right)^5 a^{2 \cdot 5} b^{3 \cdot 5} c^5$

$= \dfrac{1}{1024} a^{10} b^{15} c^5$

85. $5(3a)^2 (2b)^3 = 5 \cdot 3^2 \cdot a^2 \cdot 2^3 \cdot b^3$

$= 5 \cdot 9 \cdot a^2 \cdot 8 \cdot b^3$

$= 360 a^2 b^3$

87. $-3\left(-2a^4 b^2\right)^3 \left(3a^3 b\right)^2$

$= -3(-2)^3 a^{4 \cdot 3} b^{2 \cdot 3} (3)^2 a^{3 \cdot 2} b^2$

$= -3(-8)(9) a^{12} b^6 a^6 b^2$

$= 216 a^{12+6} b^{6+2}$

$= 216 a^{18} b^8$

89. $(6u)^2 \left(u^2 v\right)(-3uv)^2 = 6^2 u^2 \cdot u^2 v \cdot (-3)^2 u^2 v^2$

$= 36 \cdot 9 \cdot u^{2+2+2} v^{1+2}$

$= 324 u^6 v^3$

91. $(0.6u)^2 (u^2v)(-3uv)^2$

$= (0.6)^2 u^2 \cdot u^2 v \cdot (-3)^2 u^2 v^2$

$= 0.36 \cdot 9 \cdot u^{2+2+2} v^{1+2}$

$= 3.24 u^6 v^3$

93. **Mistake:** The exponents were added instead of multiplied.

Correct: $(5x^3)^2 = 5^2 x^{3\cdot 2} = 25x^6$

95. **Mistake:** The coefficient was multiplied by 4 instead of being raised to the 4th power. Also, the exponents were not multiplied properly.

Correct: $(2x^4 y)^4 = 2^4 x^{4\cdot 4} y^4 = 16 x^{16} y^4$

Review Exercises:

1. $2(3x+4) = 2 \cdot 3x + 2 \cdot 4$

$= 6x + 8$

2. $-0.3t^2 u \cdot 6t^3 u = -0.3 \cdot 6 t^{2+3} u^{1+1}$

$= -1.8 t^5 u^2$

3. $-10x - 12x = -22x$

4. $(3x + 4y - z) + (-7x - 4y - 3z) = -4x - 4z$

5. $(-4) + (-57) = -61$

6. Add the exponents on the variables: $5 + 1 = 6$

Exercise Set 5.5

1. $5(x+3) = 5 \cdot x + 5 \cdot 3$

$= 5x + 15$

3. $-6(2x - 4) = -6 \cdot 2x - 6 \cdot (-4)$

$= -12x + 24$

5. $4n(7n + 2) = 4n \cdot 7n + 4n \cdot 2$

$= 28n^2 + 8n$

7. $9a(a - b) = 9a \cdot a + 9a \cdot (-b)$

$= 9a^2 - 9ab$

9. $\dfrac{1}{8}m(2m - 5n) = \dfrac{1}{8}m \cdot 2m + \dfrac{1}{8}m \cdot (-5n)$

$= \dfrac{1}{4}m^2 - \dfrac{5}{8}mn$

11. $-0.3p^2 \left(p^3 - 2q^2\right)$

$= -0.3p^2 \cdot p^3 - 0.3p^2 \left(-2q^2\right)$

$= -0.3p^5 + 0.6p^2 q^2$

13. $5x\left(4x^2 + 2x - 5\right)$

$= 5x \cdot 4x^2 + 5x \cdot 2x + 5x \cdot (-5)$

$= 20x^3 + 10x^2 - 25x$

15. $-3x^2\left(7x^3 - 5x + 1\right)$

$= -3x^2 \cdot 7x^3 - 3x^2 \cdot (-5x) - 3x^2 \cdot 1$

$= -21x^5 + 15x^3 - 3x^2$

17. $-r^2 s\left(r^2 s^3 + 3rs^2 - s\right)$

$= -r^2 s \cdot r^2 s^3 - r^2 s \cdot 3rs^2 - r^2 s \cdot (-s)$

$= -r^4 s^4 - 3r^3 s^3 + r^2 s^2$

19. $-2a^2 b^2 \left(2a^3 - 6b + 3ab - a^2 b^2\right)$

$= -2a^2 b^2 \cdot 2a^3 - 2a^2 b^2 \cdot (-6b) - 2a^2 b^2 \cdot 3ab$

$\quad - 2a^2 b^2 \cdot \left(-a^2 b^2\right)$

$= -4a^5 b^2 + 12a^2 b^3 - 6a^3 b^3 + 2a^4 b^4$

21. $3abc\left(4a^2 b^2 c^2 - 2abc + 5\right)$

$= 3abc \cdot 4a^2 b^2 c^2 + 3abc \cdot (-2abc) + 3abc \cdot 5$

$= 12a^3 b^3 c^3 - 6a^2 b^2 c^2 + 15abc$

23. $4b\left(b^5 - \dfrac{1}{4}b^3 + \dfrac{1}{20}b^2 - \dfrac{1}{8}b + 4\right)$

$= 4b \cdot b^5 + \dfrac{\cancel{4}b}{1} \cdot \left(-\dfrac{1}{\cancel{4}}b^3\right) + \dfrac{\cancel{4}b}{1} \cdot \dfrac{1}{\cancel{20}_5}b^2$

$\quad + \dfrac{\cancel{4}b}{1} \cdot \left(-\dfrac{1}{\cancel{8}_2}b\right) + 4b \cdot 4$

$= 4b^6 - b^4 + \dfrac{1}{5}b^3 - \dfrac{1}{2}b^2 + 16b$

25. $-0.4rt^2\left(2.5r^3 t - 8r^2 t^2 + 4.1rt^3 - 2\right)$

$= -0.4rt^2 \cdot 2.5r^3 t - 0.4rt^2 \cdot \left(-8r^2 t^2\right)$

$\quad - 0.4rt^2 \cdot 4.1rt^3 - 0.4rt^2 \cdot (-2)$

$= -r^4 t^3 + 3.2r^3 t^4 - 1.64r^2 t^5 + 0.8rt^2$

27. $\dfrac{5}{6}x^2\left(30x^4y-18x^3y^2+60xy^9\right)$

$=\dfrac{5}{\cancel{6}}x^2\cdot\dfrac{\cancel{30}^5}{1}x^4y+\dfrac{5}{\cancel{6}}x^2\cdot\left(-\dfrac{\cancel{18}^3}{1}x^3y^2\right)$

$\quad+\dfrac{5}{\cancel{6}}x^2\cdot\dfrac{\cancel{60}^{10}}{1}xy^9$

$=25x^6y-15x^5y^2+50x^3y^9$

29. $(x+3)(x+4)=x\cdot x+x\cdot4+3\cdot x+3\cdot4$

$\qquad\qquad\qquad=x^2+4x+3x+12$

$\qquad\qquad\qquad=x^2+7x+12$

31. $(x-7)(x+2)=x\cdot x+x\cdot2-7\cdot x-7\cdot2$

$\qquad\qquad\qquad=x^2+2x-7x-14$

$\qquad\qquad\qquad=x^2-5x-14$

33. $(y-3)(y-6)$

$=y\cdot y+y\cdot(-6)-3\cdot y-3\cdot(-6)$

$=y^2-6y-3y+18$

$=y^2-9y+18$

35. $(2y+5)(3y+2)$

$=2y\cdot3y+2y\cdot2+5\cdot3y+5\cdot2$

$=6y^2+4y+15y+10$

$=6y^2+19y+10$

37. $(5m-3)(3m+4)$

$=5m\cdot3m+5m\cdot4-3\cdot3m-3\cdot4$

$=15m^2+20m-9m-12$

$=15m^2+11m-12$

39. $(3t-5)(4t-2)$

$=3t\cdot4t+3t\cdot(-2)-5\cdot4t-5(-2)$

$=12t^2-6t-20t+10$

$=12t^2-26t+10$

41. $(5q-3t)(3q+4t)$

$=5q\cdot3q+5q\cdot4t-3t\cdot3q-3t\cdot4t$

$=15q^2+20qt-9qt-12t^2$

$=15q^2+11qt-12t^2$

43. $(7y+3x)(2y+4x)$

$=7y\cdot2y+7y\cdot4x+3x\cdot2y+3x\cdot4x$

$=14y^2+28xy+6xy+12x^2$

$=14y^2+34xy+12x^2$

45. $\left(a^2+b\right)\left(a^2-2b\right)$

$=a^2\cdot a^2+a^2\cdot(-2b)+b\cdot a^2+b\cdot(-2b)$

$=a^4-2a^2b+a^2b-2b^2$

$=a^4-a^2b-2b^2$

47. a) $x+5$ b) $2x$ c) $2x(x+5)$

    d) $x^2+x^2+5x+5x=2x^2+10x$

    e) They describe the same area.

49. a) $x+3$ b) $x+2$

    c) $(x+2)(x+3)$

    d) $x^2+3x+2x+6=x^2+5x+6$

    e) They describe the same area.

51. $(x+2)\left(x^2-2x+1\right)$

$=x\cdot x^2+x\cdot(-2x)+x\cdot1+2\cdot x^2$

$\quad+2\cdot(-2x)+2\cdot1$

$=x^3-2x^2+x+2x^2-4x+2$

$=x^3-3x+2$

53. $(a-1)\left(3a^2-a-2\right)$

$=a\cdot3a^2+a\cdot(-a)+a\cdot(-2)-1\cdot3a^2$

$\quad-1\cdot(-a)-1\cdot(-2)$

$=3a^3-a^2-2a-3a^2+a+2$

$=3a^3-4a^2-a+2$

55. $(3c+2)\left(2c^2-c-3\right)$

$=3c\cdot2c^2+3c\cdot(-c)+3c\cdot(-3)+2\cdot2c^2$

$\quad+2\cdot(-c)+2\cdot(-3)$

$=6c^3-3c^2-9c+4c^2-2c-6$

$=6c^3+c^2-11c-6$

57. $(2f+3g)\left(2f^2-3fg-9g^2\right)$

$=2f\cdot2f^2+2f\cdot(-3fg)+2f\cdot\left(-9g^2\right)$

$\quad+3g\cdot2f^2+3g\cdot(-3fg)+3g\cdot\left(-9g^2\right)$

$=4f^3-6f^2g-18fg^2+6f^2g-9fg^2-27g^3$

$=4f^3-27fg^2-27g^3$

59. $(3m-2n)\left(9m^2+6mn+4n^2\right)$

$= 3m\left(9m^2+6mn+4n^2\right)$

$\quad -2n\left(9m^2+6mn+4n^2\right)$

$= 27m^3+18m^2n+12mn^2$

$\quad -18m^2n-12mn^2-8n^3$

$= 27m^3-8n^3$

61. $(2a+5b)\left(4a^2-10ab+25b^2\right)$

$= 2a\left(4a^2-10ab+25b^2\right)$

$\quad +5b\left(4a^2-10ab+25b^2\right)$

$= 8a^3-20a^2b+50ab^2+20a^2b-50ab^2+125b^3$

$= 8a^3+125b^3$

63. $\left(x^2+xy+y^2\right)\left(x^2+2xy-y^2\right)$

$= x^2\left(x^2+2xy-y^2\right)+xy\left(x^2+2xy-y^2\right)$

$\quad +y^2\left(x^2+2xy-y^2\right)$

$= x^4+2x^3y-x^2y^2+x^3y+2x^2y^2$

$\quad -xy^3+x^2y^2+2xy^3-y^4$

$= x^4+3x^3y+2x^2y^2+xy^3-y^4$

65. $\left(x^2+10x+25\right)\left(x^2+4x+4\right)$

$= x^2\left(x^2+4x+4\right)+10x\left(x^2+4x+4\right)$

$\quad +25\left(x^2+4x+4\right)$

$= x^4+4x^3+4x^2+10x^3+40x^2+40x$

$\quad +25x^2+100x+100$

$= x^4+14x^3+69x^2+140x+100$

67. The conjugate of $x+3$ is $x-3$.

69. The conjugate of $4x-2y$ is $4x+2y$.

71. The conjugate of $4d+3c$ is $4d-3c$.

73. The conjugate of $-3j-k$ is $-3j+k$.

75. $(x-5)(x+5) = x^2-5^2$

$\qquad = x^2-25$

77. $(2m-5)(2m+5) = (2m)^2-5^2$

$\qquad\qquad = 4m^2-25$

79. $(x+y)(x-y) = x^2-y^2$

81. $(8r-10s)(8r+10s) = (8r)^2-(10s)^2$

$\qquad\qquad\qquad = 64r^2-100s^2$

83. $(-2x-3)(-2x+3) = (-2x)^2-(3)^2$

$\qquad\qquad\qquad = 4x^2-9$

85. $(x+3)^2 = (x)^2+2(x)(3)+(3)^2$

$\qquad = x^2+6x+9$

87. $(4t-1)^2 = (4t)^2-2(4t)(1)+(1)^2$

$\qquad = 16t^2-8t+1$

89. $(m+n)^2 = (m)^2+2(m)(n)+(n)^2$

$\qquad = m^2+2mn+n^2$

91. $(2u+3v)^2 = (2u)^2+2(2u)(3v)+(3v)^2$

$\qquad = 4u^2+12uv+9v^2$

93. $(9w-4z)^2 = (9w)^2-2(9w)(4z)+(4z)^2$

$\qquad = 81w^2-72wz+16z^2$

95. $(9-5y)^2 = (9)^2-2(9)(5y)+(5y)^2$

$\qquad = 81-90y+25y^2$

97. $A = lw = 2yz(5x+y) = 10xyz+2y^2z$

99. $A = \dfrac{1}{2}h(b_1+b_2)$

$\quad = \dfrac{1}{2}(h)\left[(h+1)+(h+7)\right]$

$\quad = \dfrac{1}{2}(h)(2h+8)$

$\quad = h^2+4h$

101. Let the length be represented by $w+3$.

$A = lw = (w+3)w = w^2+3w$

103. $V = lwh$

$\quad = (x+5)(x+1)4x$

$\quad = \left(x^2+x+5x+5\right)4x$

$\quad = \left(x^2+6x+5\right)4x$

$\quad = 4x^3+24x^2+20x$

105. The length is $2w$, so the height is $2w+3$.

$$V = lwh$$
$$= (2w)(w)(2w+3)$$
$$= 2w^2(2w+3)$$
$$= 4w^3 + 6w^2$$

107. Let r be the radius. Then the height is $r+2$.

$$V = \pi r^2 h$$
$$= \pi r^2 (r+2)$$
$$= \pi r^2 \cdot r + \pi r^2 \cdot 2$$
$$= \pi r^3 + 2\pi r^2$$
$$\approx 3.14 r^3 + 6.28 r^2$$

Review Exercises

1. $1.08 \div 2.4 = 0.45$

2. $-6 - (-1) = -6 + 1 = -5$

3. $6.304 \times 10^6 = 6,304,000$

4. $x^2 \cdot x^5 = x^{2+5} = x^7$

5.
$$\frac{5-d}{r} = m$$
$$r \cdot \frac{5-d}{r} = r \cdot m$$
$$5-d = rm$$
$$\frac{5-d}{m} = \frac{rm}{m}$$
$$\frac{5-d}{m} = r$$

6. smaller number $= x$ larger number $= 4x$
$$4x - x = 27$$
$$\frac{3x}{3} = \frac{27}{3}$$
$$x = 9$$
The numbers are 9 and 36.

Exercise Set 5.6

1. $a^{-3} = \dfrac{1}{a^3}$

3. $2^{-5} = \dfrac{1}{2^5}$

5. $\dfrac{y^8}{y^3} = y^{8-3} = y^5$

7. $\dfrac{3^4}{3^3} = 3^{4-3} = 3$

9. $\dfrac{4^3}{4^9} = 4^{3-9} = 4^{-6} = \dfrac{1}{4^6}$

11. $\dfrac{x^3}{x^5} = x^{3-5} = x^{-2} = \dfrac{1}{x^2}$

13. $a^8 \div a^6 = \dfrac{a^8}{a^6} = a^{8-6} = a^2$

15. $w^7 \div w^{10} = \dfrac{w^7}{w^{10}} = w^{7-10} = w^{-3} = \dfrac{1}{w^3}$

17. $\dfrac{a^{-2}}{a^5} = a^{-2-5} = a^{-7} = \dfrac{1}{a^7}$

19. $\dfrac{r^6}{r^{-4}} = r^{6-(-4)} = r^{6+4} = r^{10}$

21. $\dfrac{y^{-7}}{y^{-15}} = y^{-7-(-15)} = y^{-7+15} = y^8$

23. $\dfrac{p^{-5}}{p^{-2}} = p^{-5-(-2)} = p^{-5+2} = p^{-3} = \dfrac{1}{p^3}$

25. $\dfrac{t^{-4}}{t^{-4}} = t^{-4-(-4)} = t^{-4+4} = t^0 = 1$

27. $\dfrac{8.32 \times 10^5}{3.2 \times 10^6} = \dfrac{8.32}{3.2} \times \dfrac{10^5}{10^6}$
$$= 2.6 \times 10^{5-6}$$
$$= 2.6 \times 10^{-1}$$

29. $\dfrac{1.26 \times 10^{-4}}{2.1 \times 10^3} = \dfrac{1.26}{2.1} \times \dfrac{10^{-4}}{10^3}$
$$= 0.6 \times 10^{-4-3}$$
$$= 0.6 \times 10^{-7}$$
$$= 6 \times 10^{-8}$$

31. $\dfrac{9.088 \times 10^1}{1.28 \times 10^{-9}} = \dfrac{9.088}{1.28} \times \dfrac{10^1}{10^{-9}}$

$\qquad = 7.1 \times 10^{1-(-9)}$

$\qquad = 7.1 \times 10^{1+9}$

$\qquad = 7.1 \times 10^{10}$

33. $\dfrac{8.64 \times 10^{-3}}{3.2 \times 10^{-7}} = \dfrac{8.64}{3.2} \times \dfrac{10^{-3}}{10^{-7}}$

$\qquad = 2.7 \times 10^{-3-(-7)}$

$\qquad = 2.7 \times 10^{-3+7}$

$\qquad = 2.7 \times 10^4$

35. $\dfrac{1.5 \times 10^{11}}{3 \times 10^8} = 0.5 \times 10^{11-8}$

$\qquad = 0.5 \times 10^3$

$\qquad = 5 \times 10^2$

It takes 500 sec. or $8\dfrac{1}{3}$ min .

37. $\dfrac{1.596 \times 10^{13}}{3.14 \times 10^8} = \dfrac{1.596}{3.14} \times 10^{13-8}$

$\qquad \approx 0.5082803 \times 10^5$

$\qquad \approx 5.082803 \times 10^4$

Each person would contribute approximately $\$50,828.03$.

39. Solve $E = mc^2$ for m:

$E = mc^2$

$\dfrac{E}{c^2} = \dfrac{mc^2}{c^2}$

$\dfrac{E}{c^2} = m$

$m = \dfrac{E}{c^2}$

$\quad = \dfrac{8.4 \times 10^{13}}{\left(3 \times 10^8\right)^2}$

$\quad = \dfrac{8.4 \times 10^{13}}{9 \times 10^{16}}$

$\quad = \dfrac{8.4}{9} \times 10^{13-16}$

$\quad \approx 0.93 \times 10^{-3}$

About 9.3×10^{-4} kg kg of mass were converted to energy.

41. $\dfrac{15x^5}{3x^2} = \dfrac{15}{3} \cdot \dfrac{x^5}{x^2} = 5x^{5-2} = 5x^3$

43. $\dfrac{24x^2}{-6x^5} = \dfrac{24}{-6} \cdot \dfrac{x^2}{x^5} = -4x^{2-5} = -4x^{-3} = \dfrac{-4}{x^3}$

45. $\dfrac{9m^3n^5}{-3mn} = \dfrac{9}{-3} \cdot \dfrac{m^3}{m} \cdot \dfrac{n^5}{n}$

$\qquad = -3 \cdot m^{3-1} \cdot n^{5-1}$

$\qquad = -3 \cdot m^2 \cdot n^4$

$\qquad = -3m^2n^4$

47. $\dfrac{-24p^3q^5}{15p^3q^2} = \dfrac{-24}{15} \cdot \dfrac{p^3}{p^3} \cdot \dfrac{q^5}{q^2}$

$\qquad = \dfrac{-8}{5} \cdot p^{3-3} \cdot q^{5-2}$

$\qquad = \dfrac{-8}{5} \cdot p^0 \cdot q^3$

$\qquad = \dfrac{-8}{5} \cdot 1 \cdot q^3$

$\qquad = -\dfrac{8}{5}q^3$

49. $\dfrac{56x^2y^6}{42x^7y^2} = \dfrac{56}{42} \cdot \dfrac{x^2}{x^7} \cdot \dfrac{y^6}{y^2}$

$\qquad = \dfrac{4}{3} \cdot x^{2-7} \cdot y^{6-2}$

$\qquad = \dfrac{4}{3} \cdot x^{-5} \cdot y^4$

$\qquad = \dfrac{4}{3} \cdot \dfrac{1}{x^5} \cdot \dfrac{y^4}{1}$

$\qquad = \dfrac{4y^4}{3x^5}$

51. $\dfrac{12a^4bc^3}{9a^6bc^2} = \dfrac{12}{9} \cdot \dfrac{a^4}{a^6} \cdot \dfrac{b}{b} \cdot \dfrac{c^3}{c^2}$

$\qquad = \dfrac{4}{3} \cdot a^{4-6} \cdot b^{1-1} \cdot c^{3-2}$

$\qquad = \dfrac{4}{3} \cdot a^{-2} \cdot b^0 \cdot c^1$

$\qquad = \dfrac{4}{3} \cdot \dfrac{1}{a^2} \cdot 1 \cdot \dfrac{c}{1}$

$\qquad = \dfrac{4c}{3a^2}$

53. $\dfrac{7a+14b}{7} = \dfrac{7a}{7} + \dfrac{14b}{7}$

$\qquad = a + 2b$

55. $\dfrac{12x^3 - 6x^2}{3x} = \dfrac{12x^3}{3x} - \dfrac{6x^2}{3x}$

$= 4x^{3-1} - 2x^{2-1}$

$= 4x^2 - 2x$

57. $\dfrac{12x^3 + 8x}{4x^2} = \dfrac{12x^3}{4x^2} + \dfrac{8x}{4x^2}$

$= 3x^{3-2} + 2x^{1-2}$

$= 3x + 2x^{-1}$

$= 3x + \dfrac{2}{x}$

59. $\dfrac{5x^2 y^2 - 15xy^3}{5xy} = \dfrac{5x^2 y^2}{5xy} - \dfrac{15xy^3}{5xy}$

$= x^{2-1} y^{2-1} - 3x^{1-1} y^{3-1}$

$= xy - 3y^2$

61. $\dfrac{6abc^2 - 24a^2 b^2 c}{-3abc}$

$= \dfrac{6abc^2}{-3abc} - \dfrac{24a^2 b^2 c}{-3abc}$

$= -2a^{1-1} b^{1-1} c^{2-1} + 8a^{2-1} b^{2-1} c^{1-1}$

$= -2a^0 b^0 c^1 + 8a^1 b^1 c^0$

$= -2c + 8ab$

63. $\dfrac{24x^3 + 16x^2 - 8x}{8x} = \dfrac{24x^3}{8x} + \dfrac{16x^2}{8x} - \dfrac{8x}{8x}$

$= 3x^{3-1} + 2x^{2-1} - 1$

$= 3x^2 + 2x - 1$

65. $\dfrac{6x^3 - 12x^2 + 9x}{3x^2} = \dfrac{6x^3}{3x^2} - \dfrac{12x^2}{3x^2} + \dfrac{9x}{3x^2}$

$= 2x^{3-2} - 4x^{2-2} + 3x^{1-2}$

$= 2x - 4 + 3x^{-1}$

$= 2x - 4 + \dfrac{3}{x}$

67. $\dfrac{36u^3 v^4 + 12uv^5 - 15u^2 v^2}{3u^2 v}$

$= \dfrac{36u^3 v^4}{3u^2 v} + \dfrac{12uv^5}{3u^2 v} - \dfrac{15u^2 v^2}{3u^2 v}$

$= 12u^{3-2} v^{4-1} + 4u^{1-2} v^{5-1} - 5u^{2-2} v^{2-1}$

$= 12uv^3 + 4u^{-1} v^4 - 5u^0 v$

$= 12uv^3 + \dfrac{4v^4}{u} - 5v$

69. $\dfrac{y^5 + y^7 - 3y^8 + y}{y^3} = \dfrac{y^5}{y^3} + \dfrac{y^7}{y^3} - \dfrac{3y^8}{y^3} + \dfrac{y}{y^3}$

$= y^{5-3} + y^{7-3} - 3y^{8-3} + y^{1-3}$

$= y^2 + y^4 - 3y^5 + y^{-2}$

$= y^2 + y^4 - 3y^5 + \dfrac{1}{y^2}$

71. $lh = A$

$\dfrac{lh}{l} = \dfrac{A}{l}$

$h = \dfrac{A}{l}$

$h = \dfrac{35mn^2}{14n} = \dfrac{5mn}{2}$

73. a) $lw = A$

$\dfrac{lw}{l} = \dfrac{A}{l}$

$w = \dfrac{A}{l}$

$w = \dfrac{9x^2 - 15x + 12}{3x}$

$= \dfrac{9x^2}{3x} - \dfrac{15x}{3x} + \dfrac{12}{3x}$

$= 3x - 5 + \dfrac{4}{x}$

b) $w = 3(2) - 5 + \dfrac{4}{(2)}$

$= 6 - 5 + 2$

$= 3$

$A = 9 \cdot 2^2 - 15 \cdot 2 + 12 \qquad l = \dfrac{A}{w}$

$= 9 \cdot 4 - 30 + 12 \qquad\qquad = \dfrac{18}{3}$

$= 36 - 30 + 12$

$= 18 \qquad\qquad\qquad\qquad = 6$

75. $x + 3 \overline{)\, x^2 + 7x + 12}$ Answer: $x + 4$

with quotient $x + 4$

$\underline{x^2 + 3x}$

$4x + 12$

$\underline{4x + 12}$

0

77. $m-3\overline{)m^2-8m+15}$ with quotient $m-5$

$$\begin{array}{r} m-5 \\ m-3\overline{)m^2-8m+15} \\ \underline{m^2-3m} \\ -5m+15 \\ \underline{-5m+15} \\ 0 \end{array}$$

Answer: $m-5$

79.
$$\begin{array}{r} x-12 \\ x-5\overline{)x^2-17x+64} \\ \underline{x^2-5x} \\ -12x+64 \\ \underline{-12x+60} \\ 4 \end{array}$$

Answer: $x-12+\dfrac{4}{x-5}$

81.
$$\begin{array}{r} 3x^2+15x+2 \\ x-5\overline{)3x^3+0x^2-73x-10} \\ \underline{3x^3-15x^2} \\ 15x^2-73x \\ \underline{15x^2-75x} \\ 2x-10 \\ \underline{2x-10} \\ 0 \end{array}$$

Answer: $3x^2+15x+2$

83.
$$\begin{array}{r} 2x^2+3x+6 \\ x-2\overline{)2x^3-x^2+0x+5} \\ \underline{2x^3-4x^2} \\ 3x^2+0x \\ \underline{3x^2-6x} \\ 6x+5 \\ \underline{6x-12} \\ 17 \end{array}$$

Answer: $2x^2+3x+6+\dfrac{17}{x-2}$

85.
$$\begin{array}{r} p^2-5p+25 \\ p+5\overline{)p^3+0p^2+0p+125} \\ \underline{p^3+5p^2} \\ -5p^2+0p \\ \underline{-5p^2-25p} \\ 25p+125 \\ \underline{25p+125} \\ 0 \end{array}$$

Answer: $p^2-5p+25$

87.
$$\begin{array}{r} 3x+2 \\ 4x-5\overline{)12x^2-7x-10} \\ \underline{12x^2-15x} \\ 8x-10 \\ \underline{8x-10} \\ 0 \end{array}$$

Answer: $3x+2$

89.
$$\begin{array}{r} 2y-2 \\ 7y+3\overline{)14y^2-8y+17} \\ \underline{14y^2+6y} \\ -14y+17 \\ \underline{-14y-6} \\ 23 \end{array}$$

Answer: $2y-2+\dfrac{23}{7y+3}$

91.
$$\begin{array}{r} 4k^2-1 \\ k+2\overline{)4k^3+8k^2-k+6} \\ \underline{4k^3+8k^2} \\ 0k^2-k+6 \\ \underline{-k-2} \\ 8 \end{array}$$

Answer: $4k^2-1+\dfrac{8}{k+2}$

93.
$$\begin{array}{r} 7u^2 - 11u + 12 \\ 3u+2{\overline{\smash{\big)}\,21u^3 - 19u^2 + 14u - 6}} \\ \underline{21u^3 + 14u^2} \\ -33u^2 + 14u \\ \underline{-33u^2 - 22u} \\ 36u - 6 \\ \underline{36u + 24} \\ -30 \end{array}$$

Answer: $7u^2 - 11u + 12 - \dfrac{30}{3u+2}$

95.
$$\begin{array}{r} b^2 - 3b + 5 \\ b-3{\overline{\smash{\big)}\,b^3 - 6b^2 + 14b - 12}} \\ \underline{b^3 - 3b^2} \\ -3b^2 + 14b \\ \underline{-3b^2 + 9b} \\ 5b - 12 \\ \underline{5b - 15} \\ 3 \end{array}$$

Answer: $b^2 - 3b + 5 + \dfrac{3}{b-3}$

97.
$$\begin{array}{r} y^2 - 2y + 3 \\ y+2{\overline{\smash{\big)}\,y^3 + 0y^2 - y + 6}} \\ \underline{y^3 + 2y^2} \\ -2y^2 - y \\ \underline{-2y^2 - 4y} \\ 3y + 6 \\ \underline{3y + 6} \\ 0 \end{array}$$

Answer: $y^2 - 2y + 3$

99. a) We need to note that $h = \dfrac{V}{wl}$ and
$wl = 6(x+3) = 6x + 18$. Then substitute.

$$\begin{array}{r} x + 4 \\ 6x+18{\overline{\smash{\big)}\,6x^2 + 42x + 72}} \\ \underline{6x^2 + 18x} \\ 24x + 72 \\ \underline{24x + 72} \\ 0 \end{array}$$
height: $x + 4$

b) length: $4 + 3 = 7$ ft.; height: $4 + 4 = 8$ ft.;
volume: $V = 6 \cdot 4^2 + 42 \cdot 4 + 72$
$\qquad = 6 \cdot 16 + 168 + 72$
$\qquad = 336$ ft.^{3}

101. $\left(6m^2 n^{-2}\right)^2 = 6^2 m^{2 \cdot 2} n^{-2 \cdot 2} = 36 m^4 n^{-4} = \dfrac{36 m^4}{n^4}$

103. $\left(4x^{-3} y^2\right)^{-3} = \dfrac{1}{\left(4x^{-3} y^2\right)^3}$

$\qquad = \dfrac{1}{4^3 x^{-3 \cdot 3} y^{2 \cdot 3}}$

$\qquad = \dfrac{1}{64 x^{-9} y^6}$

$\qquad = \dfrac{x^9}{64 y^6}$

105. $\left(4rs^3\right)^3 \left(2r^3 s^{-1}\right)^2 = \left(4^3 r^3 s^{3 \cdot 3}\right)\left(2^2 r^{3 \cdot 2} s^{-1 \cdot 2}\right)$

$\qquad = \left(64 r^3 s^9\right)\left(4 r^6 s^{-2}\right)$

$\qquad = 256 r^{3+6} s^{9+(-2)}$

$\qquad = 256 r^9 s^7$

107. $\left(\dfrac{m^6}{m^2}\right)^4 = \left(m^{6-2}\right)^4 = \left(m^4\right)^4 = m^{4 \cdot 4} = m^{16}$

109. $\left(\dfrac{x^{-3}}{x^4}\right)^2 = \left(x^{-3-4}\right)^2$

$\qquad = \left(x^{-7}\right)^2$

$\qquad = x^{-7 \cdot 2}$

$\qquad = x^{-14}$

$\qquad = \dfrac{1}{x^{14}}$

111. $\left(\dfrac{x^2 y^7}{x^5 y^{-2}}\right)^3 = \left(x^{2-5} y^{7-(-2)}\right)^3$

$\qquad = \left(x^{-3} y^9\right)^3$

$\qquad = x^{-3 \cdot 3} y^{9 \cdot 3}$

$\qquad = x^{-9} y^{27}$

$\qquad = \dfrac{y^{27}}{x^9}$

113. $\left(\dfrac{x^{-3}}{x^4}\right)^{-2} = \left(x^{-3-4}\right)^{-2}$

$= \left(x^{-7}\right)^{-2}$

$= x^{-7 \cdot -2}$

$= x^{14}$

115. $\left(\dfrac{3x^{-1}y^2}{z^2}\right)^3 = \dfrac{3^3\,x^{-1\cdot3}\,y^{2\cdot3}}{z^{2\cdot3}} = \dfrac{27x^{-3}y^6}{z^6} = \dfrac{27y^6}{x^3 z^6}$

117. $\dfrac{\left(x^4\right)^{-3}}{x^{-5}\cdot x^3} = \dfrac{x^{4\cdot(-3)}}{x^{-5+3}}$

$= \dfrac{x^{-12}}{x^{-2}}$

$= x^{-12-(-2)}$

$= x^{-10}$

$= \dfrac{1}{x^{10}}$

119. $\dfrac{4xy^{-2}z^2}{x^{-3}y^3z^{-1}} = \dfrac{4x^{1-(-3)}\,y^{-2-3}\,z^{2-(-1)}}{1}$

$= \dfrac{4x^4 y^{-5} z^3}{1}$

$= \dfrac{4x^4 z^3}{y^5}$

121. $\dfrac{(2ab)^3}{12a^4b^5} = \dfrac{2^3 a^3 b^3}{12a^4b^5}$

$= \dfrac{8a^{3-4} b^{3-5}}{12}$

$= \dfrac{2a^{-1}b^{-2}}{3}$

$= \dfrac{2}{3ab^2}$

123. $\dfrac{\left(8x^3\right)^3}{\left(4x^4\right)^5} = \dfrac{8^3\,x^{3\cdot3}}{4^5\,x^{4\cdot5}}$

$= \dfrac{512x^9}{1024x^{20}}$

$= \dfrac{1}{2}x^{9-20}$

$= \dfrac{1}{2}x^{-11}$

$= \dfrac{1}{2x^{11}}$

Review Exercises

1. $3024 = 2\cdot1512$

$= 2\cdot2\cdot756$

$= 2\cdot2\cdot2\cdot378$

$= 2\cdot2\cdot2\cdot2\cdot189$

$= 2\cdot2\cdot2\cdot2\cdot3\cdot63$

$= 2\cdot2\cdot2\cdot2\cdot3\cdot3\cdot21$

$= 2\cdot2\cdot2\cdot2\cdot3\cdot3\cdot3\cdot7$

$= 2^4\cdot3^3\cdot7$

2. $-3x^2y\cdot6x^2y^2 = -18x^{2+2}y^{1+2} = -18x^4y^3$

3. $y(3x-4) = 3xy-4y$

4. Let x be one number and let y be the other number.

$x+\dfrac{1}{2}y$

5. $A = lw$

$= 17\cdot20$

$= 340 \text{ ft.}^2$

6. $3 - y = 27$ Check: $3-(-24) = 27$

    $-y = 24$ $3+24 = 27$

      $y = -24$ $27 = 27$

Chapter 5 Review Exercises

1. $\left(\dfrac{2}{5}\right)^3 = \dfrac{2^3}{5^3} = \dfrac{8}{125}$

2. $-4^2 = -(4\cdot4) = -16$

3. $5^{-2} = \dfrac{1}{5^2} = \dfrac{1}{25}$

4. $\dfrac{1}{6^{-2}} = 6^2 = 36$

5. $4.5 \times 10^9 = 4,500,000,000$

6. 1.663×10^{-24}
    $= 0.000000000000000000000001663$

7. $0.000000000000000000000001661$
    $= 1.661 \times 10^{-24}$

8. $300,000,000 = 3 \times 10^8$

9. Yes, $3xy^5$ is a monomial because it is the product of a constant and two variables with whole-number exponents.

10. Yes, $-\dfrac{1}{2}x$ is a monomial because it is the product of a constant and a variable with a whole-number exponent.

11. No, $\dfrac{4}{ab^3}$ is not a monomial because if it is rewritten as a product, both of the variables would have a negative exponent.

12. No, $2x^2 - 9$ is not a monomial because it is not a product of a constant and variables.

13. $6x^4$
    Coefficient: 6
    Degree: 4

14. 27
    Coefficient: 27
    Degree: 0

15. $-2.6xy^3$
    Coefficient: -2.6
    Degree: $1 + 3 = 4$

16. $-m$
    Coefficient: -1
    Degree: 1

17. $4x^2 - 25$ is a binomial because it has two terms.

18. $-st^5$ is a monomial because it has one term.

19. $3x - \dfrac{4}{y^2}$ is not a polynomial because $\dfrac{4}{y^2}$ is not a monomial.

20. $5x^3 - 6x^2 - 3x + 11$ has no special polynomial name because it has more than three terms.

21. $-2x^9 + 21x^5 - 19x^3 - 3x + 15$ has degree 9 because it is the greatest degree of all the terms.

22. $-j^4 + 22j^2 + 5j - 19$ has degree 4 because it is the greatest degree of all the terms.

23. $8y^5 - 3y^4 + 2y - 4y^5 - 2y + 8 + 7y^4$
    $= 8y^5 - 4y^5 - 3y^4 + 7y^4 + 2y - 2y + 8$
    $= 4y^5 + 4y^4 + 8$

24. $3m - 4 - m^2 - 2m^3 + 7 - 2m + 8m^2$
    $= -2m^3 - m^2 + 8m^2 + 3m - 2m - 4 + 7$
    $= -2m^3 + 7m^2 + m + 3$

25. $5xyz^2 - 4x^2yz - 7xyz^2 - 5x^2yz - 2x^2zy$
    $= -4x^2yz - 5x^2yz - 2x^2zy + 5xyz^2 - 7xyz^2$
    $= -11x^2yz - 2xyz^2$

26. $4a^2bc + 5abc^3 - 8abc - 9a^5 - 2abc^3$
    $\quad - 4a^2bc + 8 + a^5$
    $= -9a^5 + a^5 + 5abc^3 - 2abc^3 + 4a^2bc - 4a^2bc$
    $\quad - 8abc + 8$
    $= -8a^5 + 3abc^3 - 8abc + 8$

27. $18jk - 2j^4 + 12jk^3 - j^4 - 8jk - 9jk^3$
    $\quad + 12 + k^2$
    $= -2j^4 - j^4 + 12jk^3 - 9jk^3 + k^2 + 18jk$
    $\quad - 8jk + 12$
    $= -3j^4 + 3jk^3 + k^2 + 10jk + 12$

28. $4m^2 - 3mn + 2mn^2 - 8n^2 - 6mn - 4m^2$
    $\quad + 7 + mn^2$
    $= 2mn^2 + mn^2 + 4m^2 - 4m^2 - 8n^2 - 3mn$
    $\quad - 6mn + 7$
    $= 3mn^2 - 8n^2 - 9mn + 7$

29. $\left(8x^2 - 3\right) + \left(2x^2 + 3x - 1\right) = 8x^2 + 2x^2 + 3x - 3 - 1$
    $\qquad\qquad\qquad\qquad = 10x^2 + 3x - 4$

30. $\left(5n + 8\right) - \left(2n^3 - 3n^2 + n + 5\right)$
    $= \left(5n + 8\right) + \left(-2n^3 + 3n^2 - n - 5\right)$
    $= -2n^3 + 3n^2 + 5n - n + 8 - 5$
    $= -2n^3 + 3n^2 + 4n + 3$

31. $(2y-4)+(4y+8)-(6y-5)$
$=(2y-4)+(4y+8)+(-6y+5)$
$=2y+4y-6y-4+8+5$
$=9$

32. $(3x^3-2x+8)+(4x^2-3x-1)$
$\quad -(4x^3-5x-10)$
$=(3x^3-2x+8)+(4x^2-3x-1)$
$\quad +(-4x^3+5x+10)$
$=3x^3-4x^3+4x^2-2x-3x+5x+8-1+10$
$=-x^3+4x^2+17$

33. $(5x^2-2xy+y^2)+(3x^2+xy+5y^2)$
$=5x^2+3x^2-2xy+xy+y^2+5y^2$
$=8x^2-xy+6y^2$

34. $(m^2+5mn+6)-(4m^2-3mn+8)$
$=(m^2+5mn+6)+(-4m^2+3mn-8)$
$=m^2-4m^2+5mn+3mn+6-8$
$=-3m^2+8mn-2$

35. $m\cdot m^4=m^{1+4}=m^5$

36. $2a\cdot 5a^8=2\cdot 5\cdot a^{1+8}$
$\quad =10a^9$

37. $(-3x^2y)(2x^4y^5)=-3\cdot 2\cdot x^{2+4}y^{1+5}$
$\quad =-6x^6y^6$

38. $(5x^2y)^2=5^2x^{2\cdot 2}y^{1\cdot 2}=25x^4y^2$

39. $(-6u^3)^2(2u^4)=(-6)^2u^{3\cdot 2}\cdot 2u^4$
$\quad =36\cdot 2\cdot u^{6+4}$
$\quad =72u^{10}$

40. $A=lw$
$\quad =(9\times 10^3)(2.5\times 10^4)$
$\quad =9\cdot 2.5\times 10^{3+4}$
$\quad =22.5\times 10^7$
$\quad =2.25\times 10^8$
The area of the region is 2.25×10^8 m^2.

41. Voltage = Current × Resistance
$\quad =(3.5\times 10^{-3})\times(8.4\times 10^{-2})$
$\quad =(3.5)(8.4)\times 10^{-3+(-2)}$
$\quad =29.4\times 10^{-5}$
$\quad =2.94\times 10^{-4}$
The voltage is 2.94×10^{-4} V.

42. $4a(a-3)=4a\cdot a+4a\cdot(-3)$
$\quad =4a^2-12a$

43. $-6b(2b^2-4b-1)$
$=-6b\cdot 2b^2-6b\cdot(-4b)-6b\cdot(-1)$
$=-12b^3+24b^2+6b$

44. $-4abc^2(3a-4ab^3+2abc^2+8)$
$=-4abc^2\cdot 3a-4abc^2\cdot(-4ab^3)$
$\quad -4abc^2\cdot 2abc^2-4abc^2\cdot 8$
$=-12a^2bc^2+16a^2b^4c^2-8a^2b^2c^4-32abc^2$

45. a) $\dfrac{1}{2}(8x)(5x-7)=4x(5x-7)$
$\quad\quad\quad\quad\quad\quad =20x^2-28x$

b) $20x^2-28x=20(3)^2-28(3)$
$\quad\quad\quad\quad\quad =20(9)-84$
$\quad\quad\quad\quad\quad =180-84$
$\quad\quad\quad\quad\quad =96$ ft.^{2}

46. $(x+5)(x-1)=x\cdot x+x\cdot(-1)+5\cdot x+5\cdot(-1)$
$\quad\quad\quad\quad\quad =x^2-x+5x-5$
$\quad\quad\quad\quad\quad =x^2+4x-5$

47. $(2m+5)(6m-1)$
$=2m\cdot 6m+2m\cdot(-1)+5\cdot 6m+5\cdot(-1)$
$=12m^2-2m+30m-5$
$=12m^2+28m-5$

48. $(5a+2b)(3a-2b)$
$=5a\cdot 3a+5a\cdot(-2b)+2b\cdot 3a+2b\cdot(-2b)$
$=15a^2-10ab+6ab-4b^2$
$=15a^2-4ab-4b^2$

49. $(3y-1)(y^2-2y+4)$

$\quad = 3y(y^2-2y+4)-1(y^2-2y+4)$

$\quad = 3y^3-6y^2+12y-y^2+2y-4$

$\quad = 3y^3-7y^2+14y-4$

50. $(a^2+ab+b^2)(a^2-2ab-b^2)$

$\quad = a^2(a^2-2ab-b^2)+ab(a^2-2ab-b^2)$

$\quad\quad +b^2(a^2-2ab-b^2)$

$\quad = a^4-2a^3b-a^2b^2+a^3b-2a^2b^2-ab^3$

$\quad\quad +a^2b^2-2ab^3-b^4$

$\quad = a^4-2a^3b+a^3b-a^2b^2-2a^2b^2+a^2b^2$

$\quad\quad -ab^3-2ab^3-b^4$

$\quad = a^4-a^3b-2a^2b^2-3ab^3-b^4$

51. $(2x+1)(2x-1)=(2x)^2-1^2=4x^2-1$

52. $(4-x)(4+x)=4^2-x^2=16-x^2$

53. $(x-6)^2=x^2+2(x)(-6)+6^2=x^2-12x+36$

54. $(3r+5)^2=(3r)^2+2(3r)(5)+(5)^2$

$\quad\quad\quad = 9r^2+30r+25$

55. $\dfrac{x^3}{x^{-5}}=x^{3-(-5)}=x^8$

56. $\dfrac{u^{-1}}{u^{-8}}=u^{-1-(-8)}=u^7$

57. $\dfrac{s^{-5}}{s}=s^{-5-1}=s^{-6}=\dfrac{1}{s^6}$

58. $\left(\dfrac{1}{x^3}\right)^{-2}=(x^3)^2=x^{2\cdot3}=x^6$

59. $\dfrac{x^3}{x^7}=x^{3-7}=x^{-4}=\dfrac{1}{x^4}$

60. $\dfrac{28z^6}{4z^2}=\dfrac{28}{4}\cdot\dfrac{z^6}{z^2}=7z^{6-2}=7z^4$

61. $\dfrac{-48x^4y^2}{6x^3y^5}=\dfrac{-48}{6}\cdot\dfrac{x^4}{x^3}\cdot\dfrac{y^2}{y^5}$

$\quad\quad = -8x^{4-3}y^{2-5}$

$\quad\quad = -8x^1y^{-3}$

$\quad\quad = \dfrac{-8}{1}\cdot\dfrac{x}{1}\cdot\dfrac{1}{y^3}$

$\quad\quad = -\dfrac{8x}{y^3}$

62. $\dfrac{24t^3u^6}{-15t^3u^4}=\dfrac{24}{-15}\cdot\dfrac{t^3}{t^3}\cdot\dfrac{u^6}{u^4}$

$\quad\quad = -\dfrac{8}{5}t^{3-3}u^{6-4}$

$\quad\quad = -\dfrac{8}{5}t^0u^2$

$\quad\quad = -\dfrac{8}{5}\cdot1\cdot u^2$

$\quad\quad = -\dfrac{8u^2}{5}$

63. $\dfrac{28a^4b^3z}{42ab^3z^4}=\dfrac{28}{42}\cdot\dfrac{a^4}{a}\cdot\dfrac{b^3}{b^3}\cdot\dfrac{z}{z^4}$

$\quad\quad = \dfrac{2}{3}a^{4-1}b^{3-3}z^{1-4}$

$\quad\quad = \dfrac{2}{3}a^3b^0z^{-3}$

$\quad\quad = \dfrac{2}{3}\cdot\dfrac{a^3}{1}\cdot1\cdot\dfrac{1}{z^{-3}}$

$\quad\quad = \dfrac{2a^3}{3z^3}$

64. $(2a^3b^{-2})^{-3}=2^{-3}a^{3\cdot(-3)}b^{-2(-3)}=\dfrac{1}{8}a^{-9}b^6=\dfrac{b^6}{8a^9}$

65. $\dfrac{4hj^4k^{-2}}{(3hj^{-2})^2}=\dfrac{4hj^4k^{-2}}{3^2h^2j^{-4}}$

$\quad\quad = \dfrac{4h^{1-2}j^{4-(-4)}k^{-2}}{9}$

$\quad\quad = \dfrac{4h^{-1}j^8k^{-2}}{9}$

$\quad\quad = \dfrac{4j^8}{9hk^2}$

66. $8x^0 - (2x)^0 = 8 \cdot 1 - 1$
$= 8 - 1$
$= 7$

67. $\left(-18x^4 y^{-5}\right)^0 = 1$

68. $\dfrac{4x - 20}{4} = \dfrac{4x}{4} - \dfrac{20}{4} = x - 5$

69. $\dfrac{5y^2 - 10y + 15}{-5} = \dfrac{5y^2}{-5} - \dfrac{10y}{-5} + \dfrac{15}{-5}$
$= -y^2 + 2y - 3$

70. $\dfrac{2st + 20s^2 t^4 - 4st^5}{2st} = \dfrac{2st}{2st} + \dfrac{20s^2 t^4}{2st} - \dfrac{4st^5}{2st}$
$= 1 + 10st^3 - 2t^4$

71. $\dfrac{a^3 bc^3 - abc^2 + 2a^2 bc^4}{abc^3}$
$= \dfrac{a^3 bc^3}{abc^3} - \dfrac{abc^2}{abc^3} + \dfrac{2a^2 bc^4}{abc^3}$
$= a^2 - \dfrac{1}{c} + 2ac$

72. $\dfrac{6a^2 b^2 + 3a^2 b^3}{3a^2 b^2} = \dfrac{6a^2 b^2}{3a^2 b^2} + \dfrac{3a^2 b^3}{3a^2 b^2} = 2 + b$

73. $\dfrac{2xyz^2 - 5x^2 y^2 z^2 + 10xy^2 z}{5xy^3 z}$
$= \dfrac{2xyz^2}{5xy^3 z} - \dfrac{5x^2 y^2 z^2}{5xy^3 z} + \dfrac{10xy^2 z}{5xy^3 z}$
$= \dfrac{2z}{5y^2} - \dfrac{xz}{y} + \dfrac{2}{y}$

74. a) To find the height of the parallelogram, divide the area, $54x^2 - 60x$, by the base, $6x$.
$\dfrac{54x^2 - 60x}{6x} = \dfrac{54x^2}{6x} - \dfrac{60x}{6x} = 9x - 10$

   b) Base: $6x = 6(4) = 24$ inches
   Height: $9x - 10 = 9(4) - 10$
   $= 36 - 10$
   $= 26$ inches
   Area: $54x^2 - 60x = 54(4)^2 - 60(4)$
   $= 54(16) - 240$
   $= 864 - 240$
   $= 624$ in.^{2}

75. To find how long it takes the light from Proxima Centauri to reach Earth, divide the distance by the rate.
$\dfrac{2.47 \times 10^{13}}{5.881 \times 10^{12}} = \dfrac{2.47}{5.881} \times \dfrac{10^{13}}{10^{12}}$
$\approx 0.42 \times 10^{13-12}$
$\approx 0.42 \times 10^1$
≈ 4.2 years

76. $\begin{array}{r} z + 5 \\ z + 3 \overline{) z^2 + 8z + 15} \end{array}$
$\underline{z^2 + 3z}$
$5z + 15$
$\underline{5z + 15}$
0
Answer: $z + 5$

77. $\begin{array}{r} 2m - 2 \\ 3m - 2 \overline{) 6m^2 - 10m + 5} \end{array}$
$\underline{6m^2 - 4m}$
$-6m + 5$
$\underline{-6m + 4}$
1
Answer: $2m - 2 + \dfrac{1}{3m - 2}$

78. $\begin{array}{r} 4x^2 + 6x + 9 \\ 2x - 3 \overline{) 8x^3 + 0x^2 + 0x - 27} \end{array}$
$\underline{8x^3 - 12x^2}$
$12x^2 + 0x$
$\underline{12x^2 - 18x}$
$18x - 27$
$\underline{18x - 27}$
0
Answer: $4x^2 + 6x + 9$

79. $\begin{array}{r} 2s^2 + 2s \\ s - 1 \overline{) 2s^3 + 0s^2 - 2s - 3} \end{array}$
$\underline{2s^3 - 2s^2}$
$2s^2 - 2s$
$\underline{2s^2 - 2s}$
$0 - 3$
Answer: $2s^2 + 2s - \dfrac{3}{s - 1}$

80. To find the width, divide the area,
$4x^2 + 23x - 35$, by the length, $x + 7$.

$$
\begin{array}{r}
4x - 5 \\
x+7\overline{)4x^2 + 23x - 35} \\
\underline{4x^2 + 28x} \\
-5x - 35 \\
\underline{-5x - 35} \\
0
\end{array}
$$

Answer: $4x - 5$

Chapter 5 Practice Test

1. $2^{-3} = \dfrac{1}{2^3} = \dfrac{1}{8}$

2. $\left(\dfrac{2}{3}\right)^{-2} = \left(\dfrac{3}{2}\right)^2 = \dfrac{3^2}{2^2} = \dfrac{9}{4}$

3. $6.201 \times 10^{-3} = \dfrac{6.201}{10^3} = 0.006201$

4. $275{,}000{,}000 = 2.75 \times 10^8$

5.   For $-7x^2 y = -7x^2 y^1$, the degree of the monomial is the sum of the variables' exponents. So, the degree is $2 + 1 = 3$.

6.   For $4x^2 - 9x^4 + 8x - 7$, the degree is 4 because it is the greatest degree of all the terms.

7.   Let $m = 4$ and $n = -2$.
$$
\begin{aligned}
-6mn - n^3 &= -6(4)(-2) - (-2)^3 \\
&= 48 - (-8) \\
&= 48 + 8 \\
&= 56
\end{aligned}
$$

8. $-5x^4 + 7x^2 + 6x^2 - 5x^4 + 12 - 6x^3 + x^2$
$$
\begin{aligned}
&= -5x^4 - 5x^4 - 6x^3 + 7x^2 + 6x^2 + x^2 + 12 \\
&= -10x^4 - 6x^3 + 14x^2 + 12
\end{aligned}
$$

9. $\left(3x^2 + 4x - 2\right) + \left(5x^2 - 3x - 2\right)$
$$
\begin{aligned}
&= 3x^2 + 5x^2 + 4x - 3x - 2 - 2 \\
&= 8x^2 + x - 4
\end{aligned}
$$

10. $\left(7x^4 - 3x^2 + 4x + 1\right) - \left(2x^4 + 5x - 7\right)$
$$
\begin{aligned}
&= \left(7x^4 - 3x^2 + 4x + 1\right) + \left(-2x^4 - 5x + 7\right) \\
&= 7x^4 - 2x^4 - 3x^2 + 4x - 5x + 1 + 7 \\
&= 5x^4 - 3x^2 - x + 8
\end{aligned}
$$

11. $\left(4x^2\right)\left(3x^5\right) = 4 \cdot 3 x^{2+5} = 12x^7$

12. $\left(2ab^3 c^7\right)\left(-a^5 b\right) = 2 \cdot (-1) a^{1+5} b^{3+1} c^7$
$$
= -2a^6 b^4 c^7
$$

13. $\left(4xy^3\right)^2 = 4^2 x^2 y^{3 \cdot 2} = 16x^2 y^6$

14. $3x\left(x^2 - 4x + 5\right) = 3x \cdot x^2 + 3x \cdot (-4x) + 3x \cdot 5$
$$
= 3x^3 - 12x^2 + 15x
$$

15. $-6t^2 u\left(4t^3 - 8tu^2\right) = -6t^2 u \cdot 4t^3 - 6t^2 u \cdot \left(-8tu^2\right)$
$$
= -24t^5 u + 48t^3 u^3
$$

16. $(n - 1)(n + 4) = n \cdot n + n \cdot 4 + (-1) \cdot n + (-1) \cdot 4$
$$
\begin{aligned}
&= n^2 + 4n - n - 4 \\
&= n^2 + 3n - 4
\end{aligned}
$$

17. $(2x - 3)^2 = (2x)^2 - 2(2x)(3) + (3)^2$
$$
= 4x^2 - 12x + 9
$$

18. $(x + 2)\left(x^2 - 4x + 3\right)$
$$
\begin{aligned}
&= x \cdot x^2 + x \cdot (-4x) + x \cdot 3 + 2 \cdot x^2 + 2 \cdot (-4x) + 2 \cdot 3 \\
&= x^3 - 4x^2 + 3x + 2x^2 - 8x + 6 \\
&= x^3 - 2x^2 - 5x + 6
\end{aligned}
$$

19. $A = lw$
$$
\begin{aligned}
&= (2n - 5)(3n + 4) \\
&= 2n \cdot 3n + 2n \cdot 4 + (-5) \cdot 3n + (-5) \cdot 4 \\
&= 6n^2 + 8n - 15n - 20 \\
&= 6n^2 - 7n - 20
\end{aligned}
$$

20. $\dfrac{x^9}{x^4} = x^{9-4} = x^5$

21. $\dfrac{\left(x^3\right)^{-2}}{x^4 \cdot x^{-5}} = \dfrac{x^{3 \cdot (-2)}}{x^{4 + (-5)}} = \dfrac{x^{-6}}{x^{-1}} = x^{-6 - (-1)} = x^{-5} = \dfrac{1}{x^5}$

22. $\dfrac{(3y)^{-2}}{\left(x^3 y^2\right)^{-3}} = \dfrac{\left(x^3 y^2\right)^3}{(3y)^2}$

$= \dfrac{x^{3\cdot 3} y^{2\cdot 3}}{3^2 y^2}$

$= \dfrac{x^9 y^6}{9 y^2}$

$= \dfrac{x^9 y^4}{9}$

23. $\dfrac{24x^5 + 18x^3}{6x^2} = \dfrac{24x^5}{6x^2} + \dfrac{18x^3}{6x^2} = 4x^3 + 3x$

24. $x + 3 \overline{\smash{)}x^2 - x - 12}$ Answer: $x - 4$

$\underline{x^2 + 3x}$

$-4x - 12$

$\underline{-4x - 12}$

0

Quotient: $x - 4$

25. $3x - 2 \overline{\smash{)}15x^2 - 22x + 14}$

$\underline{15x^2 - 10x}$

$-12x + 14$

$\underline{-12x + 8}$

6

Quotient: $5x - 4$

Answer: $5x - 4 + \dfrac{6}{3x - 2}$

Chapters 1–5 Cumulative Review

1. False;

$3x - 4y = -24$ $3x - 4y = -24$

$3(0) - 4y = -24$ $3x - 4(0) = -24$

$-4y = -24$ $3x = -24$

$\dfrac{-4y}{-4} = \dfrac{-24}{-4}$ $\dfrac{3x}{3} = \dfrac{-24}{3}$

$y = 6$ $x = -8$

$(0, 6)$ $(-8, 0)$

The x-intercept is $(-8, 0)$ and the y-intercept is $(0, 6)$.

2. False; "6 less than a number" is translated as $x - 6$.

3. True

4. False; a linear equation in two variables may have no solution, one solution, or infinitely many solutions.

5. b. coefficient; coefficient

6. $(0, 0)$ or the origin

7. $5 + 2\left(6 - 3^2\right)^2 = 5 + 2(6 - 9)^2$

$= 5 + 2(-3)^2$

$= 5 + 2(9)$

$= 5 + 18$

$= 23$

8. $|5 - 10| + 3(2) \div (-1) = |-5| + 3(2) \div (-1)$

$= 5 + 3(2) \div (-1)$

$= 5 + 6 \div (-1)$

$= 5 + (-6)$

$= -1$

9. Let n be the number.

$0.302 \cdot n = 18.12$

$0.302n = 18.12$

$n = 60$

The number is 60.

10. $\left(3y^2 - 2y + 10\right) + \left(2y^2 - 2y - 11\right)$

$= 3y^2 + 2y^2 - 2y - 2y + 10 - 11$

$= 5y^2 - 4y - 1$

11. $(2x + 1) - (-4x + 7) = (2x + 1) + (4x - 7)$

$= 2x + 4x + 1 - 7$

$= 6x - 6$

12. $(2x + 3)(x - 7) = 2x \cdot x + 2x \cdot (-7) + 3 \cdot x + 3 \cdot (-7)$

$= 2x^2 - 14x + 3x - 21$

$= 2x^2 - 11x - 21$

13. $(2x - 3)^2 = (2x)^2 - 2(2x)(3) + (3)^2$

$= 4x^2 - 12x + 9$

14. $\left(3m^2 - 6m + 1\right) \div (3m) = \dfrac{3m^2 - 6m + 1}{3m}$

$= \dfrac{3m^2}{3m} - \dfrac{6m}{3m} + \dfrac{1}{3m}$

$= m - 2 + \dfrac{1}{3m}$

15.
$$\begin{array}{r} x+3 \\ x+2\overline{)x^2+5x+7} \\ \underline{x^2+2x} \\ 3x+7 \\ \underline{3x+6} \\ 1 \end{array}$$

Answer: $x+3+\dfrac{1}{x+2}$

16. $\left(3x^{-2}y^3\right)^{-3} = \dfrac{1}{\left(3x^{-2}y^3\right)^3}$

$= \dfrac{1}{3^3 x^{-2\cdot3} y^{3\cdot3}}$

$= \dfrac{1}{27x^{-6}y^9}$

$= \dfrac{x^6}{27y^9}$

17. $\left(\dfrac{2x^3}{y}\right)^2 = \dfrac{2^2 x^{3\cdot2}}{y^2} = \dfrac{4x^6}{y^2}$

18. $\left(2x^2 y^{-3}\right)^2 \left(-3x^4 y^2\right)^3$

$= 2^2 x^{2\cdot2} y^{-3\cdot2} \cdot (-3)^3 x^{4\cdot3} y^{2\cdot3}$

$= 4x^4 y^{-6} \cdot (-27) x^{12} y^6$

$= -108 x^{4+12} y^{-6+6}$

$= -108 x^{16} y^0$

$= -108 x^{16}$

19. $3x+2y=-6$

$2y = -3x-6$

$y = -\dfrac{3}{2}x-3$

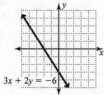

20. $y > \dfrac{2}{3}x-2$

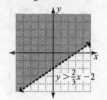

21. $10x-3+5 = 8x+26+4x$

$10x+2 = 12x+26$

$10x = 12x+24$

$-2x = 24$

$x = -12$

22. $3(x-2) > 4(x+1)$

$3x-6 > 4x+4$

$3x > 4x+10$

$-x > 10$

$x < -10$

23. $\dfrac{16}{9} = \dfrac{x}{20.25}$

$9x = 16(20.25)$

$9x = 324$

$\dfrac{9x}{9} = \dfrac{324}{9}$

$x = 36$

24. $\qquad -\dfrac{3}{7}x = \dfrac{12}{35}$

$-\dfrac{\cancel{7}}{\cancel{3}} \cdot \left(-\dfrac{\cancel{3}}{\cancel{7}}x\right) = -\dfrac{\cancel{7}}{\cancel{3}} \cdot \left(\dfrac{\overset{4}{\cancel{12}}}{\underset{5}{\cancel{35}}}\right)$

$\qquad\qquad x = -\dfrac{4}{5}$

25. $\qquad P = 2l+2w$

$P-2w = 2l$

$\dfrac{P-2w}{2} = l$

26. $\begin{cases} 2x - 3y = 14 \\ y = 3x - 14 \end{cases}$

$$2x - 3y = 14 \qquad\qquad y = 3x - 14$$
$$2x - 3(3x - 14) = 14 \qquad y = 3(4) - 14$$
$$2x - 9x + 42 = 14 \qquad\quad y = 12 - 14$$
$$-7x + 42 = 14 \qquad\qquad y = -2$$
$$-7x = -28$$
$$x = 4$$

Solution: $(4, -2)$

27. $\begin{cases} 3x - 4y = 24 \\ 5x + 3y = 11 \end{cases}$

$$\begin{array}{l} 3(3x - 4y = 24) \\ \underline{4(5x + 3y = 11)} \\ 9x - 12y = 72 \\ \underline{20x + 12y = 44} \\ 29x \qquad\quad = 116 \\ x = 4 \end{array} \qquad \begin{array}{l} 3x - 4y = 24 \\ 3(4) - 4y = 24 \\ 12 - 4y = 24 \\ -4y = 12 \\ y = -3 \end{array}$$

Solution: $(4, -3)$

28. Let x represent the principal invested at 9% annual interest. The amount invested at 13% annual interest is $3x$. We are given that the total interest after one year is \$5400. Using the relationship $I = Pr$, complete a table.

Accounts	Rate	Principal	Interest
1st account	0.09	x	$0.09x$
2nd account	0.13	$3x$	$0.13(3x)$

Translate to an equation.
$$0.09x + 0.13(3x) = 5400$$
$$0.09x + 0.39x = 5400$$
$$0.48x = 5400$$
$$\frac{0.48x}{0.48} = \frac{5400}{0.48}$$
$$x = 11,250$$

\$11,250 is invested 9% and 3(\$11,250) = \$33,750 is invested at 13% APR.

29. This situation is the same as "30% of 1600 is what number?" Translate word for word.

$$0.30(1600) = n$$
$$480 = n$$

We could expect 480 of the parents polled to think that teens should not work.

30. Find the decrease in temperature: $68 - 31.5 = 36.5$. The initial temperature is 68 degrees. Let p represent the unknown percent.

$$\frac{p}{100} = \frac{\text{Amount of decrease}}{\text{Initial amount}}$$
$$\frac{p}{100} = \frac{36.5}{68}$$
$$68p = 3650$$
$$\frac{68p}{68} = \frac{3650}{68}$$
$$p \approx 53.7$$

The percent of decrease in temperature from July to January is approximately 53.7%.

Chapter 6

Factoring

Exercise Set 6.1

1. The natural number factors of 9 are 1, 3, 9.

3. The natural number factors of 33 are 1, 3, 11, 33.

5. The natural number factors of 18 are 1, 2, 3, 6, 9, 18.

7. The natural number factors of 16 are 1, 2, 4, 8, 16.

9. The natural number factors of 44 are 1, 2, 4, 11, 22, 44.

11. The natural number factors of 60 are 1, 2, 3, 4, 5, 6, 10, 12, 15, 20, 30, 60.

13. The natural number factors of 56 are 1, 2, 4, 7, 8, 14, 28, 56.

15. The natural number factors of 90 are 1, 2, 3, 5, 6, 9, 10, 15, 18, 30, 45, 90.

17. $21 : 1, 3, 7, 21$
$30 : 1, 2, 3, 5, 6, 10, 15, 30$
$\text{GCF} = 3$

19. $72 : 1, 2, 3, 4, 6, 8, 9, 12, 18, 24, 36, 72$
$80 : 1, 2, 4, 5, 8, 10, 16, 20, 40, 80$
$\text{GCF} = 8$

21. $12 : 1, 2, 3, 4, 6, 12$
$42 : 1, 2, 3, 6, 7, 14, 21, 42$
$60 : 1, 2, 3, 4, 5, 6, 10, 12, 15, 20, 30, 60$
$\text{GCF} = 6$

23. $4xy : 2^2 \cdot x \cdot y$
$6xy : 2 \cdot 3 \cdot x \cdot y$
$\text{GCF} = 2xy$

25. $25h^2 : 5^2 \cdot h^2$
$60h^4 : 2^2 \cdot 3 \cdot 5 \cdot h^4$
$\text{GCF} = 5h^2$

27. $6a^5b : 2 \cdot 3 \cdot a^5 \cdot b$
$15a^4b^2 : 3 \cdot 5 \cdot a^4 \cdot b^2$
$\text{GCF} = 3a^4b$

29. $5c - 20 = 5\left(\dfrac{5c - 20}{5}\right)$
$= 5\left(\dfrac{5c}{5} - \dfrac{20}{5}\right)$
$= 5(c - 4)$

31. $8x - 12 = 4\left(\dfrac{8x - 12}{4}\right)$
$= 4\left(\dfrac{8x}{4} - \dfrac{12}{4}\right)$
$= 4(2x - 3)$

33. $x^2 - x = x\left(\dfrac{x^2 - x}{x}\right)$
$= x\left(\dfrac{x^2}{x} - \dfrac{x}{x}\right)$
$= x(x - 1)$

35. $18z^6 - 12z^4 = 6z^4\left(\dfrac{18z^6 - 12z^4}{6z^4}\right)$
$= 6z^4\left(\dfrac{18z^6}{6z^4} - \dfrac{12z^4}{6z^4}\right)$
$= 6z^4\left(3z^2 - 2\right)$

37. $18p^3 - 15p^5 = 3p^3\left(\dfrac{18p^3 - 15p^5}{3p^3}\right)$
$= 3p^3\left(\dfrac{18p^3}{3p^3} - \dfrac{15p^5}{3p^3}\right)$
$= 3p^3\left(6 - 5p^2\right)$

39. $6a^2b - 3ab^2 = 3ab\left(\dfrac{6a^2b - 3ab^2}{3ab}\right)$
$= 3ab\left(\dfrac{6a^2b}{3ab} - \dfrac{3ab^2}{3ab}\right)$
$= 3ab(2a - b)$

41. $14uv^2 - 7uv = 7uv\left(\dfrac{14uv^2 - 7uv}{7uv}\right)$
$= 7uv\left(\dfrac{14uv^2}{7uv} - \dfrac{7uv}{7uv}\right)$
$= 7uv(2v - 1)$

43. $25xy - 50xz + 100x$
$= 25x\left(\dfrac{25xy - 50xz + 100x}{25x}\right)$
$= 25x\left(\dfrac{25xy}{25x} - \dfrac{50xz}{25x} + \dfrac{100x}{25x}\right)$
$= 25x(y - 2z + 4)$

45. $x^2y + xy^2 + x^3y^3$

$$= xy\left(\frac{x^2y + xy^2 + x^3y^3}{xy}\right)$$

$$= xy\left(\frac{x^2y}{xy} + \frac{xy^2}{xy} + \frac{x^3y^3}{xy}\right)$$

$$= xy\left(x + y + x^2y^2\right)$$

47. $28ab^3c - 36a^2b^2c$

$$= 4ab^2c\left(\frac{28ab^3c - 36a^2b^2c}{4ab^2c}\right)$$

$$= 4ab^2c\left(\frac{28ab^3c}{4ab^2c} - \frac{36a^2b^2c}{4ab^2c}\right)$$

$$= 4ab^2c\left(7b - 9a\right)$$

49. $20p^2q + 24pq - 16pq^2$

$$= 4pq\left(\frac{20p^2q + 24pq - 16pq^2}{4pq}\right)$$

$$= 4pq\left(\frac{20p^2q}{4pq} + \frac{24pq}{4pq} - \frac{16pq^2}{4pq}\right)$$

$$= 4pq\left(5p + 6 - 4q\right)$$

51. $3mn^5p^2 + 18mn^3p - 6mnp$

$$= 3mnp\left(\frac{3mn^5p^2 + 18mn^3p - 6mnp}{3mnp}\right)$$

$$= 3mnp\left(\frac{3mn^5p^2}{3mnp} + \frac{18mn^3p}{3mnp} - \frac{6mnp}{3mnp}\right)$$

$$= 3mnp\left(n^4p + 6n^2 - 2\right)$$

53. $105a^3b^2 - 63a^2b^3 + 84a^6b^4$

$$= 21a^2b^2\left(\frac{105a^3b^2 - 63a^2b^3 + 84a^6b^4}{21a^2b^2}\right)$$

$$= 21a^2b^2\left(\frac{105a^3b^2}{21a^2b^2} - \frac{63a^2b^3}{21a^2b^2} + \frac{84a^6b^4}{21a^2b^2}\right)$$

$$= 21a^2b^2\left(5a - 3b + 4a^4b^2\right)$$

55. $18x^4 - 9x^3 + 30x^2 - 12x$

$$= 3x\left(\frac{18x^4 - 9x^3 + 30x^2 - 12x}{3x}\right)$$

$$= 3x\left(\frac{18x^4}{3x} - \frac{9x^3}{3x} + \frac{30x^2}{3x} - \frac{12x}{3x}\right)$$

$$= 3x\left(6x^3 - 3x^2 + 10x - 4\right)$$

57. $-3x + 6y = -3\left(\frac{-3x + 6y}{-3}\right)$

$$= -3\left(\frac{-3x}{-3} + \frac{6y}{-3}\right)$$

$$= -3\left(x - 2y\right)$$

59. $-20a^2 - 15a = -5a\left(\frac{-20a^2 - 15a}{-5a}\right)$

$$= -5a\left(\frac{-20a^2}{-5a} - \frac{15a}{-5a}\right)\backslash$$

$$= -5a\left(4a + 3\right)$$

61. $-12a^4b + 20a^3b^3 = -4a^3b\left(\frac{-12a^4b + 20a^3b^3}{-4a^3b}\right)$

$$= -4a^3b\left(\frac{-12a^4b}{-4a^3b} + \frac{20a^3b^3}{-4a^3b}\right)$$

$$= -4a^3b\left(3a - 5b^2\right)$$

63. $-4x^2 - 8x + 16 = -4\left(\frac{-4x^2 - 8x + 16}{-4}\right)$

$$= -4\left(\frac{-4x^2}{-4} - \frac{8x}{-4} + \frac{16}{-4}\right)$$

$$= -4\left(x^2 + 2x - 4\right)$$

65. $-24x^3y^2z - 30x^2y^3z^4 + 12x^4y^5z^2$

$$= -6x^2y^2z\left(\frac{-24x^3y^2z - 30x^2y^3z^4 + 12x^4y^5z^2}{-6x^2y^2z}\right)$$

$$= -6x^2y^2z\left(\frac{-24x^3y^2z}{-6x^2y^2z} - \frac{30x^2y^3z^4}{-6x^2y^2z} + \frac{12x^4y^5z^2}{-6x^2y^2z}\right)$$

$$= -6x^2y^2z\left(4x + 5yz^3 - 2x^2y^3z\right)$$

67. $y(a-3)+2(a-3)$

$$= (a-3)\left(\frac{y(a-3)+2(a-3)}{(a-3)}\right)$$

$$= (a-3)\left(\frac{y(a-3)}{(a-3)}+\frac{2(a-3)}{(a-3)}\right)$$

$$= (a-3)(y+2)$$

69. $4a(2m+3n)+b(2m+3n)$

$$= (2m+3n)\left(\frac{4a(2m+3n)+b(2m+3n)}{(2m+3n)}\right)$$

$$= (2m+3n)\left(\frac{4a(2m+3n)}{(2m+3n)}+\frac{b(2m+3n)}{(2m+3n)}\right)$$

$$= (2m+3n)(4a+b)$$

71. $5x(4m-5)-2(4m-5)$

$$= (4m-5)\left(\frac{5x(4m-5)-2(4m-5)}{(4m-5)}\right)$$

$$= (4m-5)\left(\frac{5x(4m-5)}{(4m-5)}-\frac{2(4m-5)}{(4m-5)}\right)$$

$$= (4m-5)(5x-2)$$

73. $2r(6p+5q)-4s(6p+5q)$

$$= 2(6p+5q)\left(\frac{2r(6p+5q)-4s(6p+5q)}{2(6p+5q)}\right)$$

$$= 2(6p+5q)\left(\frac{2r(6p+5q)}{2(6p+5q)}-\frac{4s(6p+5q)}{2(6p+5q)}\right)$$

$$= 2(6p+5q)(r-2s)$$

75. $bx+2b+cx+2c = (bx+2b)+(cx+2c)$

$$= b(x+2)+c(x+2)$$

$$= (x+2)(b+c)$$

77. $am-an-bm+bn = (am-an)-(bm-bn)$

$$= a(m-n)-b(m-n)$$

$$= (m-n)(a-b)$$

79. $x^3+2x^2-3x-6 = x^2(x+2)-3(x+2)$

$$= (x+2)(x^2-3)$$

81. $1-m+m^2-m^3 = 1(1-m)+m^2(1-m)$

$$= (1-m)(1+m^2)$$

83. $3xy+5y+6x+10 = (3xy+5y)+(6x+10)$

$$= y(3x+5)+2(3x+5)$$

$$= (3x+5)(y+2)$$

85. $4b^2-b+4b-1 = b(4b-1)+1(4b-1)$

$$= (4b-1)(b+1)$$

87. $6+2b-3a-ab = (6+2b)-(3a+ab)$

$$= 2(3+b)-a(3+b)$$

$$= (3+b)(2-a)$$

89. $3ax+6ay+4bx+8by$

$$= 3a(x+2y)+4b(x+2y)$$

$$= (x+2y)(3a+4b)$$

91. $x^2y-x^2s-ry+rs = (x^2y-x^2s)-(ry-rs)$

$$= x^2(y-s)-r(y-s)$$

$$= (y-s)(x^2-r)$$

93. $w^2+3wz+5w+15z = (w^2+3wz)+(5w+15z)$

$$= w(w+3z)+5(w+3z)$$

$$= (w+3z)(w+5)$$

95. $3st+3ty-2s-2y = (3st+3ty)-(2s+2y)$

$$= 3t(s+y)-2(s+y)$$

$$= (s+y)(3t-2)$$

97. $ax^2-5y^2+ay^2-5x^2$

$$= ax^2-5x^2+ay^2-5y^2$$

$$= (ax^2-5x^2)+(ay^2-5y^2)$$

$$= x^2(a-5)+y^2(a-5)$$

$$= (a-5)(x^2+y^2)$$

99. $2a^2+2ab+6a+6b = 2(a^2+ab+3a+3b)$

$$= 2[a(a+b)+3(a+b)]$$

$$= 2(a+b)(a+3)$$

101. $12ab+18a+16b+24$

$$= 2(6ab+9a+8b+12)$$

$$= 2[(6ab+9a)+(8b+12)]$$

$$= 2[3a(2b+3)+4(2b+3)]$$

$$= 2(2b+3)(3a+4)$$

103. $24xz - 12xw - 6yz + 3yw$

$= 3(8xz - 4xw - 2yz + yw)$

$= 3\big[(8xz - 4xw) - (2yz - yw)\big]$

$= 3\big[4x(2z - w) - y(2z - w)\big]$

$= 3(2z - w)(4x - y)$

105. $3a^2 y - 12a^2 + 9ay - 36a$

$= 3a(ay - 4a + 3y - 12)$

$= 3a\big[a(y - 4) + 3(y - 4)\big]$

$= 3a(y - 4)(a + 3)$

107. $2x^3 + 6x^2 y + 10x^2 + 30xy$

$= 2x(x^2 + 3xy + 5x + 15y)$

$= 2x\big[x(x + 3y) + 5(x + 3y)\big]$

$= 2x(x + 3y)(x + 5)$

109. Subtract the area of the inside rectangle from the area of the outside rectangle.

$A = 9x(x + 4) - 3x \cdot x$

$= 9x^2 + 36x - 3x^2$

$= 6x^2 + 36x$

$= 6x(x + 6)$

111. Add the area of the two rectangles. Then subtract the area of the hearth.

$A = \big[3x \cdot x + (3x + 4)(6x + 1)\big] - 4$

$= 3x^2 + 18x^2 + 3x + 24x + 4 - 4$

$= 21x^2 + 27x$

$= 3x(7x + 9)$

113. Add the volume of the sphere to the volume of the cylinder.

$V = \dfrac{4}{3}\pi(3r)^3 + 24\pi r^2$

$= \dfrac{4}{3}\pi(27r^3) + 24\pi r^2$

$= 36\pi r^3 + 24\pi r^2$

$= 12\pi r^2(3r + 2)$

Review Exercises

1. $3.74 \times 10^9 = 3{,}740{,}000{,}000$

2. $45{,}600{,}000 = 4.56 \times 10^7$

3. $(x + 2)(x + 5) = x \cdot x + x \cdot 5 + 2 \cdot x + 2 \cdot 5$

$= x^2 + 5x + 2x + 10$

$= x^2 + 7x + 10$

4. $(x - 4)(x + 3) = x \cdot x + x \cdot 3 - 4 \cdot x - 4 \cdot 3$

$= x^2 + 3x - 4x - 12$

$= x^2 - x - 12$

5. $(x - 5)(x - 3) = x \cdot x + x \cdot (-3) - 5 \cdot x - 5 \cdot (-3)$

$= x^2 - 3x - 5x + 15$

$= x^2 - 8x + 15$

6. $x(x - 7) = x \cdot x + x \cdot (-7)$

$= x^2 - 7x$

Exercise Set 6.2

1. 5   3. 2   5. 6   7. 2

9. To factor this trinomial, we must find a pair of numbers whose product is 3 and whose sum is 4. These numbers are 3 and 1.

$r^2 + 4r + 3 = (r + 3)(r + 1)$

11. To factor this trinomial, we must find a pair of numbers whose product is 7 and whose sum is -8. Because the product is positive and the sum is negative, both numbers must be negative. These numbers are -7 and -1.

$x^2 - 8x + 7 = (x - 7)(x - 1)$

13. To factor this trinomial, we must find a pair of numbers whose product is -3 and whose sum is -2. Because the product is negative, the two numbers must have different signs. Because the sum is also negative, the number with the greater absolute value will be negative. These numbers are -3 and 1.

$z^2 - 2z - 3 = (z - 3)(z + 1)$

15. To factor this trinomial, we must find a pair of numbers whose product is 6 and whose sum is 5. These numbers are 2 and 3.

$y^2 + 5y + 6 = (y + 2)(y + 3)$

17. To factor this trinomial, we must find a pair of numbers whose product is 8 and whose sum is -6. Because the product is positive and the sum is negative, both numbers must be negative. These numbers are -2 and -4.

$u^2 - 6u + 8 = (u - 2)(u - 4)$

19. To factor this trinomial, we must find a pair of numbers whose product is -6 and whose sum is 1. Because the product is negative, the two numbers must have different signs. Because the sum is positive, the number with the lesser absolute value will be negative. These numbers are 3 and -2.

$$u^2 + u - 6 = (u+3)(u-2)$$

21. To factor this trinomial, we must find a pair of numbers whose product is 12 and whose sum is 7. These numbers are 3 and 4.

$$a^2 + 7a + 12 = (a+3)(a+4)$$

23. To factor this trinomial, we must find a pair of numbers whose product is 9 and whose sum is -6. Because the product is positive and the sum is negative, both numbers must be negative. These numbers are -3 and -3.

$$y^2 - 6y + 9 = (y-3)(y-3) = (y-3)^2$$

25. To factor this trinomial, we must find a pair of numbers whose product is -12 and whose sum is -1. Because the product is negative, the two numbers must have different signs. Because the sum is also negative, the number with the greater absolute value will be negative. These numbers are -4 and 3.

$$w^2 - w - 12 = (w-4)(w+3)$$

27. To factor this trinomial, we must find a pair of numbers whose product is -30 and whose sum is -1. Because the product is negative, the two numbers must have different signs. Because the sum is also negative, the number with the greater absolute value will be negative. These numbers are -6 and 5.

$$x^2 - x - 30 = (x-6)(x+5)$$

29. To factor this trinomial, we must find a pair of numbers whose product is 30 and whose sum is -11. Because the product is positive and the sum is negative, both numbers must be negative. These numbers are -6 and -5.

$$n^2 - 11n + 30 = (n-6)(n-5)$$

31. To factor this trinomial, we must find a pair of numbers whose product is 18 and whose sum is -9. Because the product is positive and the sum is negative, both numbers must be negative. These numbers are -3 and -6.

$$r^2 - 9r + 18 = (r-3)(r-6)$$

33. To factor this trinomial, we must find a pair of numbers whose product is -24 and whose sum is -5. Because the product is negative, the two numbers must have different signs. Because the sum is also negative, the number with the greater absolute value will be negative. These numbers are -8 and 3.

$$x^2 - 5x - 24 = (x-8)(x+3)$$

35. To factor this trinomial, we must find a pair of numbers whose product is -36 and whose sum is -5. Because the product is negative, the two numbers must have different signs. Because the sum is also negative, the number with the greater absolute value will be negative. These numbers are -9 and 4.

$$p^2 - 5p - 36 = (p-9)(p+4)$$

37. To factor this trinomial, we must find a pair of numbers whose product is 6 and whose sum is -4. There are no combinations of factors of 6 that satisfy these conditions. Therefore, $m^2 - 4m + 6$ cannot be factored and is considered to be prime.

39. To factor this trinomial, we must find a pair of numbers whose product is -8 and whose sum is -6. There are no combinations of factors of -8 that satisfy these conditions. Therefore, $x^2 - 6x - 8$ cannot be factored and is considered to be prime.

41. To factor this trinomial, we must find a pair of terms whose product is $21q^2$ and whose sum is $-10q$. These terms would have to be $-3q$ and $-7q$.

$$p^2 - 10pq + 21q^2 = (p-3q)(p-7q)$$

43. To factor this trinomial, we must find a pair of terms whose product is $-27b^2$ and whose sum is $-6b$. These terms would have to be $-9b$ and $3b$.

$$a^2 - 6ab - 27b^2 = (a-9b)(a+3b)$$

45. To factor this trinomial, we must find a pair of terms whose product is $24y^2$ and whose sum is $-14y$. These terms would have to be $-12y$ and $-2y$.

$$x^2 - 14xy + 24y^2 = (x-12y)(x-2y)$$

47. To factor this trinomial, we must find a pair of terms whose product is $-30s^2$ and whose sum is $-s$. These terms would have to be $-6s$ and $5s$.
$$r^2 - rs - 30s^2 = (r - 6s)(r + 5s)$$

49. Factor out the GCF of the terms. Then factor the trinomial to two binomials. We must find a pair of numbers whose product is 21 and whose sum is -10.
$$4x^2 - 40x + 84 = 4(x^2 - 10x + 21)$$
$$= 4(x - 3)(x - 7)$$

51. Factor out the GCF of the terms. Then factor the trinomial to two binomials. We must find a pair of numbers whose product is 6 and whose sum is -7.
$$2m^3 - 14m^2 + 12m = 2m(m^2 - 7m + 6)$$
$$= 2m(m - 1)(m - 6)$$

53. Factor out the GCF of the terms. Then factor the trinomial to two binomials. We must find a pair of numbers whose product is -24 and whose sum is -5.
$$3a^2b - 15ab - 72b = 3b(a^2 - 5a - 24)$$
$$= 3b(a - 8)(a + 3)$$

55. Factor out the GCF of the terms. Then factor the trinomial to two binomials. We must find a pair of numbers whose product is 9 and whose sum is -6.
$$4x^2 - 24x + 36 = 4(x^2 - 6x + 9)$$
$$= 4(x - 3)(x - 3)$$
$$= 4(x - 3)^2$$

57. Factor out the GCF of the terms. Then factor the trinomial to two binomials. We must find a pair of numbers whose product is 6 and whose sum is 5.
$$n^4 + 5n^3 + 6n^2 = n^2(n^2 + 5n + 6)$$
$$= n^2(n + 2)(n + 3)$$

59. Factor out the GCF of the terms. Then factor the trinomial to two binomials. We must find a pair of numbers whose product is 5 and whose sum is 6.
$$7u^4 + 42u^3 + 35u^2 = 7u^2(u^2 + 6u + 5)$$
$$= 7u^2(u + 5)(u + 1)$$

61. Factor out the GCF of the terms. Then factor the trinomial to two binomials. We must find a pair of numbers whose product is 8 and whose sum is -6.
$$6a^2b^2c - 36ab^2c + 48b^2c = 6b^2c(a^2 - 6a + 8)$$
$$= 6b^2c(a - 2)(a - 4)$$

63. Factor out the GCF of the terms. Then factor the trinomial to two binomials. We must find a pair of numbers whose product is $10y^2$ and whose sum is $-7y$.
$$3x^2 - 21xy + 30y^2 = 3(x^2 - 7xy + 10y^2)$$
$$= 3(x - 2y)(x - 5y)$$

65. Factor out the GCF of the terms. Then factor the trinomial to two binomials. We must find a pair of numbers whose product is $-18b^2$ and whose sum is $-3b$.
$$2a^2b - 6ab^2 - 36b^3 = 2b(a^2 - 3ab - 18b^2)$$
$$= 2b(a - 6b)(a + 3b)$$

67. Factor out the GCF of the terms. When we try to factor the trinomial further, we find that there are no factor pairs of $-15y^2$ that yield a sum of $-3y$. The trinomial cannot be factored further.
$$4x^4y - 12x^3y^2 - 60x^2y^3$$
$$= 4x^2y(x^2 - 3xy - 15y^2)$$

69. The mistake is $(-3)(-2) = 6$, not -6. The correct factored form is
$$x^2 - 5x - 6 = (x - 6)(x + 1).$$

71. The mistake is $4x + (-1x) = 3x$, not $-3x$. The correct factored form is
$$x^2 - 3x - 4 = (x - 4)(x + 1).$$

73. $x^2 + bx - 21$
To make the trinomial factorable, the value of b must be the sum of factor pairs of -21. Because we wish b to be a natural number, we can disregard all negative values of b. Factor pairs of -21, and their resulting sums, are:
$$-1 + 21 = 20$$
$$1 + (-21) = -20$$
$$-3 + 7 = 4$$
$$3 + (-7) = -4$$
The natural number values of b that make the trinomial factorable are 4 and 20.

75. $x^2 + bx + 12$

To make the trinomial factorable, the value of b must be the sum of factor pairs of 12. Because b must be a natural number, both factors of 12 must be positive. Factor pairs of 12, and their resulting sums, are:

$1 + 12 = 13$

$2 + 6 = 8$

$3 + 4 = 7$

The natural number values of b that make the trinomial factorable are 7, 8, and 13.

77. $x^2 - 7x + c$

To make the trinomial factorable, the value of c must be the product of pairs of numbers whose sum is -7. Because c must be a natural number, the value of c is positive, which means both numbers must be negative. Possibilities are

$-1(-6) = 6$

$-2(-5) = 10$

$-3(-4) = 12$

The natural number values of c that make the trinomial factorable are 6, 10, and 12.

79. $x^2 + 10x + c$

To make the trinomial factorable, the value of c must be the product of pairs of numbers whose sum is 10. Because c must be a natural number, the value of c must be positive, which means both numbers must be positive. Possibilities are

$1(9) = 9$

$2(8) = 16$

$3(7) = 21$

$4(6) = 24$

$5(5) = 25$

The natural number values of c that make the trinomial factorable are 9, 16, 21, 24, and 25.

81. Factor the expression for area.

$h^2 + 6h + 8 = (h + 4)(h + 2)$

length: $h + 4$ width: $h + 2$

83. Factor the expression for the viewing area.

$w^2 - 16w + 60 = (w - 6)(w - 10)$

length: $w - 6$ width: $w - 10$

Review Exercises

1. $6x:\ \ 2 \cdot 3 \cdot x$

   $35y:\ \ 5 \cdot 7 \cdot y$

   These two expressions have no common factor other than 1. The GCF is 1.

2. $(2x + 3)(5x + 1) = 2x \cdot 5x + 2x \cdot 1 + 3 \cdot 5x + 3 \cdot 1$

   $\qquad\qquad\qquad = 10x^2 + 2x + 15x + 3$

   $\qquad\qquad\qquad = 10x^2 + 17x + 3$

3. $(3y - 4)(5y + 6) = 3y \cdot 5y + 3y \cdot 6 - 4 \cdot 5y - 4 \cdot 6$

   $\qquad\qquad\qquad = 15y^2 + 18y - 20y - 24$

   $\qquad\qquad\qquad = 15y^2 - 2y - 24$

4. $4n(6n - 1)(3n - 2)$

   $= 4n(6n \cdot 3n + 6n \cdot (-2) - 1 \cdot 3n - 1 \cdot (-2))$

   $= 4n(18n^2 - 12n - 3n + 2)$

   $= 4n(18n^2 - 15n + 2)$

   $= 72n^3 - 60n^2 + 8n$

5. $10x^2y^3 - 5x^3y - 20x^2y^2 = 5x^2y(2y^2 - x - 4y)$

6. $8a^2b^4 - 16ab^2 - 12a^2b^3 = 4ab^2(2ab^2 - 4 - 3ab)$

7. $x(x + 2) - 3(x + 2) = (x + 2)(x - 3)$

8. $z(z + 5) - 6(z + 5) = (z + 5)(z - 6)$

9. $15ac - 5ad - 3bc + bd$

   $= (15ac - 5ad) - (3bc - bd)$

   $= 5a(3c - d) - b(3c - d)$

   $= (3c - d)(5a - b)$

10. $6ac + 4ad - 9bc - 6bd$

    $= (6ac + 4ad) - (9bc + 6bd)$

    $= 2a(3c + 2d) - 3b(3c + 2d)$

    $= (3c + 2d)(2a - 3b)$

Exercise Set 6.3

1. $2, 4$ 3. $t, 4t$ 5. $7, 3x$

7. To factor this trinomial, we need to find a pair of first terms whose product equals $2j^2$ and a pair of last terms whose product is 2.
$$2j^2 + 5j + 2 = (2j+1)(j+2)$$

9. To factor this trinomial, we need to find a pair of first terms whose product equals $2y^2$ and a pair of last terms whose product is -5.
$$2y^2 - 3y - 5 = (2y-5)(y+1)$$

11. To factor this trinomial, we need to find a pair of first terms whose product equals $3m^2$ and a pair of last terms whose product is 8.
$$3m^2 - 10m + 8 = (3m-4)(m-2)$$

13. To factor this trinomial, we need to find a pair of first terms whose product equals $6a^2$ and a pair of last terms whose product is 7.
$$6a^2 + 13a + 7 = (6a+7)(a+1)$$

15. To factor this trinomial, we need to find a pair of first terms whose product equals $6p^2$ and a pair of last terms whose product is 1. There are no such pairs that yield a middle term of $2p$.
$$6p^2 + 2p + 1 \text{ is prime.}$$

17. To factor this trinomial, we need to find a pair of first terms whose product equals $4a^2$ and a pair of last terms whose product is 12.
$$4a^2 - 19a + 12 = (4a-3)(a-4)$$

19. To factor this trinomial, we need to find a pair of first terms whose product equals $6x^2$ and a pair of last terms whose product is 15.
$$6x^2 + 19x + 15 = (2x+3)(3x+5)$$

21. To factor this trinomial, we need to find a pair of first terms whose product equals $16d^2$ and a pair of last terms whose product is -15.
$$16d^2 - 14d - 15 = (2d-3)(8d+5)$$

23. To factor this trinomial, we need to find a pair of first terms whose product equals $3p^2$ and a pair of last terms whose product is $4q^2$.
$$3p^2 + 13pq + 4q^2 = (3p+q)(p+4q)$$

25. To factor this trinomial, we need to find a pair of first terms whose product equals $5k^2$ and a pair of last terms whose product is $-12h^2$.
$$5k^2 - 7kh - 12h^2 = (k+h)(5k-12h)$$

27. To factor this trinomial, we need to find a pair of first terms whose product equals $12a^2$ and a pair of last terms whose product is $25b^2$.
$$12a^2 - 40ab + 25b^2 = (2a-5b)(6a-5b)$$

29. To factor this trinomial, we need to find a pair of first terms whose product equals $8m^2$ and a pair of last terms whose product is $9n^2$.
$$8m^2 - 27mn + 9n^2 = (8m-3n)(m-3n)$$

31. Factor out the GCF, y. The remaining trinomial cannot be factored further.
$$8x^2 - 4xy - y = y(8x^2 - 4x - 1)$$

33. To factor this trinomial, first factor out the GCF, 3. Then find a pair of first terms whose product equals $4y^2$ and a pair of last terms whose product is 3.
$$12y^2 + 24y + 9 = 3(4y^2 + 8y + 3)$$
$$= 3(2y+1)(2y+3)$$

35. To factor this trinomial, first factor out the GCF, $2a$. Then find a pair of first terms whose product equals $3b^2$ and a pair of last terms whose product is 7.
$$6ab^2 - 20ab + 14a = 2a(3b^2 - 10b + 7)$$
$$= 2a(3b-7)(b-1)$$

37. To factor this trinomial, first factor out the GCF, $2w$. Then find a pair of first terms whose product equals $3w^2$ and a pair of last terms whose product is $4v^2$.
$$6w^3 + 16w^2v + 8wv^2 = 2w(3w^2 + 8wv + 4v^2)$$
$$= 2w(3w+2v)(w+2v)$$

39. $3a^2 + 4a + 1 = 3a^2 + 3a + a + 1$
$$= 3a(a+1) + 1(a+1)$$
$$= (a+1)(3a+1)$$

41. $2t^2 - 3t + 1 = 2t^2 - t - 2t + 1$
$$= t(2t-1) - 1(2t-1)$$
$$= (2t-1)(t-1)$$

43. $3x^2 - 4x - 7 = 3x^2 + 3x - 7x - 7$
$$= 3x(x+1) - 7(x+1)$$
$$= (x+1)(3x-7)$$

45. $2y^2 - y - 6 = 2y^2 - 4y + 3y - 6$
$$= 2y(y-2) + 3(y-2)$$
$$= (y-2)(2y+3)$$

47. $10a^2 - 19a + 7 = 10a^2 - 14a - 5a + 7$
$$= 2a(5a-7) - 1(5a-7)$$
$$= (5a-7)(2a-1)$$

49. $8r^2 - 6r - 9 = 8r^2 + 6r - 12r - 9$
$$= 2r(4r+3) - 3(4r+3)$$
$$= (4r+3)(2r-3)$$

51. $20x^2 - 23x + 6 = 20x^2 - 8x - 15x + 6$
$$= 4x(5x-2) - 3(5x-2)$$
$$= (5x-2)(4x-3)$$

53. $6k^2 + 7jk + 2j^2 = 6k^2 + 3jk + 4jk + 2j^2$
$$= 3k(2k+j) + 2j(2k+j)$$
$$= (2k+j)(3k+2j)$$

55. $5x^2 - 26xy + 5y^2 = 5x^2 - 25xy - xy + 5y^2$
$$= 5x(x-5y) - y(x-5y)$$
$$= (x-5y)(5x-y)$$

57. $10s^2 + st - 2t^2 = 10s^2 - 4st + 5st - 2t^2$
$$= 2s(5s-2t) + t(5s-2t)$$
$$= (5s-2t)(2s+t)$$

59. $6u^2 + 5uv - 6v^2 = 6u^2 - 4uv + 9uv - 6v^2$
$$= 2u(3u-2v) + 3v(3u-2v)$$
$$= (3u-2v)(2u+3v)$$

61. $15y^2 - 13y - 8$ cannot be factored any further. This trinomial is prime.

63. $4k^2 - 14k + 12 = 2(2k^2 - 7k + 6)$
$$= 2(2k^2 - 4k - 3k + 6)$$
$$= 2[2k(k-2) - 3(k-2)]$$
$$= 2(k-2)(2k-3)$$

65. $12m^3 + 10m^2 - 12m$
$$= 2m(6m^2 + 5m - 6)$$
$$= 2m(6m^2 + 9m - 4m - 6)$$
$$= 2m[3m(2m+3) - 2(2m+3)]$$
$$= 2m(2m+3)(3m-2)$$

67. $24x^3 - 64x^2y - 24xy^2$
$$= 8x(3x^2 - 8xy - 3y^2)$$
$$= 8x(3x^2 + xy - 9xy - 3y^2)$$
$$= 8x[x(3x+y) - 3y(3x+y)]$$
$$= 8x(3x+y)(x-3y)$$

69. $24x^3 - 20x^2y - 24xy^2$
$$= 4x(6x^2 - 5xy - 6y^2)$$
$$= 4x(6x^2 - 9xy + 4xy - 6y^2)$$
$$= 4x[3x(2x-3y) + 2y(2x-3y)]$$
$$= 4x(2x-3y)(3x+2y)$$

71. Mistake: Using FOIL to check, we see that the first and last terms check but the inner and outer terms combine to give $31x$ instead of $13x$.
Correct: $(2x+1)(3x+5)$

73. Mistake: Did not factor completely.
Correct: $4(n+2)(n+1)$

75. $15x^2 - 11x + 2 = (5x-2)(3x-1)$
Possible expressions for the length and width are $5x-2$ and $3x-1$.

77. $6w^2 - w - 2 = (2w+1)(3w-2)$
Possible expressions for the length and width are $2w+1$ and $3w-2$.

79. $2x^2 + bx + 6$
The possibilities for b are
$2 \cdot 6 + 1 \cdot 1 = 12 + 1 = 13$
$2 \cdot 1 + 6 \cdot 1 = 2 + 6 = 8$
$2 \cdot 3 + 2 \cdot 1 = 6 + 2 = 8$
$2 \cdot 2 + 3 \cdot 1 = 4 + 3 = 7$
The values of b that make the expression factorable are 7, 8, or 13

81. $5x^2 + bx - 9$

The possibilities for b are

$5 \cdot (-9) + 1 \cdot 1 = -45 + 1 = -44$

$5 \cdot 9 + (-1) \cdot 1 = 45 - 1 = 44$

$5 \cdot (-1) + 9 \cdot 1 = -5 + 9 = 4$

$5 \cdot 1 + (-9) \cdot 1 = 5 - 9 = -4$

$5 \cdot 3 + (-3) \cdot 1 = 15 - 3 = 12$

$5 \cdot (-3) + 3 \cdot 1 = -15 + 3 = -12$

We want only the natural numbers that can replace b. The values of b that make the expression factorable are 4, 12, or 44.

83. $6x^2 + bx - 7$

The possibilities for b are

$6 \cdot (-7) + 1 \cdot 1 = -42 + 1 = -41$

$6 \cdot 7 + (-1) \cdot 1 = 42 - 1 = 41$

$6 \cdot (-1) + 7 \cdot 1 = -6 + 7 = 1$

$6 \cdot 1 + (-7) \cdot 1 = 6 - 7 = -1$

$3 \cdot (-7) + 1 \cdot 2 = -21 + 2 = -19$

$3 \cdot 7 + (-1) \cdot 2 = 21 - 2 = 19$

$3 \cdot (-1) + 7 \cdot 2 = -3 + 14 = 11$

$3 \cdot 1 + (-7) \cdot 2 = 3 - 14 = -11$

$2 \cdot (-7) + 1 \cdot 3 = -14 + 3 = -11$

$2 \cdot 7 + (-1) \cdot 3 = 14 - 3 = 11$

$2 \cdot (-1) + 7 \cdot 3 = -2 + 21 = 19$

$2 \cdot 1 + (-7) \cdot 3 = 2 - 21 = -19$

We want only the natural numbers that can replace b. The values of b that make the expression factorable are 1, 11, 19, or 41

Review Exercises

1. $(3x+5)(3x-5) = (3x)^2 - 5^2$
$$= 9x^2 - 25$$

2. $(x-2)(x^2 + 2x + 4)$
$$= x \cdot x^2 + x \cdot 2x + x \cdot 4 - 2 \cdot x^2 - 2 \cdot 2x - 2 \cdot 4$$
$$= x^3 + 2x^2 + 4x - 2x^2 - 4x - 8$$
$$= x^3 - 8$$

3. $(2x-5)^2 = 4x^2 - 20x + 25$

4. $16c^3 d^2 - 24c^2 d^4 + 36cd^2$
$$= 4cd^2 (4c^2 - 6cd^2 + 9)$$

5. $ab + 5a + 3b + 15 = (ab + 5a) + (3b + 15)$
$$= a(b+5) + 3(b+5)$$
$$= (b+5)(a+3)$$

6. $m^2 + 2m - 24 = (m-4)(m+6)$

7. $3x^3 + 9x^2 - 30x = 3x(x^2 + 3x - 10)$
$$= 3x(x-2)(x+5)$$

Exercise Set 6.4

1. $x^2 + 14x + 49 = (x)^2 + 2(7)(x) + (7)^2$
$$= (x+7)^2$$

3. $b^2 - 8b + 16 = (b)^2 - 2(4)(b) + (4)^2$
$$= (b-4)^2$$

5. $n^2 + 12n + 144$ is not a perfect square.

7. $25u^2 - 30u + 9 = (5u)^2 - 2(3)(5u) + (3)^2$
$$= (5u-3)^2$$

9. $100w^2 + 20w + 1 = (10w)^2 + 2(1)(10w) + (1)^2$
$$= (10w+1)^2$$

11. $y^2 + 2yz + z^2 = (y)^2 + 2(y)(z) + (z)^2$
$$= (y+z)^2$$

13. $4p^2 - 28pq + 49q^2 = (2p)^2 - 2(2p)(7q) + (7q)^2$
$$= (2p-7q)^2$$

15. $16g^2 + 24gh + 9h^2 = (4g)^2 + 2(4g)(3h) + (3h)^2$
$$= (4g+3h)^2$$

17. $16t^2 + 80t + 100 = 4(4t^2 + 20t + 25)$
$$= 4\left[(2t)^2 + 2(5)(2t) + (5)^2 \right]$$
$$= 4(2t+5)^2$$

19. $x^2 - 4 = (x+2)(x-2)$

21. $16 - y^2 = (4+y)(4-y)$

23. $p^2 - q^2 = (p+q)(p-q)$

25. $25u^2 - 16 = (5u)^2 - (4)^2$
$= (5u + 4)(5u - 4)$

27. $9x^2 + b^2$ is prime.

29. $64m^2 - 25n^2 = (8m)^2 - (5n)^2$
$= (8m + 5n)(8m - 5n)$

31. $50x^2 - 32y^2 = 2(25x^2 - 16y^2)$
$= 2[(5x)^2 - (4y)^2]$
$= 2(5x + 4y)(5x - 4y)$

33. $4x^2 + 100y^2 = 4(x^2 + 25y^2)$

35. $x^4 - y^4 = (x^2)^2 - (y^2)^2$
$= (x^2 + y^2)(x^2 - y^2)$
$= (x^2 + y^2)[(x)^2 - (y)^2]$
$= (x^2 + y^2)(x + y)(x - y)$

37. $x^4 - 16 = (x^2)^2 - (4)^2$
$= (x^2 + 4)(x^2 - 4)$
$= (x^2 + 4)[(x)^2 - (2)^2]$
$= (x^2 + 4)(x + 2)(x - 2)$

39. $n^3 - 27 = (n - 3)(n^2 + 3n + 9)$

41. $x^3 + 27 = (x + 3)(x^2 - 3x + 9)$

43. $x^3 - 1 = (x - 1)(x^2 + x + 1)$

45. $m^3 + n^3 = (m + n)(m^2 - mn + n^2)$

47. $27k^3 - 8 = (3k)^3 - (2)^3$
$= (3k - 2)(9k^2 + 6k + 4)$

49. $27k^3 + 8 = (3k)^3 + (2)^3$
$= (3k + 2)(9k^2 - 6k + 4)$

51. $c^3 - 64d^3 = (c)^3 - (4d)^3$
$= (c - 4d)(c^2 + 4cd + 16d^2)$

53. $125x^3 + 64y^3 = (5x)^3 + (4y)^3$
$= (5x + 4y)(25x^2 - 20xy + 16y^2)$

55. $27x^3 - 64y^3 = (3x)^3 - (4y)^3$
$= (3x - 4y)(9x^2 + 12xy + 16y^2)$

57. $8p^3 + q^3z^3 = (2p)^3 + (qz)^3$
$= (2p + qz)(4p^2 - 2pqz + q^2z^2)$

59. $2x^2 - 50 = 2(x^2 - 25)$
$= 2(x + 5)(x - 5)$

61. $16x^2 - \dfrac{25}{49} = \left(4x + \dfrac{5}{7}\right)\left(4x - \dfrac{5}{7}\right)$

63. $2u^3 - 2u = 2u(u^2 - 1) = 2u(u + 1)(u - 1)$

65. $y^5 - 16y^3b^2 = y^3(y^2 - 16b^2)$
$= y^3(y + 4b)(y - 4b)$

67. $50x^3 + 2x = 2x(25x^2 + 1)$

69. $3y^3 - 24z^3 = 3(y^3 - 8z^3)$
$= 3(y - 2z)(y^2 + 2yz + 4z^2)$

71. $c^3 - \dfrac{8}{27} = \left(c - \dfrac{2}{3}\right)\left(c^2 + \dfrac{2}{3}c + \dfrac{4}{9}\right)$

73. $16c^4 + 2cd^3 = 2c(8c^3 + d^3)$
$= 2c(2c + d)(4c^2 - 2cd + d^2)$

75. $(2a - b)^2 - c^2 = ((2a - b) + c)((2a - b) - c)$
$= (2a - b + c)(2a - b - c)$

77. $16 - 9(x - y)^2 = (4 - 3(x - y))(4 + 3(x - y))$
$= (4 - 3x + 3y)(4 + 3x - 3y)$

79. $x^3 + 27(y + z)^3$
$= (x + 3(y + z))(x^2 - 3x(y + z) + 9(y + z)^2)$
$= (x + 3y + 3z)(x^2 - 3xy - 3xz + 9(y^2 + 2yz + z^2))$
$= (x + 3y + 3z)(x^2 - 3xy - 3xz + 9y^2 + 18yz + 9z^2)$

81. $(x+y)^3 - 64d^3$

$$= \big((x+y) - 4d\big)\big((x+y)^2 + 4d(x+y) + 16d^2\big)$$

$$= (x+y-4d)\big(x^2 + 2xy + y^2 + 4dx + 4dy + 16d^2\big)$$

83. $16x^2 + bx + 9 = (4x)^2 + bx + (3)^2$

This trinomial will be a perfect square trinomial if $b = 2(4)(3) = 24$.

85. $4x^2 - bx + 81 = (2x)^2 - bx + (9)^2$

This trinomial will be a perfect square trinomial if $b = 2(2)(9) = 36$.

87. $x^2 + 10x + c = (1x)^2 + 10x + (b)^2$

Let c be equal to the perfect square b^2. This trinomial will be a perfect square trinomial if
$$2ab = 10$$
$$2(1)b = 10$$
$$2b = 10$$
$$b = 5$$

The trinomial will be a perfect square if $c = 5^2 = 25$.

89. $9x^2 - 24x + c = (3x)^2 - 24x + (b)^2$

Let c be equal to the perfect square b^2. This trinomial will be a perfect square trinomial if
$$2ab = 24$$
$$2(3)b = 24$$
$$6b = 24$$
$$b = 4$$

The trinomial will be a perfect square if $c = 4^2 = 16$.

91. Subtract the volume of the inner box from the volume of the outer box.
$$A = 10x \cdot x - 8 \cdot 5$$
$$= 10x^2 - 40$$
$$= 10(x^2 - 4)$$
$$= 10(x+2)(x-2)$$

93. Subtract the volume of the inner rectangle from the area of the outer rectangle.
$$V = 4x \cdot 2x \cdot x - 9 \cdot 3 \cdot 1$$
$$= 8x^3 - 27$$
$$= (2x-3)(4x^2 + 6x + 9)$$

Review Exercises

1. The natural number factors of 36 are: 1, 2, 3, 4, 6, 9, 12, 18, 36

2. $100 = 2 \cdot 2 \cdot 5 \cdot 5$

3. $14x^3 y^3 - 21x^2 y^4 - 7xy^2$
$$= 7xy^2\big(2x^2 y - 3xy^2 - 1\big)$$

4. $x^2 + 4xy + 3x + 12y$
$$= \big(x^2 + 4xy\big) + \big(3x + 12y\big)$$
$$= x(x+4y) + 3(x+4y)$$
$$= (x+4y)(x+3)$$

5. $a^2 - 3ab - 18b^2 = (a+3b)(a-6b)$

6. $3a^2 b + 6ab - 45b = 3b\big(a^2 + 2a - 15\big)$
$$= 3b(a+5)(a-3)$$

7. $6n^2 + 5n - 6 = (3n-2)(2n+3)$

8. $12x^3 y - 38x^2 y^2 + 20xy^3$
$$= 2xy\big(6x^2 - 19xy + 10y^2\big)$$
$$= 2xy(2x-5y)(3x-2y)$$

Exercise Set 6.5

1. $3xy^2 + 6x^2 y = 3xy(y+2x)$

3. $7a(x+y) - b(x+y) = (x+y)(7a-b)$

5. $2x^2 - 32 = 2\big(x^2 - 16\big) = 2(x+4)(x-4)$

7. $ax + ay + bx + by = (ax+ay) + (bx+by)$
$$= a(x+y) + b(x+y)$$
$$= (x+y)(a+b)$$

9. $12a^3 b^2 c + 3a^2 b^2 c^2 + 5abc^3$
$$= abc\big(12a^2 b + 3abc + 5c^2\big)$$

11. $x^2 + 8x + 15 = (x+3)(x+5)$

13. $x^4 - 16 = \big(x^2 + 4\big)\big(x^2 - 4\big)$
$$= \big(x^2 + 4\big)(x+2)(x-2)$$

15. $x^2 + 25$ is prime.

17. $15x^2 + 7x - 2 = (5x-1)(3x+2)$

19. $ax^2 + 4ax + 4a = a(x^2 + 4x + 4)$
$$= a(x+2)^2$$

21. $6ab - 36ab^2 = 6ab(1-6b)$

23. $x^2 + x + 2$ is prime.

25. $x^2 - 49 = (x+7)(x-7)$

27. $p^2 + p - 30 = (p+6)(p-5)$

29. $u^3 - u = u(u^2 - 1)$
$$= u(u+1)(u-1)$$

31. $2b^2 + 14b + 24 = 2(b^2 + 7b + 12)$
$$= 2(b+3)(b+4)$$

33. $6r^2 - 15r^3 = 3r^2(2-5r)$

35. $14u^2 + 7u - 105 = 7(2u^2 + u - 15)$
$$= 7(2u-5)(u+3)$$

37. $4h^2 + 12h + 4 = 4(h^2 + 3h + 1)$

39. $5p^2 - 80 = 5(p^2 - 16)$
$$= 5(p+4)(p-4)$$

41. $3w^2 + 5w + 2 = (3w+2)(w+1)$

43. $8q^2 - 10q - 3 = (2q-3)(4q+1)$

45. $12v^2 + 23v + 10 = (3v+2)(4v+5)$

47. $2 - 50x^2 = 2(1 - 25x^2)$
$$= 2(1+5x)(1-5x)$$

49. $80k^2 - 20k^2l^2 = 20k^2(4-l^2)$
$$= 20k^2(2+l)(2-l)$$

51. $2j^2 + j + 3$ is prime.

53. $50 - 20t + 2t^2 = 2(25 - 10t + t^2)$
$$= 2(5-t)^2$$

55. $3ax - 6ay - 8by + 4bx$
$$= 3ax - 6ay + 4bx - 8by$$
$$= 3a(x-2y) + 4b(x-2y)$$
$$= (x-2y)(3a+4b)$$

57. $x^3 - x^2 - x + 1 = x^2(x-1) - 1(x-1)$
$$= (x-1)(x^2-1)$$
$$= (x-1)(x+1)(x-1)$$
$$= (x+1)(x-1)^2$$

59. $4x^2 - 28x + 49 = (2x-7)^2$

61. $2x^4 - 162 = 2(x^4 - 81)$
$$= 2(x^2+9)(x^2-9)$$
$$= 2(x^2+9)(x+3)(x-3)$$

63. $a^2 + 2ab + ab + 2b^2$
$$= a(a+2b) + b(a+2b)$$
$$= (a+2b)(a+b)$$

65. $9 - 4m^2 = (3+2m)(3-2m)$

67. $x^5 - 4xy^2 = x(x^4 - 4y^2)$
$$= x(x^2+2y)(x^2-2y)$$

69. $20x^2 + 3xy + 2y^2$ is prime.

71. $b^3 + 125 = (b+5)(b^2 - 5b + 25)$

73. $3y^3 - 24 = 3(y^3 - 8)$
$$= 3(y-2)(y^2 + 2y + 4)$$

75. $54x - 2xy^3 = 2x(27 - y^3)$
$$= 2x(3-y)(9+3y+y^2)$$

77. $A = 6x^2 - 11x - 7$
$$= (2x+1)(3x-7)$$
$l = 2x+1$
$w = 3x-7$

79. $V = 18x^3 + 60x^2 + 50x$

   $V = 2x(9x^2 + 30x + 25)$

   $V = 2x(3x+5)(3x+5)$

   $l = 2x$

   $w = 3x+5$

   $h = 3x+5$

81. $100 - 16t^2 = 4(25 - 4t^2) = 4(5+2t)(5-2t)$

83. $6ir + 15i + 8r + 20$

   $= 3i(2r+5) + 4(2r+5)$

   $= (2r+5)(3i+4)$

   current: $3i+4$

   resistance: $2r+5$

Review Exercises

1. $\begin{array}{r} x-4 \\ 2x+1 \overline{)2x^2 - 7x - 8} \end{array}$ so: $x - 4 - \dfrac{4}{2x+1}$

   $\underline{2x^2 + x}$

   $-8x - 8$

   $\underline{-8x - 4}$

   -4

2. $6y^2 + 23y - 4 = (6y-1)(y+4)$

3. $A = 14 \cdot 16$

   $= 224$

   The area is 224 ft.^{2}

4. $4(x+3) = 20$

   $4x + 12 = 20$

   $4x + 12 - 12 = 20 - 12$

   $4x = 8$

   $x = 2$

   The number is 2.

5. Let the first integer be x. Then the second consecutive integer is $(x + 1)$, and the third consecutive integer is $(x + 2)$. Translate to the equation $x + (x+1) + (x+2) = 54$ and solve for x.

   $x + (x+1) + (x+2) = 54$

   $3x + 3 = 54$

   $3x = 51$

   $x = 17$

   The integers are 17, 18, and 19.

Exercise Set 6.6

1. $(x+5)(x+2) = 0$

   $x+5 = 0$ or $x+2 = 0$

   $x = -5$ $x = -2$

3. $(a+3)(a-4) = 0$

   $a+3 = 0$ or $a-4 = 0$

   $a = -3$ $a = 4$

5. $(3x-4)(2x+3) = 0$

   $3x-4 = 0$ or $2x+3 = 0$

   $3x = 4$ $2x = -3$

   $x = \dfrac{4}{3}$ $x = -\dfrac{3}{2}$

7. $x(x-7) = 0$

   $x = 0$ or $x-7 = 0$

   $x = 7$

9. $m(m-1)(m+2) = 0$

   $m = 0$ or $m-1 = 0$ or $m+2 = 0$

   $m = 1$ $m = -2$

11. $b(b-2)^2 = 0$

   $b = 0$ or $b-2 = 0$

   $b = 2$

13. $y^2 - 4y = 0$

   $y(y-4) = 0$

   $y = 0$ or $y-4 = 0$

   $y = 4$

15. $6r^2 + 10r = 0$

   $2r(3r+5) = 0$

   $2r = 0$ or $3r+5 = 0$

   $r = 0$ $r = -\dfrac{5}{3}$

17. $x^2 - 4 = 0$

   $(x+2)(x-2) = 0$

   $x+2 = 0$ or $x-2 = 0$

   $x = -2$ $x = 2$

19. $x^2 + x - 6 = 0$

   $(x+3)(x-2) = 0$

   $x+3 = 0$ or $x-2 = 0$

   $x = -3$ $x = 2$

21. $n^2 - 6n + 9 = 0$

    $(n-3)^2 = 0$

    $n - 3 = 0$

    $n = 3$

23. $3a^2 - 11a - 4 = 0$

    $(a-4)(3a+1) = 0$

    $a - 4 = 0$ or $3a + 1 = 0$

    $a = 4$ $3a = -1$

    $a = -\dfrac{1}{3}$

25. $6r^2 + 11r - 10 = 0$

    $(2r+5)(3r-2) = 0$

    $2r + 5 = 0$ or $3r - 2 = 0$

    $2r = -5$ $3r = 2$

    $r = -\dfrac{5}{2}$ $r = \dfrac{2}{3}$

27. $x^2 - 4x = 21$

    $x^2 - 4x - 21 = 0$

    $(x-7)(x+3) = 0$

    $x - 7 = 0$ or $x + 3 = 0$

    $x = 7$ $x = -3$

29. $p^2 = 3p - 2$

    $p^2 - 3p + 2 = 0$

    $(p-2)(p-1) = 0$

    $p - 2 = 0$ or $p - 1 = 0$

    $p = 2$ $p = 1$

31. $v^2 = 9$

    $v^2 - 9 = 0$

    $(v+3)(v-3) = 0$

    $v + 3 = 0$ or $v - 3 = 0$

    $v = -3$ $v = 3$

33. $2a^2 + 18a + 28 = 0$

    $2(a^2 + 9a + 14) = 0$

    $2(a+2)(a+7) = 0$

    $a + 2 = 0$ or $a + 7 = 0$

    $a = -2$ $a = -7$

35. $x^3 - x^2 - 12x = 0$

    $x(x^2 - x - 12) = 0$

    $x(x-4)(x+3) = 0$

    $x = 0$ or $x - 4 = 0$ or $x + 3 = 0$

    $x = 4$ $x = -3$

37. $12r^3 + 22r^2 - 20r = 0$

    $2r(6r^2 + 11r - 10) = 0$

    $2r(3r-2)(2r+5) = 0$

    $2r = 0$ or $3r - 2 = 0$ or $2r + 5 = 0$

    $r = 0$ $r = \dfrac{2}{3}$ $r = -\dfrac{5}{2}$

39. $3y^2 + 8 = 14y$

    $3y^2 - 14y + 8 = 0$

    $(3y-2)(y-4) = 0$

    $3y - 2 = 0$ or $y - 4 = 0$

    $y = \dfrac{2}{3}$ $y = 4$

41. $4x^2 = 60 - x$

    $4x^2 + x - 60 = 0$

    $(4x-15)(x+4) = 0$

    $4x - 15 = 0$ or $x + 4 = 0$

    $x = \dfrac{15}{4}$ $x = -4$

43. $7t = 12 + t^2$

    $t^2 - 7t + 12 = 0$

    $(t-4)(t-3) = 0$

    $t - 4 = 0$ or $t - 3 = 0$

    $t = 4$ $t = 3$

45. $b(b-5) = 14$

    $b^2 - 5b = 14$

    $b^2 - 5b - 14 = 0$

    $(b-7)(b+2) = 0$

    $b - 7 = 0$ or $b + 2 = 0$

    $b = 7$ $b = -2$

47.
$$4x(x+7) = -49$$
$$4x^2 + 28x = -49$$
$$4x^2 + 28x + 49 = 0$$
$$(2x+7)^2 = 0$$
$$2x+7 = 0$$
$$x = -\frac{7}{2}$$

49. $(x+1)(x+5) = -3$
$$x^2 + 6x + 5 = -3$$
$$x^2 + 6x + 8 = 0$$
$$(x+2)(x+4) = 0$$
$$x+2 = 0 \quad \text{or} \quad x+4 = 0$$
$$x = -2 \qquad x = -4$$

51. $(a-1)(a+4) = 14$
$$a^2 + 3a - 4 = 14$$
$$a^2 + 3a - 18 = 0$$
$$(a-3)(a+6) = 0$$
$$a-3 = 0 \quad \text{or} \quad a+6 = 0$$
$$a = 3 \qquad a = -6$$

53. Let x represent one of the numbers. Translate to an equation and solve.
$$x^2 + 55 = 16x$$
$$x^2 - 16x + 55 = 0$$
$$(x-5)(x-11) = 0$$
$$x-5 = 0 \quad \text{or} \quad x-11 = 0$$
$$x = 5 \qquad x = 11$$
The numbers are 5 and 11.

55. Let x represent the first odd natural number. Then the second consecutive odd natural number is represented by $x+2$. Translate to an equation and solve.
$$x(x+2) = 143$$
$$x^2 + 2x = 143$$
$$x^2 + 2x - 143 = 0$$
$$(x+13)(x-11) = 0$$
$$x+13 = 0 \quad \text{or} \quad x-11 = 0$$
$$x = \cancel{-13} \qquad x = 11$$
Because we are considering only natural numbers, 11 is the only solution. The second consecutive odd natural number is 11 + 2 = 13.

57. Let x represent the first natural number. Then the second consecutive natural number is represented by $x+1$. Translate to an equation and solve.
$$x^2 + (x+1)^2 = 365$$
$$x^2 + x^2 + 2x + 1 - 365 = 0$$
$$2x^2 + 2x - 364 = 0$$
$$2(x^2 + x - 182) = 0$$
$$(x-13)(x+14) = 0$$
$$x-13 = 0 \qquad x+14 = 0$$
$$x = 13 \qquad x = \cancel{-14}$$
Because we are considering only natural numbers, 13 is the only solution. The second consecutive natural number is 13 + 1 = 14.

59. Let w represent the width. Then the length is represented by $w+9$. Substitute into the formula and solve.
$$lw = A$$
$$(w+9)w = 252$$
$$w^2 + 9w - 252 = 0$$
$$(w-12)(w+21) = 0$$
$$w-12 = 0 \quad \text{or} \quad w+21 = 0$$
$$w = 12 \qquad w = \cancel{-21}$$
Because lengths can only be positive, disregard the negative solution. The width is 12 m and the length is 12 + 9 = 21 m. The dimensions of the garden are 12 m by 21 m.

61. Let w represent the width. Then the length is represented by $3w+10$. Substitute into the formula and solve.
$$lw = A$$
$$(3w+10)w = 9288$$
$$3w^2 + 10w = 9288$$
$$3w^2 + 10w - 9288 = 0$$
$$(3w+172)(w-54) = 0$$
$$3w+172 = 0 \quad \text{or} \quad w-54 = 0$$
$$w = \cancel{-57\frac{1}{3}} \qquad w = 54$$
Because lengths can only be positive, disregard the negative solution. The width is 54 ft. and the length is 3(54) + 10 = 172 ft. The dimensions are 172 ft. by 54 ft.

63. The area of the rectangle is $22 \cdot 28 = 616 \text{ ft.}^2$
Substitute into the formula and solve.

$$A = \pi r^2$$

$$616 = \frac{22}{7} r^2$$

$$\frac{7}{22} \cdot 616 = \frac{7}{22} \cdot \frac{22}{7} r^2$$

$$196 = r^2$$

$$0 = r^2 - 196$$

$$0 = (r - 14)(r + 14)$$

$$r - 14 = 0 \qquad \text{or} \quad r + 14 = 0$$

$$r = 14 \text{ ft.} \qquad\qquad r = \cancel{-14}$$

Because values of the radius can only be positive, disregard the negative solution. The radius of the building should be 14 ft.

65. Let h represent the height of the triangle. Then $h - 6$ represents the length of the base of the triangle.

$$A = \frac{1}{2} bh$$

$$216 = \frac{1}{2}(h - 6)h$$

$$2 \cdot 216 = 2 \cdot \frac{1}{2}(h - 6)h$$

$$432 = h^2 - 6h$$

$$0 = h^2 - 6h - 432$$

$$0 = (h - 24)(h + 18)$$

$$h - 24 = 0 \quad \text{or} \quad h + 18 = 0$$

$$h = 24 \qquad\qquad h = \cancel{-18}$$

Because lengths can only be positive, disregard the negative solution. The height of the triangle is 24 cm and the base is $24 - 6 = 18$ cm.

67. Let h represent the height of the trapezoid. Substitute into the formula and solve.

$$A = \frac{1}{2} h(a + b)$$

$$85.5 = \frac{1}{2} h(h + 10)$$

$$2 \cdot 85.5 = 2 \cdot \frac{1}{2} h(h + 10)$$

$$171 = h(h + 10)$$

$$171 = h^2 + 10h$$

$$0 = h^2 + 10h - 171$$

$$0 = (h + 19)(h - 9)$$

$$h - 9 = 0 \qquad \text{or} \qquad h + 19 = 0$$

$$h = 9 \text{ in.} \qquad\qquad h = \cancel{-19}$$

Because lengths can only be positive, disregard the negative solution. The height of the trapezoid is 9 in.

69. Let x represent the first even integer. The second consecutive even integer is $x + 2$, and the third consecutive even integer is $x + 4$. Translate to an equation and solve.

$$(x + 4)(x + 2) = 168$$

$$x^2 + 6x + 8 = 168$$

$$x^2 + 6x - 160 = 0$$

$$(x + 16)(x - 10) = 0$$

$$x + 16 = 0 \qquad \text{or} \quad x - 10 = 0$$

$$x = -16 \qquad\qquad x = 10$$

Beginning with the positive solution, the first integer is 10. The second consecutive even integer is $10 + 2 = 12$, and the third consecutive even integer is $10 + 4 = 14$. Now consider the negative solution. Then the integers are -16, -14, and -12.

71. Let $h = 9$, $v_0 = 4$, and $h_0 = 29$.

$$h = -16t^2 + v_0 t + h_0$$
$$9 = -16t^2 + 4t + 29$$
$$16t^2 - 4t - 20 = 0$$
$$4\left(4t^2 - t - 5\right) = 0$$
$$\left(4t^2 - t - 5\right) = 0$$
$$(4t - 5)(t + 1) = 0$$
$$4t - 5 = 0 \qquad t + 1 = 0$$
$$4t = 5 \qquad\quad t = \cancel{-1}$$
$$t = \frac{5}{4}$$
$$t = 1.25 \text{ sec.}$$

Because time can only be positive, disregard the negative solution. The ball will be 9 feet above the ground after 1.25 sec.

73. Let $B = 4840$, $P = 4000$, $t = 2$, and $n = 1$.

$$B = P\left(1 + \frac{r}{n}\right)^{nt}$$
$$4840 = 4000\left(1 + \frac{r}{1}\right)^{1 \cdot 2}$$
$$4840 = 4000(1 + r)^2$$
$$4840 = 4000\left(1 + 2r + r^2\right)$$
$$4840 = 4000 + 8000r + 4000r^2$$
$$0 = 4000r^2 + 8000r - 840$$
$$0 = 40\left(100r^2 + 200r - 21\right)$$
$$0 = \left(100r^2 + 200r - 21\right)$$
$$0 = (10r - 1)(10r + 21)$$
$$10r - 1 = 0 \qquad \text{or} \quad 10r + 21 = 0$$
$$10r = 1 \qquad\qquad 10r = -21$$
$$r = \frac{1}{10} \qquad\qquad r = \cancel{-\frac{21}{10}}$$
$$r = 0.10$$

Disregard the negative interest rate. The interest rate on the account is 0.10 = 10%.

75.
$$9^2 + 12^2 = c^2$$
$$81 + 144 = c^2$$
$$225 = c^2$$
$$0 = c^2 - 225$$
$$0 = (c + 15)(c - 15)$$
$$c + 15 = 0 \qquad \text{or} \quad c - 15 = 0$$
$$c = \cancel{-15} \qquad\qquad c = 15$$

77.
$$24^2 + 7^2 = c^2$$
$$576 + 49 = c^2$$
$$625 = c^2$$
$$0 = c^2 - 625$$
$$0 = (c + 25)(c - 25)$$
$$c + 25 = 0 \qquad \text{or} \quad c - 25 = 0$$
$$c = \cancel{-25} \qquad\qquad c = 25$$

79.
$$12^2 + 35^2 = c^2$$
$$144 + 1225 = c^2$$
$$1369 = c^2$$
$$0 = c^2 - 1369$$
$$0 = (c + 37)(c - 37)$$
$$c + 37 = 0 \qquad \text{or} \quad c - 37 = 0$$
$$c = \cancel{-37} \qquad\qquad c = 37$$

The third rod must be 37 cm long.

81.
$$18^2 + 24^2 = c^2$$
$$324 + 576 = c^2$$
$$900 = c^2$$
$$0 = c^2 - 900$$
$$0 = (c + 30)(c - 30)$$
$$c + 30 = 0 \qquad \text{or} \quad c - 30 = 0$$
$$c = \cancel{-30} \qquad\qquad c = 30$$

The shortest distance is 30 blocks.

83.
$$24^2 + b^2 = 51^2$$
$$576 + b^2 = 2601$$
$$b^2 - 2025 = 0$$
$$(b + 45)(b - 45) = 0$$
$$b + 45 = 0 \qquad \text{or} \quad b - 45 = 0$$
$$b = \cancel{-45} \qquad\qquad b = 45$$

The window is 45 in. long.

Review Exercises

1. No, it is not linear because the variable x has an exponent other than 1.

2. $4(1) + 3(-4)? - 8$

    $4 + (-12)? - 8$

    $-8 = -8$

    Yes, $(1, -4)$ is a solution.

3. To find the x-intercept, set $y = 0$ and solve for x.

    $2x - 5(0) = -20$ x-intercept: $(-10, 0)$

    $2x = -20$

    $x = -10$

    To find the y-intercept, set $x = 0$ and solve for y.

    $2(0) - 5y = -20$ y-intercept: $(0, 4)$

    $-5y = -20$

    $y = 4$

4. Find the slope between $(-1, 4)$ and $(-3, -2)$.

    $m = \dfrac{-2 - 4}{-3 - (-1)} = \dfrac{-6}{-3 + 1} = \dfrac{-6}{-2} = 3$

    Use the point-slope equation:

    $y - 4 = 3(x - (-1))$

    $y - 4 = 3(x + 1)$

    $y - 4 = 3x + 3$

    $y = 3x + 7$

    $Ax + By = C$

    $3x - y = -7$

5. $f(x) = x^2 + 2$

    a) $f(0) = 0^2 + 2 = 2$

    b) $f(-2) = (-2)^2 + 2 = 4 + 2 = 6$

6. $f(x) = 2x - 3$

    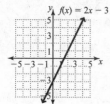

Exercise Set 6.7

1. The vertex is the lowest point of the parabola, $(2, -1)$.

3. The vertex is the highest point of the parabola, $(3, 0)$.

5. $y = 2x^2$

x	y
-2	8
-1	2
0	0
1	2
2	8

    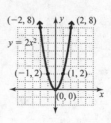

7. $y = -x^2$

x	y
-2	-4
-1	-1
0	0
1	-1
2	-4

    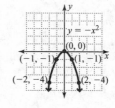

9. $y = x^2 - 3$

x	y
-2	1
-1	-2
0	-3
1	-2
2	1

    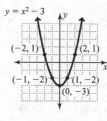

11. $y = -x^2 + 2x$

x	y
-1	-3
0	0
1	1
2	0
3	-3

    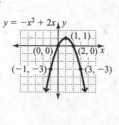

13. $y = 2x^2 - 4x + 1$

x	y
-1	7
0	1
1	-1
2	1
3	7

    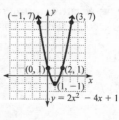

15. $y = -3x^2 + 6x + 4$

x	y
-1	-5
0	4
1	7
2	4
3	-5

$(1, 7)$
$(0, 4)$ $(2, 4)$
$(-1, -5)$ $(3, -5)$
$y = -3x^2 + 6x + 4$

17. $y = x^2 + 1$

x	y
-2	5
-1	2
0	1
1	2
2	5

$y = x^2 + 1$

19. $y = -x^2 + 3$

x	y
-2	-1
-1	2
0	3
1	2
2	-1

$y = -x^2 + 3$

21. $y = 2x^2 - 5$

x	y
-2	3
-1	-3
0	-5
1	-3
2	3

$y = 2x^2 - 5$

23. $y = -3x^2 + 2$

x	y
-2	-10
-1	-1
0	2
1	-1
2	-10

$y = -3x^2 + 2$

25. $y = x^2 + 4x - 1$

x	y
-4	-1
-3	-4
-2	-5
-1	-4
0	-1

$y = x^2 + 4x - 1$

27. $y = -x^2 + 6x - 5$

x	y
1	0
2	3
3	4
4	3
5	0

$y = -x^2 + 6x - 5$

29. $y = 2x^2 - 4x - 3$

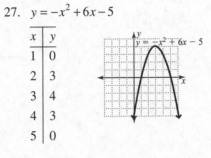

x	y
-1	3
0	-3
1	-5
2	-3
3	3

$y = 2x^2 - 4x - 3$

31. $y = -2x^2 + 8x - 5$

x	y
0	-5
1	1
2	3
3	1
4	-5

$y = -2x^2 + 8x - 5$

33. $f(x) = 3x^2$

x	y
-2	12
-1	3
0	0
1	3
2	12

$f(x) = 3x^2$

35. $f(x) = -x^2 + 1$

x	y
-2	-3
-1	0
0	1
1	0
2	-3

$f(x) = -x^2 + 1$

37. $f(x) = x^2 + 4x - 1$

x	y
-4	-1
-3	-4
-2	-5
-1	-4
0	-1

$f(x) = x^2 + 4x - 1$

39. $f(x) = -x^2 - 4x - 3$

x	y
-4	-3
-3	0
-2	1
-1	0
0	-3

$f(x) = -x^2 - 4x - 3$

41. $f(x) = 3x^2 - 12x + 5$

x	y
0	5
1	-4
2	-7
3	-4
4	5

$f(x) = 3x^2 - 12x + 5$

43. $f(x) = -2x^2 - 4x - 1$

x	y
-3	-7
-2	-1
-1	1
0	-1
1	-7

$f(x) = -2x^2 - 4x - 1$

45. This is a function because any vertical line intersects the graph in at most one point.
Domain: all real numbers or $(-\infty, \infty)$
Range: $\{y \mid y \geq 0\}$ or $[0, \infty)$

47. This is not a function because a vertical line can be drawn that intersects the graph in more than one point.
Domain: $\{x \mid x \geq 0\}$ or $[0, \infty)$
Range: all real numbers or $(-\infty, \infty)$

49. This is a function because any vertical line intersects the graph in at most one point.
Domain: all real numbers or $(-\infty, \infty)$
Range: $\{y \mid y \leq 2\}$ or $(-\infty, 2]$

51. This is not a function because a vertical line can be drawn that intersects the graph in more than one point.
Domain: $\{x \mid x \leq -1\}$ or $(-\infty, -1]$
Range: all real numbers or $(-\infty, \infty)$

53. a) $(0, 0)$ b) $(0, 2)$
c) The original graph is moved upward c units.
d) The original graph moved downward 3 units.

55. a) $y = x^2$ opens up and $y = -x^2$ opens down.
The sign of a indicates whether the parabola opens up or down.
b) Yes
c) These graphs are "wider."
d) The smaller the absolute value of a, the wider the parabola.

Review Exercises

1. $3mn(m^2 - 4mn + 5) = 3m^3n - 12m^2n^2 + 15mn$

2. $(x-3)(2x+1) = x \cdot 2x + x \cdot 1 + (-3) \cdot 2x + (-3) \cdot 1$
$$= 2x^2 + x - 6x - 3$$
$$= 2x^2 - 5x - 3$$

3. $3n^2 - 17n + 20 = (3n - 5)(n - 4)$

4. $x^2 - 25 = (x + 5)(x - 5)$

5. $h(h - 3) = 0$
$h = 0$ or $h - 3 = 0$
$h = 3$

6. $y^2 + 3y - 18 = 0$
$(y + 6)(y - 3) = 0$
$y + 6 = 0$ or $y - 3 = 0$
$y = -6$ $y = 3$

Chapter 6 Review Exercises

1. The natural number factors of 9 are 1, 3, and 9.

2. The natural number factors of 33 are 1, 3, 11, and 33.

3. The natural number factors of 81 are 1, 3, 9, 27, and 81.

4. The natural number factors of 45 are 1, 3, 5, 9, 15, and 45.

5. The natural number factors of 30 are 1, 2, 3, 5, 6, 10, 15, and 30.

6. The natural number factors of 60 are 1, 2, 3, 4, 5, 6, 10, 12, 15, 20, 30, and 60.

7. $21 : 1, 3, 7, 21$
   $30 : 1, 2, 3, 5, 6, 10, 15, 30$
   $\text{GCF} = 3$

8. $12 : 1, 2, 3, 4, 6, 12$
   $42 : 1, 2, 3, 6, 7, 14, 21, 42$
   $60 : 1, 2, 3, 4, 5, 6, 10, 12, 15, 20, 30, 60$
   $\text{GCF} = 6$

9. $10y^4 : 2 \cdot 5 \cdot y^4$
   $25y^2 : 5^2 \cdot y^2$
   $\text{GCF} = 5y^2$

10. $15m^2 n : 3 \cdot 5 \cdot m^2 \cdot n$
    $3mn^5 : 3 \cdot m \cdot n^5$
    $\text{GCF} = 3 \cdot m \cdot n = 3mn$

11. $4x - 2 = 2\left(\dfrac{4x-2}{2}\right) = 2\left(\dfrac{4x}{2} - \dfrac{2}{2}\right) = 2(2x-1)$

12. $5m - 35m^3 = 5m\left(\dfrac{5m - 35m^3}{5m}\right)$
    $= 5m\left(\dfrac{5m}{5m} - \dfrac{35m^3}{5m}\right)$
    $= 5m\left(1 - 7m^2\right)$

13. $x^2 y + xy^2 + x^3 y^3$
    $= xy\left(\dfrac{x^2 y + xy^2 + x^3 y^3}{xy}\right)$
    $= xy\left(\dfrac{x^2 y}{xy} + \dfrac{xy^2}{xy} + \dfrac{x^3 y^3}{xy}\right)$
    $= xy\left(x + y + x^2 y^2\right)$

14. $105a^3 b^2 - 63a^2 b^3 + 84a^6 b^4$
    $= 21a^2 b^2\left(\dfrac{105a^3 b^2 - 63a^2 b^3 + 84a^6 b^4}{21a^2 b^2}\right)$
    $= 21a^2 b^2\left(\dfrac{105a^3 b^2}{21a^2 b^2} - \dfrac{63a^2 b^3}{21a^2 b^2} + \dfrac{84a^6 b^4}{21a^2 b^2}\right)$
    $= 21a^2 b^2\left(5a - 3b + 4a^4 b^2\right)$

15. $18ab^3 c - 36a^2 b^2 c$
    $= 18ab^2 c\left(\dfrac{18ab^3 c - 36a^2 b^2 c}{18ab^2 c}\right)$
    $= 18ab^2 c\left(\dfrac{18ab^3 c}{18ab^2 c} - \dfrac{36a^2 b^2 c}{18ab^2 c}\right)$
    $= 18ab^2 c(b - 2a)$

16. $100k^4 + 120k^5 - 10k + 40k^3$
    $= 10k\left(\dfrac{100k^4 + 120k^5 - 10k + 40k^3}{10k}\right)$
    $= 10k\left(\dfrac{100k^4}{10k} + \dfrac{120k^5}{10k} - \dfrac{10k}{10k} + \dfrac{40k^3}{10k}\right)$
    $= 10k\left(10k^3 + 12k^4 - 1 + 4k^2\right)$

17. $ax + ay + bx + by = a(x+y) + b(x+y)$
    $\qquad\qquad\qquad\quad = (x+y)(a+b)$

18. $ax + 2a + bx + 2b = a(x+2) + b(x+2)$
    $\qquad\qquad\qquad\quad = (x+2)(a+b)$

19. $y^3 + 2y^2 + 3y + 6 = y^2(y+2) + 3(y+2)$
    $\qquad\qquad\qquad\quad = (y+2)(y^2+3)$

20. $y^4 + 4y^3 - by - 4b = y^3(y+4) - b(y+4)$
    $\qquad\qquad\qquad\quad = (y+4)(y^3-b)$

21. $xy + y + x + 1 = y(x+1) + 1(x+1)$
    $\qquad\qquad\qquad = (x+1)(y+1)$

22. $ax^2 - 5y^2 + ay^2 - 5x^2 = ax^2 + ay^2 - 5x^2 - 5y^2$
    $\qquad\qquad\qquad\qquad = a(x^2+y^2) - 5(x^2+y^2)$
    $\qquad\qquad\qquad\qquad = (x^2+y^2)(a-5)$

23. $2b^3 - 2b^2 + b - 1 = 2b^2(b-1) + 1(b-1)$
$$= (b-1)(2b^2 + 1)$$

24. $u^2 - 3u + 4uv - 12v = u(u-3) + 4v(u-3)$
$$= (u-3)(u+4v)$$

25. To factor this trinomial, we must find a pair of numbers whose product is -12 and whose sum is -1. Because the product is negative, the two numbers must have different signs. Because the sum is also negative, the number with the greater absolute value will be negative. These numbers are -4 and 3.
$$x^2 - x - 12 = (x-4)(x+3)$$

26. To factor this trinomial, we must find a pair of numbers whose product is 45 and whose sum is 14. These numbers are 5 and 9.
$$x^2 + 14x + 45 = (x+5)(x+9)$$

27. To factor this trinomial, we must find a pair of numbers whose product is 8 and whose sum is -6. Because the product is positive and the sum is negative, both numbers must be negative. These numbers are -2 and -4.
$$n^2 - 6n + 8 = (n-2)(n-4)$$

28. To factor this trinomial, we must find a pair of numbers whose product is -20 and whose sum is 3. There are no combinations of factors of -20 that satisfy these conditions. Therefore, $a^2 + 3a - 20$ cannot be factored and is considered to be prime.

29. To factor this trinomial, we must find a pair of numbers whose product is 144 and whose sum is 51. These numbers are 3 and 48.
$$h^2 + 51h + 144 = (h+3)(h+48)$$

30. To factor this trinomial, we must find a pair of numbers whose product is -24 and whose sum is -10. Because the product is negative, the two numbers must have different signs. Because the sum is also negative, the number with the greater absolute value will be negative. These numbers are -12 and 2.
$$y^2 - 10y - 24 = (y-12)(y+2)$$

31. Factor out the GCF of the terms. Then factor the trinomial to two binomials. We must find a pair of numbers whose product is 9 and whose sum is -6.
$$4x^2 - 24x + 36 = 4(x^2 - 6x + 9)$$
$$= 4(x-3)(x-3)$$
$$= 4(x-3)^2$$

32. Factor out the GCF of the terms. Then factor the trinomial to two binomials. We must find a pair of numbers whose product is 18 and whose sum is -11.
$$3m^2 - 33m + 54 = 3(m^2 - 11m + 18)$$
$$= 3(m-9)(m-2)$$

33. To factor this trinomial, we need to find a pair of first terms whose product equals $6x^2$ and a pair of last terms whose product is -7. There are no such pairs that yield a middle term of $3x$.
$$6x^2 + 3x - 7 \text{ is prime}$$

34. To factor this trinomial, we need to find a pair of first terms whose product equals $2u^2$ and a pair of last terms whose product is 2.
$$2u^2 + 5u + 2 = (2u+1)(u+2)$$

35. To factor this trinomial, we need to find a pair of first terms whose product equals $3m^2$ and a pair of last terms whose product is 3.
$$3m^2 - 10m + 3 = (3m-1)(m-3)$$

36. To factor this trinomial, we need to find a pair of first terms whose product equals $5k^2$ and a pair of last terms whose product is $-12h^2$.
$$5k^2 - 7kh - 12h^2 = (5k-12h)(k+h)$$

37. To factor this trinomial, first factor out the GCF. Then find a pair of first terms whose product equals $3a^2$ and a pair of last terms whose product is 8.
$$6a^2 - 20a + 16 = 2(3a^2 - 10a + 8)$$
$$= 2(3a-4)(a-2)$$

38. Factor out the GCF. The remaining trinomial cannot be factored further.
$$8x^2y - 4xy - y = y(8x^2 - 4x - 1)$$

39. To factor this trinomial, we need to find a pair of first terms whose product equals $3p^2$ and a pair of last terms whose product is $4q^2$.

$$3p^2 - 13pq + 4q^2 = (3p - q)(p - 4q)$$

40. To factor this trinomial, first factor out the GCF. Then find a pair of first terms whose product equals $3x^2$ and a pair of last terms whose product is $-3y^2$.

$$24x^2 - 64xy - 24y^2 = 8(3x^2 - 8xy - 3y^2)$$
$$= 8(3x + y)(x - 3y)$$

41. $v^2 - 8v + 16 = (v)^2 - 2(4)(v) + (4)^2$
$$= (v - 4)^2$$

42. $u^2 + 6u + 9 = (u)^2 + 2(3)(u) + (3)^2$
$$= (u + 3)^2$$

43. $4x^2 + 20x + 25 = (2x)^2 + 2(5)(2x) + (5)^2$
$$= (2x + 5)^2$$

44. $9y^2 - 12y + 4 = (3y)^2 - 2(2)(3y) + (2)^2$
$$= (3y - 2)^2$$

45. $x^2 - 4 = (x + 2)(x - 2)$

46. $25 - y^2 = (5 + y)(5 - y)$

47. $x^3 - 1 = (x - 1)(x^2 + x + 1)$

48. $x^3 + 27 = (x + 3)(x^2 - 3x + 9)$

49. $6b^2 + b - 2 = (3b + 2)(2b - 1)$

50. $4ab - 24ab^2 = 4ab(1 - 6b)$

51. $y^2 + 25$ is prime.

52. $3x^2 - 3y^2 = 3(x^2 - y^2)$
$$= 3(x + y)(x - y)$$

53. $x^4 - 81 = (x^2 + 9)(x^2 - 9)$
$$= (x^2 + 9)(x + 3)(x - 3)$$

54. $7u^2 - 14u - 105 = 7(u^2 - 2u - 15)$
$$= 7(u - 5)(u + 3)$$

55. $8x^3 - 27y^3 = (2x - 3y)(4x^2 + 6xy + 9y^2)$

56. $2 - 50y^2 = 2(1 - 25y^2)$
$$= 2(1 + 5y)(1 - 5y)$$

57. $3am - 6an - 8bn + 4bm$
$$= 3am - 6an + 4bm - 8bn$$
$$= 3a(m - 2n) + 4b(m - 2n)$$
$$= (m - 2n)(3a + 4b)$$

58. $3m^2 + 9m + 27 = 3(m^2 + 3m + 9)$

59. $(m + 3)(m - 4) = 0$
$m + 3 = 0$ or $m - 4 = 0$
$m = -3$ $\qquad m = 4$

60. $\qquad y^2 - 4 = 0$
$(y + 2)(y - 2) = 0$
$y + 2 = 0$ or $y - 2 = 0$
$y = -2$ $\qquad y = 2$

61. $\qquad x^2 - 5x + 6 = 0$
$(x - 2)(x - 3) = 0$
$x - 2 = 0$ or $x - 3 = 0$
$x = 2$ $\qquad x = 3$

62. $\qquad x^2 - 4x = 21$
$x^2 - 4x - 21 = 0$
$(x - 7)(x + 3) = 0$
$x - 7 = 0$ or $x + 3 = 0$
$x = 7$ $\qquad x = -3$

63. $\qquad m^2 = 9$
$m^2 - 9 = 0$
$(m + 3)(m - 3) = 0$
$m + 3 = 0$ or $m - 3 = 0$
$m = -3$ $\qquad m = 3$

64. $\qquad x^2 + 3x = 18$
$x^2 + 3x - 18 = 0$
$(x + 6)(x - 3) = 0$
$x + 6 = 0$ or $x - 3 = 0$
$x = -6$ $\qquad x = 3$

65. $y(y+6) = -9$

$y^2 + 6y + 9 = 0$

$(y+3)^2 = 0$

$y + 3 = 0$

$y = -3$

66. $3n^2 - 11n = 4$

$3n^2 - 11n - 4 = 0$

$(3n+1)(n-4) = 0$

$3n + 1 = 0$ or $n - 4 = 0$

$n = -\dfrac{1}{3}$ $n = 4$

67. Let x represent one of the numbers. Translate to an equation and solve.

$5x^2 = 2x$

$5x^2 - 2x = 0$

$x(5x-2) = 0$

$x = 0$ or $5x - 2 = 0$

$x = \dfrac{2}{5}$

68. Let x represent the first natural number. Then the second consecutive natural number is represented by $x + 1$. Translate to an equation and solve.

$x^2 + (x+1)^2 = 61$

$x^2 + x^2 + 2x + 1 = 61$

$2x^2 + 2x - 60 = 0$

$2(x^2 + x - 30) = 0$

$x^2 + x - 30 = 0$

$(x+6)(x-5) = 0$

$x + 6 = 0$ or $x - 5 = 0$

$x = \cancel{-6}$ $x = 5$

Because we are considering only natural numbers, 5 is the only solution. The two numbers are 5 and $5 + 1 = 6$.

69. Let x represent the number. Translate to an equation and solve.

$x \cdot 4x = 100$

$4x^2 - 100 = 0$

$4(x^2 - 25) = 0$

$x^2 - 25 = 0$

$(x+5)(x-5) = 0$

$x + 5 = 0$ or $x - 5 = 0$

$x = -5$ $x = 5$

The numbers are -5 and 5.

70. Let the width be w, then the length will be $(w + 6)$. Translate to the equation

$w(w+6) = 91$

$w^2 + 6w - 91 = 0$

$(w+13)(w-7) = 0$

$w + 13 = 0$ or $w - 7 = 0$

$w = \cancel{-13}$ $w = 7$

Because lengths can only be positive, disregard the negative solution. The dimensions are 7 in. by $7 + 6 = 13$ in.

71. Let x represent the first natural number. Then the second consecutive natural number is represented by $x + 1$. Translate to an equation and solve.

$x(x+1) = 110$

$x^2 + x - 110 = 0$

$(x+11)(x-10) = 0$

$x + 11 = 0$ or $x - 10 = 0$

$x = \cancel{-11}$ $x = 10$

Because we are considering only natural numbers, 10 is the only solution. The numbers are 10 and $10 + 1 = 11$.

72. $3^2 + 4^2 = c^2$

$9 + 16 = c^2$

$25 = c^2$

$0 = c^2 - 25$

$0 = (c+5)(c-5)$

$c + 5 = 0$ or $c - 5 = 0$

$c = \cancel{-5}$ $c = 5$

73. $y = -5x^2$

x	y
-2	-20
-1	-5
0	0
1	-5
2	-20

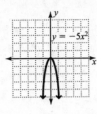

74. $y = x^2 - 3$

x	y
-2	1
-1	-2
0	-3
1	-2
2	1

75. $f(x) = x^2 + 2$

x	y
-2	6
-1	3
0	2
1	3
2	6

76. $f(x) = -3x^2 + 6x + 2$

x	y
-1	-7
0	2
1	5
2	2
3	-7

Chapter 6 Practice Test

1. The natural number factors of 80 are:
1, 2, 4, 5, 8, 10, 16, 20, 40, 80

2. Prime factorization of $25m^2 = 5^2 \cdot m^2$
Prime factorization of $40m = 2^3 \cdot 5 \cdot m$
GCF $= 5m$

3. The GCF of $5y$ and 30 is 5.
$$5y - 30 = 5\left(\frac{5y - 30}{5}\right)$$
$$= 5\left(\frac{5y}{5} - \frac{30}{5}\right)$$
$$= 5(y - 6)$$

4. The GCF of $6x^2y$ and $2y^2$ is $2y$.
$$6x^2y - 2y^2 = 2y\left(\frac{6x^2y - 2y^2}{2y}\right)$$
$$= 2y\left(\frac{6x^2y}{2y} - \frac{2y^2}{2y}\right)$$
$$= 2y\left(3x^2 - y\right)$$

5. Use the method of factoring by grouping.
$$ax - ay - bx + by = (ax - ay) + (-bx + by)$$
$$= a(x - y) - b(x - y)$$
$$= (x - y)(a - b)$$

6. To factor this trinomial, we must find a pair of numbers whose product is 12 and whose sum is 7. These numbers are 3 and 4.
$$m^2 + 7m + 12 = (m + 3)(m + 4)$$

7. Notice that this trinomial is a perfect square because the first and last terms are perfect squares and twice the product of their roots is the middle term.
$$r^2 - 8r + 16 = (r - 4)^2$$

8. Notice that this trinomial is a perfect square because the first and last terms are perfect squares and twice the product of their roots is the middle term.
$$4y^2 + 20y + 25 = (2y + 5)^2$$

9. In the factored form, the first terms must multiply to equal $3q^2$. These must be q and $3q$. The last terms must multiply to equal $+8$, which could be 1 and 8, or 2 and 4. Because the sum of the outside and inside products is negative, both signs of the binomial must be negative. By trial and error, we find the correct combination to be
$$3q^2 - 10q + 8 = (q - 2)(3q - 4)$$

10. In the factored form, the first terms must multiply to equal $6x^2$. These could be x and $6x$, or $2x$ and $3x$. The last terms must multiply to equal +20, which could be 1 and 20, or 2 and 10, or 4 and 5. Because the sum of the outside and inside products is negative, both signs of the binomial must be negative. By trial and error, we find the correct combination to be

$$6x^2 - 23x + 20 = (3x - 4)(2x - 5)$$

11. Begin by factoring out a as a common factor. Then use trial and error to factor the resulting polynomial.

$$ax^2 - 5ax - 24a = a(x^2 - 5x - 24)$$
$$= a(x - 8)(x + 3)$$

12. Begin by factoring out $2n$ as a common factor. Then use trial and error to factor the resulting polynomial.

$$10n^3 + 38n^2 - 8n = 2n(5n^2 + 19n - 4)$$
$$= 2n(5n - 1)(n + 4)$$

13. This binomial is the difference of squares because $c^2 - 25 = (c)^2 - (5)^2$.

$$c^2 - 25 = (c + 5)(c - 5)$$

14. $x^2 + 4$ is the sum of squares, which cannot be factored. Therefore, $x^2 + 4$ is prime.

15. Factor out the common factor 2 and then factor the resulting difference of squares.

$$2 - 50u^2 = 2(1 - 25u^2)$$
$$= 2(1 + 5u)(1 - 5u)$$

16. Factor out the common factor -4 and then factor the resulting difference of squares.

$$-4x^2 + 16 = -4(x^2 - 4)$$
$$= -4(x + 2)(x - 2)$$

17. This binomial is a sum of cubes because $m^3 + 125 = (m)^3 + (5)^3$.

$$m^3 + 125 = (m + 5)(m^2 - 5m + 25)$$

18. This binomial is a difference of cubes because $x^3 - 8 = (x)^3 - (2)^3$.

$$x^3 - 8 = (x - 2)(x^2 + 2x + 4)$$

19. $a(a + 3) = 0$

$a = 0$ or $a + 3 = 0$
$\qquad\qquad\qquad a = -3$

The solutions are 0 and -3.

20. $\quad x^2 - 4x - 12 = 0$

$(x - 6)(x + 2) = 0$

$x - 6 = 0$ or $x + 2 = 0$
$\quad x = 6 \qquad\qquad x = -2$

The solutions are 6 and -2.

21. $\quad 2n^2 + 7n - 15 = 0$

$(2n - 3)(n + 5) = 0$

$2n - 3 = 0$ or $n + 5 = 0$

$\quad n = \dfrac{3}{2} \qquad\qquad n = -5$

The solutions are $\dfrac{3}{2}$ and -5.

22. Let the first number be n. Then the next consecutive number is $n + 1$. We are given that the product of the two numbers is 72. Translate to an equation.

$$n(n + 1) = 72$$
$$n^2 + n - 72 = 0$$
$$(n + 9)(n - 8) = 0$$

$n + 9 = 0$ or $n - 8 = 0$
$\quad n = \cancel{-9} \qquad\quad n = 8$

The numbers must be natural numbers, so -9 is not possible. The numbers are 8 and $8 + 1 = 9$.

23. Let l represent the length of the rectangle. Because the width is 5 feet less than the length, the width is $l - 5$. We are given that the area is 36 square feet. Use the formula for the area of a rectangle: $lw = A$. Translate to an equation.

$$lw = A$$
$$l(l - 5) = 36$$
$$l^2 - 5l - 36 = 0$$
$$(l - 9)(l + 4) = 0$$

$l - 9 = 0$ or $l + 4 = 0$
$\quad l = 9 \qquad\qquad l = \cancel{-4}$

Length must be a positive number, so -4 is not a possible answer. The length is 9 ft. and the width is $9 - 5 = 4$ ft.

24. $y = -2x^2 + 4$

x	y
-2	-4
-1	2
0	4
1	2
2	-4

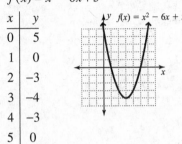

$y = -2x^2 + 4$

25. $f(x) = x^2 - 6x + 5$

x	y
0	5
1	0
2	-3
3	-4
4	-3
5	0

$f(x) = x^2 - 6x + 5$

Chapters 1–6 Cumulative Review Exercises

1. False; $x^3 \cdot x^4 = x^{3+4} = x^7$

2. True

3. False; unlike terms cannot be combined.

4. False; $4x + 12 = 4(x + 3)$ is factored completely.

5. $(a + b)(a - b)$

6. standard, factored

7. $14 - 2 \cdot 6 \cdot 3 + 8^2 = 14 - 2 \cdot 6 \cdot 3 + 64$
$= 14 - 36 + 64$
$= 42$

8. $-6|5 + 3^2| - \sqrt{14 + 11} = -6|5 + 9| - \sqrt{25}$
$= -6 \cdot 14 - 5$
$= -84 - 5$
$= -89$

9. $(5x^2 + 6x - 1) + (-2x + 3) = 5x^2 + 4x + 2$

10. $(6x^2)(3x^2 y) = 18x^4 y$

11. $(3x)(4x^3)^2 = 3x \cdot 16x^6 = 48x^7$

12. $(y - 8)(2y + 1) = y \cdot 2y + y \cdot 1 - 8 \cdot 2y - 8 \cdot 1$
$= 2y^2 + y - 16y - 8$
$= 2y^2 - 15y - 8$

13. $\dfrac{16h^3 + 4h^2 - 8h}{4h^2} = \dfrac{16h^3}{4h^2} + \dfrac{4h^2}{4h^2} - \dfrac{8h}{4h^2}$
$= 4h + 1 - \dfrac{2}{h}$

14.
$$\begin{array}{r} x + 4 \\ x + 3 \overline{) x^2 + 7x + 12} \\ \underline{x^2 + 3x} \\ 4x + 12 \\ \underline{4x + 12} \\ 0 \end{array}$$

15. $x^2 - 121 = (x + 11)(x - 11)$

16. $10x^2 + 40 = 10\left(\dfrac{10x^2}{10} + \dfrac{40}{10}\right)$
$= 10(x^2 + 4)$

17. $x^2 - 8x + 16 = (x)^2 - 2(4)(x) + (4)^2 = (x - 4)^2$

18. $2x^2 + 5x - 12 = (2x - 3)(x + 4)$

19. $x^3 - 5x^2 + 5x - 25 = x^2(x - 5) + 5(x - 5)$
$= (x - 5)(x^2 + 5)$

20.
$$\dfrac{3}{4} = x - \dfrac{2}{3}$$
$$\dfrac{3}{4} + \dfrac{2}{3} = x$$
$$\dfrac{9}{12} + \dfrac{8}{12} = x$$
$$\dfrac{17}{12} = x$$

21. $2.6 + 7a + 5 = 8a - 5.6$
$7.6 + 7a = 8a - 5.6$
$7.6 = a - 5.6$
$13.2 = a$

22. $x^2 - 3x - 28 = 0$
$(x - 7)(x + 4) = 0$
$x - 7 = 0 \quad \text{or} \quad x + 4 = 0$
$x = 7 \qquad\qquad x = -4$

23. $2x - 3y = 7$

$$2x = 7 + 3y$$

$$x = \frac{7 + 3y}{2}$$

24. $y = 2x - 6$

$\underline{\text{y-intercept:}}$ $\underline{\text{x-intercept:}}$

$x = 0: y = 2 \cdot 0 - 6$ $y = 0: 0 = 2x - 6$

      $y = 0 - 6$ $6 = 2x$

      $y = -6$ $3 = x$

     $(0, -6)$ $(3, 0)$

25. $(x_1, y_1) = (6, 2)$ and $(x_2, y_2) = (5, 2)$

$$m = \frac{y_2 - y_1}{x_2 - x_1}$$

$$= \frac{2 - 2}{5 - 6}$$

$$= \frac{0}{-1}$$

$$= 0$$

26. $y = 3x + 2$

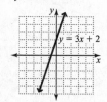

27. a) $7x - 5 \le 3x + 11$ b) $(-\infty, 4]$

     $4x - 5 \le 11$

       $4x \le 16$

        $x \le 4$

      $\{x | x \le 4\}$

c) ◄─┼─┼─┼─┼─┼─┼─┼─┼─┼─┼─┼─►
     $-5\ -4\ -3\ -2\ -1\ \ 0\ \ 1\ \ 2\ \ 3\ \ 4\ \ 5$

28. Solve the first equation for y.

$3x + y = 2$

     $y = -3x + 2$

Now substitute this value of y into the second equation and solve for x.

$6x + 5(-3x + 2) = 7$

$6x - 15x + 10 = 7$

        $-9x = -3$

           $x = \dfrac{1}{3}$

Use this value to find y.

$y = -3x + 2$

$$= -3\left(\frac{1}{3}\right) + 2$$

$$= -1 + 2$$

$$= 1$$

Solution: $\left(\dfrac{1}{3}, 1\right)$

29. Let the first integer be x. Then the second even integer is $(x + 2)$, and the third even integer is $(x + 4)$. Translate to the equation $x + (x + 2) + (x + 4) = 24$ and solve.

$x + (x + 2) + (x + 4) = 24$

         $3x + 6 = 24$

            $3x = 18$

             $x = 6$

Answer: 6, 8, 10

30. Use a table to solve the problem:

Categories	Amount	Rate	Interest
1st account	$8000 - x$	0.05	$0.05(8000 - x)$
2nd account	x	0.08	$0.08x$

Now, translate to an equation and solve for x.

$0.05(8000 - x) + 0.08x = 565$

$400 - 0.05x + 0.08x = 565$

     $400 + 0.03x = 565$

          $0.03x = 165$

            $x = \$5500$

There is $8000 - \$5500 = \$2500$ in the first account at a rate of 5% and $5500 in the second account at a rate of 8%.

Chapter 7
Rational Expressions
and Equations

Exercise Set 7.1

1. a) $x = -2, y = 5$:

    $$\frac{4x^2}{9y} = \frac{4(-2)^2}{9(5)} = \frac{4 \cdot 4}{45} = \frac{16}{45}$$

    b) $x = 3, y = -4$:

    $$\frac{4x^2}{9y} = \frac{4(3)^2}{9(-4)} = \frac{4 \cdot 9}{-36} = \frac{36}{-36} = -1$$

    c) $x = 0, y = 7$:

    $$\frac{4x^2}{9y} = \frac{4(0)^2}{9(7)} = \frac{4 \cdot 0}{63} = \frac{0}{63} = 0$$

3. a) $x = 4$: $\dfrac{2x}{x+4} = \dfrac{2(4)}{4+4} = \dfrac{8}{8} = 1$

    b) $x = -4$: $\dfrac{2x}{x+4} = \dfrac{2(-4)}{-4+4} = \dfrac{-8}{0}$ is undefined

    c) $x = -2.4$: $\dfrac{2x}{x+4} = \dfrac{2(-2.4)}{-2.4+4} = \dfrac{-4.8}{1.6} = -3$

5. a) $x = 3$: $\dfrac{2x+5}{3x-1} = \dfrac{2(3)+5}{3(3)-1} = \dfrac{6+5}{9-1} = \dfrac{11}{8}$

    b) $x = -2$: $\dfrac{2x+5}{3x-1} = \dfrac{2(-2)+5}{3(-2)-1} = \dfrac{-4+5}{-6-1} = -\dfrac{1}{7}$

    c) $x = -1.3$:

    $$\frac{2x+5}{3x-1} = \frac{2(-1.3)+5}{3(-1.3)-1} = -\frac{24}{49} \approx -0.49$$

7. a) $x = 0$: $\dfrac{x^2-4}{x+3} = \dfrac{(0)^2-4}{(0)+3} = \dfrac{-4}{3}$

    b) $x = -1$: $\dfrac{x^2-4}{x+3} = \dfrac{(-1)^2-4}{(-1)+3} = \dfrac{1-4}{2} = -\dfrac{3}{2}$

    c) $x = -4.2$:

    $$\frac{x^2-4}{x+3} = \frac{(-4.2)^2-4}{(-4.2)+3} = -\frac{341}{30} = -11.3\overline{6}$$

9. $x - 3 = 0$

    $x = 3$

    The expression $\dfrac{2x}{x-3}$ is undefined if x is replaced by 3.

11. $x^2 - 9 = 0$

    $(x+3)(x-3) = 0$

    $x+3 = 0$ or $x-3 = 0$

    $x = -3$ $x = 3$

    The expression $\dfrac{3}{x^2-9}$ is undefined if x is replaced by -3 or 3.

13. $x^2 - 14x + 45 = 0$

    $(x-5)(x-9) = 0$

    $x-5 = 0$ or $x-9 = 0$

    $x = 5$ $x = 9$

    The expression $\dfrac{6}{x^2-14x+45}$ is undefined if x is replaced by 5 or 9.

15. $m^2 - 3m = 0$

    $m(m-3) = 0$

    $m = 0$ or $m-3 = 0$

    $m = 3$

    The expression $\dfrac{2m}{m^2-3m}$ is undefined if m is replaced by 0 or 3.

17. $2a^2 - 3a - 5 = 0$

    $(2a-5)(a+1) = 0$

    $2a-5 = 0$ or $a+1 = 0$

    $2a = 5$ $a = -1$

    $a = \dfrac{5}{2}$

    The expression $\dfrac{4a}{2a^2-3a-5}$ is undefined if a is replaced by -1 or $\dfrac{5}{2}$.

19. $x^2 + 16 \neq 0$ for all real numbers. So $\dfrac{x-5}{x^2+16}$ is defined for all real numbers.

21. $\dfrac{9x^2}{12xy} = \dfrac{\cancel{3} \cdot 3 \cdot \cancel{x} \cdot x}{\cancel{3} \cdot 4 \cdot \cancel{x} \cdot y} = \dfrac{3x}{4y}$

23. $-\dfrac{4m^3n}{mn^2} = -\dfrac{4 \cdot \cancel{m} \cdot m \cdot m \cdot \cancel{n}}{\cancel{m} \cdot n \cdot \cancel{n}} = -\dfrac{4m^2}{n}$

25. $\dfrac{54x^4y}{36x^6yz^2}$

$= \dfrac{\cancel{2} \cdot \cancel{3} \cdot \cancel{3} \cdot 3 \cdot \cancel{x} \cdot \cancel{x} \cdot \cancel{x} \cdot \cancel{x} \cdot \cancel{y}}{\cancel{2} \cdot 2 \cdot \cancel{3} \cdot \cancel{3} \cdot \cancel{x} \cdot \cancel{x} \cdot \cancel{x} \cdot \cancel{x} \cdot x \cdot x \cdot \cancel{y} \cdot z \cdot z}$

$= \dfrac{3}{2x^2z^2}$

27. $-\dfrac{15m^2n}{5m+10n} = -\dfrac{3 \cdot \cancel{5} \cdot m^2n}{\cancel{5}(m+2n)} = -\dfrac{3m^2n}{m+2n}$

29. $\dfrac{7}{21(x+5)} = \dfrac{\cancel{7}}{3 \cdot \cancel{7} \cdot (x+5)} = \dfrac{1}{3(x+5)}$

31. $\dfrac{3\cancel{(m-2)}}{5\cancel{(m-2)}} = \dfrac{3}{5}$

33. $\dfrac{10x-20}{15x-30} = \dfrac{10\cancel{(x-2)}}{15\cancel{(x-2)}} = \dfrac{2}{3}$

35. $\dfrac{30a+15b}{18a+9b} = \dfrac{15\cancel{(2a+b)}}{9\cancel{(2a+b)}} = \dfrac{5}{3}$

37. $\dfrac{x^2+5x}{4x+20} = \dfrac{x\cancel{(x+5)}}{4\cancel{(x+5)}} = \dfrac{x}{4}$

39. $\dfrac{m^2-6m}{m-6} = \dfrac{m\cancel{(m-6)}}{\cancel{(m-6)}} = m$

41. $\dfrac{5y^2-10y-15}{xy^2-2xy-3x} = \dfrac{5\cancel{(y^2-2y-3)}}{x\cancel{(y^2-2y-3)}} = \dfrac{5}{x}$

43. $\dfrac{x^2-y^2}{x^2+2xy+y^2} = \dfrac{\cancel{(x+y)}(x-y)}{\cancel{(x+y)}(x+y)} = \dfrac{x-y}{x+y}$

45. $\dfrac{x^2-5x}{x^2-7x+10} = \dfrac{x\cancel{(x-5)}}{\cancel{(x-5)}(x-2)} = \dfrac{x}{x-2}$

47. $\dfrac{t^2-9}{t^2+5t+6} = \dfrac{\cancel{(t+3)}(t-3)}{(t+2)\cancel{(t+3)}} = \dfrac{t-3}{t+2}$

49. $\dfrac{4a^2-4a-3}{6a^2-a-2} = \dfrac{(2a-3)\cancel{(2a+1)}}{(3a-2)\cancel{(2a+1)}} = \dfrac{2a-3}{3a-2}$

51. $\dfrac{6b^2+b-12}{10b^2+13b-3} = \dfrac{\cancel{(2b+3)}(3b-4)}{\cancel{(2b+3)}(5b-1)} = \dfrac{3b-4}{5b-1}$

53. $\dfrac{x^3-16x}{x^3+6x^2+8x} = \dfrac{x(x^2-16)}{x(x^2+6x+8)}$

$= \dfrac{\cancel{x}\cancel{(x+4)}(x-4)}{\cancel{x}\cancel{(x+4)}(x+2)}$

$= \dfrac{x-4}{x+2}$

55. $\dfrac{px+py+qx+qy}{mx-nx+my-ny} = \dfrac{p(x+y)+q(x+y)}{x(m-n)+y(m-n)}$

$= \dfrac{\cancel{(x+y)}(p+q)}{(m-n)\cancel{(x+y)}}$

$= \dfrac{p+q}{m-n}$

57. $\dfrac{6u^3-4u^2-3u+2}{3u^2+u-2} = \dfrac{2u^2(3u-2)-(3u-2)}{(3u-2)(u+1)}$

$= \dfrac{\cancel{(3u-2)}(2u^2-1)}{\cancel{(3u-2)}(u+1)}$

$= \dfrac{2u^2-1}{u+1}$

59. $\dfrac{x-3}{3-x} = \dfrac{\cancel{(x-3)}}{-1\cancel{(x-3)}} = \dfrac{1}{-1} = -1$

61. $\dfrac{4x-4}{1-x} = \dfrac{4\cancel{(x-1)}}{-1\cancel{(x-1)}} = -4$

63. $\dfrac{y^2-4}{2-y} = \dfrac{(y+2)\cancel{(y-2)}}{-1\cancel{(y-2)}}$

$= -(y+2)$

$= -y-2$

65. $\dfrac{3-w}{w^2-2w-3} = \dfrac{-1\cancel{(w-3)}}{\cancel{(w-3)}(w+1)} = -\dfrac{1}{w+1}$

67. $\dfrac{x^2-4}{x^3+8} = \dfrac{(x-2)\cancel{(x+2)}}{\cancel{(x+2)}(x^2-2x+4)} = \dfrac{x-2}{x^2-2x+4}$

69. $\dfrac{x^3+y^3}{x^3-x^2y+xy^2} = \dfrac{(x+y)\cancel{(x^2-xy+y^2)}}{x\cancel{(x^2-xy+y^2)}} = \dfrac{x+y}{x}$

71. $\dfrac{3-m}{m^3-27} = \dfrac{-1\cancel{(m-3)}}{\cancel{(m-3)}(m^2+3m+9)}$

$= -\dfrac{1}{m^2+3m+9}$

73. Mistake: Placed remaining x in numerator.

Correct: $-\dfrac{5x^2y}{10x^3y} = -\dfrac{\cancel{5}\cdot\cancel{x}\cdot x\cdot\cancel{y}}{2\cdot\cancel{5}\cdot\cancel{x}\cdot\cancel{x}\cdot x\cdot\cancel{y}} = -\dfrac{1}{2x}$

75. Mistake: Divided out a part of a multi-term polynomial.
Correct: Can't simplify.

77. Mistake: $y-x$ and $x-y$ are not identical factors until -1 is factored out of one of them.

Correct: $\dfrac{-x\cancel{(x-y)}}{\cancel{x-y}} = -x$

79. a) $BMI = \dfrac{704.5w}{h^2}$

$5'4'' = 64'', \ w = 138:$

$BMI = \dfrac{704.5(138)}{64^2} \approx 23.7$

$5'9'' = 69'', \ w = 155:$

$BMI = \dfrac{704.5(155)}{69^2} \approx 22.9$

$6'2'' = 74'', \ w = 220:$

$BMI = \dfrac{704.5(220)}{74^2} \approx 28.3$

b) The first two are normal weight and the third is overweight.

81. $C = \dfrac{\pi dN}{12}$

$d = 8; N = 800: C = \dfrac{\pi\cdot 8\cdot 800}{12} = 533.\overline{3}\pi$

$d = 8; N = 1000: C = \dfrac{\pi\cdot 8\cdot 1000}{12} = 666.\overline{6}\pi$

$d = 10; N = 800: C = \dfrac{\pi\cdot 10\cdot 800}{12} = 666.\overline{6}\pi$

$d = 10; N = 1000: C = \dfrac{\pi\cdot 10\cdot 1000}{12} = 833.\overline{3}\pi$

83. a) $V = \dfrac{1}{3}\pi r^2 h$

$3\cdot V = 3\cdot\dfrac{1}{3}\pi r^2 h$

$3V = \pi r^2 h$

$\dfrac{3V}{\pi r^2} = h$

b) $V = 150.8, r = 4$

$\dfrac{3(150.8)}{\pi\cdot 4^2} = h$

$9 \text{ cm} \approx h$

85. $\dfrac{s^2-9}{s^2+7s+12} = \dfrac{\cancel{(s+3)}(s-3)}{\cancel{(s+3)}(s+4)} = \dfrac{s-3}{s+4}$

Review Exercises

1. $\dfrac{2}{3}\cdot\dfrac{6}{4} = \dfrac{12}{12} = 1$　　2. $-\dfrac{15}{16}\cdot\dfrac{4}{5} = -\dfrac{60}{80} = -\dfrac{3}{4}$

3. $\dfrac{3}{8}\div\dfrac{5}{6} = \dfrac{3}{8}\cdot\dfrac{6}{5} = \dfrac{18}{40} = \dfrac{9}{20}$

4. $-\dfrac{7}{9}\div\left(-\dfrac{5}{12}\right) = -\dfrac{7}{9}\cdot\left(-\dfrac{12}{5}\right) = \dfrac{84}{45} = \dfrac{28}{15}$

5. $4y^2-25 = (2y+5)(2y-5)$

6. $3x^2+13x-10 = (3x-2)(x+5)$

Exercise Set 7.2

1. $\dfrac{m}{n}\cdot\dfrac{2m}{5n} = \dfrac{2m^2}{5n^2}$

3. $\dfrac{r^3}{s^3}\cdot\dfrac{s^5}{r^6} = \dfrac{\cancel{r}\cdot\cancel{r}\cdot\cancel{r}}{\cancel{s}\cdot\cancel{s}\cdot\cancel{s}}\cdot\dfrac{\cancel{s}\cdot\cancel{s}\cdot\cancel{s}\cdot s\cdot s}{\cancel{r}\cdot\cancel{r}\cdot\cancel{r}\cdot r\cdot r\cdot r} = \dfrac{s^2}{r^3}$

5. $\dfrac{3}{y}\cdot\dfrac{y^3}{3} = \dfrac{\cancel{3}}{\cancel{y}}\cdot\dfrac{\cancel{y}\cdot y\cdot y}{\cancel{3}} = y^2$

7. $-\dfrac{n}{rs^2}\cdot\dfrac{rs}{mn^3} = -\dfrac{\cancel{n}}{\cancel{r}\cdot\cancel{s}\cdot s}\cdot\dfrac{\cancel{r}\cdot\cancel{s}}{m\cdot\cancel{n}\cdot n\cdot n} = -\dfrac{1}{mn^2s}$

9. $-\dfrac{6r^2s}{5r^2s^2}\cdot\left(-\dfrac{15r^4s^2}{2r^3s^2}\right)$

$=-\dfrac{2\cdot3\cdot r\cdot r\cdot s}{5\cdot r\cdot r\cdot s\cdot s}\cdot\left(-\dfrac{3\cdot5\cdot r\cdot r\cdot r\cdot r\cdot s\cdot s}{2\cdot r\cdot r\cdot r\cdot s\cdot s}\right)$

$=\dfrac{9r}{s}$

11. $\dfrac{5n^2}{15m}\cdot\dfrac{n}{2}\cdot\dfrac{3m^2n}{n^2}=\dfrac{5\cdot n\cdot n}{3\cdot5\cdot m}\cdot\dfrac{n}{2}\cdot\dfrac{3\cdot m\cdot m\cdot n}{n\cdot n}$

$=\dfrac{mn^2}{2}$

13. $\dfrac{3b-12}{14}\cdot\dfrac{21}{5b-20}=\dfrac{3(b-4)}{2\cdot7}\cdot\dfrac{3\cdot7}{5(b-4)}=\dfrac{9}{10}$

15. $\dfrac{2x-6}{3y+12}\cdot\dfrac{2y+8}{5x-15}=\dfrac{2(x-3)}{3(y+4)}\cdot\dfrac{2(y+4)}{5(x-3)}=\dfrac{4}{15}$

17. $\dfrac{4x^2-25}{3x-4}\cdot\dfrac{9x^2-16}{2x+5}$

$=\dfrac{(2x+5)(2x-5)}{(3x-4)}\cdot\dfrac{(3x+4)(3x-4)}{(2x+5)}$

$=(2x-5)(3x+4)$

19. $\dfrac{r^2+3r}{r^2-36}\cdot\dfrac{r^2-6r}{r^2-9}$

$=\dfrac{r(r+3)}{(r+6)(r-6)}\cdot\dfrac{r(r-6)}{(r+3)(r-3)}$

$=\dfrac{r^2}{(r+6)(r-3)}$

21. $\dfrac{12q}{q^2-2q+1}\cdot\dfrac{q^2-1}{24q^2}$

$=\dfrac{12\cdot q}{(q-1)(q-1)}\cdot\dfrac{(q+1)(q-1)}{2\cdot12\cdot q\cdot q}$

$=\dfrac{q+1}{2q(q-1)}$

23. $\dfrac{w^2-4w-5}{w^2-1}\cdot\dfrac{w^2-w-20}{w^2-10w+25}$

$=\dfrac{(w+1)(w-5)}{(w+1)(w-1)}\cdot\dfrac{(w-5)(w+4)}{(w-5)(w-5)}$

$=\dfrac{w+4}{w-1}$

25. $\dfrac{n^2-6n+9}{n^2-9}\cdot\dfrac{n^2+4n+3}{n^2-3n}$

$=\dfrac{(n-3)(n-3)}{(n+3)(n-3)}\cdot\dfrac{(n+3)(n+1)}{n(n-3)}$

$=\dfrac{n+1}{n}$

27. $\dfrac{6x^2+x-12}{2x^2-5x-12}\cdot\dfrac{3x^2-14x+8}{9x^2-18x+8}$

$=\dfrac{(3x-4)(2x+3)}{(2x+3)(x-4)}\cdot\dfrac{(3x-2)(x-4)}{(3x-2)(3x-4)}$

$=1$

29. $\dfrac{ac+3a+2c+6}{ad+a+2d+2}\cdot\dfrac{ad-5a+2d-10}{ac+2c+3a+6}$

$=\dfrac{a(c+3)+2(c+3)}{a(d+1)+2(d+1)}\cdot\dfrac{a(d-5)+2(d-5)}{c(a+2)+3(a+2)}$

$=\dfrac{(c+3)(a+2)}{(d+1)(a+2)}\cdot\dfrac{(d-5)(a+2)}{(a+2)(c+3)}$

$=\dfrac{d-5}{d+1}$

31. $\dfrac{x^3+27}{x^3-3x^2+9x}\cdot\dfrac{x^2-7x+10}{x^2-2x-15}$

$=\dfrac{(x+3)(x^2-3x+9)}{x(x^2-3x+9)}\cdot\dfrac{(x-5)(x-2)}{(x-5)(x+3)}$

$=\dfrac{x-2}{x}$

33. $\dfrac{x^2+7xy+10y^2}{x^2+6xy+5y^2}\cdot\dfrac{x+2y}{y}\cdot\dfrac{x+y}{x^2+4xy+4y^2}$

$=\dfrac{(x+5y)(x+2y)}{(x+5y)(x+y)}\cdot\dfrac{x+2y}{y}\cdot\dfrac{x+y}{(x+2y)(x+2y)}$

$=\dfrac{1}{y}$

35. $\dfrac{4x^2y^3}{2x-6}\cdot\dfrac{12-4x}{6x^3y}$

$=\dfrac{2\cdot 2\cdot \cancel{x}\cdot \cancel{x}\cdot \cancel{y}\cdot y\cdot y}{\cancel{2}\,\cancel{(x-3)}}\cdot\dfrac{-2\cdot 2\cdot \cancel{(x-3)}}{\cancel{2}\cdot 3\cdot \cancel{x}\cdot \cancel{x}\cdot x\cdot \cancel{y}}$

$=-\dfrac{4y^2}{3x}$

37. $\dfrac{2x^2-11x+12}{2x^2+11x-21}\cdot\dfrac{3x^2+20x-7}{4-x}$

$=\dfrac{\cancel{(2x-3)}\,\cancel{(x-4)}}{\cancel{(2x-3)}\,\cancel{(x+7)}}\cdot\dfrac{(3x-1)\,\cancel{(x+7)}}{-1\,\cancel{(x-4)}}$

$=-(3x-1)$

$=1-3x$

39. $\dfrac{x}{y^4}\div\dfrac{x^3}{y^2}=\dfrac{x}{y^4}\cdot\dfrac{y^2}{x^3}$

$\qquad=\dfrac{\cancel{x}}{\cancel{y}\cdot \cancel{y}\cdot y\cdot y}\cdot\dfrac{\cancel{y}\cdot \cancel{y}}{x\cdot x\cdot x}$

$\qquad=\dfrac{1}{x^2y^2}$

41. $\dfrac{2m}{3n^2}\div\dfrac{8m^3}{15n}=\dfrac{2m}{3n^2}\cdot\dfrac{15n}{8m^3}$

$=\dfrac{\cancel{2}\cdot \cancel{m}}{\cancel{3}\cdot \cancel{n}\cdot n}\cdot\dfrac{\cancel{3}\cdot 5\cdot \cancel{n}}{\cancel{2}\cdot 2\cdot 2\cdot \cancel{m}\cdot m\cdot m}$

$=\dfrac{5}{4m^2n}$

43. $-\dfrac{12w^2y}{5z^2}\div\left(-\dfrac{12w^2}{y}\right)$

$=-\dfrac{12w^2y}{5z^2}\cdot\left(-\dfrac{y}{12w^2}\right)$

$=\dfrac{\cancel{12}\cdot \cancel{w}\cdot \cancel{w}\cdot y}{5\cdot z\cdot z}\cdot\dfrac{y}{\cancel{12}\cdot \cancel{w}\cdot \cancel{w}}$

$=\dfrac{y^2}{5z^2}$

45. $\dfrac{8x^7g^3}{15x^2g^4}\div\left(-\dfrac{3xg^2}{4x^2g^3}\right)=\dfrac{8x^7g^3}{15x^2g^4}\cdot\dfrac{-4x^2g^3}{3xg^2}$

$=\dfrac{2\cdot 2\cdot 2\cdot \cancel{x}\cdot \cancel{x}\cdot x\cdot x\cdot x\cdot x\cdot x\cdot \cancel{g}\cdot \cancel{g}\cdot \cancel{g}}{3\cdot 5\cdot \cancel{x}\cdot \cancel{x}\cdot \cancel{g}\cdot \cancel{g}\cdot \cancel{g}\cdot \cancel{g}}$

$\cdot\dfrac{-2\cdot 2\cdot \cancel{x}\cdot x\cdot \cancel{g}\cdot \cancel{g}\cdot \cancel{g}}{3\cdot \cancel{x}\cdot \cancel{g}\cdot \cancel{g}}$

$=-\dfrac{32x^6}{45}$

47. $\dfrac{4a-8}{15a}\div\dfrac{3a-6}{5a^2}=\dfrac{4a-8}{15a}\cdot\dfrac{5a^2}{3a-6}$

$=\dfrac{4\,\cancel{(a-2)}}{3\cdot \cancel{5}\cdot \cancel{a}}\cdot\dfrac{\cancel{5}\cdot \cancel{a}\cdot a}{3\,\cancel{(a-2)}}$

$=\dfrac{4a}{9}$

49. $\dfrac{4p+12q}{3p-6q}\div\dfrac{5p+15q}{6p-12q}$

$=\dfrac{4p+12q}{3p-6q}\cdot\dfrac{6p-12q}{5p+15q}$

$=\dfrac{2\cdot 2\cdot \cancel{(p+3q)}}{\cancel{3}\,\cancel{(p-2q)}}\cdot\dfrac{2\cdot \cancel{3}\cdot \cancel{(p-2q)}}{5\,\cancel{(p+3q)}}$

$=\dfrac{8}{5}$

51. $\dfrac{a^2-b^2}{x^2-y^2}\div\dfrac{a+b}{x-y}=\dfrac{a^2-b^2}{x^2-y^2}\cdot\dfrac{x-y}{a+b}$

$=\dfrac{\cancel{(a+b)}\,(a-b)}{(x+y)\,\cancel{(x-y)}}\cdot\dfrac{\cancel{(x-y)}}{\cancel{(a+b)}}$

$=\dfrac{a-b}{x+y}$

53. $\dfrac{x^2+16x+64}{x^2-16x+64}\div\dfrac{x+8}{x-8}$

$=\dfrac{x^2+16x+64}{x^2-16x+64}\cdot\dfrac{x-8}{x+8}$

$=\dfrac{(x+8)\,(x+8)}{\cancel{(x-8)}\,(x-8)}\cdot\dfrac{\cancel{(x-8)}}{\cancel{(x+8)}}$

$=\dfrac{x+8}{x-8}$

55. $\dfrac{x^2+7x+10}{x^2+2x} \div \dfrac{x+5}{x^2-4x}$

$= \dfrac{x^2+7x+10}{x^2+2x} \cdot \dfrac{x^2-4x}{x+5}$

$= \dfrac{(x+5)(x+2)}{x(x+2)} \cdot \dfrac{x(x-4)}{(x+5)} = x-4$

57. $\dfrac{a^2-b^2}{a^2+2ab+b^2} \div \dfrac{a^2-3ab+2b^2}{a^2+3ab+2b^2}$

$= \dfrac{a^2-b^2}{a^2+2ab+b^2} \cdot \dfrac{a^2+3ab+2b^2}{a^2-3ab+2b^2}$

$= \dfrac{(a+b)(a-b)}{(a+b)(a+b)} \cdot \dfrac{(a+2b)(a+b)}{(a-2b)(a-b)}$

$= \dfrac{a+2b}{a-2b}$

59. $\dfrac{3k+2}{3k^4+2k^3} \div \dfrac{10k^2+3k-1}{5k^3-k^2}$

$= \dfrac{3k+2}{3k^4+2k^3} \cdot \dfrac{5k^3-k^2}{10k^2+3k-1}$

$= \dfrac{(3k+2)}{k \cdot k \cdot k \cdot (3k+2)} \cdot \dfrac{k \cdot k \cdot (5k-1)}{(5k-1)(2k+1)}$

$= \dfrac{1}{k(2k+1)}$

61. $\dfrac{8x^3-27}{3x^2+20x-7} \div \dfrac{2x^2-7x+6}{x^2+5x-14}$

$= \dfrac{8x^3-27}{3x^2+20x-7} \cdot \dfrac{x^2+5x-14}{2x^2-7x+6}$

$= \dfrac{(2x-3)(4x^2+6x+9)}{(3x-1)(x+7)} \cdot \dfrac{(x-2)(x+7)}{(2x-3)(x-2)}$

$= \dfrac{4x^2+6x+9}{3x-1}$

63. $\dfrac{x^2-2x-8}{x^2+6x+8} \div \left(x^2-3x-4 \right)$

$= \dfrac{x^2-2x-8}{x^2+6x+8} \div \dfrac{x^2-3x-4}{1}$

$= \dfrac{x^2-2x-8}{x^2+6x+8} \cdot \dfrac{1}{x^2-3x-4}$

$= \dfrac{(x-4)(x+2)}{(x+4)(x+2)} \cdot \dfrac{1}{(x-4)(x+1)}$

$= \dfrac{1}{(x+4)(x+1)}$

65. $\dfrac{ab+3a+2b+6}{bc+4b+3c+12} \div \dfrac{ac-3a+2c-6}{bc+4b-4c-16}$

$= \dfrac{ab+3a+2b+6}{bc+4b+3c+12} \cdot \dfrac{bc+4b-4c-16}{ac-3a+2c-6}$

$= \dfrac{a(b+3)+2(b+3)}{b(c+4)+3(c+4)} \cdot \dfrac{b(c+4)-4(c+4)}{a(c-3)+2(c-3)}$

$= \dfrac{(b+3)(a+2)}{(c+4)(b+3)} \cdot \dfrac{(c+4)(b-4)}{(c-3)(a+2)}$

$= \dfrac{b-4}{c-3}$

67. $\dfrac{b^2-16}{b^2+b-12} \div \dfrac{4-b}{b^2+2b-15}$

$= \dfrac{b^2-16}{b^2+b-12} \cdot \dfrac{b^2+2b-15}{4-b}$

$= \dfrac{(b+4)(b-4)}{(b+4)(b-3)} \cdot \dfrac{(b+5)(b-3)}{-1(b-4)}$

$= -(b+5)$

$= -b-5$

69. $\dfrac{y^2+y-6}{2y^2-3y-2} \cdot \dfrac{2y^2+9y+4}{y^2+7y+12} \div \dfrac{4y+y^2}{7+y}$

$= \dfrac{y^2+y-6}{2y^2-3y-2} \cdot \dfrac{2y^2+9y+4}{y^2+7y+12} \cdot \dfrac{7+y}{4y+y^2}$

$= \dfrac{(y+3)(y-2)}{(2y+1)(y-2)} \cdot \dfrac{(2y+1)(y+4)}{(y+4)(y+3)} \cdot \dfrac{7+y}{y(4+y)}$

$= \dfrac{7+y}{y(y+4)}$

71. $\dfrac{6x^2-5x-6}{x^4-81}\div\dfrac{3x-2x^2}{x^3+3x^2+9x+27}\div\dfrac{3x+2}{x^4}$

$=\dfrac{6x^2-5x-6}{x^4-81}\cdot\dfrac{x^3+3x^2+9x+27}{3x-2x^2}\cdot\dfrac{x^4}{3x+2}$

$=\dfrac{(3x+2)(2x-3)}{(x^2+9)(x^2-9)}\cdot\dfrac{x^2(x+3)+9(x+3)}{x(3-2x)}\cdot\dfrac{x^4}{3x+2}$

$=\dfrac{\cancel{(3x+2)}\,\cancel{(2x-3)}}{\cancel{(x^2+9)}\,(x+3)(x-3)}\cdot\dfrac{\cancel{(x^2+9)}\,\cancel{(x+3)}}{-x\cancel{(2x-3)}}\cdot\dfrac{x^4}{\cancel{3x+2}}$

$=-\dfrac{x^4}{x(x-3)}$

$=-\dfrac{\cancel{x}\cdot x\cdot x\cdot x}{\cancel{x}(x-3)}$

$=-\dfrac{x^3}{(x-3)}$

73. $\dfrac{6x^2}{x^2-9}\cdot\dfrac{3x^2+x-24}{20y-6xy}\cdot\dfrac{x-3}{3x-8}\div\dfrac{3x^2+9x}{3x^2-x-30}$

$=\dfrac{6x^2}{x^2-9}\cdot\dfrac{3x^2+x-24}{20y-6xy}\cdot\dfrac{x-3}{3x-8}\cdot\dfrac{3x^2-x-30}{3x^2+9x}$

$=\dfrac{\cancel{2}\cdot\cancel{3}\cdot x\cdot x}{(x+3)\cancel{(x-3)}}\cdot\dfrac{\cancel{(3x-8)}\,\cancel{(x+3)}}{-1\cdot\cancel{2}\cdot y(3x-10)}\cdot\dfrac{\cancel{x-3}}{\cancel{3x-8}}$

$\cdot\dfrac{(3x-10)\,\cancel{(x+3)}}{\cancel{3}\cdot x\cdot(x+3)}$

$=-\dfrac{x}{y}$

75. Mistake: Did not invert the divisor.

Correct: $\dfrac{5\cancel{y}}{9}\cdot\dfrac{10}{3\cancel{y}}=\dfrac{50}{27}$

77. Mistake: Divided out y, which is not allowed because it is not a factor in $y-2$.

Correct: $\dfrac{(x+3)(x-3)}{15y}\cdot\dfrac{y-2}{x+3}=\dfrac{(x-3)(y-2)}{15y}$

79. 18.5 ft. $=\dfrac{18.5\text{ ft.}}{1}\cdot\dfrac{12\text{ in.}}{1\text{ ft.}}=222$ in.

81. 6 mi. $=\dfrac{6\text{ mi.}}{1}\cdot\dfrac{5280\text{ ft.}}{1\text{ mi.}}\cdot\dfrac{1\text{ yd.}}{3\text{ ft.}}=10{,}560$ yd.

83. 8 yd. $=\dfrac{8\text{ yd.}}{1}\cdot\dfrac{3\text{ ft.}}{1\text{ yd.}}\cdot\dfrac{12\text{ in.}}{1\text{ ft.}}=288$ in.

85. 5.5 lb. $=\dfrac{5.5\text{ lb.}}{1}\cdot\dfrac{16\text{ oz.}}{1\text{ lb.}}=88$ oz.

87. $25{,}400$ lb. $=\dfrac{25{,}400\text{ lb.}}{1}\cdot\dfrac{1\text{ T}}{2000\text{ lb.}}=12.7$ T

89. 3 tons $=\dfrac{3\text{ tons}}{1}\cdot\dfrac{2000\text{ pounds}}{1\text{ ton}}\cdot\dfrac{16\text{ oz.}}{1\text{ pound}}$

$=96{,}000$ oz.

91. $\dfrac{763\text{ miles}}{1\text{ hour}}\cdot\dfrac{1\text{ hour}}{60\text{ min.}}\cdot\dfrac{1\text{ min.}}{60\text{ sec.}}\cdot\dfrac{5280\text{ ft.}}{1\text{ mile}}$

$=1119.0\overline{6}$ ft./sec.

93. $\dfrac{40\text{ yd.}}{4.3\text{ sec.}}\cdot\dfrac{3\text{ ft.}}{1\text{ yd.}}\cdot\dfrac{1\text{ mi.}}{5280\text{ ft.}}\cdot\dfrac{60\text{ sec.}}{1\text{ min.}}$

$\cdot\dfrac{60\text{ min.}}{1\text{ hr.}}\approx19.03$ mi./hr.

95. $\dfrac{29.78\text{ km}}{1\text{ sec.}}\cdot\dfrac{60\text{ sec.}}{1\text{ min.}}\cdot\dfrac{60\text{ min.}}{1\text{ hr.}}$

$\cdot\dfrac{24\text{ hr.}}{1\text{ day}}\cdot\dfrac{365.25\text{ days}}{1\text{ yr.}}$

$=939{,}785{,}328$ km/yr.

97. $\dfrac{186{,}000\text{ mi.}}{1\text{ sec.}}\cdot\dfrac{60\text{ sec.}}{1\text{ min.}}\cdot\dfrac{60\text{ min.}}{1\text{ hr.}}\cdot\dfrac{24\text{ hr.}}{1\text{ day}}$

$\cdot\dfrac{365.25\text{ days}}{1\text{ yr.}}\approx5.87\times10^{12}$ mi./yr.

99. $\dfrac{\$500}{1}\cdot\dfrac{0.6235\text{ £}}{\$1}=311.75$ £

101. $\dfrac{450\text{ €}}{1}\cdot\dfrac{\$1.2979}{1\text{ €}}=\$584.06$

103. $\dfrac{650\text{ £}}{1}\cdot\dfrac{1.2338\text{ €}}{1\text{ £}}=801.97$ €

Review Exercises

1. $\dfrac{1}{3}+\dfrac{2}{3}=\dfrac{3}{3}=1$

2. $\dfrac{1}{4}-\dfrac{3}{4}=-\dfrac{2}{4}=-\dfrac{1}{2}$

3. $\dfrac{1}{2}x+\dfrac{3}{2}x=\dfrac{4}{2}x=2x$

4. $\dfrac{3}{8}y - 5 - \dfrac{1}{8}y + 1 = \dfrac{2}{8}y - 4 = \dfrac{1}{4}y - 4$

5. $\left(6x^2 + x - 1\right) + \left(2x^2 - 7x - 3\right) = 8x^2 - 6x - 4$

6. $\left(3n^2 - 5n + 7\right) - \left(4n^2 - 5n + 3\right)$
$= 3n^2 - 5n + 7 - 4n^2 + 5n - 3$
$= -n^2 + 4$

Exercise Set 7.3

1. $\dfrac{3x}{10} + \dfrac{2x}{10} = \dfrac{3x + 2x}{10} = \dfrac{5x}{10} = \dfrac{\cancel{5} \cdot x}{2 \cdot \cancel{5}} = \dfrac{x}{2}$

3. $\dfrac{4a^2}{9} + \left(-\dfrac{a^2}{9}\right) = \dfrac{4a^2 - a^2}{9} = \dfrac{3a^2}{9} = \dfrac{\cancel{3}a^2}{\cancel{3} \cdot 3} = \dfrac{a^2}{3}$

5. $\dfrac{6}{a} - \left(-\dfrac{3}{a}\right) = \dfrac{6}{a} + \dfrac{3}{a} = \dfrac{6 + 3}{a} = \dfrac{9}{a}$

7. $\dfrac{6x}{7y^2} - \dfrac{2x}{7y^2} - \dfrac{5x}{7y^2} = \dfrac{6x - 2x - 5x}{7y^2} = -\dfrac{x}{7y^2}$

9. $\dfrac{n}{n+5} + \dfrac{3}{n+5} = \dfrac{n+3}{n+5}$

11. $\dfrac{2r}{r+2} + \dfrac{r}{r+2} = \dfrac{2r + r}{r+2} = \dfrac{3r}{r+2}$

13. $\dfrac{1 - 2q}{3q} + \dfrac{5q + 2}{3q} = \dfrac{1 - 2q + 5q + 2}{3q}$
$= \dfrac{3q + 3}{3q}$
$= \dfrac{\cancel{3}(q + 1)}{\cancel{3}q}$
$= \dfrac{q + 1}{q}$

15. $\dfrac{19u + 6}{5u} - \dfrac{4u + 6}{5u} = \dfrac{19u + 6 - \left(4u + 6\right)}{5u}$
$= \dfrac{19u + 6 - 4u - 6}{5u}$
$= \dfrac{15u}{5u}$
$= \dfrac{3 \cdot \cancel{5} \cdot \cancel{u}}{\cancel{5} \cdot \cancel{u}}$
$= 3$

17. $\dfrac{2 - 2a}{2a - 3} + \dfrac{8a - 11}{2a - 3} = \dfrac{2 - 2a + 8a - 11}{2a - 3}$
$= \dfrac{6a - 9}{2a - 3}$
$= \dfrac{3\cancel{(2a - 3)}}{\cancel{2a - 3}}$
$= 3$

19. $\dfrac{w^2 + 5}{w + 1} + \dfrac{6 - w^2}{w + 1} = \dfrac{w^2 + 5 + 6 - w^2}{w + 1}$
$= \dfrac{11}{w + 1}$

21. $\dfrac{4m + 3}{m - 1} - \dfrac{m + 4}{m - 1} = \dfrac{4m + 3 + \left(-m - 4\right)}{m - 1} = \dfrac{3m - 1}{m - 1}$

23. $\dfrac{1 - 3t}{2t - 3} - \dfrac{8t - 2}{2t - 3} = \dfrac{1 - 3t + \left(-8t + 2\right)}{2t - 3} = \dfrac{-11t + 3}{2t - 3}$

25. $\dfrac{2w + 3}{w + 5} + \dfrac{5w - 2}{w + 5} - \dfrac{w + 6}{w + 5}$
$= \dfrac{2w + 3 + 5w - 2 + \left(-w - 6\right)}{w + 5}$
$= \dfrac{6w - 5}{w + 5}$

27. $\dfrac{g}{g^2 - 9} - \dfrac{3}{g^2 - 9} = \dfrac{g - 3}{g^2 - 9}$
$= \dfrac{\cancel{g - 3}}{(g + 3)\cancel{(g - 3)}}$
$= \dfrac{1}{g + 3}$

29. $\dfrac{z^2+2z}{z-7}+\dfrac{21-12z}{z-7}=\dfrac{z^2+2z+21-12z}{z-7}$

$=\dfrac{z^2-10z+21}{z-7}$

$=\dfrac{\cancel{(z-7)}\,(z-3)}{\cancel{z-7}}$

$=z-3$

31. $\dfrac{2s}{s^2+4s+4}-\dfrac{s-2}{s^2+4s+4}=\dfrac{2s+(-s+2)}{s^2+4s+4}$

$=\dfrac{\cancel{s+2}}{\cancel{(s+2)}\,(s+2)}$

$=\dfrac{1}{s+2}$

33. $\dfrac{y^2}{y^2+3y}-\dfrac{9}{y^2+3y}=\dfrac{y^2-9}{y^2+3y}$

$=\dfrac{\cancel{(y+3)}\,(y-3)}{y\cancel{(y+3)}}$

$=\dfrac{y-3}{y}$

35. $\dfrac{x^2+3x}{x^2-x-12}+\dfrac{2x+6}{x^2-x-12}=\dfrac{x^2+3x+2x+6}{x^2-x-12}$

$=\dfrac{x^2+5x+6}{x^2-x-12}$

$=\dfrac{(x+2)\cancel{(x+3)}}{(x-4)\cancel{(x+3)}}$

$=\dfrac{x+2}{x-4}$

37. $\dfrac{t^2-4t}{t^2+10t+21}-\dfrac{-6t+3}{t^2+10t+21}=\dfrac{t^2-4t+(6t-3)}{t^2+10t+21}$

$=\dfrac{t^2+2t-3}{t^2+10t+21}$

$=\dfrac{\cancel{(t+3)}\,(t-1)}{\cancel{(t+3)}\,(t+7)}$

$=\dfrac{t-1}{t+7}$

39. $\dfrac{3x^3}{x^2-2x+4}+\dfrac{24}{x^2-2x+4}$

$=\dfrac{3x^3+24}{x^2-2x+4}$

$=\dfrac{3\left(x^3+8\right)}{x^2-2x+4}$

$=\dfrac{3(x+2)\cancel{\left(x^2-2x+4\right)}}{\cancel{\left(x^2-2x+4\right)}}$

$=3(x+2)$

41. $\dfrac{x-5}{x^2-25}+\dfrac{2x+10}{x^2-25}+\dfrac{-2x}{x^2-25}$

$=\dfrac{x-5+2x+10-2x}{x^2-25}$

$=\dfrac{x+5}{x^2-25}$

$=\dfrac{\cancel{x+5}}{\cancel{(x+5)}\,(x-5)}$

$=\dfrac{1}{x-5}$

43. $\dfrac{3x^2+2x}{x^2-2x-15}+\dfrac{3x-8}{x^2-2x-15}+\dfrac{2x+2}{x^2-2x-15}$

$=\dfrac{3x^2+2x+3x-8+2x+2}{x^2-2x-15}$

$=\dfrac{3x^2+7x-6}{x^2-2x-15}$

$=\dfrac{(3x-2)\cancel{(x+3)}}{(x-5)\cancel{(x+3)}}$

$=\dfrac{3x-2}{x-5}$

45. $\dfrac{2x^2+x-3}{x^2+6x+5}+\dfrac{x^2-2x+4}{x^2+6x+5}-\dfrac{2x^2+x+4}{x^2+6x+5}$

$=\dfrac{2x^2+x-3+x^2-2x+4+\left(-2x^2-x-4\right)}{x^2+6x+5}$

$=\dfrac{x^2-2x-3}{x^2+6x+5}$

$=\dfrac{(x-3)\,\cancel{(x+1)}}{(x+5)\,\cancel{(x+1)}}$

$=\dfrac{x-3}{x+5}$

47. Mistake: Combined terms that are not like.

Correct: $\dfrac{4y+1}{5}$

49. Mistake: Did not change sign of –5.

Correct: $\dfrac{(x+1)+(-x+5)}{3}=\dfrac{6}{3}=2$

51. $\cancel{2}\cdot\dfrac{3x}{\cancel{4}}+\cancel{2}\cdot\dfrac{2x+9}{\cancel{4}}=\dfrac{3x}{2}+\dfrac{2x+9}{2}$

$=\dfrac{3x+2x+9}{2}$

$=\dfrac{5x+9}{2}$

53. $\dfrac{x}{4}+\dfrac{3x-5}{4}+\dfrac{2x-1}{4}=\dfrac{x+3x-5+2x-1}{4}$

$=\dfrac{6x-6}{4}$

$=\dfrac{6(x-1)}{4}$

$=\dfrac{3(x-1)}{2}$

$=\dfrac{3x-3}{2}$

Review Exercises

1. $\dfrac{1}{2}+\dfrac{3}{5}=\dfrac{5}{10}+\dfrac{6}{10}=\dfrac{5+6}{10}=\dfrac{11}{10}$

2. $\dfrac{5}{6}-\dfrac{3}{8}=\dfrac{20}{24}-\dfrac{9}{24}=\dfrac{20-9}{24}=\dfrac{11}{24}$

3. $\dfrac{4}{7}-\left(-\dfrac{3}{5}\right)=\dfrac{4}{7}+\dfrac{3}{5}$

$=\dfrac{20}{35}+\dfrac{21}{35}$

$=\dfrac{20+21}{35}$

$=\dfrac{41}{35}$

$=1\dfrac{6}{35}$

4. $\dfrac{1}{3}x-\dfrac{1}{6}+\dfrac{2}{5}x+\dfrac{3}{4}=\dfrac{1}{3}x+\dfrac{2}{5}x-\dfrac{1}{6}+\dfrac{3}{4}$

$=\dfrac{5}{15}x+\dfrac{6}{15}x-\dfrac{2}{12}+\dfrac{9}{12}$

$=\dfrac{11}{15}x+\dfrac{7}{12}$

5. $\left(x^2-9x-\dfrac{5}{8}\right)+\left(x^2-3x+\dfrac{1}{6}\right)$

$=x^2+x^2-9x-3x-\dfrac{5}{8}+\dfrac{1}{6}$

$=2x^2-12x-\dfrac{15}{24}+\dfrac{4}{24}$

$=2x^2-12x-\dfrac{11}{24}$

6. $\left(3y^2+\dfrac{3}{4}y-8\right)-\left(y^2+\dfrac{4}{5}y+2\right)$

$=\left(3y^2+\dfrac{15}{20}y-8\right)+\left(-y^2-\dfrac{16}{20}y-2\right)$

$=3y^2-y^2+\dfrac{15}{20}y-\dfrac{16}{20}y-8-2$

$=2y^2-\dfrac{1}{20}y-10$

Exercise Set 7.4

1. The LCD is mn.

$\dfrac{2}{n}=\dfrac{2\cdot m}{n\cdot m}=\dfrac{2m}{mn}$

$\dfrac{3}{m}=\dfrac{3\cdot n}{m\cdot n}=\dfrac{3n}{mn}$

3. The LCD is a^3b^5.

$$\frac{2t}{a^2b^5} = \frac{2t \cdot a}{a^2b^5 \cdot a} = \frac{2at}{a^3b^5}$$

$$\frac{3z}{a^3b^3} = \frac{3z \cdot b^2}{a^3b^3 \cdot b^2} = \frac{3b^2z}{a^3b^5}$$

5. The LCD is $15a^2$.

$$\frac{2}{3a} = \frac{2 \cdot 5a}{3a \cdot 5a} = \frac{10a}{15a^2}$$

$$\frac{3}{5a^2} = \frac{3 \cdot 3}{5a^2 \cdot 3} = \frac{9}{15a^2}$$

7. The LCD is $14x^2y^2$.

$$\frac{3}{7x^2y} = \frac{3 \cdot 2y}{7x^2y \cdot 2y} = \frac{6y}{14x^2y^2}$$

$$\frac{5}{14xy^2} = \frac{5 \cdot x}{14xy^2 \cdot x} = \frac{5x}{14x^2y^2}$$

9. $15m^3n^3 = 3 \cdot 5 \cdot m^3 \cdot n^3$

$18m^2n = 2 \cdot 3^2 \cdot m^2 \cdot n$

The LCD is $2 \cdot 3^2 \cdot 5 \cdot m^3 \cdot n^3 = 90m^3n^3$.

$$\frac{8}{15m^3n^3} = \frac{8 \cdot 6}{15m^3n^3 \cdot 6} = \frac{48}{90m^3n^3}$$

$$\frac{7}{18m^2n} = \frac{7 \cdot 5mn^2}{18m^2n \cdot 5mn^2} = \frac{35mn^2}{90m^3n^3}$$

11. The LCD is $(y+2)(y-2)$.

$$\frac{3}{(y+2)} = \frac{3(y-2)}{(y+2)(y-2)} = \frac{3y-6}{(y+2)(y-2)}$$

$$\frac{4y}{(y-2)} = \frac{4y(y+2)}{(y-2)(y+2)} = \frac{4y^2+8y}{(y+2)(y-2)}$$

13. $4y - 20 = 4(y-5) = 2^2 \cdot (y-5)$

$6y - 30 = 6(y-5) = 2 \cdot 3 \cdot (y-5)$

The LCD is $2^2 \cdot 3 \cdot (y-5) = 12(y-5)$.

$$\frac{3y}{4(y-5)} = \frac{3y \cdot 3}{4(y-5) \cdot 3} = \frac{9y}{12(y-5)}$$

$$\frac{7y}{6(y-5)} = \frac{7y \cdot 2}{6(y-5) \cdot 2} = \frac{14y}{12(y-5)}$$

15. $6x + 18 = 6(x+3)$

$x^2 + 3x = x(x+3)$

The LCD is $6x(x+3)$.

$$\frac{5}{6(x+3)} = \frac{5 \cdot x}{6(x+3) \cdot x} = \frac{5x}{6x(x+3)}$$

$$\frac{2}{x(x+3)} = \frac{2 \cdot 6}{x(x+3) \cdot 6} = \frac{12}{6x(x+3)}$$

17. $p^2 - 4 = (p+2)(p-2)$

$p + 2 = (p+2)$

The LCD is $(p+2)(p-2) = p^2 - 4$.

$$\frac{2}{p^2-4}$$

$$\frac{(3+p)}{(p+2)} = \frac{(3+p)(p-2)}{(p+2)(p-2)} = \frac{p^2+p-6}{p^2-4}$$

19. The LCD is $3(u+2)^2$.

$$\frac{4u}{(u+2)^2} = \frac{4u \cdot 3}{(u+2)^2 \cdot 3} = \frac{12u}{3(u+2)^2};$$

$$\frac{2}{3(u+2)} = \frac{2 \cdot (u+2)}{3(u+2) \cdot (u+2)} = \frac{2u+4}{3(u+2)^2}$$

21. $t^2 - 36 = (t+6)(t-6)$

$t^2 - 7t + 6 = (t-6)(t-1)$

The LCD is $(t+6)(t-6)(t-1)$.

$$\frac{2t-4}{t^2-36} = \frac{2t-4}{(t+6)(t-6)}$$

$$= \frac{(2t-4)(t-1)}{(t+6)(t-6)(t-1)}$$

$$= \frac{2t^2-6t+4}{(t+6)(t-6)(t-1)}$$

$$\frac{3t+2}{t^2-7t+6} = \frac{3t+2}{(t-6)(t-1)}$$

$$= \frac{(3t+2)(t+6)}{(t-6)(t-1)(t+6)}$$

$$= \frac{3t^2+20t+12}{(t+6)(t-6)(t-1)}$$

23. $x^2 + 3x - 4 = (x+4)(x-1)$

$x^2 + 6x + 8 = (x+4)(x+2)$

The LCD is $(x+4)(x-1)(x+2)$.

$$\frac{x+2}{x^2+3x-4} = \frac{x+2}{(x+4)(x-1)}$$

$$= \frac{(x+2)\cdot(x+2)}{(x+4)(x-1)\cdot(x+2)}$$

$$= \frac{x^2+4x+4}{(x+4)(x-1)(x+2)}$$

$$\frac{x-3}{x^2+6x+8} = \frac{x-3}{(x+4)(x+2)}$$

$$= \frac{(x-3)\cdot(x-1)}{(x+4)(x+2)\cdot(x-1)}$$

$$= \frac{x^2-4x+3}{(x+4)(x-1)(x+2)}$$

25. $x - 1 = (x-1)$

$x^2 - 3x + 2 = (x-1)(x-2)$

$x - 2 = (x-2)$

The LCD is $(x-1)(x-2)$.

$$\frac{2}{x-1} = \frac{2\cdot(x-2)}{(x-1)\cdot(x-2)} = \frac{2x-4}{(x-1)(x-2)}$$

$$\frac{3}{x^2-3x+2} = \frac{3}{(x-1)(x-2)}$$

$$\frac{5x}{x-2} = \frac{5x\cdot(x-1)}{(x-2)\cdot(x-1)} = \frac{5x^2-5x}{(x-1)(x-2)}$$

27. The LCD is 18.

$$\frac{3a-5}{9} - \frac{2a+7}{6} = \frac{3a-5}{9}\cdot\frac{2}{2} - \frac{2a+7}{6}\cdot\frac{3}{3}$$

$$= \frac{6a-10}{18} - \frac{6a+21}{18}$$

$$= \frac{6a-10+(-6a-21)}{18}$$

$$= -\frac{31}{18}$$

29. The LCD is $6c$.

$$\frac{5}{2c} + \frac{1}{6c} = \frac{5}{2c}\cdot\frac{3}{3} + \frac{1}{6c}$$

$$= \frac{15}{6c} + \frac{1}{6c}$$

$$= \frac{15+1}{6c}$$

$$= \frac{16}{6c}$$

$$= \frac{8}{3c}$$

31. The LCD is x^3.

$$\frac{5}{x} - \frac{2}{x^3} = \frac{5}{x}\cdot\frac{x^2}{x^2} - \frac{2}{x^3}$$

$$= \frac{5x^2}{x^3} - \frac{2}{x^3}$$

$$= \frac{5x^2-2}{x^3}$$

33. The LCD is $(y+2)(y-4)$.

$$\frac{3}{y+2} + \frac{5}{y-4}$$

$$= \frac{3}{y+2}\cdot\frac{y-4}{y-4} + \frac{5}{y-4}\cdot\frac{y+2}{y+2}$$

$$= \frac{3y-12}{(y+2)(y-4)} + \frac{5y+10}{(y+2)(y-4)}$$

$$= \frac{3y-12+5y+10}{(y+2)(y-4)}$$

$$= \frac{8y-2}{(y+2)(y-4)}$$

35. The LCD is $(x-2)(x+2)$.

$$\frac{3}{(x-2)} - \frac{2}{(x+2)}$$

$$= \frac{3(x+2)}{(x+2)(x-2)} - \frac{2(x-2)}{(x+2)(x-2)}$$

$$= \frac{3x+6-2x+4}{(x+2)(x-2)}$$

$$= \frac{x+10}{(x+2)(x-2)}$$

37. The LCD is $(x+5)(x-5)$.

$$\frac{4}{x^2-25}+\frac{2}{x-5}$$

$$=\frac{4}{(x+5)(x-5)}+\frac{2}{x-5}$$

$$=\frac{4}{(x+5)(x-5)}+\frac{2\cdot(x+5)}{(x-5)\cdot(x+5)}$$

$$=\frac{4+2x+10}{(x+5)(x-5)}$$

$$=\frac{2x+14}{(x+5)(x-5)}$$

39. $p^2-2p+1=(p-1)(p-1)$

The LCD is $(p-1)^2$.

$$\frac{2p}{p^2-2p+1}+\frac{8}{p-1}$$

$$=\frac{2p}{(p-1)^2}+\frac{8(p-1)}{(p-1)(p-1)}$$

$$=\frac{2p+8p-8}{p^2-2p+1}$$

$$=\frac{10p-8}{p^2-2p+1}$$

41. $z^2+z-12=(z+4)(z-3)$

The LCD is $(z+4)(z-3)$.

$$\frac{z+5}{z^2+z-12}-\frac{z+2}{z-3}$$

$$=\frac{z+5}{(z+4)(z-3)}-\frac{(z+2)\cdot(z+4)}{(z-3)\cdot(z+4)}$$

$$=\frac{z+5-z^2-6z-8}{(z+4)(z-3)}$$

$$=\frac{-z^2-5z-3}{(z+4)(z-3)}$$

43. The LCD is $(n-4)(n+3)$.

$$\frac{n+2}{n-4}+\frac{n-3}{n+3}$$

$$=\frac{(n+2)\cdot(n+3)}{(n-4)\cdot(n+3)}+\frac{(n-3)\cdot(n-4)}{(n+3)\cdot(n-4)}$$

$$=\frac{n^2+5n+6+n^2-7n+12}{(n-4)(n+3)}$$

$$=\frac{2n^2-2n+18}{(n-4)(n+3)}$$

45. The LCD is $(t+4)(t+6)$.

$$\frac{t+2}{t+4}-\frac{2t-1}{t+6}$$

$$=\frac{t+2}{t+4}\cdot\frac{t+6}{t+6}-\frac{2t-1}{t+6}\cdot\frac{t+4}{t+4}$$

$$=\frac{t^2+8t+12}{(t+4)(t+6)}-\frac{2t^2+7t-4}{(t+4)(t+6)}$$

$$=\frac{t^2+8t+12+(-2t^2-7t+4)}{(t+4)(t+6)}$$

$$=\frac{-t^2+t+16}{(t+4)(t+6)}$$

47. $c^2+10c+25=(c+5)^2$

$2c+10=2(c+5)$

The LCD is $2(c+5)^2$.

$$\frac{c+5}{c^2+10c+25}+\frac{3}{2c+10}$$

$$=\frac{c+5}{(c+5)^2}+\frac{3}{2(c+5)}$$

$$=\frac{2\cdot(c+5)}{2\cdot(c+5)^2}+\frac{3\cdot(c+5)}{2(c+5)\cdot(c+5)}$$

$$=\frac{2c+10}{2(c+5)^2}+\frac{3c+15}{2(c+5)^2}$$

$$=\frac{5c+25}{2(c+5)^2}$$

$$=\frac{5\cancel{(c+5)}}{2(c+5)\cancel{(c+5)}}$$

$$=\frac{5}{2(c+5)}$$

49. The LCD is $y(y+3)$.

$$\frac{5y+6}{y^2+3y}+\frac{y}{y+3}=\frac{5y+6}{y(y+3)}+\frac{y}{y+3}\cdot\frac{y}{y}$$

$$=\frac{5y+6}{y(y+3)}+\frac{y^2}{y(y+3)}$$

$$=\frac{y^2+5y+6}{y(y+3)}$$

$$=\frac{(y+2)\cancel{(y+3)}}{y\cancel{(y+3)}}$$

$$=\frac{y+2}{y}$$

51. $r^2-1=(r+1)(r-1)$

$r^2-2r+1=(r-1)(r-1)$

The LCD is $(r+1)(r-1)^2$.

$$\frac{3r}{r^2-1}-\frac{2r+1}{r^2-2r+1}$$

$$=\frac{3r}{(r+1)(r-1)}-\frac{2r+1}{(r-1)(r-1)}$$

$$=\frac{3r}{(r+1)(r-1)}\cdot\frac{(r-1)}{(r-1)}-\frac{2r+1}{(r-1)(r-1)}\cdot\frac{(r+1)}{(r+1)}$$

$$=\frac{3r^2-3r}{(r+1)(r-1)^2}-\frac{2r^2+3r+1}{(r+1)(r-1)^2}$$

$$=\frac{3r^2-3r+\left(-2r^2-3r-1\right)}{(r+1)(r-1)^2}$$

$$=\frac{r^2-6r-1}{(r+1)(r-1)^2}$$

53. $q^2-9=(q+3)(q-3)$

$q^2+3q=q(q+3)$

The LCD is $q(q+3)(q-3)$.

$$\frac{q+5}{q^2-9}-\frac{1}{q^2+3q}$$

$$=\frac{q+5}{(q+3)(q-3)}-\frac{1}{q(q+3)}$$

$$=\frac{q+5}{(q+3)(q-3)}\cdot\frac{q}{q}-\frac{1}{q(q+3)}\cdot\frac{(q-3)}{(q-3)}$$

$$=\frac{q^2+5q}{q(q+3)(q-3)}-\frac{q-3}{q(q+3)(q-3)}$$

$$=\frac{q^2+5q+(-q+3)}{q(q+3)(q-3)}$$

$$=\frac{q^2+4q+3}{q(q+3)(q-3)}$$

$$=\frac{\cancel{(q+3)}(q+1)}{q\cancel{(q+3)}(q-3)}$$

$$=\frac{q+1}{q(q-3)}$$

55. $w^2+4w-5=(w+5)(w-1)$

$w^2+3w-10=(w+5)(w-2)$

The LCD is $(w+5)(w-1)(w-2)$.

$$\frac{w-7}{w^2+4w-5}-\frac{w-9}{w^2+3w-10}$$

$$=\frac{w-7}{(w+5)(w-1)}-\frac{w-9}{(w+5)(w-2)}$$

$$=\frac{w-7}{(w+5)(w-1)}\cdot\frac{(w-2)}{(w-2)}-\frac{w-9}{(w+5)(w-2)}\cdot\frac{(w-1)}{(w-1)}$$

$$=\frac{w^2-9w+14}{(w+5)(w-1)(w-2)}-\frac{w^2-10w+9}{(w+5)(w-1)(w-2)}$$

$$=\frac{w^2-9w+14+\left(-w^2+10w-9\right)}{(w+5)(w-1)(w-2)}$$

$$=\frac{\cancel{w+5}}{\cancel{(w+5)}(w-1)(w-2)}$$

$$=\frac{1}{(w-1)(w-2)}$$

57. $v^2 - 16 = (v+4)(v-4)$

$v^2 - v - 12 = (v-4)(v+3)$

The LCD is $(v+4)(v-4)(v+3)$.

$\dfrac{2v+5}{v^2-16} - \dfrac{v-9}{v^2-v-12}$

$= \dfrac{2v+5}{(v+4)(v-4)} - \dfrac{v-9}{(v-4)(v+3)}$

$= \dfrac{(2v+5)\cdot(v+3)}{(v+4)(v-4)\cdot(v+3)} - \dfrac{(v-9)\cdot(v+4)}{(v-4)(v+3)\cdot(v+4)}$

$= \dfrac{2v^2+11v+15}{(v+4)(v-4)(v+3)} - \dfrac{v^2-5v-36}{(v-4)(v+3)(v+4)}$

$= \dfrac{2v^2+11v+15-v^2+5v+36}{(v+4)(v-4)(v+3)}$

$= \dfrac{v^2+16v+51}{(v+4)(v-4)(v+3)}$

59. $\dfrac{3}{t-3} + \dfrac{6}{3-t} = \dfrac{3}{t-3} + \dfrac{6}{3-t}\cdot\dfrac{-1}{-1}$

$= \dfrac{3}{t-3} + \dfrac{-6}{t-3}$

$= \dfrac{-3}{t-3}$

61. $\dfrac{4}{2x-1} + \dfrac{x}{1-2x} = \dfrac{4}{2x-1} + \dfrac{x}{1-2x}\cdot\dfrac{-1}{-1}$

$= \dfrac{4}{2x-1} + \dfrac{-x}{2x-1}$

$= \dfrac{4-x}{2x-1}$

63. $3y - 6 = 3(y-2)$

$2 - y = -1(y-2)$

$\dfrac{y}{3y-6} - \dfrac{2}{2-y} = \dfrac{y}{3(y-2)} - \dfrac{2}{2-y}\cdot\dfrac{-3}{-3}$

$= \dfrac{y}{3(y-2)} - \dfrac{-6}{3(y-2)}$

$= \dfrac{y+6}{3(y-2)}$

65. Mistake: Did not find a common denominator.

Correct: $\dfrac{3}{x}\cdot\dfrac{2}{2} - \dfrac{x}{2}\cdot\dfrac{x}{x} = \dfrac{6}{2x} - \dfrac{x^2}{2x} = \dfrac{6-x^2}{2x}$

67. Mistake: Did not change the sign of 7.

Correct: $\dfrac{3x}{x+2} - \dfrac{4x+7}{x+2} = \dfrac{3x+(-4x-7)}{x+2}$

$= \dfrac{-x-7}{x+2}$

69. Mistake: Added denominators.

Correct: $\dfrac{2w+5w}{x} = \dfrac{7w}{x}$

71. $\dfrac{t}{2} + \dfrac{t}{3} = \dfrac{t\cdot 3}{2\cdot 3} + \dfrac{t\cdot 2}{3\cdot 2} = \dfrac{3t+2t}{6} = \dfrac{5t}{6}$

73. The perimeter is given by the sum of twice the length and twice the width:

$\dfrac{2}{x+3} + \dfrac{2}{x} = \dfrac{2\cdot x}{(x+3)\cdot x} + \dfrac{2\cdot(x+3)}{x\cdot(x+3)}$

$= \dfrac{2x+2x+6}{x(x+3)}$

$= \dfrac{4x+6}{x^2+3x}$

75. area of trapezoid − area of triangle

$= \dfrac{1}{2}h(b_1+b_2) - \dfrac{1}{2}bh$

$= \dfrac{1}{2}h(a+2b) - \dfrac{1}{2}b\left(\dfrac{3}{4}h\right)$

$= \dfrac{h(a+2b)}{2} - \dfrac{3bh}{8}$

$= \dfrac{h(a+2b)\cdot 4}{2\cdot 4} - \dfrac{3bh}{8}$

$= \dfrac{4ah+8bh}{8} - \dfrac{3bh}{8}$

$= \dfrac{4ah+5bh}{8}$

77. $\dfrac{x}{4x^2-1} + \dfrac{2}{4x-2} - \dfrac{3x+1}{1-2x}$

$= \dfrac{x}{(2x-1)(2x+1)} + \dfrac{\cancel{2}}{\cancel{2}(2x-1)} - \dfrac{3x+1}{1-2x}$

$= \dfrac{x}{(2x-1)(2x+1)} + \dfrac{1}{2x-1}\cdot\dfrac{2x+1}{2x+1} - \dfrac{3x+1}{1-2x}\cdot\dfrac{-1(2x+1)}{-1(2x+1)}$

$= \dfrac{x}{(2x-1)(2x+1)} + \dfrac{2x+1}{(2x-1)(2x+1)} - \dfrac{-6x^2-5x-1}{(2x-1)(2x+1)}$

$= \dfrac{x+2x+1+(6x^2+5x+1)}{(2x-1)(2x+1)} = \dfrac{6x^2+8x+2}{(2x-1)(2x+1)}$

79. $(x+y) \div \left(\dfrac{1}{x} + \dfrac{1}{y} \right) = (x+y) \div \left(\dfrac{1 \cdot y}{x \cdot y} + \dfrac{1 \cdot x}{y \cdot x} \right)$

$\qquad = (x+y) \div \left(\dfrac{x+y}{xy} \right)$

$\qquad = \cancel{(x+y)} \cdot \dfrac{xy}{\cancel{(x+y)}}$

$\qquad = xy$

81. $\dfrac{2}{a} + \dfrac{3}{4a+8} \cdot \dfrac{4a}{5} = \dfrac{2}{a} + \dfrac{3}{\cancel{4}(a+2)} \cdot \dfrac{\cancel{4}a}{5}$

$\qquad = \dfrac{2}{a} + \dfrac{3a}{5(a+2)}$

$\qquad = \dfrac{2 \cdot 5(a+2)}{a \cdot 5(a+2)} + \dfrac{3a \cdot a}{5(a+2) \cdot a}$

$\qquad = \dfrac{10(a+2)}{5a(a+2)} + \dfrac{3a^2}{5a(a+2)}$

$\qquad = \dfrac{10a+20}{5a(a+2)} + \dfrac{3a^2}{5a(a+2)}$

$\qquad = \dfrac{3a^2 + 10a + 20}{5a(a+2)}$

83. $\left(\dfrac{a+b}{a-b} + \dfrac{a-b}{a+b} \right) \div \dfrac{a+b}{a-b}$

$= \left(\dfrac{a+b}{a-b} \cdot \dfrac{a+b}{a+b} + \dfrac{a-b}{a+b} \cdot \dfrac{a-b}{a-b} \right) \div \dfrac{a+b}{a-b}$

$= \left(\dfrac{a^2 + 2ab + b^2}{(a+b)(a-b)} + \dfrac{a^2 - 2ab + b^2}{(a+b)(a-b)} \right) \div \dfrac{a+b}{a-b}$

$= \dfrac{a^2 + 2ab + b^2 + a^2 - 2ab + b^2}{(a+b)(a-b)} \div \dfrac{a+b}{a-b}$

$= \dfrac{2a^2 + 2b^2}{(a+b)\cancel{(a-b)}} \cdot \dfrac{\cancel{a-b}}{a+b}$

$= \dfrac{2a^2 + 2b^2}{(a+b)^2}$

85. $\left(\dfrac{1}{x+2} - \dfrac{3}{x^2-4} \right) \div \dfrac{3}{x-2}$

$= \left(\dfrac{1}{x+2} - \dfrac{3}{(x+2)(x-2)} \right) \div \dfrac{3}{x-2}$

$= \left(\dfrac{1 \cdot (x-2)}{(x+2) \cdot (x-2)} - \dfrac{3}{(x+2)(x-2)} \right) \div \dfrac{3}{x-2}$

$= \left(\dfrac{x-2-3}{(x+2)(x-2)} \right) \div \dfrac{3}{x-2}$

$= \dfrac{x-5}{(x+2)\cancel{(x-2)}} \cdot \dfrac{\cancel{x-2}}{3}$

$= \dfrac{x-5}{3(x+2)}$

Review Exercises

1. $\dfrac{1}{4} \div \dfrac{2}{3} = \dfrac{1}{4} \cdot \dfrac{3}{2} = \dfrac{3}{8}$

2. $\dfrac{m}{6} \div \dfrac{18}{m} = \dfrac{m}{6} \cdot \dfrac{m}{18} = \dfrac{m^2}{108}$

3. $4x^2 - 16 = 4\left(x^2 - 4\right) = 4(x+2)(x-2)$

4. $6y^3 - 27y^2 - 15y = 3y\left(2y^2 - 9y - 5\right)$

$\qquad\qquad\qquad\qquad = 3y(2y+1)(y-5)$

5. $\dfrac{x^2 - 7x + 6}{x^2 - 4x - 12} = \dfrac{\cancel{(x-6)}(x-1)}{\cancel{(x-6)}(x+2)} = \dfrac{x-1}{x+2}$

6. $\dfrac{x+3}{x^2-9} = \dfrac{\cancel{x+3}}{\cancel{(x+3)}(x-3)} = \dfrac{1}{x-3}$

Exercise Set 7.5

1. $\dfrac{\frac{3}{4}}{\frac{3}{2}} = \dfrac{3}{4} \div \dfrac{3}{2} = \dfrac{3}{4} \cdot \dfrac{2}{3} = \dfrac{6}{12} = \dfrac{1}{2}$

3. $\dfrac{\frac{m}{n}}{\frac{q}{p}} = \dfrac{m}{n} \div \dfrac{q}{p} = \dfrac{m}{n} \cdot \dfrac{p}{q} = \dfrac{mp}{nq}$

5. $\dfrac{\dfrac{1}{2}+\dfrac{3}{5}}{\dfrac{3}{2}-\dfrac{1}{5}}=\dfrac{\dfrac{1\cdot 5}{2\cdot 5}+\dfrac{3\cdot 2}{5\cdot 2}}{\dfrac{3\cdot 5}{2\cdot 5}-\dfrac{1\cdot 2}{5\cdot 2}}$

$=\dfrac{\dfrac{5}{10}+\dfrac{6}{10}}{\dfrac{15}{10}-\dfrac{2}{10}}$

$=\dfrac{\dfrac{11}{10}}{\dfrac{13}{10}}$

$=\dfrac{11}{10}\div\dfrac{13}{10}$

$=\dfrac{11}{10}\cdot\dfrac{10}{13}$

$=\dfrac{11}{13}$

7. $\dfrac{5+\dfrac{2}{a}}{3+\dfrac{4}{a}}=\dfrac{a\cdot\left(5+\dfrac{2}{a}\right)}{a\cdot\left(3+\dfrac{4}{a}\right)}$

$=\dfrac{a\cdot 5+a\cdot\dfrac{2}{a}}{a\cdot 3+a\cdot\dfrac{4}{a}}$

$=\dfrac{5a+2}{3a+4}$

9. $\dfrac{\dfrac{1}{y}-1}{\dfrac{1}{y}+1}=\dfrac{y\cdot\left(\dfrac{1}{y}-1\right)}{y\cdot\left(\dfrac{1}{y}+1\right)}=\dfrac{y\cdot\dfrac{1}{y}-y\cdot 1}{y\cdot\dfrac{1}{y}+y\cdot 1}=\dfrac{1-y}{1+y}$

11. $\dfrac{r-\dfrac{3}{4}}{\dfrac{1}{2}-r}=\dfrac{\left(r-\dfrac{3}{4}\right)\cdot 4}{\left(\dfrac{1}{2}-r\right)\cdot 4}=\dfrac{r\cdot 4-\dfrac{3}{4}\cdot 4}{\dfrac{1}{2}\cdot 4-r\cdot 4}=\dfrac{4r-3}{2-4r}$

13. $\dfrac{\dfrac{3}{x}-1}{\dfrac{9}{x^2}-1}=\dfrac{\left(\dfrac{3}{x}-1\right)\cdot x^2}{\left(\dfrac{9}{x^2}-1\right)\cdot x^2}$

$=\dfrac{\dfrac{3}{x}\cdot x^2-1\cdot x^2}{\dfrac{9}{x^2}\cdot x^2-1\cdot x^2}$

$=\dfrac{3x-x^2}{9-x^2}$

$=\dfrac{x\,(3-x)}{(3+x)\,(3-x)}$

$=\dfrac{x}{3+x}$

15. $\dfrac{x+y}{\dfrac{1}{x}+\dfrac{1}{y}}=(x+y)\div\left(\dfrac{1\cdot y}{x\cdot y}+\dfrac{1\cdot x}{y\cdot x}\right)$

$=(x+y)\div\left(\dfrac{x+y}{xy}\right)$

$=(x+y)\cdot\dfrac{xy}{x+y}$

$=xy$

17. $\dfrac{x+2}{\dfrac{x^2-4}{x}}=(x+2)\div\dfrac{(x+2)(x-2)}{x}$

$=(x+2)\cdot\dfrac{x}{(x+2)(x-2)}$

$=\dfrac{x}{x-2}$

19. $\dfrac{\dfrac{6}{3x-2}}{\dfrac{x}{2x+1}}=\dfrac{6}{3x-2}\div\dfrac{x}{2x+1}$

$=\dfrac{6}{3x-2}\cdot\dfrac{2x+1}{x}$

$=\dfrac{6(2x+1)}{x(3x-2)}$

21. $\dfrac{\dfrac{a}{a^2-b^2}}{\dfrac{b}{a+b}} = \dfrac{a}{(a+b)(a-b)} \div \dfrac{b}{a+b}$

$= \dfrac{a}{(a+b)(a-b)} \cdot \dfrac{a+b}{b}$

$= \dfrac{a}{b(a-b)}$

23. $\dfrac{\dfrac{y^2-25}{5y}}{\dfrac{y^2+5y}{2y^3}} = \dfrac{y^2-25}{5y} \div \dfrac{y^2+5y}{2y^3}$

$= \dfrac{y^2-25}{5y} \cdot \dfrac{2y^3}{y^2+5y}$

$= \dfrac{(y+5)(y-5)}{5y} \cdot \dfrac{2y^3}{y(y+5)}$

$= \dfrac{2y^3(y-5)}{5y^2}$

$= \dfrac{2y(y-5)}{5}$

25. $\dfrac{2+\dfrac{2}{u-1}}{\dfrac{2}{u-1}} = \dfrac{(u-1)\cdot 2 + (u-1)\cdot\dfrac{2}{u-1}}{(u-1)\cdot\dfrac{2}{u-1}}$

$= \dfrac{2u-2+2}{2}$

$= \dfrac{2u}{2}$

$= u$

27. $\dfrac{\dfrac{a}{a-2}-1}{1+\dfrac{a}{a-2}} = \dfrac{(a-2)\cdot\left(\dfrac{a}{a-2}-1\right)}{(a-2)\cdot\left(1+\dfrac{a}{a-2}\right)}$

$= \dfrac{(a-2)\cdot\dfrac{a}{a-2}-(a-2)\cdot 1}{(a-2)\cdot 1+(a-2)\cdot\dfrac{a}{a-2}}$

$= \dfrac{a-(a-2)}{a-2+a}$

$= \dfrac{a-a+2}{a-2+a}$

$= \dfrac{2}{2a-2}$

$= \dfrac{2\cdot 1}{2(a-1)}$

$= \dfrac{1}{a-1}$

29. $\dfrac{\dfrac{x-3}{x}-1}{\dfrac{x^2-9}{x}-x} = \dfrac{\left(\dfrac{x-3}{x}-1\right)\cdot x}{\left(\dfrac{x^2-9}{x}-x\right)\cdot x}$

$= \dfrac{\dfrac{x-3}{x}\cdot x - 1\cdot x}{\dfrac{x^2-9}{x}\cdot x - x\cdot x}$

$= \dfrac{x-3-x}{x^2-9-x^2}$

$= \dfrac{-3}{-9}$

$= \dfrac{1}{3}$

31. $\dfrac{\dfrac{1}{x-1}-\dfrac{1}{x+1}}{\dfrac{1}{x^2-1}}$

$=\dfrac{(x+1)(x-1)\cdot\left(\dfrac{1}{x-1}-\dfrac{1}{x+1}\right)}{(x+1)(x-1)\cdot\left(\dfrac{1}{(x+1)(x-1)}\right)}$

$=\dfrac{(x+1)(x-1)\cdot\dfrac{1}{x-1}-(x+1)(x-1)\cdot\dfrac{1}{x+1}}{(x+1)(x-1)\cdot\dfrac{1}{(x+1)(x-1)}}$

$=\dfrac{x+1-(x-1)}{1}$

$=x+1-x+1$

$=2$

33. $\dfrac{1+\dfrac{1}{y}-\dfrac{2}{y^2}}{\dfrac{1}{y}+\dfrac{2}{y^2}}=\dfrac{\left(1+\dfrac{1}{y}-\dfrac{2}{y^2}\right)\cdot y^2}{\left(\dfrac{1}{y}+\dfrac{2}{y^2}\right)\cdot y^2}$

$=\dfrac{1\cdot y^2+\dfrac{1}{y}\cdot y^2-\dfrac{2}{y^2}\cdot y^2}{\dfrac{1}{y}\cdot y^2+\dfrac{2}{y^2}\cdot y^2}$

$=\dfrac{y^2+y-2}{y+2}$

$=\dfrac{(y+2)(y-1)}{y+2}$

$=y-1$

35. $\dfrac{1-\dfrac{1}{n}}{n-2+\dfrac{1}{n}}=\dfrac{\left(1-\dfrac{1}{n}\right)\cdot n}{\left(n-2+\dfrac{1}{n}\right)\cdot n}$

$=\dfrac{1\cdot n-\dfrac{1}{n}\cdot n}{n\cdot n-2\cdot n+\dfrac{1}{n}\cdot n}$

$=\dfrac{n-1}{n^2-2n+1}$

$=\dfrac{n-1}{(n-1)(n-1)}$

$=\dfrac{1}{n-1}$

37. $\dfrac{1-\dfrac{1}{2y+1}}{\dfrac{1}{2y^2-5y-3}-\dfrac{1}{y-3}}$

$=\dfrac{1-\dfrac{1}{2y+1}}{\dfrac{1}{(2y+1)(y-3)}-\dfrac{1}{y-3}}$

$=\dfrac{(2y+1)(y-3)\cdot\left(1-\dfrac{1}{2y+1}\right)}{(2y+1)(y-3)\cdot\left(\dfrac{1}{(2y+1)(y-3)}-\dfrac{1}{y-3}\right)}$

$=\dfrac{(2y+1)(y-3)\cdot 1-\dfrac{(2y+1)(y-3)}{2y+1}}{\dfrac{(2y+1)(y-3)}{(2y+1)(y-3)}-\dfrac{(2y+1)(y-3)}{y-3}}$

$=\dfrac{(2y+1)(y-3)-(y-3)}{1-(2y+1)}$

$=\dfrac{(2y^2-5y-3)+(-y+3)}{1+(-2y-1)}$

$=\dfrac{2y^2-5y-3-y+3}{1-2y-1}$

$=\dfrac{2y^2-6y}{-2y}$

$=\dfrac{-2y(-y+3)}{-2y}$

$=3-y$

39. $\dfrac{\dfrac{x^2-2x-8}{x^2-16}}{\dfrac{x^2+2x}{x^2+3x-4}}=\dfrac{x^2-2x-8}{x^2-16}\div\dfrac{x^2+2x}{x^2+3x-4}$

$=\dfrac{x^2-2x-8}{x^2-16}\cdot\dfrac{x^2+3x-4}{x^2+2x}$

$=\dfrac{(x-4)(x+2)}{(x+4)(x-4)}\cdot\dfrac{(x+4)(x-1)}{x(x+2)}$

$=\dfrac{x-1}{x}$

41. $\dfrac{\dfrac{2}{x+h}-\dfrac{2}{x}}{h} = \dfrac{x(x+h)\cdot\left(\dfrac{2}{x+h}-\dfrac{2}{x}\right)}{x(x+h)\cdot(h)}$

$= \dfrac{x\cancel{(x+h)}\cdot\dfrac{2}{\cancel{x+h}}-\cancel{x}(x+h)\cdot\dfrac{2}{\cancel{x}}}{x(x+h)\cdot h}$

$= \dfrac{2x-2(x+h)}{x(x+h)h}$

$= \dfrac{2x-2x-2h}{xh(x+h)}$

$= \dfrac{-2\cancel{h}}{x\cancel{h}(x+h)}$

$= \dfrac{-2}{x(x+h)}$

43. Mistake: Divided out $\dfrac{1}{2}$, which is not a common factor.

Correct: $\dfrac{2\cdot x-2\cdot\dfrac{1}{2}}{2\cdot y-2\cdot\dfrac{1}{2}} = \dfrac{2x-1}{2y-1}$

45. Mistake: Did not write as an equivalent multiplication.

Correct: $\dfrac{m}{6}\div\dfrac{3}{m} = \dfrac{m}{6}\cdot\dfrac{m}{3} = \dfrac{m^2}{18}$

47. Mistake: Did not distribute c in the numerator.

Correct: $\dfrac{c\cdot c+c\cdot\dfrac{1}{c}}{c\cdot\dfrac{1}{c}} = \dfrac{c^2+1}{1} = c^2+1$

49. Mistake: Added the x's instead of multiplying them.

Correct: $\dfrac{3}{x}\div\dfrac{x}{6} = \dfrac{3}{x}\cdot\dfrac{6}{x} = \dfrac{18}{x^2}$

51. $\dfrac{10+10}{\dfrac{1}{3}+\dfrac{1}{4}} = \dfrac{20}{\dfrac{1\cdot4}{3\cdot4}+\dfrac{1\cdot3}{4\cdot3}} = \dfrac{20}{\dfrac{4}{12}+\dfrac{3}{12}}$

$= \dfrac{20}{\dfrac{7}{12}} = 20\div\dfrac{7}{12} = 20\cdot\dfrac{12}{7} = \dfrac{240}{7} = 34\dfrac{2}{7}$ mph

53. $\dfrac{\dfrac{5n-2}{8}}{\dfrac{n-1}{6}} = \dfrac{5n-2}{8}\div\dfrac{n-1}{6}$

$= \dfrac{5n-2}{\cancel{8}_{4}}\cdot\dfrac{\cancel{6}^{3}}{n-1}$

$= \dfrac{3(5n-2)}{4(n-1)}$

55. a) $\dfrac{1}{\dfrac{1}{R_1}+\dfrac{1}{R_2}} = \dfrac{R_1R_2\cdot 1}{\cancel{R_1}R_2\cdot\dfrac{1}{\cancel{R_1}}+R_1\cancel{R_2}\cdot\dfrac{1}{\cancel{R_2}}}$

$= \dfrac{R_1R_2}{R_2+R_1}$

b) $\dfrac{R_1R_2}{R_2+R_1} = \dfrac{100\cdot10}{100+10} = \dfrac{1000}{110} \approx 9.1\,\Omega$

c) $\dfrac{16\cdot16}{16+16} = \dfrac{256}{32} = 8\,\Omega$

Review Exercises

1. $-6x = 42$

$x = -7$

2. $9t-8 = 19$

$9t = 27$

$t = 3$

3. $\dfrac{x}{4} = \dfrac{5}{8}$

$8\cdot x = 4\cdot5$

$8x = 20$

$x = \dfrac{20}{8} = 2.5$

4. $\dfrac{3}{4}y-\dfrac{1}{3} = \dfrac{1}{6}y+2$

$12\cdot\left(\dfrac{3}{4}y-\dfrac{1}{3}\right) = 12\cdot\left(\dfrac{1}{6}y+2\right)$

$12\cdot\dfrac{3}{4}y-12\cdot\dfrac{1}{3} = 12\cdot\dfrac{1}{6}y+12\cdot2$

$\dfrac{36}{4}y-\dfrac{12}{3} = \dfrac{12}{6}y+24$

$9y-4 = 2y+24$

$7y-4 = 24$

$7y = 28$

$y = 4$

5. You must change the minutes to hours. Fifteen minutes is the same as 0.25 hours.

$$d = rt$$
$$1.2 = r(0.25)$$
$$1.2 = 0.25r$$
$$4.8 = r$$

The runner's average rate is 4.8 mph.

6. Complete a table.

Categories	Rate	Time	Distance
Eastbound	40	t	$40t$
Westbound	60	t	$60t$

$$40t + 60t = 9$$
$$100t = 9$$
$$t = 0.09$$

In 0.09 hours or 5.4 minutes the cars will be 9 miles apart.

Exercise Set 7.6

1.
$$\frac{3}{x} + \frac{2}{3} = \frac{5}{6}$$
$$\frac{3}{9} + \frac{2}{3} \overset{?}{=} \frac{5}{6}$$
$$\frac{3}{9} + \frac{6}{9} \overset{?}{=} \frac{5}{6}$$
$$\frac{9}{9} \overset{?}{=} \frac{5}{6}$$
$$1 \neq \frac{5}{6}$$

No, $x = 9$ is not a solution.

3.
$$\frac{4}{m} - \frac{2}{5} = \frac{3}{4m}$$
$$\frac{4}{\frac{65}{8}} - \frac{2}{5} \overset{?}{=} \frac{3}{4\left(\frac{65}{8}\right)}$$
$$\frac{32}{65} - \frac{2}{5} \overset{?}{=} \frac{6}{65}$$
$$\frac{32}{65} - \frac{26}{65} \overset{?}{=} \frac{6}{65}$$
$$\frac{6}{65} = \frac{6}{65}$$

Yes, $m = \frac{65}{8}$ is a solution.

5.
$$\frac{1}{t+2} + \frac{1}{4} = \frac{t}{16}$$
$$\frac{1}{-4+2} + \frac{1}{4} \overset{?}{=} \frac{-4}{16}$$
$$-\frac{1}{2} + \frac{1}{4} \overset{?}{=} -\frac{1}{4}$$
$$-\frac{2}{4} + \frac{1}{4} \overset{?}{=} -\frac{1}{4}$$
$$-\frac{1}{4} = -\frac{1}{4}$$

Yes, $t = -4$ is a solution.

7. The LCD is 6.
$$\frac{2n}{3} + \frac{3n}{2} = \frac{13}{3}$$
$$6 \cdot \frac{2n}{3} + 6 \cdot \frac{3n}{2} = 6 \cdot \frac{13}{3}$$
$$4n + 9n = 26$$
$$13n = 26$$
$$n = 2$$

Check:
$$\frac{2n}{3} + \frac{3n}{2} = \frac{13}{3}$$
$$\frac{2(2)}{3} + \frac{3(2)}{2} \overset{?}{=} \frac{13}{3}$$
$$\frac{4}{3} + \frac{6}{2} \overset{?}{=} \frac{13}{3}$$
$$\frac{8}{6} + \frac{18}{6} \overset{?}{=} \frac{13}{3}$$
$$\frac{26}{6} \overset{?}{=} \frac{13}{3}$$
$$\frac{13}{3} = \frac{13}{3}$$

9. The LCD is 5.
$$3y - \frac{4y}{5} = 22$$
$$5 \cdot 3y - 5 \cdot \frac{4y}{5} = 5 \cdot 22$$
$$15y - 4y = 110$$
$$11y = 110$$
$$y = 10$$

Check:

$$3y - \frac{4y}{5} = 22$$

$$3(10) - \frac{4(10)}{5} \overset{?}{=} 22$$

$$30 - \frac{40}{5} \overset{?}{=} 22$$

$$30 - 8 \overset{?}{=} 22$$

$$22 = 22$$

11. The LCD is 18.

$$\frac{r-7}{2} - 1 = \frac{r+9}{9} + \frac{1}{3}$$

$$18 \cdot \frac{r-7}{2} - 18 \cdot 1 = 18 \cdot \frac{r+9}{9} + 18 \cdot \frac{1}{3}$$

$$9(r-7) - 18 = 2(r+9) + 6$$

$$9r - 63 - 18 = 2r + 18 + 6$$

$$9r - 81 = 2r + 24$$

$$7r = 105$$

$$r = 15$$

Check:

$$\frac{r-7}{2} - 1 = \frac{r+9}{9} + \frac{1}{3}$$

$$\frac{15-7}{2} - 1 \overset{?}{=} \frac{15+9}{9} + \frac{1}{3}$$

$$\frac{8}{2} - 1 \overset{?}{=} \frac{24}{9} + \frac{1}{3}$$

$$4 - 1 \overset{?}{=} \frac{24}{9} + \frac{3}{9}$$

$$3 \overset{?}{=} \frac{27}{9}$$

$$3 = 3$$

13. The LCD is $6x$.

$$\frac{4}{x} + \frac{3}{2x} = \frac{11}{6}$$

$$6x \cdot \frac{4}{x} + 6x \cdot \frac{3}{2x} = 6x \cdot \frac{11}{6}$$

$$6 \cdot 4 + 3 \cdot 3 = x \cdot 11$$

$$24 + 9 = 11x$$

$$33 = 11x$$

$$3 = x$$

Check:

$$\frac{4}{x} + \frac{3}{2x} = \frac{11}{6}$$

$$\frac{4}{3} + \frac{3}{2(3)} \overset{?}{=} \frac{11}{6}$$

$$\frac{8}{6} + \frac{3}{6} \overset{?}{=} \frac{11}{6}$$

$$\frac{11}{6} = \frac{11}{6}$$

15. The LCD is $10t$.

$$\frac{(4-6t)}{2t} + \frac{3}{5} = -\frac{2}{5t}$$

$$10t \cdot \frac{(4-6t)}{2t} + 10t \cdot \frac{3}{5} = 10t \cdot -\frac{2}{5t}$$

$$5 \cdot (4-6t) + 2t \cdot 3 = 2 \cdot -2$$

$$20 - 30t + 6t = -4$$

$$20 - 24t = -4$$

$$-24t = -24$$

$$t = 1$$

Check:

$$\frac{(4-6t)}{2t} + \frac{3}{5} = -\frac{2}{5t}$$

$$\frac{(4-6(1))}{2(1)} + \frac{3}{5} \overset{?}{=} -\frac{2}{5(1)}$$

$$\frac{-2}{2} + \frac{3}{5} \overset{?}{=} -\frac{2}{5}$$

$$-1 + \frac{3}{5} \overset{?}{=} -\frac{2}{5}$$

$$-\frac{2}{5} = -\frac{2}{5}$$

17.

$$\frac{5}{20} = \frac{x-7}{x+2}$$

$$5(x+2) = 20(x-7)$$

$$5x + 10 = 20x - 140$$

$$150 = 15x$$

$$10 = x$$

Check:

$$\frac{5}{20} = \frac{x-7}{x+2}$$

$$\frac{5}{20} \overset{?}{=} \frac{10-7}{10+2}$$

$$\frac{5}{20} \overset{?}{=} \frac{3}{12}$$

$$\frac{1}{4} = \frac{1}{4}$$

19. $\dfrac{6}{y-2} = \dfrac{5}{y-3}$

$6(y-3) = 5(y-2)$

$6y - 18 = 5y - 10$

$y = 8$

Check:

$\dfrac{6}{y-2} = \dfrac{5}{y-3}$

$\dfrac{6}{8-2} \overset{?}{=} \dfrac{5}{8-3}$

$\dfrac{6}{6} \overset{?}{=} \dfrac{5}{5}$

$1 = 1$

21. The LCD is $x+2$.

$$\dfrac{x+4}{x+2} - 5 = \dfrac{6}{x+2}$$

$$(x+2) \cdot \dfrac{x+4}{x+2} - (x+2) \cdot 5 = (x+2) \cdot \dfrac{6}{x+2}$$

$$x + 4 - 5x - 10 = 6$$

$$-4x - 6 = 6$$

$$-4x = 12$$

$$x = -3$$

Check:

$\dfrac{x+4}{x+2} - 5 = \dfrac{6}{x+2}$

$\dfrac{-3+4}{-3+2} - 5 \overset{?}{=} \dfrac{6}{-3+2}$

$\dfrac{1}{-1} - 5 \overset{?}{=} \dfrac{6}{-1}$

$-1 - 5 \overset{?}{=} -6$

$-6 = -6$

23. The LCD is $6(2f+5)$.

$$\dfrac{6f-5}{6} = \dfrac{2f-1}{2} - \dfrac{f+2}{2f+5}$$

$$6(2f+5) \cdot \dfrac{6f-5}{6} = 6(2f+5)\left(\dfrac{2f-1}{2} - \dfrac{f+2}{2f+5}\right)$$

$$(2f+5)(6f-5) = 3(2f+5)(2f-1) - 6(f+2)$$

$$12f^2 + 20f - 25 = 12f^2 + 24f - 15 + (-6f - 12)$$

$$20f - 25 = 18f - 27$$

$$2f = -2$$

$$f = -1$$

Check:

$\dfrac{6f-5}{6} = \dfrac{2f-1}{2} - \dfrac{f+2}{2f+5}$

$\dfrac{6(-1)-5}{6} \overset{?}{=} \dfrac{2(-1)-1}{2} - \dfrac{(-1)+2}{2(-1)+5}$

$\dfrac{-6-5}{6} \overset{?}{=} \dfrac{-2-1}{2} - \dfrac{1}{-2+5}$

$\dfrac{-11}{6} \overset{?}{=} \dfrac{-3}{2} - \dfrac{1}{3}$

$\dfrac{-11}{6} \overset{?}{=} \dfrac{-9}{6} - \dfrac{2}{6}$

$\dfrac{-11}{6} = \dfrac{-11}{6}$

25. The LCD is $(x-3)(x+1)$.

$$\dfrac{2}{x^2-2x-3} - \dfrac{3}{x-3} = \dfrac{2}{x+1}$$

$$(x-3)(x+1) \cdot \dfrac{2}{(x-3)(x+1)} - (x-3)(x+1) \cdot \dfrac{3}{x-3}$$

$$= (x-3)(x+1) \cdot \dfrac{2}{x+1}$$

$$2 - 3(x+1) = 2(x-3)$$

$$2 - 3x - 3 = 2x - 6$$

$$-3x - 1 = 2x - 6$$

$$-1 = 5x - 6$$

$$5 = 5x$$

$$1 = x$$

Check:

$\dfrac{2}{x^2-2x-3} - \dfrac{3}{x-3} = \dfrac{2}{x+1}$

$\dfrac{2}{1^2-2(1)-3} - \dfrac{3}{1-3} \overset{?}{=} \dfrac{2}{1+1}$

$\dfrac{2}{1-2-3} - \dfrac{3}{-2} \overset{?}{=} \dfrac{2}{2}$

$\dfrac{2}{-4} - \dfrac{3}{-2} \overset{?}{=} \dfrac{2}{2}$

$-\dfrac{1}{2} + \dfrac{3}{2} \overset{?}{=} \dfrac{2}{2}$

$\dfrac{2}{2} = \dfrac{2}{2}$

27. The LCD is $(u+4)(u-4)$.

$$\frac{1}{u-4}+\frac{2}{u^2-16}=\frac{3}{u+4}$$

$$(u+4)(u-4)\cdot\frac{1}{u-4}+(u+4)(u-4)\cdot\frac{2}{u^2-16}$$

$$=(u+4)(u-4)\cdot\frac{3}{u+4}$$

$$u+4+2=3(u-4)$$

$$u+6=3u-12$$

$$18=2u$$

$$9=u$$

Check:

$$\frac{1}{u-4}+\frac{2}{u^2-16}=\frac{3}{u+4}$$

$$\frac{1}{9-4}+\frac{2}{9^2-16}\overset{?}{=}\frac{3}{9+4}$$

$$\frac{1}{5}+\frac{2}{65}\overset{?}{=}\frac{3}{13}$$

$$\frac{13}{65}+\frac{2}{65}\overset{?}{=}\frac{3}{13}$$

$$\frac{15}{65}\overset{?}{=}\frac{3}{13}$$

$$\frac{3}{13}=\frac{3}{13}$$

29.
$$\frac{x}{2-3x}=\frac{1}{3x+2}$$

$$x(3x+2)=(2-3x)1$$

$$3x^2+2x=2-3x$$

$$3x^2+5x-2=0$$

$$(3x-1)(x+2)=0$$

$$3x-1=0 \quad\text{or}\quad x+2=0$$

$$x=\frac{1}{3} \qquad\qquad x=-2$$

Check:

$$x=\frac{1}{3}:\qquad \frac{x}{2-3x}=\frac{1}{3x+2}$$

$$\frac{\frac{1}{3}}{2-3\left(\frac{1}{3}\right)}\overset{?}{=}\frac{1}{3\left(\frac{1}{3}\right)+2}$$

$$\frac{\frac{1}{3}}{2-1}\overset{?}{=}\frac{1}{1+2}$$

$$\frac{1}{3}=\frac{1}{3}$$

$$x=-2:\qquad \frac{x}{2-3x}=\frac{1}{3x+2}$$

$$\frac{-2}{2-3(-2)}\overset{?}{=}\frac{1}{3(-2)+2}$$

$$\frac{-2}{2+6}\overset{?}{=}\frac{1}{-6+2}$$

$$-\frac{1}{4}=-\frac{1}{4}$$

31. The LCD is $2x(x+8)$.

$$\frac{5}{2x}+\frac{15}{2x+16}=\frac{6}{x+8}$$

$$2x(x+8)\cdot\left(\frac{5}{2x}+\frac{15}{2(x+8)}\right)=2x(x+8)\cdot\frac{6}{x+8}$$

$$5(x+8)+15x=2x\cdot6$$

$$5x+40+15x=12x$$

$$20x+40=12x$$

$$40=-8x$$

$$-5=x$$

Check:

$$\frac{5}{2x}+\frac{15}{2x+16}=\frac{6}{x+8}$$

$$\frac{5}{2(-5)}+\frac{15}{2(-5)+16}\overset{?}{=}\frac{6}{-5+8}$$

$$\frac{5}{-10}+\frac{15}{-10+16}\overset{?}{=}\frac{6}{3}$$

$$-\frac{1}{2}+\frac{15}{6}\overset{?}{=}2$$

$$-\frac{1}{2}+\frac{5}{2}\overset{?}{=}2$$

$$\frac{4}{2}\overset{?}{=}2$$

$$2=2$$

33.
$$\frac{2}{x^2-1}=\frac{-3}{7x+7}$$

$$2(7x+7)=-3(x^2-1)$$

$$14x+14=-3x^2+3$$

$$3x^2+14x+11=0$$

$$(3x+11)(x+1)=0$$

$$3x+11=0 \quad \text{or} \quad x+1=0$$

$$3x=-11 \qquad\qquad x=-1$$

$$x=-\frac{11}{3}$$

Check:

$$\frac{2}{x^2-1}=\frac{-3}{7x+7}$$

$$\frac{2}{\left(-\frac{11}{3}\right)^2-1} \overset{?}{=} \frac{-3}{7\left(-\frac{11}{3}\right)+7}$$

$$\frac{2}{\left(\frac{121}{9}\right)-1} \overset{?}{=} \frac{-3}{\left(-\frac{77}{3}\right)+7}$$

$$\frac{2}{\frac{112}{9}} \overset{?}{=} \frac{-3}{-\frac{56}{3}}$$

$$\frac{9}{56}=\frac{9}{56}$$

In the original equation, note that both rational expressions are undefined when $x=-1$; so -1 is an extraneous solution. Consequently, $-\frac{11}{3}$ is the only solution.

35. The LCD is $2x(x+5)$.

$$\frac{3}{x+5}-\frac{x-1}{2x}=\frac{-3}{2x^2+10x}$$

$$2x(x+5)\left(\frac{3}{x+5}-\frac{x-1}{2x}\right)=2x(x+5)\left(\frac{-3}{2x(x+5)}\right)$$

$$2x\cdot 3-(x+5)(x-1)=-3$$

$$6x-x^2-4x+5=-3$$

$$0=x^2-2x-8$$

$$0=(x-4)(x+2)$$

$$x-4=0 \quad \text{or} \quad x+2=0$$

$$x=4 \qquad\qquad x=-2$$

Check:

$$x=4: \quad \frac{3}{x+5}-\frac{x-1}{2x}=\frac{-3}{2x^2+10x}$$

$$\frac{3}{4+5}-\frac{4-1}{2(4)} \overset{?}{=} \frac{-3}{2(4)^2+10(4)}$$

$$\frac{3}{9}-\frac{3}{8} \overset{?}{=} \frac{-3}{32+40}$$

$$\frac{24}{72}-\frac{27}{72} \overset{?}{=} -\frac{3}{72}$$

$$-\frac{3}{72}=-\frac{3}{72}$$

$$x=-2:$$

$$\frac{3}{x+5}-\frac{x-1}{2x}=\frac{-3}{2x^2+10x}$$

$$\frac{3}{-2+5}-\frac{-2-1}{2(-2)} \overset{?}{=} \frac{-3}{2(-2)^2+10(-2)}$$

$$\frac{3}{3}-\frac{-3}{-4} \overset{?}{=} \frac{-3}{8-20}$$

$$\frac{3}{3}-\frac{-3}{-4} \overset{?}{=} \frac{3}{12}$$

$$\frac{12}{12}-\frac{9}{12} \overset{?}{=} \frac{3}{12}$$

$$\frac{3}{12}=\frac{3}{12}$$

37.
$$\frac{3m+1}{m^2-9}-\frac{m+3}{m-3}=\frac{1-5m}{m+3}$$

$$(m+3)(m-3)\cdot\frac{3m+1}{m^2-9}-(m+3)(m-3)\cdot\frac{m+3}{m-3}$$

$$=(m+3)(m-3)\cdot\frac{1-5m}{m+3}$$

$$3m+1-(m+3)(m+3)=(m-3)(1-5m)$$

$$3m+1-(m^2+6m+9)=m-5m^2-3+15m$$

$$3m+1-m^2-6m-9=-5m^2+16m-3$$

$$4m^2-19m-5=0$$

$$(4m+1)(m-5)=0$$

$$4m+1=0 \quad \text{or} \quad m-5=0$$

$$m=-\frac{1}{4} \qquad\qquad m=5$$

Check:

$m = -\dfrac{1}{4}:$

$$\dfrac{3m+1}{m^2-9} - \dfrac{m+3}{m-3} = \dfrac{1-5m}{m+3}$$

$$\dfrac{3\left(-\dfrac{1}{4}\right)+1}{\left(-\dfrac{1}{4}\right)^2 - 9} - \dfrac{-\dfrac{1}{4}+3}{-\dfrac{1}{4}-3} \overset{?}{=} \dfrac{1-5\left(-\dfrac{1}{4}\right)}{-\dfrac{1}{4}+3}$$

$$\dfrac{\dfrac{1}{4}}{-\dfrac{143}{16}} - \dfrac{\dfrac{11}{4}}{-\dfrac{13}{4}} \overset{?}{=} \dfrac{\dfrac{9}{4}}{\dfrac{11}{4}}$$

$$\dfrac{-4}{143} + \dfrac{11}{13} \overset{?}{=} \dfrac{9}{11}$$

$$\dfrac{9}{11} = \dfrac{9}{11}$$

$m = 5:$

$$\dfrac{3m+1}{m^2-9} - \dfrac{m+3}{m-3} = \dfrac{1-5m}{m+3}$$

$$\dfrac{3(5)+1}{(5)^2-9} - \dfrac{5+3}{5-3} \overset{?}{=} \dfrac{1-5(5)}{5+3}$$

$$\dfrac{16}{16} - \dfrac{8}{2} \overset{?}{=} \dfrac{-24}{8}$$

$$1 - 4 \overset{?}{=} -3$$

$$-3 = -3$$

39. The LCD is $3(x-1)(x+3)$.

$$\dfrac{x+3}{x-1} + \dfrac{2}{x+3} = \dfrac{5}{3}$$

$$3(x-1)(x+3)\left(\dfrac{x+3}{x-1} + \dfrac{2}{x+3}\right) = 3(x-1)(x+3)\dfrac{5}{3}$$

$$3(x+3)(x+3) + 3(x-1)\cdot 2 = (x-1)(x+3)\cdot 5$$

$$3(x^2+6x+9) + 6x - 6 = 5(x^2+2x-3)$$

$$3x^2 + 18x + 27 + 6x - 6 = 5x^2 + 10x - 15$$

$$3x^2 + 24x + 21 = 5x^2 + 10x - 15$$

$$0 = 2x^2 - 14x - 36$$

$$0 = x^2 - 7x - 18$$

$$0 = (x-9)(x+2)$$

$$x - 9 = 0 \quad \text{or} \quad x + 2 = 0$$
$$x = 9 \qquad\qquad x = -2$$

Check:

$x = 9:$ $\qquad\qquad\qquad$ $x = -2:$

$$\dfrac{x+3}{x-1} + \dfrac{2}{x+3} = \dfrac{5}{3}$$

$$\dfrac{9+3}{9-1} + \dfrac{2}{9+3} \overset{?}{=} \dfrac{5}{3} \qquad \dfrac{x+3}{x-1} + \dfrac{2}{x+3} = \dfrac{5}{3}$$

$$\dfrac{12}{8} + \dfrac{2}{12} \overset{?}{=} \dfrac{5}{3} \qquad \dfrac{-2+3}{-2-1} + \dfrac{2}{-2+3} \overset{?}{=} \dfrac{5}{3}$$

$$\dfrac{3}{2} + \dfrac{1}{6} \overset{?}{=} \dfrac{5}{3} \qquad\qquad \dfrac{1}{-3} + \dfrac{2}{1} \overset{?}{=} \dfrac{5}{3}$$

$$\dfrac{10}{6} \overset{?}{=} \dfrac{5}{3} \qquad\qquad\qquad \dfrac{5}{3} = \dfrac{5}{3}$$

$$\dfrac{5}{3} = \dfrac{5}{3}$$

41. The LCD is $(x-2)(x+2)$.

$$1 - \dfrac{3}{x-2} = -\dfrac{12}{x^2-4}$$

$$(x-2)(x+2)\left(1 - \dfrac{3}{x-2}\right) = (x-2)(x+2)\left(-\dfrac{12}{x^2-4}\right)$$

$$(x^2-4) - 3(x+2) = -12$$

$$x^2 - 4 - 3x - 6 = -12$$

$$x^2 - 3x + 2 = 0$$

$$(x-1)(x-2) = 0$$

$$x - 1 = 0 \quad \text{or} \quad x - 2 = 0$$
$$x = 1 \qquad\qquad x = 2$$

Check:

$x = 1:$ $\quad 1 - \dfrac{3}{x-2} = -\dfrac{12}{x^2-4}$

$$1 - \dfrac{3}{1-2} \overset{?}{=} -\dfrac{12}{1^2-4}$$

$$1 - \dfrac{3}{-1} \overset{?}{=} -\dfrac{12}{-3}$$

$$1 + 3 \overset{?}{=} 4$$

$$4 = 4$$

In the original equation, note that both rational expressions are undefined when $x = 2$; so 2 is an extraneous solution. Consequently, 1 is the only solution.

43. The LCD is $(r-2)(r-3)$.

$$\frac{2r}{r-2} - \frac{4r}{r-3} = -\frac{7r}{r^2 - 5r + 6}$$

$$(r-2)(r-3)\left(\frac{2r}{r-2} - \frac{4r}{r-3}\right)$$
$$= (r-2)(r-3)\left(-\frac{7r}{(r-2)(r-3)}\right)$$

$$2r(r-3) - 4r(r-2) = -7r$$

$$2r^2 - 6r - 4r^2 + 8r = -7r$$

$$-2r^2 + 9r = 0$$

$$r(-2r+9) = 0$$

$$r = 0 \quad \text{or} \quad -2r + 9 = 0$$
$$-2r = -9$$
$$r = \frac{9}{2}$$

Check:

$r = 0$:

$$\frac{2r}{r-2} - \frac{4r}{r-3} = -\frac{7r}{r^2 - 5r + 6}$$

$$\frac{2(0)}{(0)-2} - \frac{4(0)}{(0)-3} \overset{?}{=} -\frac{7(0)}{(0)^2 - 5(0) + 6}$$

$$\frac{0}{-2} - \frac{0}{-3} \overset{?}{=} -\frac{0}{6}$$

$$0 = 0$$

$r = \frac{9}{2}$:

$$\frac{2r}{r-2} - \frac{4r}{r-3} = -\frac{7r}{r^2 - 5r + 6}$$

$$\frac{2\left(\frac{9}{2}\right)}{\left(\frac{9}{2}\right)-2} - \frac{4\left(\frac{9}{2}\right)}{\left(\frac{9}{2}\right)-3} \overset{?}{=} -\frac{7\left(\frac{9}{2}\right)}{\left(\frac{9}{2}\right)^2 - 5\left(\frac{9}{2}\right) + 6}$$

$$\frac{9}{2.5} - \frac{18}{1.5} \overset{?}{=} -\frac{31.5}{3.75}$$

$$-8.4 = -8.4$$

45.

$$\frac{4u^2 + 3u + 4}{u^2 + u - 2} - \frac{3u}{u+2} = \frac{-2u-1}{u-1}$$

$$\frac{4u^2 + 3u + 4}{(u+2)(u-1)} - \frac{3u}{u+2} = \frac{-2u-1}{u-1}$$

$$(u+2)(u-1)\left(\frac{4u^2 + 3u + 4}{(u+2)(u-1)} - \frac{3u}{u+2}\right)$$
$$= (u+2)(u-1)\left(\frac{-2u-1}{u-1}\right)$$

$$4u^2 + 3u + 4 - 3u(u-1) = (-2u-1)(u+2)$$

$$4u^2 + 3u + 4 - 3u^2 + 3u = -2u^2 - 4u - u - 2$$

$$u^2 + 6u + 4 = -2u^2 - 5u - 2$$

$$3u^2 + 11u + 6 = 0$$

$$(3u+2)(u+3) = 0$$

$$3u + 2 = 0 \quad \text{or} \quad u + 3 = 0$$
$$3u = -2 \qquad\qquad u = -3$$
$$u = -\frac{2}{3}$$

Check:

$u = -\frac{2}{3}$:

$$\frac{4u^2 + 3u + 4}{u^2 + u - 2} - \frac{3u}{u+2} = \frac{-2u-1}{u-1}$$

$$\frac{4\left(-\frac{2}{3}\right)^2 + 3\left(-\frac{2}{3}\right) + 4}{\left(-\frac{2}{3}\right)^2 + \left(-\frac{2}{3}\right) - 2} - \frac{3\left(-\frac{2}{3}\right)}{\left(-\frac{2}{3}\right)+2} \overset{?}{=} \frac{-2\left(-\frac{2}{3}\right)-1}{\left(-\frac{2}{3}\right)-1}$$

$$\frac{\frac{34}{9}}{-\frac{20}{9}} - \frac{-2}{\frac{4}{3}} \overset{?}{=} \frac{\frac{1}{3}}{-\frac{5}{3}}$$

$$-\frac{1}{5} = -\frac{1}{5}$$

$u = -3$:

$$\frac{4u^2 + 3u + 4}{u^2 + u - 2} - \frac{3u}{u+2} = \frac{-2u-1}{u-1}$$

$$\frac{4(-3)^2 + 3(-3) + 4}{(-3)^2 + (-3) - 2} - \frac{3(-3)}{(-3)+2} \overset{?}{=} \frac{-2(-3)-1}{(-3)-1}$$

$$\frac{31}{4} - \frac{-9}{-1} \overset{?}{=} \frac{5}{-4}$$

$$-\frac{5}{4} = -\frac{5}{4}$$

47. The LCD is $(k+7)(k+3)$.

$$\frac{2k}{k+7}-1=\frac{1}{k^2+10k+21}+\frac{k}{k+3}$$

$$\frac{2k}{k+7}-1=\frac{1}{(k+7)(k+3)}+\frac{k}{k+3}$$

$$(k+7)(k+3)\cdot\frac{2k}{k+7}-(k+7)(k+3)\cdot1$$

$$=(k+7)(k+3)\cdot\frac{1}{(k+7)(k+3)}$$

$$+(k+7)(k+3)\cdot\frac{k}{k+3}$$

$$2k(k+3)-(k+7)(k+3)=1+k(k+7)$$

$$2k^2+6k-\left(k^2+10k+21\right)=1+k^2+7k$$

$$2k^2+6k-k^2-10k-21=k^2+7k+1$$

$$-22=11k$$

$$-2=k$$

Check:

$$\frac{2k}{k+7}-1=\frac{1}{k^2+10k+21}+\frac{k}{k+3}$$

$$\frac{2(-2)}{-2+7}-1\stackrel{?}{=}\frac{1}{(-2)^2+10(-2)+21}+\frac{-2}{-2+3}$$

$$\frac{-4}{5}-1\stackrel{?}{=}\frac{1}{4-20+21}+\frac{-2}{1}$$

$$\frac{-9}{5}\stackrel{?}{=}\frac{1}{5}+\frac{-2}{1}$$

$$-\frac{9}{5}=-\frac{9}{5}$$

49.

$$\frac{6}{t^2+t-12}=\frac{4}{t^2-t-6}+\frac{1}{t^2+6t+8}$$

$$\frac{6}{(t+4)(t-3)}=\frac{4}{(t-3)(t+2)}+\frac{1}{(t+4)(t+2)}$$

The LCD is $(t+4)(t-3)(t+2)$.

$$(t+4)(t-3)(t+2)\cdot\left(\frac{6}{(t+4)(t-3)}\right)$$

$$=(t+4)(t-3)(t+2)\cdot$$

$$\left(\frac{4}{(t-3)(t+2)}+\frac{1}{(t+4)(t+2)}\right)$$

$$6(t+2)=4(t+4)+1(t-3)$$

$$6t+12=4t+16+t-3$$

$$6t+12=5t+13$$

$$t=1$$

Check:

$$\frac{6}{t^2+t-12}=\frac{4}{t^2-t-6}+\frac{1}{t^2+6t+8}$$

$$\frac{6}{1^2+1-12}\stackrel{?}{=}\frac{4}{1^2-1-6}+\frac{1}{1^2+6(1)+8}$$

$$\frac{6}{-10}\stackrel{?}{=}\frac{4}{-6}+\frac{1}{15}$$

$$-\frac{6}{10}\stackrel{?}{=}-\frac{20}{30}+\frac{2}{30}$$

$$-\frac{6}{10}\stackrel{?}{=}-\frac{18}{30}$$

$$-\frac{6}{10}=-\frac{6}{10}$$

51.

$$\frac{x+1}{x^2-2x-15}-\frac{6}{x^2-3x-10}=\frac{2}{x^2+5x+6}$$

$$\frac{x+1}{(x+3)(x-5)}-\frac{6}{(x-5)(x+2)}=\frac{2}{(x+3)(x+2)}$$

The LCD is $(x-5)(x+3)(x+2)$.

$$(x-5)(x+3)(x+2)\left(\frac{x+1}{(x+3)(x-5)}-\frac{6}{(x-5)(x+2)}\right)$$

$$=(x-5)(x+3)(x+2)\cdot\frac{2}{(x+3)(x+2)}$$

$$(x+2)(x+1)-6(x+3)=2(x-5)$$

$$x^2+3x+2-6x-18=2x-10$$

$$x^2-3x-16=2x-10$$

$$x^2-5x-6=0$$

$$(x-6)(x+1)=0$$

$$x-6=0\quad\text{or}\quad x+1=0$$

$$x=6\qquad\qquad x=-1$$

Check:

$x=6$:

$$\frac{x+1}{x^2-2x-15}-\frac{6}{x^2-3x-10}=\frac{2}{x^2+5x+6}$$

$$\frac{6+1}{6^2-2(6)-15}-\frac{6}{6^2-3(6)-10}\stackrel{?}{=}\frac{2}{6^2+5(6)+6}$$

$$\frac{7}{36-12-15}-\frac{6}{36-18-10}\stackrel{?}{=}\frac{2}{36+30+6}$$

$$\frac{7}{9}-\frac{6}{8}\stackrel{?}{=}\frac{2}{72}$$

$$\frac{56}{72}-\frac{54}{72}\stackrel{?}{=}\frac{2}{72}$$

$$\frac{2}{72}=\frac{2}{72}$$

$x = -1$:

$$\frac{x+1}{x^2 - 2x - 15} - \frac{6}{x^2 - 3x - 10} = \frac{2}{x^2 + 5x + 6}$$

$$\frac{-1+1}{(-1)^2 - 2(-1) - 15} - \frac{6}{(-1)^2 - 3(-1) - 10} \overset{?}{=} \frac{2}{(-1)^2 + 5(-1) + 6}$$

$$\frac{0}{1+2-15} - \frac{6}{1+3-10} \overset{?}{=} \frac{2}{1-5+6}$$

$$\frac{0}{-12} - \frac{6}{-6} \overset{?}{=} \frac{2}{2}$$

$$1 = 1$$

53.

$$\frac{1}{d^2 - 5d + 6} + \frac{d-2}{d^2 - d - 6} - \frac{d}{d^2 - 4} = 0$$

$$\frac{1}{(d-2)(d-3)} + \frac{d-2}{(d-3)(d+2)} - \frac{d}{(d+2)(d-2)} = 0$$

The LCD is $(d-2)(d-3)(d+2)$.

$$(d-2)(d-3)(d+2) \cdot \frac{1}{(d-2)(d-3)}$$

$$+ (d-2)(d-3)(d+2) \cdot \frac{d-2}{(d-3)(d+2)}$$

$$- (d-2)(d-3)(d+2) \cdot \frac{d}{(d+2)(d-2)} = 0$$

$$(d+2) + (d-2)(d-2) - d(d-3) = 0$$

$$d + 2 + (d^2 - 4d + 4) - d^2 + 3d = 0$$

$$d + 2 + d^2 - 4d + 4 - d^2 + 3d = 0$$

$$6 \neq 0$$

This equation has no solution.

55.

$$\frac{1}{a-3} - \frac{6}{a^2 - 1} = \frac{12}{a^3 - 3a^2 - a + 3}$$

$$\frac{1}{a-3} - \frac{6}{(a+1)(a-1)} = \frac{12}{(a-3)(a+1)(a-1)}$$

The LCD is $(a-3)(a+1)(a-1)$.

$$(a-3)(a+1)(a-1) \cdot \frac{1}{a-3}$$

$$- (a-3)(a+1)(a-1) \cdot \frac{6}{(a+1)(a-1)}$$

$$= (a-3)(a+1)(a-1) \cdot \frac{12}{(a-3)(a+1)(a-1)}$$

$$(a+1)(a-1) - 6(a-3) = 12$$

$$a^2 - 1 - 6a + 18 = 12$$

$$a^2 - 6a + 5 = 0$$

$$(a-5)(a-1) = 0$$

$$a - 5 = 0 \quad \text{or} \quad a - 1 = 0$$

$$a = 5 \qquad\qquad a = 1$$

Check:

$a = 5$:

$$\frac{1}{a-3} - \frac{6}{a^2 - 1} = \frac{12}{a^3 - 3a^2 - a + 3}$$

$$\frac{1}{5-3} - \frac{6}{5^2 - 1} \overset{?}{=} \frac{12}{5^3 - 3(5)^2 - 5 + 3}$$

$$\frac{1}{2} - \frac{6}{24} \overset{?}{=} \frac{12}{125 - 75 - 5 + 3}$$

$$\frac{1}{2} - \frac{1}{4} \overset{?}{=} \frac{12}{48}$$

$$\frac{1}{4} = \frac{1}{4}$$

In the original equation, note that two rational expressions are undefined when $x = 1$; so 1 is an extraneous solution. Consequently, 5 is the only solution.

57. Mistake: 3 is extraneous.
Correct: −3 is the only answer.

59. In an expression, the LCD is used to combine the numerator over a single denominator. In an equation, the LCD is used to eliminate the denominators.

61. Let x represent the value of the other resistor.

$$\frac{1}{R} = \frac{1}{R_1} + \frac{1}{R_2}$$

$$\frac{1}{80} = \frac{1}{x} + \frac{1}{100}$$

$$400x \cdot \frac{1}{80} = 400x \cdot \frac{1}{x} + 400x \cdot \frac{1}{100}$$

$$5x = 400 + 4x$$

$$x = 400$$

The value of the other resistor is 400Ω.

63. Let x represent the value of one resistor and let $x + 10$ represent the value of the other resistor.

$$\frac{1}{R} = \frac{1}{R_1} + \frac{1}{R_2}$$

$$\frac{1}{12} = \frac{1}{x} + \frac{1}{x+10}$$

$$12x(x+10) \cdot \frac{1}{12}$$

$$= 12x(x+10) \cdot \frac{1}{x} + 12x(x+10) \cdot \frac{1}{x+10}$$

$$x(x+10) = 12(x+10) + 12x$$

$$x^2 + 10x = 12x + 120 + 12x$$

$$x^2 - 14x - 120 = 0$$

$$(x-20)(x+6) = 0$$

$$x - 20 = 0 \quad \text{or} \quad x + 6 = 0$$

$$x = 20 \qquad\qquad x = -6$$

Disregard the negative solution because it doesn't make sense in the context of the problem. The value of one resistor is 20Ω and the value of the other resistor is $20 + 10 = 30\Omega$.

65.

$$\frac{1}{o} + \frac{1}{i} = \frac{1}{f}$$

$$\frac{1}{40} + \frac{1}{i} = \frac{1}{2}$$

$$40i \cdot \frac{1}{40} + 40i \cdot \frac{1}{i} = 40i \cdot \frac{1}{2}$$

$$i + 40 = 20i$$

$$40 = 19i$$

$$2.1 \approx i$$

The object is approximately 2.1 feet from the lens.

67.

$$\frac{1}{o} + \frac{1}{i} = \frac{1}{f}$$

$$\frac{1}{60} + \frac{1}{i} = \frac{1}{i-5}$$

$$\frac{60i(i-5)}{60} + \frac{60i(i-5)}{i} = \frac{60i(i-5)}{i-5}$$

$$i(i-5) + 60(i-5) = 60i$$

$$i^2 - 5i + 60i - 300 = 60i$$

$$i^2 - 5i - 300 = 0$$

$$(i-20)(i+15) = 0$$

$$i - 20 = 0 \quad \text{or} \quad i + 15 = 0$$

$$i = 20 \qquad\qquad i = -15$$

Disregard the negative solution because it doesn't make sense in the context of the problem. The object is $i = 20$ mm from the lens, and the focal length is $f = 20 - 5 = 15$ mm.

Review Exercises

1.
$$rt = d$$

$$r\left(\frac{7}{4}\right) = 120$$

$$\frac{4}{7} \cdot r\left(\frac{7}{4}\right) = \frac{4}{7} \cdot 120$$

$$r = \frac{480}{7} \approx 68.6 \text{ mph}$$

2. $\dfrac{0.88 \div 10.5}{10.5 \div 10.5} \approx \dfrac{\$0.084}{1 \text{ oz.}}$

Each ounce costs 8.4 cents.

3. $\dfrac{800}{4} = \dfrac{1200}{x}$

$$800x = 4800$$

$$x = 6 \text{ hr.}$$

4. Let one number be x and the other number is $(x + 2)$. Translate to the equation:

$$x + (x+2) = 146$$

$$2x = 144$$

$$x = 72$$

The numbers are 72 and 72 + 2 = 74.

5. The width of the eraser is x and the length is $(2x - 3)$. Translate to the equation:

$$2(2x-3) + 2x = 36$$

$$4x - 6 + 2x = 36$$

$$6x - 6 = 36$$

$$6x = 42$$

$$x = 7$$

The width of the eraser is 7 mm, and the length is 2(7) – 3 = 11 mm.

6. Complete a table.

Categories	Rate (mph)	Time (hr.)	Distance
1st jogger	6	x	$6x$
2nd jogger	4	x	$4x$

$$6x + 4x = 1.5$$

$$10x = 1.5$$

$$x = 0.15 \text{ hr. or 9 min.}$$

Exercise Set 7.7

1. Complete a table.

Categories	Rate of Work	Time at Work	Amt of Task Completed
Jason	$\dfrac{1}{4}$	t	$\dfrac{t}{4}$
sister	$\dfrac{1}{6}$	t	$\dfrac{t}{6}$

$$\frac{t}{4}+\frac{t}{6}=1$$
$$24\left(\frac{t}{4}+\frac{t}{6}\right)=24\cdot 1$$
$$6t+4t=24$$
$$10t=24$$
$$t=2\frac{2}{5}$$

Working together, it takes Jason and his sister $2\frac{2}{5}$ hours to wash and wax the car.

3. Complete a table.

Categories	Rate of Work	Time at Work	Amt of Task Completed
Alicia	$\dfrac{1}{25}$	t	$\dfrac{t}{25}$
Geraldine	$\dfrac{1}{35}$	t	$\dfrac{t}{35}$

$$\frac{t}{25}+\frac{t}{35}=1$$
$$175\left(\frac{t}{25}+\frac{t}{35}\right)=175\cdot 1$$
$$7t+5t=175$$
$$12t=175$$
$$t=14\frac{7}{12}$$

Working together it will take Alicia and Geraldine $14\frac{7}{12}$ days to make the quilt.

5. Complete a table.

Categories	Rate of Work	Time at Work	Amt of Task Completed
1st roofer	$\dfrac{1}{6}$	4	$\dfrac{2}{3}$
2nd roofer	$\dfrac{1}{t}$	4	$\dfrac{4}{t}$

$$\frac{2}{3}+\frac{4}{t}=1$$
$$3t\left(\frac{2}{3}+\frac{4}{t}\right)=3t\cdot 1$$
$$2t+12=3t$$
$$12=t$$

It would take the second roofer 12 hr. to do the job alone.

7. Complete a table.

Categories	Rate of Work	Time at Work	Amt of Task Completed
cold	$\dfrac{1}{15}$	9	$\dfrac{3}{5}$
hot	$\dfrac{1}{t}$	9	$\dfrac{9}{t}$

$$\frac{3}{5}+\frac{9}{t}=1$$
$$5t\left(\frac{3}{5}+\frac{9}{t}\right)=5t\cdot 1$$
$$3t+45=5t$$
$$45=2t$$
$$22.5=t$$

It would take the hot water faucet 22.5 minutes to fill the tub alone.

9. Complete a table.

Categories	Rate	Time	Distance
northbound bus	63	$\dfrac{x}{63}$	x
southbound car	72	$\dfrac{675-x}{72}$	$675-x$

$$\frac{x}{63}=\frac{675-x}{72}\qquad \text{time}=\frac{x}{63}=\frac{315}{63}=5\text{ hrs.}$$
$$72x=42{,}525-63x$$
$$135x=42{,}525$$
$$x=315$$

They will be 675 miles apart after 5 hours, at 3 P.M.

11. Complete a table.

Categories	Rate	Time	Distance
freight train	$\dfrac{3}{5}r$	3	$\dfrac{9}{5}r$
passenger train	r	3	$3r$

$$\frac{9}{5}r + 3r = 360$$

$$\frac{24}{5}r = 360$$

$$r = 75$$

The passenger train travels at 75 mph.

13. Complete a table.

Categories	Rate	Time	Distance
Jack	45	$\dfrac{x}{45}$	x
Frances	60	$\dfrac{x}{60}$	x

$$\frac{x}{45} - \frac{x}{60} = 2$$

$$180\left(\frac{x}{45} - \frac{x}{60}\right) = 180 \cdot 2$$

$$4x - 3x = 360$$

$$x = 360$$

Frances' time is $\dfrac{x}{60} = \dfrac{360}{60} = 6$.

It takes Frances 6 hours to catch Jack.

15. Complete a table.

Categories	Rate	Time	Distance
against	$r-21$	5.5	$5.5(r-21)$
with	$r+21$	5	$5(r+21)$

$$5.5(r-21) = 5(r+21)$$

$$5.5r - 115.5 = 5r + 105$$

$$0.5r = 220.5$$

$$r = 441$$

The speed of the airliner in still air is 441 mph.

17. Translate "a varies directly as b."

$a = k \cdot b$ Now, $a = \dfrac{4}{9}b$

$4 = k \cdot 9$

$\dfrac{4}{9} = k$ $a = \dfrac{4}{9}(27)$

$a = 12$

19. Translate "y varies directly as the square of x."

$y = k \cdot x^2$ Now, $y = 4x^2$

$100 = k \cdot 5^2$ $y = 4(3)^2$

$100 = 25k$ $y = 36$

$4 = k$

21. Translate "m varies directly as n."

$m = k \cdot n$ Now, $m = \dfrac{3}{4}n$

$6 = k \cdot 8$

$\dfrac{3}{4} = k$ $9 = \dfrac{3}{4}n$

$12 = n$

23. Translating "The cost increases with the quantity purchased," we write $c = kq$, where c represents cost and q represents the quantity.

$c = kq$ Now, $c = 6.5q$

$16.25 = k \cdot 2.5$ $c = 6.5(6)$

$6.5 = k$ $c = 39$

The salmon will cost $39 for 6 pounds.

25. Translating "the volume of a gas varies directly with temperature," we write $v = kt$ where v represents the volume and t represents the temperature.

$v = kt$ Now, $v = 3.6t$

$288 = k \cdot 80$ $v = 3.6(50)$

$3.6 = k$ $v = 180$

The volume is 180 cubic cm.

27. $d = k \cdot t^2$ Now, $d = 16t^2$

$144 = k \cdot 3^2$ $400 = 16t^2$

$\dfrac{144}{9} = k$ $25 = t^2$

$16 = k$ $5 = t$

It will take 5 seconds for it to fall 400 feet.

29. $d = k \cdot t^2$ Now, $d = 1.62t^2$

$6.48 = k \cdot 2^2$ $d = 1.62(5)^2$

$\dfrac{6.48}{4} = k$ $d = 40.5$

$1.62 = k$

In 5 seconds, it will fall 40.5 meters.

31. Translating "The circumference varies directly with its diameter," we write $c = kd$, where c represents circumference and d represents the diameter.

$c = k \cdot d$	Now, $c = 3.14d$
$12.56 = k \cdot 4$	$21.98 = 3.14d$
$3.14 = k$	$7 = d$

The diameter will be 7 ft. when the circumference is 21.98 feet.

33. Translate "a varies inversely as b."

$a = \dfrac{k}{b}$	Now, $a = \dfrac{16}{b}$
$3.2 = \dfrac{k}{5}$	$8 = \dfrac{16}{b}$
$16 = k$	$8b = 16$
	$b = 2$

35. Translate "m varies inversely as n."

$m = \dfrac{k}{n}$	Now, $m = \dfrac{66}{n}$
$11 = \dfrac{k}{6}$	$8 = \dfrac{66}{n}$
$66 = k$	$8n = 66$
	$n = 8.25$

37. Translating "The pressure that a gas exerts against the walls of a container is inversely proportional to the volume of the container," we write $p = \dfrac{k}{v}$ where p represents the pressure the gas exerts and v is the volume of the container.

$p = \dfrac{k}{v}$	Now, $p = \dfrac{800}{v}$
$40 = \dfrac{k}{20}$	$p = \dfrac{800}{16}$
$800 = k$	$p = 50$

The pressure is 50 psi.

39. Translating "the current I varies inversely as the resistance R," we write $I = \dfrac{k}{R}$ where I represents the current in amperes and R represents the resistance in ohms.

$I = \dfrac{k}{R}$	Now, $I = \dfrac{150}{R}$
$10 = \dfrac{k}{15}$	$25 = \dfrac{150}{R}$
$150 = k$	$25R = 150$
	$R = 6$

The resistance is 6 ohms.

41.

$l = \dfrac{k}{f}$	Now, $l = \dfrac{360,000}{f}$
$400 = \dfrac{k}{900}$	$l = \dfrac{360,000}{600}$
$360,000 = k$	$l = 600$

The wavelength is 600 m.

43.

$f = \dfrac{k}{a}$	Now, $f = \dfrac{280}{a}$
$5.6 = \dfrac{k}{50}$	$2.8 = \dfrac{280}{a}$
$280 = k$	$2.8a = 280$
	$a = 100$

The aperture is 100 mm.

45. Translate "a varies jointly with b and c."

$a = kbc$	Now, $a = 4bc$
$96 = k \cdot 6 \cdot 4$	$a = 4(2)(8)$
$96 = 24k$	$a = 64$
$4 = k$	

47. Translate "a varies jointly as the square of b and c."

$a = k \cdot b^2 \cdot c$	Now, $a = 4b^2c$
$96 = k \cdot 2^2 \cdot 6$	$a = 4(3)^2(2)$
$96 = 24k$	$a = 72$
$4 = k$	

49. Translating "the volume of a rectangular solid varies jointly with the length and the height," we write $v = klh$, where v represents volume, l represents length, and h represents height.

$v = k \cdot l \cdot h$	Now, $v = 4lh$
$192 = k \cdot 8 \cdot 6$	$v = 4(7)(12)$
$192 = 48k$	$v = 336$
$4 = k$	

The volume is 336 cubic inches.

51. Translating "the volume of a right circular cylinder varies jointly as the square of the radius and the height," we write $v = kr^2h$, where v represents volume, r represents radius, and h represents height.

$v = k \cdot r^2 \cdot h$	Now, $v = 3.141\overline{6} \cdot r^2 \cdot h$
$301.6 = k \cdot 4^2 \cdot 6$	$v = 3.141\overline{6} \cdot 3^2 \cdot 6$
$301.6 = 96k$	$v = 169.65$
$3.141\overline{6} = k$	

The volume is 169.65 cubic cm.

53. Translate "y varies directly with x and inversely as z."

$$y = \frac{kx}{z} \qquad \text{Now,} \quad y = \frac{12x}{z}$$

$$8 = \frac{k \cdot 4}{6} \qquad\qquad y = \frac{12 \cdot 5}{10}$$

$$48 = 4k \qquad\qquad\qquad y = 6$$

$$12 = k$$

55. Translate "y varies jointly with x and z and inversely with n."

$$y = \frac{kxz}{n} \qquad \text{Now,} \quad y = \frac{18xz}{n}$$

$$81 = \frac{k(4)(9)}{8} \qquad\quad 8 = \frac{18(6)(12)}{n}$$

$$648 = 36k \qquad\qquad 8n = 1296$$

$$18 = k \qquad\qquad\quad n = 162$$

57. Translating "the resistance of a wire varies directly with the length and inversely with the square of the diameter," we write $R = \dfrac{kl}{d^2}$, where R represents resistance, l represents length, and d represents diameter.

$$R = \frac{kl}{d^2} \qquad \text{Now,} \qquad R = \frac{0.0005l}{d^2}$$

$$7.5 = \frac{k \cdot 6}{0.02^2} \qquad\qquad R = \frac{0.0005 \cdot 10}{0.04^2}$$

$$0.003 = 6k \qquad\qquad 0.0016R = 0.005$$

$$0.0005 = k \qquad\qquad\quad R = 3.125$$

The resistance is 3.125 ohms.

59. Translating "the force of attraction between two bodies is jointly proportional to their masses and inversely proportional to the square of the distance between them," we write $f = \dfrac{k \cdot m_1 \cdot m_2}{d^2}$, where f is the force of attraction, m_1 is the mass of one body, m_2 is the mass of the other body, and d is the distance between them.

$$f = \frac{k \cdot m_1 \cdot m_2}{d^2} \quad \text{Now,} \quad f = \frac{18m_1m_2}{d^2}$$

$$48 = \frac{k \cdot 4 \cdot 6}{3^2} \qquad\qquad f = \frac{18(2)(12)}{6^2}$$

$$432k = 24k \qquad\qquad\quad f = 12$$

$$18 = k$$

The force of attraction is 12 dynes.

61. a)

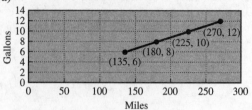

b) They are directly proportional. As the number of miles increases, so does the number of gallons.

c) $\quad d = k \cdot g$

$\quad 135 = k \cdot 6$

$\quad 22.5 = k$

This represents the miles per gallon the car gets.

d) Yes, the data represents a function because for any given number of miles, there will be one quantity of gasoline (assuming that the miles per gallon stays constant).

Review Exercises

1. $|-4.3| = 4.3$

2. $\left|\dfrac{5}{8}\right| = \dfrac{5}{8}$

3. $x \geq 2$

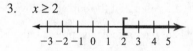

4. $x < -3$

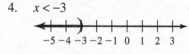

5. $2x + 11 > 3$

$$2x > -8$$

$$\frac{2x}{2} > \frac{-8}{2}$$

$$x > -4$$

6. $-4x - 3 \leq -11$

$$-4x \leq -8$$

$$\frac{-4x}{-4} \geq \frac{-8}{-4}$$

$$x \geq 2$$

Chapter 7 Review Exercises

1. a) $x = 3$:

$$\frac{4x+5}{2x} = \frac{4(3)+5}{2(3)} = \frac{12+5}{6} = \frac{17}{6}$$

b) $x = -1.3$:

$$\frac{4x+5}{2x} = \frac{4(-1.3)+5}{2(-1.3)} = \frac{-5.2+5}{-2.6} \approx 0.077$$

c) $x = -2$:

$$\frac{4x+5}{2x} = \frac{4(-2)+5}{2(-2)} = \frac{-8+5}{-4} = \frac{3}{4}$$

2. a) $x = 3$:

$$\frac{5x}{x+2} = \frac{5(3)}{3+2} = \frac{15}{5} = 3$$

b) $x = -3$:

$$\frac{5x}{x+2} = \frac{5(-3)}{-3+2} = \frac{-15}{-1} = 15$$

c) $x = 1.2$:

$$\frac{5x}{x+2} = \frac{5(1.2)}{1.2+2} = \frac{6}{3.2} = 1.875$$

3. $5 - y = 0$

$-y = -5$

$y = 5$

The expression $\dfrac{9y}{5-y}$ is undefined if y is replaced by 5.

4. $x^2 - 4 = 0$

$(x+2)(x-2) = 0$

$x+2 = 0$ or $x-2 = 0$

$x = -2$ or $\quad x = 2$

The expression $\dfrac{2x}{x^2-4}$ is undefined if x is replaced by -2 or 2.

5. $x^2 + 5x - 6 = 0$

$(x-1)(x+6) = 0$

$x-1 = 0$ or $x+6 = 0$

$x = 1 \qquad x = -6$

The expression $\dfrac{6}{x^2+5x-6}$ is undefined if x is replaced by -6 or 1.

6. $x^2 + 6x = 0$

$x(x+6) = 0$

$x+6 = 0$ or $x = 0$

$x = -6$

The expression $\dfrac{2x+3}{x^2+6x}$ is undefined if x is replaced by -6 or 0.

7. $\dfrac{3(x+5)}{9} = \dfrac{\cancel{3}(x+5)}{\cancel{3}\cdot 3} = \dfrac{x+5}{3}$

8. $\dfrac{2y+12}{3y+18} = \dfrac{2\cancel{(y+6)}}{3\cancel{(y+6)}} = \dfrac{2}{3}$

9. $\dfrac{x^2+6x}{4x+24} = \dfrac{x\cancel{(x+6)}}{4\cancel{(x+6)}} = \dfrac{x}{4}$

10. $\dfrac{xy-y^2}{3x-3y} = \dfrac{y\cancel{(x-y)}}{3\cancel{(x-y)}} = \dfrac{y}{3}$

11. $\dfrac{a^2-b^2}{a^2-2ab+b^2} = \dfrac{(a+b)\cancel{(a-b)}}{(a-b)\cancel{(a-b)}} = \dfrac{a+b}{a-b}$

12. $\dfrac{3x+9}{x^4-81} = \dfrac{3(x+3)}{(x^2+9)(x^2-9)}$

$= \dfrac{3\cancel{(x+3)}}{(x^2+9)\cancel{(x+3)}(x-3)}$

$= \dfrac{3}{(x^2+9)(x-3)}$

13. $\dfrac{x^2-y^2}{y-x} = \dfrac{(x+y)\cancel{(x-y)}}{-1\cancel{(x-y)}} = -(x+y)$

14. $\dfrac{1-w}{w^2+2w-3} = \dfrac{-1\cancel{(w-1)}}{(w+3)\cancel{(w-1)}} = -\dfrac{1}{w+3}$

15. $-\dfrac{2n^2}{8m^2n} \cdot \dfrac{24mp}{10n^2p^5}$

$= -\dfrac{\cancel{2} \cdot \cancel{n} \cdot n}{\cancel{2} \cdot \cancel{2} \cdot 2 \cdot m \cdot m \cdot \cancel{n}}$

$\qquad \cdot \dfrac{\cancel{2} \cdot \cancel{2} \cdot \cancel{2} \cdot 3 \cdot \cancel{m} \cdot \cancel{p}}{\cancel{2} \cdot 5 \cdot \cancel{m} \cdot n \cdot \cancel{p} \cdot p \cdot p \cdot p}$

$= -\dfrac{3}{5mnp^4}$

16. $\dfrac{6}{2m+4} \cdot \dfrac{3m+6}{15} = \dfrac{\cancel{2} \cdot 3}{\cancel{2}\cancel{(m+2)}} \cdot \dfrac{\cancel{3}\cancel{(m+2)}}{\cancel{3} \cdot 5} = \dfrac{3}{5}$

17. $\dfrac{n^2 - 6n + 9}{n^2 - 9} \cdot \dfrac{n^2 + 4n + 3}{n^2 - 3n}$

$= \dfrac{\cancel{(n-3)}\cancel{(n-3)}}{\cancel{(n+3)}\cancel{(n-3)}} \cdot \dfrac{\cancel{(n+3)}(n+1)}{n\cancel{(n-3)}}$

$= \dfrac{n+1}{n}$

18. $\dfrac{4r^2 + 4r + 1}{r + 2r^2} \cdot \dfrac{2r}{2r^2 - r - 1}$

$= \dfrac{\cancel{(2r+1)}\cancel{(2r+1)}}{\cancel{r}\cancel{(1+2r)}} \cdot \dfrac{2\cancel{r}}{\cancel{(2r+1)}(r-1)}$

$= \dfrac{2}{r-1}$

19. $\dfrac{4m}{m^2 - 2m + 1} \cdot \dfrac{m^2 - 1}{16m^2}$

$= \dfrac{\cancel{4}\cancel{m}}{(m-1)\cancel{(m-1)}} \cdot \dfrac{(m+1)\cancel{(m-1)}}{\cancel{4} \cdot 4 \cdot m \cdot \cancel{m}}$

$= \dfrac{m+1}{4m(m-1)}$

20. $\dfrac{w^2 - 2w - 24}{w^2 - 16} \cdot \dfrac{(w+5)(w+4)}{w^2 - w - 30}$

$= \dfrac{\cancel{(w-6)}\cancel{(w+4)}}{\cancel{(w+4)}(w-4)} \cdot \dfrac{(w+5)\cancel{(w+4)}}{\cancel{(w-6)}\cancel{(w+5)}}$

$= \dfrac{w+4}{w-4}$

21. $-\dfrac{7y^2}{10b^2} \div -\dfrac{21y^2}{25b^2} = \dfrac{7y^2}{10b^2} \cdot \dfrac{25b^2}{21y^2}$

$= \dfrac{\cancel{7} \cdot \cancel{y} \cdot \cancel{y}}{2 \cdot \cancel{5} \cdot \cancel{b} \cdot \cancel{b}} \cdot \dfrac{\cancel{5} \cdot 5 \cdot \cancel{b} \cdot \cancel{b}}{3 \cdot \cancel{7} \cdot \cancel{y} \cdot \cancel{y}}$

$= \dfrac{5}{6}$

22. $\dfrac{2a+4}{5} \div \dfrac{4a+8}{25a} = \dfrac{\cancel{2}\cancel{(a+2)}}{\cancel{5}} \cdot \dfrac{\cancel{5} \cdot 5 \cdot a}{\cancel{2} \cdot 2 \cdot \cancel{(a+2)}} = \dfrac{5a}{2}$

23. $\dfrac{x^2 - y^2}{c^2 - d^2} \div \dfrac{x-y}{c+d} = \dfrac{(x+y)\cancel{(x-y)}}{\cancel{(c+d)}(c-d)} \cdot \dfrac{\cancel{c+d}}{\cancel{x-y}}$

$\qquad = \dfrac{x+y}{c-d}$

24. $\dfrac{b^3 - 6b^2 + 8b}{6b} \div \dfrac{2b-4}{10b+40}$

$= \dfrac{\cancel{b}\cancel{(b-2)}(b-4)}{\cancel{2} \cdot 3 \cdot \cancel{b}} \cdot \dfrac{\cancel{2} \cdot 5 \cdot (b+4)}{2\cancel{(b-2)}}$

$= \dfrac{5(b-4)(b+4)}{6}$

25. $\dfrac{3j+2}{5j^2 - j} \div \dfrac{6j^2 + j - 2}{10j^2 + 3j - 1}$

$= \dfrac{\cancel{3j+2}}{j\cancel{(5j-1)}} \cdot \dfrac{\cancel{(5j-1)}(2j+1)}{\cancel{(3j+2)}(2j-1)}$

$= \dfrac{2j+1}{j(2j-1)}$

26. $\dfrac{u^2 - 2u - 8}{u^2 + 3u + 2} \div \left(u^2 - 3u - 4\right)$

$= \dfrac{\cancel{(u-4)}\cancel{(u+2)}}{\cancel{(u+2)}(u+1)} \cdot \dfrac{1}{\cancel{(u-4)}(u+1)}$

$= \dfrac{1}{(u+1)^2}$

27. $42 \text{ in.} = \dfrac{42 \cancel{\text{ in.}}}{1} \cdot \dfrac{1 \text{ ft.}}{12 \cancel{\text{ in.}}} = 3.5 \text{ ft.}$

28. $2.5 \text{ mi.} = \dfrac{2.5 \cancel{\text{ mi.}}}{1} \cdot \dfrac{5280 \text{ ft.}}{1 \cancel{\text{ mi.}}} = 13,200 \text{ ft.}$

29. $68 \text{ oz.} = \dfrac{68 \cancel{\text{ oz.}}}{1} \cdot \dfrac{1 \text{ lb.}}{16 \cancel{\text{ oz.}}} = 4.25 \text{ lb.}$

30. $4 \text{ T} = \dfrac{4 \cancel{T}}{1} \cdot \dfrac{2000 \cancel{lb.}}{1 \cancel{T}} \cdot \dfrac{16 \text{ oz.}}{1 \cancel{lb.}} = 128{,}000 \text{ oz.}$

31. $\dfrac{5x}{18} + \dfrac{x}{18} = \dfrac{5x + x}{18} = \dfrac{6x}{18} = \dfrac{x}{3}$

32. $\dfrac{8m+5}{6m^2} - \dfrac{10m+5}{6m^2} = \dfrac{8m+5+(-10m-5)}{6m^2}$

$= \dfrac{-2m}{6m^2}$

$= -\dfrac{1}{3m}$

33. $\dfrac{2x}{x+y} - \dfrac{x-y}{x+y} = \dfrac{2x+(-x+y)}{x+y} = \dfrac{\cancel{x+y}}{\cancel{x+y}} = 1$

34. $\dfrac{2r-5s}{r^2-s^2} - \dfrac{2r-6s}{r^2-s^2} = \dfrac{2r-5s+(-2r+6s)}{r^2-s^2}$

$= \dfrac{s}{r^2-s^2}$

35. $\dfrac{9x^2-24x+16}{2x+1} + \dfrac{2x^2+3x-9}{2x+1}$

$= \dfrac{9x^2-24x+16+2x^2+3x-9}{2x+1}$

$= \dfrac{11x^2-21x+7}{2x+1}$

36. $\dfrac{q^2+2q}{q-7} - \dfrac{12q-21}{q-7} = \dfrac{q^2+2q-12q+21}{q-7}$

$= \dfrac{q^2-10q+21}{q-7}$

$= \dfrac{\cancel{(q-7)}(q-3)}{\cancel{q-7}}$

$= q-3$

37. $4c = 2 \cdot 2 \cdot c$

$3c^3 = 3 \cdot c \cdot c \cdot c$

The LCD is $12c^3$.

$\dfrac{2}{4c} = \dfrac{2 \cdot 3c^2}{4c \cdot 3c^2} = \dfrac{6c^2}{12c^3}$

$\dfrac{3}{3c^3} = \dfrac{3 \cdot 4}{3c^3 \cdot 4} = \dfrac{12}{12c^3}$

38. $xy^3 = x \cdot y \cdot y \cdot y$

$x^2y^2 = x \cdot x \cdot y \cdot y$

The LCD is x^2y^3.

$\dfrac{5}{xy^3} = \dfrac{5 \cdot x}{xy^3 \cdot x} = \dfrac{5x}{x^2y^3}$

$\dfrac{3}{x^2y^2} = \dfrac{3 \cdot y}{x^2y^2 \cdot y} = \dfrac{3y}{x^2y^3}$

39. The LCD is $(m+1)(m-1)$.

$\dfrac{4}{m-1} = \dfrac{4(m+1)}{(m-1)(m+1)} = \dfrac{4m+4}{(m-1)(m+1)}$

$\dfrac{4y}{m+1} = \dfrac{4y(m-1)}{(m+1)(m-1)} = \dfrac{4my-4y}{(m-1)(m+1)}$

40. $4h-8 = 4(h-2)$

$h^2-2h = h(h-2)$

The LCD is $4h(h-2)$.

$\dfrac{3}{4h-8} = \dfrac{3}{4(h-2)} = \dfrac{3 \cdot h}{4(h-2) \cdot h} = \dfrac{3h}{4h(h-2)}$

$\dfrac{5}{h^2-2h} = \dfrac{5}{h(h-2)} = \dfrac{5 \cdot 4}{h(h-2) \cdot 4} = \dfrac{20}{4h(h-2)}$

41. The LCD is $(x+1)(x-1)$.

$\dfrac{x}{x^2-1} = \dfrac{x}{(x+1)(x-1)}$

$\dfrac{2+x}{x+1} = \dfrac{(2+x)(x-1)}{(x+1)(x-1)} = \dfrac{x^2+x-2}{(x+1)(x-1)}$

42. $p^2-4 = (p+2)(p-2)$

$p^2+4p+4 = (p+2)(p+2)$

The LCD is $(p-2)(p+2)^2$.

$\dfrac{2}{p^2-4} = \dfrac{2(p+2)}{(p-2)(p+2)(p+2)}$

$= \dfrac{2p+4}{(p-2)(p+2)^2}$

$\dfrac{p}{p^2+4p+4} = \dfrac{p(p-2)}{(p+2)^2(p-2)}$

$= \dfrac{p^2-2p}{(p-2)(p+2)^2}$

43. The LCD is 36.

$$\frac{3a-4}{18}-\frac{2a+5}{12}=\frac{(3a-4)\cdot 2}{18\cdot 2}-\frac{(2a+5)\cdot 3}{12\cdot 3}$$

$$=\frac{6a-8+(-6a-15)}{36}$$

$$=-\frac{23}{36}$$

44. The LCD is $(y-2)(y-4)$.

$$\frac{3}{y-2}+\frac{5}{y-4}$$

$$=\frac{3(y-4)}{(y-2)(y-4)}+\frac{5(y-2)}{(y-4)(y-2)}$$

$$=\frac{3y-12+5y-10}{(y-2)(y-4)}$$

$$=\frac{8y-22}{(y-2)(y-4)}$$

45. The LCD is $2(t-3)^2$.

$$\frac{6t}{(t-3)^2}-\frac{3t}{2t-6}$$

$$=\frac{6t}{(t-3)^2}-\frac{3t}{2(t-3)}$$

$$=\frac{6t\cdot 2}{(t-3)^2\cdot 2}-\frac{3t(t-3)}{2(t-3)(t-3)}$$

$$=\frac{12t-3t^2+9t}{2(t-3)^2}$$

$$=\frac{-3t^2+21t}{2(t-3)^2}$$

46. The LCD is $(a+3)(a+4)$.

$$\frac{a+6}{a^2+7a+12}-\frac{a-3}{a+3}$$

$$=\frac{a+6}{(a+3)(a+4)}-\frac{(a-3)(a+4)}{(a+3)(a+4)}$$

$$=\frac{a+6-(a^2+a-12)}{(a+3)(a+4)}$$

$$=\frac{18-a^2}{(a+3)(a+4)}$$

47. The LCD is $2y-1$.

$$\frac{10}{2y-1}-\frac{5}{1-2y}=\frac{10}{2y-1}-\frac{5(-1)}{(1-2y)(-1)}$$

$$=\frac{10}{2y-1}-\frac{-5}{2y-1}$$

$$=\frac{10+5}{2y-1}$$

$$=\frac{15}{2y-1}$$

48. The LCD is $(x+4)(x-4)$.

$$\frac{3}{3x-12}+\frac{15}{x^2-16}$$

$$=\frac{\cancel{3}}{\cancel{3}(x-4)}+\frac{15}{(x+4)(x-4)}$$

$$=\frac{1\cdot(x+4)}{(x-4)\cdot(x+4)}+\frac{15}{(x+4)(x-4)}$$

$$=\frac{x+4+15}{(x+4)(x-4)}$$

$$=\frac{x+19}{(x+4)(x-4)}$$

49. $\dfrac{\dfrac{a}{b}}{\dfrac{x}{y}}=\dfrac{a}{b}\div\dfrac{x}{y}=\dfrac{a}{b}\cdot\dfrac{y}{x}=\dfrac{ay}{bx}$

50. $\dfrac{\dfrac{1}{3}-\dfrac{1}{2}}{\dfrac{1}{2}-\dfrac{1}{3}}=\dfrac{\dfrac{2}{6}-\dfrac{3}{6}}{\dfrac{3}{6}-\dfrac{2}{6}}=\dfrac{\dfrac{-1}{6}}{\dfrac{1}{6}}=\dfrac{-1}{6}\div\dfrac{1}{6}=-\dfrac{1}{6}\cdot\dfrac{6}{1}=-1$

51. $\dfrac{\dfrac{5}{3x+5}}{\dfrac{x}{x-2}}=\dfrac{5}{3x+5}\div\dfrac{x}{x-2}$

$$=\frac{5}{3x+5}\cdot\frac{x-2}{x}$$

$$=\frac{5(x-2)}{x(3x+5)}$$

52. $\dfrac{\dfrac{x+3}{x^2-9}}{x} = (x+3) \div \dfrac{x^2-9}{x}$

$= \cancel{(x+3)} \cdot \dfrac{x}{\cancel{(x+3)}(x-3)}$

$= \dfrac{x}{x-3}$

53. $\dfrac{\dfrac{x}{4x^2-1}}{\dfrac{5}{2x+1}} = \dfrac{x}{(2x+1)(2x-1)} \div \dfrac{5}{(2x+1)}$

$= \dfrac{x}{(2x+1)(2x-1)} \cdot \dfrac{\cancel{2x+1}}{5}$

$= \dfrac{x}{5(2x-1)}$

54. $\dfrac{r - \dfrac{r^2-1}{r}}{1 - \dfrac{r-1}{r}} = \dfrac{r \cdot r - r \cdot \dfrac{r^2-1}{r}}{r \cdot 1 - r \cdot \dfrac{r-1}{r}} = \dfrac{r^2 - r^2 + 1}{r - r + 1} = \dfrac{1}{1} = 1$

55. $\dfrac{\dfrac{1}{x^2} + \dfrac{1}{y^2}}{\dfrac{7}{xy}} = \dfrac{x^2 y^2 \cdot \dfrac{1}{x^2} + x^2 y^2 \cdot \dfrac{1}{y^2}}{x^2 y^2 \cdot \dfrac{7}{xy}} = \dfrac{y^2 + x^2}{7xy}$

56. $\dfrac{\dfrac{1}{h+1} - 1}{\dfrac{1}{h+1}} = \dfrac{\cancel{(h+1)} \cdot \dfrac{1}{\cancel{h+1}} - (h+1) \cdot 1}{\cancel{(h+1)} \cdot \dfrac{1}{\cancel{h+1}}}$

$= \dfrac{1 - h - 1}{1}$

$= -h$

57. $\dfrac{3x}{2} - \dfrac{x}{5} = \dfrac{19}{5}$

$10\left(\dfrac{3x}{2} - \dfrac{x}{5}\right) = 10 \cdot \dfrac{19}{5}$

$15x - 2x = 38$

$13x = 38$

$x = \dfrac{38}{13}$

Check:

$\dfrac{3x}{2} - \dfrac{x}{5} = \dfrac{19}{5}$

$\dfrac{3(38/13)}{2} - \dfrac{(38/13)}{5} \overset{?}{=} \dfrac{19}{5}$

$\dfrac{57}{13} - \dfrac{38}{65} \overset{?}{=} \dfrac{19}{5}$

$\dfrac{285}{65} - \dfrac{38}{65} \overset{?}{=} \dfrac{19}{5}$

$\dfrac{247}{65} \overset{?}{=} \dfrac{19}{5}$

$\dfrac{19}{5} = \dfrac{19}{5}$

58. $\dfrac{y+5}{y-3} - 5 = \dfrac{4}{y-3}$

$(y-3)\left(\dfrac{y+5}{y-3} - 5\right) = (y-3) \cdot \dfrac{4}{y-3}$

$y + 5 - 5(y-3) = 4$

$y + 5 - 5y + 15 = 4$

$-4y + 20 = 4$

$-4y = -16$

$y = 4$

Check:

$\dfrac{y+5}{y-3} - 5 = \dfrac{4}{y-3}$

$\dfrac{4+5}{4-3} - 5 \overset{?}{=} \dfrac{4}{4-3}$

$\dfrac{9}{1} - 5 \overset{?}{=} \dfrac{4}{1}$

$9 - 5 \overset{?}{=} 4$

$4 = 4$

59.
$$\frac{x}{x-2} = \frac{6}{x-1}$$
$$6(x-2) = x(x-1)$$
$$6x-12 = x^2 - x$$
$$0 = x^2 - 7x + 12$$
$$0 = (x-3)(x-4)$$
$$x-3 = 0 \quad \text{or} \quad x-4 = 0$$
$$x = 3 \qquad\qquad x = 4$$

Check:

$x = 3:$

$$\frac{x}{x-2} = \frac{6}{x-1}$$
$$\frac{3}{3-2} \overset{?}{=} \frac{6}{3-1}$$
$$\frac{3}{1} \overset{?}{=} \frac{6}{2}$$
$$3 = 3$$

$x = 4:$

$$\frac{x}{x-2} = \frac{6}{x-1}$$
$$\frac{4}{4-2} \overset{?}{=} \frac{6}{4-1}$$
$$\frac{4}{2} \overset{?}{=} \frac{6}{3}$$
$$2 = 2$$

60.
$$\frac{6g-5}{6} = \frac{2g-1}{2} - \frac{g+2}{2g+5}$$
$$6(2g+5)\left(\frac{6g-5}{6}\right) = 6(2g+5)\left(\frac{2g-1}{2} - \frac{g+2}{2g+5}\right)$$
$$(2g+5)(6g-5) = 3(2g+5)(2g-1) - 6(g+2)$$
$$12g^2 + 20g - 25 = 3\left(4g^2 + 8g - 5\right) - 6(g+2)$$
$$12g^2 + 20g - 25 = 12g^2 + 24g - 15 - 6g - 12$$
$$2g = -2$$
$$g = -1$$

Check:

$$\frac{6g-5}{6} = \frac{2g-1}{2} - \frac{g+2}{2g+5}$$
$$\frac{6(-1)-5}{6} \overset{?}{=} \frac{2(-1)-1}{2} - \frac{(-1)+2}{2(-1)+5}$$
$$\frac{-6-5}{6} \overset{?}{=} \frac{-2-1}{2} - \frac{1}{-2+5}$$
$$\frac{-11}{6} \overset{?}{=} \frac{-3}{2} - \frac{1}{3}$$
$$\frac{-11}{6} \overset{?}{=} \frac{-9}{6} - \frac{2}{6}$$
$$-\frac{11}{6} = -\frac{11}{6}$$

61.
$$\frac{1}{p-4} + \frac{1}{4} = \frac{8}{p^2 - 16}$$
$$4(p+4)(p-4)\left(\frac{1}{p-4} + \frac{1}{4}\right)$$
$$= 4(p+4)(p-4)\left(\frac{8}{p^2 - 16}\right)$$
$$4(p+4) + (p+4)(p-4) = 32$$
$$4p + 16 + p^2 - 16 = 32$$
$$p^2 + 4p - 32 = 0$$
$$(p+8)(p-4) = 0$$
$$p+8 = 0 \quad \text{or} \quad p-4 = 0$$
$$p = -8 \qquad\qquad p = 4$$

Check:

$$\frac{1}{p-4} + \frac{1}{4} = \frac{8}{p^2 - 16}$$
$$\frac{1}{-8-4} + \frac{1}{4} \overset{?}{=} \frac{8}{(-8)^2 - 16}$$
$$\frac{1}{-12} + \frac{1}{4} \overset{?}{=} \frac{8}{64 - 16}$$
$$\frac{1}{-12} + \frac{1}{4} \overset{?}{=} \frac{8}{48}$$
$$-\frac{1}{12} + \frac{3}{12} \overset{?}{=} \frac{1}{6}$$
$$\frac{2}{12} \overset{?}{=} \frac{1}{6}$$
$$\frac{1}{6} = \frac{1}{6}$$

In the original equation, note that two rational expressions are undefined when $x = 4$; so 4 is an extraneous solution. Consequently, -8 is the only solution.

62. $\dfrac{3}{r-2}-\dfrac{4}{r+3}=\dfrac{-6}{r^2+r-6}$

$(r-2)(r+3)\left(\dfrac{3}{r-2}-\dfrac{4}{r+3}\right)$

$\qquad=(r-2)(r+3)\left(\dfrac{-6}{(r-2)(r+3)}\right)$

$3(r+3)-4(r-2)=-6$

$3r+9-4r+8=-6$

$-r+17=-6$

$-r=-23$

$r=23$

Check:

$\dfrac{3}{r-2}-\dfrac{4}{r+3}=\dfrac{-6}{r^2+r-6}$

$\dfrac{3}{23-2}-\dfrac{4}{23+3}\overset{?}{=}\dfrac{-6}{23^2+23-6}$

$\dfrac{3}{21}-\dfrac{4}{26}\overset{?}{=}\dfrac{-6}{529+23-6}$

$\dfrac{78}{546}-\dfrac{84}{546}\overset{?}{=}-\dfrac{6}{546}$

$-\dfrac{6}{546}=-\dfrac{6}{546}$

63. Translate "p varies directly as q."

$p=kq$ $\qquad$ Now, $p=6q$

$24=k\cdot4$ $\qquad$ $p=6(7)$

$6=k$ $\qquad$ $p=42$

64. Translate "s varies directly as the square of t."

$s=k\cdot t^2$ $\qquad$ Now, $s=8t^2$

$72=k\cdot9$ $\qquad$ $s=8(5)^2$

$8=k$ $\qquad$ $s=200$

65. Translate "y varies inversely as x."

$y=\dfrac{k}{x}$ $\qquad$ Now, $y=\dfrac{24}{x}$

$8=\dfrac{k}{3}$ $\qquad$ $4=\dfrac{24}{x}$

$24=k$ $\qquad$ $4x=24$

$\qquad\qquad\qquad$ $x=6$

66. Translate "m varies inversely as the square of n."

$m=\dfrac{k}{n^2}$ $\qquad$ Now, $m=\dfrac{144}{n^2}$

$16=\dfrac{k}{3^2}$ $\qquad$ $m=\dfrac{144}{4^2}$

$144=k$ $\qquad$ $m=9$

67. Translate "y varies jointly with x and z."

$y=kxz$ $\qquad$ Now, $y=4xz$

$40=k\cdot2\cdot5$ $\qquad$ $y=4\cdot4\cdot2$

$4=k$ $\qquad$ $y=32$

68. Translate "m varies directly with n and inversely with p."

$m=\dfrac{kn}{p}$ $\qquad$ Now, $m=\dfrac{6n}{p}$

$2=\dfrac{k\cdot3}{9}$ $\qquad$ $m=\dfrac{6\cdot6}{4}$

$18=3k$ $\qquad$ $m=9$

$6=k$

69. Translating "the distance a car can travel varies directly with the amount of gas it carries," we write $d=kg$ where d is the distance traveled and g is the amount of gas.

$d=kg$ $\qquad$ Now, $d=26g$

$156=k\cdot6$ $\qquad$ $234=26g$

$26=k$ $\qquad$ $9=g$

9 gallons are required to travel 234 miles.

70. $v=\dfrac{k}{T}$ $\qquad$ Now, $v=\dfrac{28}{T}$

$4=\dfrac{k}{7}$ $\qquad$ $v=\dfrac{28}{12}$

$28=k$ $\qquad$ $v=\dfrac{7}{3}$

The velocity is $\dfrac{7}{3}$ cm/sec.

71. $v=kr^2h$ $\qquad$ Now, $v=3.14r^2h$

$62.8=k\cdot2^2\cdot5$ $\qquad$ $v=3.14(4)^2(2)$

$62.8=20k$ $\qquad$ $v=100.48$

$3.14=k$

The volume is 100.48 in.^{3}

Chapter 7 Practice Test

1. $\dfrac{2x}{x-7}$

$x-7=0$

$x=7$

The expression is undefined if x is replaced with 7.

2. $\dfrac{8-m}{m^2-16}$

$$m^2-16=0$$
$$(m+4)(m-4)=0$$
$$m+4=0 \qquad \text{or} \qquad m-4=0$$
$$m=-4 \qquad\qquad m=4$$

The expression is undefined if m is replaced with -4 or 4.

3. $\dfrac{12-3x}{x^2-8x+16} = \dfrac{-3\,\cancel{(x-4)}}{\cancel{(x-4)}\,(x-4)}$

$$= -\dfrac{3}{x-4}$$

4. $\dfrac{x-y}{x^2-y^2} = \dfrac{\cancel{x-y}}{(x+y)\,\cancel{(x-y)}}$

$$= \dfrac{1}{x+y}$$

5.   The LCD is $f^2 g^3$.

$$\dfrac{6}{fg^3} = \dfrac{6\cdot f}{fg^3 \cdot f} = \dfrac{6f}{f^2 g^3}$$

$$\dfrac{2}{f^2} = \dfrac{2\cdot g^3}{f^2 \cdot g^3} = \dfrac{2g^3}{f^2 g^3}$$

6. $\dfrac{5x}{x^2-9} = \dfrac{5x}{(x+3)(x-3)}$

$$\dfrac{2}{x^2+6x+9} = \dfrac{2}{(x+3)(x+3)}$$

The LCD is $(x+3)^2(x-3)$.

$$\dfrac{5x}{x^2-9} = \dfrac{5x}{(x+3)(x-3)}$$

$$= \dfrac{5x(x+3)}{(x+3)(x-3)(x+3)}$$

$$= \dfrac{5x(x+3)}{(x+3)^2(x-3)}$$

$$\dfrac{2}{x^2+6x+9} = \dfrac{2}{(x+3)(x+3)}$$

$$= \dfrac{2(x-3)}{(x+3)^2(x-3)}$$

7. $\dfrac{4a^2b^3}{15x^3y} \cdot \dfrac{25x^5y}{16ab}$

$$= \dfrac{\cancel{2}\cdot\cancel{2}\cdot\cancel{a}\cdot a\cdot\cancel{b}\cdot b\cdot b}{3\cdot\cancel{5}\cdot\cancel{x}\cdot\cancel{x}\cdot\cancel{x}\cdot\cancel{y}} \cdot \dfrac{\cancel{5}\cdot 5\cdot\cancel{x}\cdot\cancel{x}\cdot\cancel{x}\cdot x\cdot x\cdot\cancel{y}}{\cancel{2}\cdot\cancel{2}\cdot 2\cdot 2\cdot\cancel{a}\cdot\cancel{b}}$$

$$= \dfrac{a\cdot b\cdot b}{3} \cdot \dfrac{5\cdot x\cdot x}{2\cdot 2}$$

$$= \dfrac{5ab^2x^2}{12}$$

8. $-\dfrac{9x^3y^4}{16ab^2} \div \dfrac{45x^5y^2}{14a^7b^9}$

$$= -\dfrac{9x^3y^4}{16ab^2} \cdot \dfrac{14a^7b^9}{45x^5y^2}$$

$$= -\dfrac{\cancel{3}\cdot\cancel{3}\cdot\cancel{x}\cdot\cancel{x}\cdot\cancel{x}\cdot\cancel{y}\cdot\cancel{y}\cdot y\cdot y}{\cancel{2}\cdot 2\cdot 2\cdot 2\cdot\cancel{a}\cdot\cancel{b}\cdot\cancel{b}}$$

$$\cdot \dfrac{\cancel{2}\cdot 7\cdot\cancel{a}\cdot a\cdot a\cdot a\cdot a\cdot a\cdot a\cdot\cancel{b}\cdot\cancel{b}\cdot b\cdot b\cdot b\cdot b\cdot b\cdot b\cdot b}{\cancel{3}\cdot\cancel{3}\cdot 5\cdot\cancel{x}\cdot\cancel{x}\cdot\cancel{x}\cdot x\cdot x\cdot\cancel{y}\cdot\cancel{y}}$$

$$= -\dfrac{y\cdot y}{2\cdot 2\cdot 2} \cdot \dfrac{7\cdot a\cdot a\cdot a\cdot a\cdot a\cdot a\cdot b\cdot b\cdot b\cdot b\cdot b\cdot b\cdot b}{5\cdot x\cdot x}$$

$$= -\dfrac{7a^6b^7y^2}{40x^2}$$

9. $\dfrac{4ab-8b}{x^2} \div \dfrac{2b-ab}{x^3} = \dfrac{4ab-8b}{x^2} \cdot \dfrac{x^3}{2b-ab}$

$$= \dfrac{4\cancel{b}\,\cancel{(a-2)}}{\cancel{x}\cdot\cancel{x}} \cdot \dfrac{\cancel{x}\cdot\cancel{x}\cdot x}{-\cancel{b}\,\cancel{(a-2)}}$$

$$= \dfrac{4}{1} \cdot \dfrac{x}{-1}$$

$$= \dfrac{4x}{-1}$$

$$= -4x$$

10. $\dfrac{12x^2-6x}{x^2+6x+5} \cdot \dfrac{2x^2+10x}{4x^2-1}$

$$= \dfrac{6x\,\cancel{(2x-1)}}{(x+5)(x+1)} \cdot \dfrac{2x\,\cancel{(x+5)}}{(2x+1)\,\cancel{(2x-1)}}$$

$$= \dfrac{6x}{(x+1)} \cdot \dfrac{2x}{(2x+1)}$$

$$= \dfrac{12x^2}{(x+1)(2x+1)}$$

11. $\dfrac{2x}{2x+3}+\dfrac{5x}{2x+3}=\dfrac{2x+5x}{2x+3}$

$\qquad\qquad\qquad = \dfrac{7x}{2x+3}$

12. $\dfrac{x}{x^2-9}-\dfrac{3}{x^2-9}=\dfrac{x-3}{x^2-9}$

$\qquad\qquad\qquad = \dfrac{\cancel{x-3}}{(x+3)\cancel{(x-3)}}$

$\qquad\qquad\qquad = \dfrac{1}{x+3}$

13. $\dfrac{6}{x^2-4}+\dfrac{x-3}{x^2-2x}=\dfrac{6}{(x+2)(x-2)}+\dfrac{x-3}{x(x-2)}$

The LCD is $x(x-2)(x+2)$.

$\dfrac{6}{x^2-4}+\dfrac{x-3}{x^2-2x}$

$=\dfrac{6}{(x+2)(x-2)}+\dfrac{x-3}{x(x-2)}$

$=\dfrac{6(x)}{(x+2)(x-2)(x)}+\dfrac{(x-3)(x+2)}{x(x-2)(x+2)}$

$=\dfrac{6x}{x(x-2)(x+2)}+\dfrac{x^2-x-6}{x(x-2)(x+2)}$

$=\dfrac{6x+x^2-x-6}{x(x+2)(x-2)}$

$=\dfrac{x^2+5x-6}{x(x-2)(x+2)}$

14. The LCD is $(x+2)(x-1)$.

$\dfrac{4}{x-1}+\dfrac{5}{x+2}=\dfrac{4(x+2)}{(x-1)(x+2)}+\dfrac{5(x-1)}{(x+2)(x-1)}$

$\qquad\qquad\qquad = \dfrac{4x+8}{(x+2)(x-1)}+\dfrac{5x-5}{(x+2)(x-1)}$

$\qquad\qquad\qquad = \dfrac{4x+8+5x-5}{(x+2)(x-1)}$

$\qquad\qquad\qquad = \dfrac{9x+3}{(x+2)(x-1)}$

15. The LCD is $(x+1)(x-4)$.

$\dfrac{5}{x-4}-\dfrac{2}{x+1}=\dfrac{5(x+1)}{(x-4)(x+1)}-\dfrac{2(x-4)}{(x+1)(x-4)}$

$\qquad\qquad\qquad = \dfrac{5x+5}{(x+1)(x-4)}-\dfrac{2x-8}{(x+1)(x-4)}$

$\qquad\qquad\qquad = \dfrac{(5x+5)-(2x-8)}{(x+1)(x-4)}$

$\qquad\qquad\qquad = \dfrac{5x+5+(-2x+8)}{(x+1)(x-4)}$

$\qquad\qquad\qquad = \dfrac{3x+13}{(x+1)(x-4)}$

16. $\dfrac{x}{2x+4}-\dfrac{2}{x^2+2x}=\dfrac{x}{2(x+2)}-\dfrac{2}{x(x+2)}$

The LCD is $2x(x+2)$.

$\dfrac{x}{2x+4}-\dfrac{2}{x^2+2x}=\dfrac{x}{2(x+2)}-\dfrac{2}{x(x+2)}$

$\qquad\qquad\qquad = \dfrac{x\cdot x}{2(x+2)\cdot x}-\dfrac{2\cdot 2}{x(x+2)\cdot 2}$

$\qquad\qquad\qquad = \dfrac{x^2}{2x(x+2)}-\dfrac{4}{2x(x+2)}$

$\qquad\qquad\qquad = \dfrac{x^2-4}{2x(x+2)}$

$\qquad\qquad\qquad = \dfrac{\cancel{(x+2)}(x-2)}{2x\cancel{(x+2)}}$

$\qquad\qquad\qquad = \dfrac{x-2}{2x}$

17. Multiply the numerator and the denominator of the rational expression by the LCD of all the rational expressions involved, y.

$\dfrac{2+\dfrac{1}{y}}{3-\dfrac{1}{y}}=\dfrac{\left(2+\dfrac{1}{y}\right)\cdot y}{\left(3-\dfrac{1}{y}\right)\cdot y}=\dfrac{\dfrac{2}{1}\cdot\dfrac{y}{1}+\dfrac{1}{\cancel{y}}\cdot\dfrac{\cancel{y}}{1}}{\dfrac{3}{1}\cdot\dfrac{y}{1}-\dfrac{1}{\cancel{y}}\cdot\dfrac{\cancel{y}}{1}}=\dfrac{2y+1}{3y-1}$

18. Multiply the numerator and the denominator of the rational expression by the LCD of all the rational expressions involved, 4.

$$\frac{\dfrac{x-y}{2}}{\dfrac{x^2-y^2}{4}} = \frac{\left(\dfrac{x-y}{2}\right)\cdot 4}{\left(\dfrac{x^2-y^2}{4}\right)\cdot 4}$$

$$= \frac{2(x-y)}{x^2-y^2}$$

$$= \frac{2\cancel{(x-y)}}{(x+y)\cancel{(x-y)}}$$

$$= \frac{2}{x+y}$$

19. $\dfrac{3x-1}{4} - \dfrac{7}{6} = \dfrac{2}{3}$

Multiply both sides by the LCD, 12.

$$12\left(\frac{3x-1}{4} - \frac{7}{6}\right) = 12\left(\frac{2}{3}\right)$$

$$12\left(\frac{3x-1}{4}\right) - 12\left(\frac{7}{6}\right) = 12\left(\frac{2}{3}\right)$$

$$3(3x-1) - 14 = 8$$

$$9x - 3 - 14 = 8$$

$$9x - 17 = 8$$

$$9x = 25$$

$$x = \frac{25}{9}$$

20. $\dfrac{4}{5m-1} = \dfrac{2}{2m-1}$

Since the equation is a proportion, we can cross multiply to solve it.

$$4(2m-1) = 2(5m-1)$$

$$8m - 4 = 10m - 2$$

$$-4 = 2m - 2$$

$$-2 = 2m$$

$$-1 = m$$

21. $\dfrac{y}{y+2} - \dfrac{y+6}{y^2-4} + \dfrac{2}{y-2} = 0$

$$\frac{y}{y+2} - \frac{y+6}{(y+2)(y-2)} + \frac{2}{y-2} = 0$$

By inspecting the denominators, notice that the solution cannot be 2 or -2. The LCD of all of the denominators is $(y+2)(y-2)$. Multiply both sides of the equation by the LCD.

$$(y+2)(y-2)\left(\frac{y}{y+2} - \frac{y+6}{y^2-4} + \frac{2}{y-2}\right) = 0\cdot(y+2)(y-2)$$

$$y(y-2) - (y+6) + 2(y+2) = 0$$

$$y^2 - 2y - y - 6 + 2y + 4 = 0$$

$$y^2 - y - 2 = 0$$

$$(y-2)(y+1) = 0$$

$$y - 2 = 0 \quad \text{or} \quad y + 1 = 0$$

$$y = 2 \qquad\qquad y = -1$$

Since 2 cannot be a solution (it causes an expression to be undefined), 2 is extraneous. The only solution is -1.

22. $\dfrac{g}{g-1} = \dfrac{8}{g+2}$

Since the equation is a proportion, we can cross multiply to solve it.

$$g(g+2) = 8(g-1)$$

$$g^2 + 2g = 8g - 8$$

$$g^2 - 6g + 8 = 0$$

$$(g-4)(g-2) = 0$$

$$g - 4 = 0 \quad \text{or} \quad g - 2 = 0$$

$$g = 4 \qquad\qquad g = 2$$

23. Erin paints $\dfrac{1}{3}$ of the room per hour. Her husband paints $\dfrac{1}{2}$ of the room per hour. Complete a table.

Category	Rate of Work	Time at Work	Amt of Task Completed
Erin	$\dfrac{1}{3}$	t	$\dfrac{t}{3}$
husband	$\dfrac{1}{2}$	t	$\dfrac{t}{2}$

The total job in this case is 1 room.
Erin's amount completed + her husband's amount completed = 1 room.

$$\frac{t}{3} + \frac{t}{2} = 1$$

$$6\left(\frac{t}{3} + \frac{t}{2}\right) = 6(1)$$

$$2t + 3t = 6$$

$$5t = 6$$

$$t = 1.2$$

Erin and her husband can paint the room in 1.2 hours if they work together.

24. Let r represent Jake's rate while running outbound from the gym. His rate on the return trip is 2 miles per hour slower, or $r-2$. He runs 3 miles each way. Complete a table. Remember that because $d = rt$, time is found from $t = \dfrac{d}{r}$.

Category	Distance	Rate	Time
outbound	3	r	$\dfrac{3}{r}$
return	3	$r-2$	$\dfrac{3}{r-2}$

The time for the total trip is $1\dfrac{1}{4}$ or $\dfrac{5}{4}$ hours.

Outbound time + return time $= \dfrac{5}{4}$ hours

$$\frac{3}{r} + \frac{3}{r-2} = \frac{5}{4}$$

$$4r(r-2)\left(\frac{3}{r} + \frac{3}{r-2}\right) = 4r(r-2)\left(\frac{5}{4}\right)$$

$$4\cancel{r}(r-2)\cdot\frac{3}{\cancel{r}} + 4r\cancel{(r-2)}\cdot\frac{3}{\cancel{r-2}} = \cancel{4}r(r-2)\cdot\frac{5}{\cancel{4}}$$

$$12(r-2) + 12r = 5r(r-2)$$

$$12r - 24 + 12r = 5r^2 - 10r$$

$$24r - 24 = 5r^2 - 10r$$

$$0 = 5r^2 - 34r + 24$$

$$0 = (5r-4)(r-6)$$

$$5r - 4 = 0 \quad \text{or} \quad r - 6 = 0$$
$$5r = 4 \qquad\qquad r = 6$$
$$r = \frac{4}{5}$$

If the outbound rate were $\dfrac{4}{5}$ mph, then the return rate would be negative, which does not make sense in this situation. Therefore, Jake runs 6 mph away from the gym and then $6 - 2 = 4$ mph returning.

25. Let w represent the object's weight and let m represent the object's mass. Translating "the weight of an object is directly proportional to its mass," we have $w = km$. We need to find the value of the constant k. We know that the object has a mass of 40 kg and a weight of 392 N.

$$w = km$$
$$392 = k \cdot 40$$
$$9.8 = k$$

Now we can replace k with 9.8 in $w = km$ so that we have $w = 9.8m$. We can now use this equation to find an object's weight when we know its mass is 54kg.

$$w = 9.8m$$
$$= 9.8(54)$$
$$= 529.2 \text{ N}$$

Chapters 1–7 Cumulative Review

1. False; the given equation is a quadratic equation.

2. False; the x-intercept for $y = 2x$ is (0, 0).

3. False; the LCD is $(x+5)(x-5)$.

4. denominator

5. There are no common factors in the numerator and denominator besides 1.

6. multiply; add

7. complex rational expression

8. $$5^2 - \left(-4^2\right) - 20 = 25 - (-16) - 20$$
$$= 25 + 16 - 20$$
$$= 41 - 20$$
$$= 21$$

9. $$16 \div (-8) + 2^3 = 16 \div (-8) + 8$$
$$= -2 + 8$$
$$= 6$$

10. $$\frac{15a^4b}{-5a^{-9}b} = \frac{15}{-5} \cdot \frac{a^4}{a^{-9}} \cdot \frac{b}{b}$$
$$= \frac{15}{-5} \cdot a^{4-(-9)} \cdot b^{1-1}$$
$$= -3a^{13}b^0$$
$$= -3a^{13}$$

11. $$(m+3)\left(m^2 - 6m + 1\right)$$
$$= m\left(m^2 - 6m + 1\right) + 3\left(m^2 - 6m + 1\right)$$
$$= m^3 - 6m^2 + m + 3m^2 - 18m + 3$$
$$= m^3 - 3m^2 - 17m + 3$$

12. $$ax - 2x + a - 2 = (ax - 2x) + (a - 2)$$
$$= x(a-2) + (a-2)$$
$$= (a-2)(x+1)$$

13. $3y^2 - 3y - 60 = 3(y^2 - y - 20)$
$= 3(y-5)(y+4)$

14. $-3h^2 + 75 = -3(h^2 - 25)$
$= -3(h+5)(h-5)$

15. $\dfrac{3x-6}{4} \cdot \dfrac{12}{x-2} = \dfrac{3\cancel{(x-2)}}{\cancel{4}} \cdot \dfrac{\cancel{4}\cdot 3}{\cancel{x-2}}$
$= 9$

16. $\dfrac{8}{m-4} \div \dfrac{4}{m^2-16} = \dfrac{8}{m-4} \cdot \dfrac{m^2-16}{4}$
$= \dfrac{\cancel{4}\cdot 2}{\cancel{m-4}} \cdot \dfrac{\cancel{(m-4)}(m+4)}{\cancel{4}}$
$= 2(m+4)$

17. $\dfrac{x}{x^2-9} - \dfrac{3}{x^2-9} = \dfrac{x-3}{x^2-9}$
$= \dfrac{x-3}{(x-3)(x+3)}$
$= \dfrac{\cancel{x-3}}{\cancel{(x-3)}(x+3)}$
$= \dfrac{1}{x+3}$

18. $\dfrac{4}{x^2-49} + \dfrac{2}{x^2-7x}$
$= \dfrac{4}{(x+7)(x-7)} + \dfrac{2}{x(x-7)}$
$= \dfrac{4x}{x(x+7)(x-7)} + \dfrac{2(x+7)}{x(x-7)(x+7)}$
$= \dfrac{4x+2(x+7)}{x(x+7)(x-7)}$
$= \dfrac{4x+2x+14}{x(x+7)(x-7)}$
$= \dfrac{6x+14}{x(x+7)(x-7)}$

19. $\dfrac{\dfrac{y^2-y-12}{4y}}{\dfrac{y-4}{8}} = \dfrac{y^2-y-12}{4y} \div \dfrac{y-4}{8}$
$= \dfrac{y^2-y-12}{4y} \cdot \dfrac{8}{y-4}$
$= \dfrac{\cancel{(y-4)}(y+3)}{\cancel{4}\cdot y} \cdot \dfrac{\cancel{4}\cdot 2}{\cancel{y-4}}$
$= \dfrac{2(y+3)}{y}$

20. $20 + 24x = 8 - 3(5-9x)$
$20 + 24x = 8 - 15 + 27x$
$20 + 24x = -7 + 27x$
$27 + 24x = 27x$
$27 = 3x$
$9 = x$
Check:
$20 + 24x = 8 - 3(5-9x)$
$20 + 24(9) \overset{?}{=} 8 - 3(5-9(9))$
$20 + 216 \overset{?}{=} 8 - 3(5-81)$
$236 \overset{?}{=} 8 - 3(-76)$
$236 \overset{?}{=} 8 + 228$
$236 = 236$

21. $y = \dfrac{2}{3}x + 1$

a) $m = \dfrac{2}{3}$

b) The y-intercept is $(0,1)$.

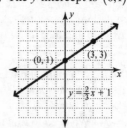

22. $2x^2 - 13x = 7$

$2x^2 - 13x - 7 = 0$

$(2x+1)(x-7) = 0$

$2x+1 = 0 \quad$ or $\quad x-7 = 0$

$\qquad 2x = -1 \qquad\qquad x = 7$

$\qquad x = -\dfrac{1}{2}$

23. $V = \dfrac{1}{3}\pi r^2 h$

$3 \cdot V = 3 \cdot \dfrac{1}{3}\pi r^2 h$

$3V = \pi r^2 h$

$\dfrac{3V}{\pi r^2} = \dfrac{\pi r^2 h}{\pi r^2}$

$\dfrac{3V}{\pi r^2} = h$

24. $\dfrac{u^2}{u-3} - \dfrac{12}{u-3} = 8$

$(u-3)\left(\dfrac{u^2}{u-3} - \dfrac{12}{u-3}\right) = (u-3)(8)$

$\qquad u^2 - 12 = 8u - 24$

$\qquad u^2 - 8u + 12 = 0$

$\qquad (u-6)(u-2) = 0$

$u-6 = 0 \quad$ or $\quad u-2 = 0$

$\qquad u = 6 \qquad\qquad u = 2$

25. $\dfrac{12}{r} = \dfrac{4}{r-4}$

$12(r-4) = 4r$

$12r - 48 = 4r$

$-48 = -8r$

$6 = r$

26. Although either the x-terms or y-terms can be eliminated, we choose to eliminate the y-terms.

$3x + 2y = 2 \quad \xrightarrow{\text{Multiply by 3}} \quad 9x + 6y = 6$

$4x - 3y = 14 \quad \xrightarrow{\text{Multiply by 2}} \quad 8x - 6y = 28$

Now add the rewritten equations to eliminate the y-terms.

$9x + 6y = 6$

$\underline{8x - 6y = 28}$

$\quad 17x = 34$

$\qquad x = 2$

Substitute $x = 2$ into the first equation and solve for y.

$3x + 2y = 2$

$3(2) + 2y = 2$

$6 + 2y = 2$

$2y = -4$

$y = -2$

Thus, the solution is $(2, -2)$.

27. Since the planes fly for the same amount of time before they are 2625 miles apart, let t represent the time spent flying for each plane. We are given that the first plane flies at a speed of 600 mph and that the second plane flies at a speed of 450 mph. We are also given that the total distance flown by the two planes is 2625 miles. Using the relationship rate $\times$ time = distance, complete a table.

Categories	Rate	Time	Distance
first plane	600	t	$600t$
second plane	450	t	$450t$

Translate to an equation.

$600t + 450t = 2625$

$1050t = 2625$

$\dfrac{1050t}{1050} = \dfrac{2625}{1050}$

$t = 2.5$

In 2.5 hours the planes will be 2625 miles apart.

28. Let x represent the principal invested at 9%. Because we know that the total amount invested is $7000, the amount invested at 7% APR is $7000 - x$. We are given that the total interest after one year is $600. Using the relationship $I = Pr$, complete a table.

Accounts	Rate	Principal	Interest
7% acct	0.07	$7000 - x$	$0.07(7000 - x)$
9% acct	0.09	x	$0.09x$

Translate to an equation.

$0.07(7000 - x) + 0.09x = 600$

$490 - 0.07x + 0.09x = 600$

$490 + 0.02x = 600$

$0.02x = 110$

$x = 5500$

$5500 was invested at 9% and $7000 - 5500 = 1500 was invested at 7%.

29. Complete a table.

Categories	Rate of Work	Time at Work	Amt of Task Completed
Jake	$\frac{1}{8}$	t	$\frac{t}{8}$
brother	$\frac{1}{6}$	t	$\frac{t}{6}$

$$\frac{t}{8}+\frac{t}{6}=1$$

$$24\left(\frac{t}{8}+\frac{t}{6}\right)=24\cdot 1$$

$$3t+4t=24$$

$$7t=24$$

$$t=\frac{24}{7}$$

$$t=3\frac{3}{7}$$

Working together, it takes Jake and his brother $3\frac{3}{7}$ hours to clean the garage.

30. $y=kx$ Now, $y=\frac{7}{2}x$

$7=k\cdot 2$ $21=\frac{7}{2}x$

$\frac{7}{2}=k$ $\frac{2}{7}\cdot 21=\frac{2}{7}\cdot\frac{7}{2}x$

                             $6=x$

Chapter 8
More on Inequalities, Absolute Value, and Functions

Exercise Set 8.1

1. $\{1, 3, 5\}$　　　　3. $\{7, 8\}$

5. $-4 < x < 5$　　　7. $0 < y \le 2$

9. $-7 < w < 3$　　　11. $0 \le u \le 2$

13. $2 < x < 7$

15. $-1 < x \le 5$

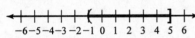

17. $1 \le x \le 10$

19. $-3 \le x < 4$

21. $-3 < x < -1$
　　a)　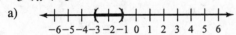

　　b) $\{x \mid -3 < x < -1\}$　c) $(-3, -1)$

23. $x + 2 > 5$　and　$x - 4 \le 2$
　　　　$x > 3$　　　　　$x \le 6$
　　a)

　　b) $\{x \mid 3 < x \le 6\}$　　c) $(3, 6]$

25. $-x > 1$　and　$-2x \le -10$
　　　$x < -1$　　　　$x \ge 5$
　　a)

　　b) $\varnothing$　　　　　　c) no interval notation

27. $3x + 8 \ge -7$　and　$4x - 7 < 5$
　　　$3x \ge -15$　　　　$4x < 12$
　　　$x \ge -5$　　　　　$x < 3$
　　a)　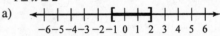

　　b) $\{x \mid -5 \le x < 3\}$　c) $[-5, 3)$

29. $-3x - 8 > 1$　and　$-4x + 5 \le -3$
　　　$-3x > 9$　　　　　$-4x \le -8$
　　　$x < -3$　　　　　$x \ge 2$
　　a)

　　b) $\varnothing$　　　　　　c) no interval notation

31. $-3 < x + 4 < 1$
　　　$-7 < x < -3$
　　a)

　　b) $\{x \mid -7 < x < -3\}$　c) $(-7, -3)$

33. $-7 \le 4x - 3 \le 5$
　　　$-4 \le 4x \le 8$
　　　$-1 \le x \le 2$
　　a)

　　b) $\{x \mid -1 \le x \le 2\}$　　c) $[-1, 2]$

35. $0 \le 2 + 3x < 8$
　　　$-2 \le 3x < 6$
　　　$-\dfrac{2}{3} \le x < 2$
　　a)

　　b) $\left\{x \mid -\dfrac{2}{3} \le x < 2\right\}$　c) $\left[-\dfrac{2}{3}, 2\right)$

37. $-1 < -2x + 5 \le 5$
　　　$-6 < -2x \le 0$
　　　$3 > x \ge 0$
　　a)

　　b) $\{x \mid 0 \le x < 3\}$　　c) $[0, 3)$

39. $3 \le 6 - x \le 6$
     $-3 \le -x \le 0$
      $3 \ge x \ge 0$

a)
     $-6\ -5\ -4\ -3\ -2\ -1\ 0\ 1\ 2\ 3\ 4\ 5\ 6$

b) $\{x \mid 0 \le x \le 3\}$ c) $[0, 3]$

41. $\{a, c, d, g, o, t\}$ 43. $\{w, x, y, z\}$

45. $x < -2$ or $x > 6$

     $-6\ -5\ -4\ -3\ -2\ -1\ 0\ 1\ 2\ 3\ 4\ 5\ 6$

47. $y < -3$ or $y \ge 0$

     $-6\ -5\ -4\ -3\ -2\ -1\ 0\ 1\ 2\ 3\ 4\ 5\ 6$

49. $w > -3$ or $w > 2$

     $-6\ -5\ -4\ -3\ -2\ -1\ 0\ 1\ 2\ 3\ 4\ 5\ 6$

51. $u \ge 0$ or $u \le 2$

     $-6\ -5\ -4\ -3\ -2\ -1\ 0\ 1\ 2\ 3\ 4\ 5\ 6$

53. $y + 2 < -7$ or $y + 2 > 7$
      $y < -9$ $y > 5$

a)
     $-9\ -8\ -7\ -6\ -5\ -4\ -3\ -2\ -1\ 0\ 1\ 2\ 3\ 4\ 5\ 6$

b) $\{y \mid y < -9 \text{ or } y > 5\}$

c) $(-\infty, -9) \cup (5, \infty)$

55. $4r - 3 < -11$ or $2r - 3 > -1$
      $4r < -8$ $2r > 2$
       $r < -2$ $r > 1$

a)
     $-6\ -5\ -4\ -3\ -2\ -1\ 0\ 1\ 2\ 3\ 4\ 5\ 6$

b) $\{r \mid r < -2 \text{ or } r > 1\}$

c) $(-\infty, -2) \cup (1, \infty)$

57. $-w + 2 \le -5$ or $-w + 2 \ge 3$
       $-w \le -7$ $-w \ge 1$
        $w \ge 7$ $w \le -1$

a)
     $-6\ -5\ -4\ -3\ -2\ -1\ 0\ 1\ 2\ 3\ 4\ 5\ 6\ 7\ 8$

b) $\{w \mid w \le -1 \text{ or } w \ge 7\}$

c) $(-\infty, -1] \cup [7, \infty)$

59. $7 - 4k \le -5$ or $6 - 2k \ge 2$
      $-4k \le -12$ $-2k \ge -4$
        $k \ge 3$ $k \le 2$

a)
     $-6\ -5\ -4\ -3\ -2\ -1\ 0\ 1\ 2\ 3\ 4\ 5\ 6$

b) $\{k \mid k \le 2 \text{ or } k \ge 3\}$

c) $(-\infty, 2] \cup [3, \infty)$

61. $\dfrac{2}{3} x - 4 \ge -2$ or $\dfrac{2}{5} x - 2 \le -4$

      $\dfrac{2}{3} x \ge 2$ $\dfrac{2}{5} x \le -2$

       $2x \ge 6$ $2x \le -10$

        $x \ge 3$ $x \le -5$

a)
     $-6\ -5\ -4\ -3\ -2\ -1\ 0\ 1\ 2\ 3\ 4\ 5\ 6$

b) $\{x \mid x \le -5 \text{ or } x \ge 3\}$

c) $(-\infty, -5] \cup [3, \infty)$

63. $-2(c - 1) < -4$ or $-2(c - 2) < -6$
      $-2c + 2 < -4$ $-2c + 4 < -6$
        $-2c < -6$ $-2c < -10$
         $c > 3$ $c > 5$

a)
     $-6\ -5\ -4\ -3\ -2\ -1\ 0\ 1\ 2\ 3\ 4\ 5\ 6$

b) $\{c \mid c > 3\}$ c) $(3, \infty)$

65. $2(x + 3) + 3 \le 1$ or $2(x + 1) + 7 \le -3$
      $2x + 6 + 3 \le 1$ $2x + 2 + 7 \le -3$
       $2x + 9 \le 1$ $2x + 9 \le -3$
        $2x \le -8$ $2x \le -12$
         $x \le -4$ $x \le -6$

a)
     $-6\ -5\ -4\ -3\ -2\ -1\ 0\ 1\ 2\ 3\ 4\ 5\ 6$

b) $\{x \mid x \le -4\}$

c) $(-\infty, -4]$

67. $-3(x-1)-1 \le -1$ or $-3(x+1)+5 \ge -2$

$\quad -3x+3-1 \le -1 \qquad\quad -3x-3+5 \ge -2$

$\qquad\quad -3x+2 \le -1 \qquad\qquad -3x+2 \ge -2$

$\qquad\qquad -3x \le -3 \qquad\qquad\quad -3x \ge -4$

$\qquad\qquad\quad x \ge 1$

$\qquad\qquad\qquad\qquad\qquad\qquad x \le \dfrac{4}{3}$

a)

b) $\{x \mid x \text{ is a real number}\}$ or $\mathbb{R}$

c) $(-\infty, \infty)$

69. $-3 < x+4$ and $x+4 < 7$

$\quad -7 < x \qquad\qquad\quad x < 3$

a)

b) $\{x \mid -7 < x < 3\}$ c) $(-7, 3)$

71. $4x+3 < -5$ or $3x+2 > 8$

$\quad 4x < -8 \qquad\qquad 3x > 6$

$\quad\quad x < -2 \qquad\qquad x > 2$

a)

b) $\{x \mid x < -2 \text{ or } x > 2\}$

c) $(-\infty, -2) \cup (2, \infty)$

73. $4 < -2x < 6$

$\quad -2 > x > -3$

a)

b) $\{x \mid -3 < x < -2\}$ c) $(-3, -2)$

75. $7 \le 4x-5 \le 19$

$\quad 12 \le 4x \le 24$

$\quad\quad 3 \le x \le 6$

a)

b) $\{x \mid 3 \le x \le 6\}$ c) $[3, 6]$

77. $-5 < 1-2x < -1$

$\quad -6 < -2x < -2$

$\quad\quad 3 > x > 1$

a)

b) $\{x \mid 1 < x < 3\}$ c) $(1, 3)$

79. $x+3 < 2x+1$ and $2x+1 < 3x$

$\quad -x < -2 \qquad\qquad\quad 1 < x$

$\quad\quad x > 2$

a)

b) $\{x \mid x > 2\}$ c) $(2, \infty)$

81. $36 \le 0.09x \le 45$

$\quad 400 \le x \le 500$

a)

b) $\{x \mid 400 \le x \le 500\}$

c) $[400, 500]$

83. $80 \le \dfrac{95+80+82+88+x}{5} < 90$

$\quad 400 \le 345+x < 450$

$\quad\quad 55 \le x < 105$

a)

b) $\{x \mid 55 \le x < 105\}$ c) $[55, 105)$

85. a)

b) $\{x \mid 68° \le x \le 78°\}$ c) $[68, 78]$

87. a)

b) $\{x \mid 72° \le x \le 80°\}$ c) $[72, 80]$

89. $6000 \le \dfrac{1}{2} \cdot 50(60+x) \le 8000$

$\quad 6000 \le 25(60+x) \le 8000$

$\quad 6000 \le 1500 + 25x \le 8000$

$\quad 4500 \le 25x \le 6500$

$\quad\quad 180 \le x \le 260$

a)

b) $\{x \mid 180 \text{ ft} \le x \le 260 \text{ ft}\}$

c) $[180, 260]$

Review Exercises

1.   No, the absolute value of zero is zero and zero is neither positive nor negative.

2. $-\left|2-3^2\right|-\left|-4\right| = -\left|2-9\right|-\left|-4\right|$
$= -\left|-7\right|-4$
$= -7-4$
$= -11$

3. $-\left|-\left|-16\right|\right| = -\left|-16\right| = -16$

4. $3x-14x = 17-12x-11$
$-11x = 6-12x$
$x = 6$

5. $\dfrac{3}{5}(25-5x) = 15-\dfrac{3}{5}$
$15-3x = \dfrac{72}{5}$
$5\cdot 15-5\cdot 3x = 5\cdot\dfrac{72}{5}$
$75-15x = 72$
$-15x = -3$
$x = \dfrac{1}{5}$

6. $\dfrac{1}{2}(x+2) = 0$
$2\cdot\dfrac{1}{2}(x+2) = 2\cdot 0$
$x+2 = 0$
$x = -2$

Exercise Set 8.2

1. $x = 2$ or $x = -2$

3. $\left|a\right| = -4$

Because the absolute value of every real number is a positive number or zero, this equation has no solution.

5. $x+3 = 8$ or $x+3 = -8$
$x = 5$ $\qquad$ $x = -11$

7. $2m-5 = 1$ or $2m-5 = -1$
$2m = 6$ $\qquad$ $2m = 4$
$m = 3$ $\qquad$ $m = 2$

9. $6-5x = 1$ or $6-5x = -1$
$-5x = -5$ $\qquad$ $-5x = -7$
$x = 1$ $\qquad$ $x = \dfrac{7}{5}$

11. $4-3w = -6$ or $4-3w = 6$
$-3w = -10$ $\qquad$ $-3w = 2$
$w = \dfrac{10}{3}$ $\qquad$ $w = -\dfrac{2}{3}$

13. $\left|4m-2\right| = -5$
Because the absolute value of every real number is a positive number or zero, this equation has no solution.

15. $4w-3 = 0$
$4w = 3$
$w = \dfrac{3}{4}$

17. $\left|2y\right|-3 = 5$
$\left|2y\right| = 8$
$2y = 8$ or $2y = -8$
$y = 4$ $\qquad$ $y = -4$

19. $\left|y+1\right|+2 = 4$
$\left|y+1\right| = 2$
$y+1 = 2$ or $y+1 = -2$
$y = 1$ $\qquad$ $y = -3$

21. $\left|b-4\right|-6 = 2$
$\left|b-4\right| = 8$
$b-4 = 8$ or $b-4 = -8$
$b = 12$ $\qquad$ $b = -4$

23. $3+\left|5x-1\right| = 7$
$\left|5x-1\right| = 4$
$5x-1 = 4$ or $5x-1 = -4$
$5x = 5$ $\qquad$ $5x = -3$
$x = 1$ $\qquad$ $x = -\dfrac{3}{5}$

25. $1-\left|2k+3\right| = -4$
$-\left|2k+3\right| = -5$
$\left|2k+3\right| = 5$
$2k+3 = 5$ or $2k+3 = -5$
$2k = 2$ $\qquad$ $2k = -8$
$k = 1$ $\qquad$ $k = -4$

27. $4 - 3|z - 2| = -8$

   $-3|z - 2| = -12$

   $|z - 2| = 4$

   $z - 2 = 4$ or $z - 2 = -4$

   $z = 6$ $z = -2$

29. $6 - 2|3 - 2w| = -18$

   $-2|3 - 2w| = -24$

   $|3 - 2w| = 12$

   $3 - 2w = 12$ or $3 - 2w = -12$

   $-2w = 9$ $-2w = -15$

   $w = -\dfrac{9}{2}$ $w = \dfrac{15}{2}$

31. $|3x - 2(x + 5)| = 10$

   $|3x - 2x - 10| = 10$

   $|x - 10| = 10$

   $x - 10 = 10$ or $x - 10 = -10$

   $x = 20$ $x = 0$

33. $2x + 1 = x + 5$ or $2x + 1 = -(x + 5)$

   $x = 4$ $2x + 1 = -x - 5$

                    $3x = -6$

                    $x = -2$

35. $x + 3 = 2x - 4$ or $x + 3 = -(2x - 4)$

   $7 = x$ $x + 3 = -2x + 4$

                    $3x = 1$

                    $x = \dfrac{1}{3}$

37. $3v + 4 = 1 - 2v$ or $3v + 4 = -(1 - 2v)$

   $5v = -3$ $3v + 4 = -1 + 2v$

   $v = -\dfrac{3}{5}$ $v = -5$

39. $2n + 3 = 3 + 2n$ or $2n + 3 = -(3 + 2n)$

   $3 = 3$ $2n + 3 = -3 - 2n$

                    $4n = -6$

                    $n = -\dfrac{3}{2}$

One equation leads to a solution with no variables yet is true. The solution to this absolute value equation is all real numbers.

41. $2k + 1 = 2k - 5$ or $2k + 1 = -(2k - 5)$

   $0 = -6$ $2k + 1 = -2k + 5$

   no solution $4k = 4$

                    $k = 1$

This absolute value equation has only one solution, 1.

43. $|10 - (5 - h)| = 8$

   $|10 - 5 + h| = 8$

   $|5 + h| = 8$

   $5 + h = 8$ or $5 + h = -8$

   $h = 3$ $h = -13$

45. $\dfrac{b}{2} - 1 = 4$ or $\dfrac{b}{2} - 1 = -4$

   $\dfrac{b}{2} = 5$ $\dfrac{b}{2} = -3$

   $b = 10$ $b = -6$

47. $\dfrac{4 - 3x}{2} = \dfrac{3}{4}$ or $\dfrac{4 - 3x}{2} = -\dfrac{3}{4}$

   $4(4 - 3x) = 2 \cdot 3$ $4(4 - 3x) = -3 \cdot 2$

   $16 - 12x = 6$ $16 - 12x = -6$

   $-12x = -10$ $-12x = -22$

   $x = \dfrac{5}{6}$ $x = \dfrac{11}{6}$

49. $\left|2y + \dfrac{3}{2}\right| - 2 = 5$

   $\left|2y + \dfrac{3}{2}\right| = 7$

   $2y + \dfrac{3}{2} = 7$ or $2y + \dfrac{3}{2} = -7$

   $2y = \dfrac{11}{2}$ $2y = -\dfrac{17}{2}$

   $y = \dfrac{11}{4}$ $y = -\dfrac{17}{4}$

51. $f(x) = |x| - 4$

Domain: $(-\infty, \infty)$ Range: $[4, \infty)$

53. $f(x) = |x+3| - 2$

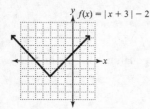

Domain: $(-\infty, \infty)$ Range: $[-2, \infty)$

55. $f(x) = 2|x-1| + 3$

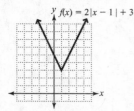

Domain: $(-\infty, \infty)$ Range: $[3, \infty)$

57. $f(x) = -|x| + 3$

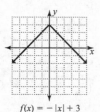

Domain: $(-\infty, \infty)$ Range: $(-\infty, 3]$

59. $f(x) = -3|x+2| - 1$

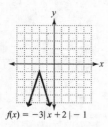

Domain: $(-\infty, \infty)$ Range: $(-\infty, -1]$

61. $f(x) = -\dfrac{1}{2}|x+1| + 1$

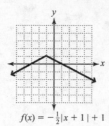

Domain: $(-\infty, \infty)$ Range: $(-\infty, 1]$

Review Exercises

1. $(-2, 0]$

2.

![number line with bracket interval from -2 to -1, marks -6 through 6]

3. $6 - 4x > -5x - 1$ 4. $-4x \le 12$
        $x > -7$ $x \ge -3$

5. $-1 < \dfrac{3x+2}{4} < 4$

     $-4 < 3x + 2 < 16$

     $-6 < 3x < 14$

     $-2 < x < \dfrac{14}{3}$

6. $n - 2 < 5$

     $n - 2 + 2 < 5 + 2$

          $n < 7$

Exercise Set 8.3

1.   a)

![number line with open interval from -5 to 5, marks -6 through 6]

     b) $\{x \mid -5 < x < 5\}$ c) $(-5, 5)$

3. $-7 \le x + 3 \le 7$
     $-10 \le x \le 4$
     a)

![number line with closed interval from -10 to 4, marks -12 through 4]

     b) $\{x \mid -10 \le x \le 4\}$ c) $[-10, 4]$

5. $|s+3|+6<9$

$\qquad |s+3|<3$

$\qquad -3<s+3<3$

$\qquad -6<s<0$

a)

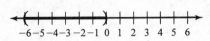

b) $\{s|-6<s<0\}$　　c) $(-6,0)$

7. $|2m-5|-3<6$

$\qquad |2m-5|<9$

$\qquad -9<2m-5<9$

$\qquad -4<2m<14$

$\qquad -2<m<7$

a)

b) $\{m|-2<m<7\}$　c) $(-2,7)$

9. $|-3k+5|+7\le 8$

$\qquad |-3k+5|\le 1$

$\qquad -1\le -3k+5\le 1$

$\qquad -6\le -3k\le -4$

$\qquad 2\ge k\ge \dfrac{4}{3}$

a)

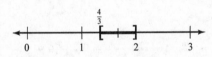

b) $\left\{k\middle|\dfrac{4}{3}\le k\le 2\right\}$　c) $\left[\dfrac{4}{3},2\right]$

11. $2|x|+7\le 3$

$\qquad 2|x|\le -4$

$\qquad |x|\le -2$

Because absolute values cannot be negative, this inequality has no solution.

a)

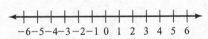

b) $\{\ \}$ or $\varnothing$　　c) no interval notation

13. $2|w-3|+4<10$

$\qquad 2|w-3|<6$

$\qquad |w-3|<3$

$\qquad -3<w-3<3$

$\qquad 0<w<6$

a)

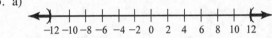

b) $\{w|0<w<6\}$　c. $(0,6)$

15. a)

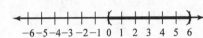

b) $\{c|c<-12 \text{ or } c>12\}$

c) $(-\infty,-12)\cup (12,\infty)$

17. $y+2\le -7$　or　$y+2\ge 7$

$\qquad y\le -9$　　　　$y\ge 5$

a)

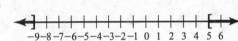

b) $\{y|y\le -9 \text{ or } y\ge 5\}$

c) $(-\infty,-9]\cup [5,\infty)$

19. $|p-6|-3>5$

$\qquad |p-6|>8$

$\qquad p-6<-8$　or　$p-6>8$

$\qquad p<-2$　　　　$p>14$

a)

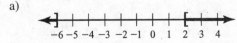

b) $\{p|p<-2 \text{ or } p>14\}$

c) $(-\infty,-2)\cup (14,\infty)$

21. $|3x+6|-3\ge 9$

$\qquad |3x+6|\ge 12$

$\qquad 3x+6\le -12$　or　$3x+6\ge 12$

$\qquad 3x\le -18$　　　$3x\ge 6$

$\qquad x\le -6$　　　　$x\ge 2$

a)

b) $\{x|x\le -6 \text{ or } x\ge 2\}$

c) $(-\infty,-6]\cup [2,\infty)$

23. $|-4n-5|+3>8$

   $|-4n-5|>5$

   $-4n-5<-5$ or $-4n-5>5$

   $-4n<0$ $-4n>10$

   $n>0$ $n<-\dfrac{5}{2}$

a)

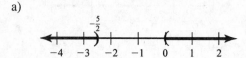

b) $\left\{n\Big|n<-\dfrac{5}{2}\text{ or }n>0\right\}$

c) $\left(-\infty,-\dfrac{5}{2}\right)\cup(0,\infty)$

25. $4|v|+3\geq 7$

   $4|v|\geq 4$

   $|v|\geq 1$

   $v\leq -1$ or $v\geq 1$

a)

b) $\left\{v\,|\,v\leq -1\text{ or }v\geq 1\right\}$

c) $(-\infty,-1]\cup[1,\infty)$

27. $4|y+2|-1>3$

   $4|y+2|>4$

   $|y+2|>1$

   $y+2<-1$ or $y+2>1$

   $y<-3$ $y>-1$

a)

b) $\left\{y\,|\,y<-3\text{ or }y>-1\right\}$

c) $(-\infty,-3)\cup(-1,\infty)$

29. $|4m+8|-2>10$

   $|4m+8|>12$

   $4m+8<-12$ or $4m+8>12$

   $4m<-20$ $4m>4$

   $m<-5$ $m>1$

a)

b) $\left\{m\,|\,m<-5\text{ or }m>1\right\}$

c) $(-\infty,-5)\cup(1,\infty)$

31. $-6<-3x+6<6$

   $-12<-3x<0$

   $4>x>0$

   $0<x<4$

a)

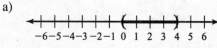

b) $\left\{x\,|\,0<x<4\right\}$ c) $(0,4)$

33. $|2r-3|>-3$

This inequality indicates that the absolute value is greater than a negative number. Because the absolute value of every real number is either positive or 0, the solution set is $\mathbb{R}$.

a)

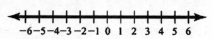

b) $\left\{r\,|\,r\text{ is a real number}\right\}$

c) $(-\infty,\infty)$

35. $4-2|x+3|>2$

   $-2|x+3|>-2$

   $|x+3|<1$

   $-1<x+3<1$

   $-4<x<-2$

a)

b) $\left\{x\,|-4<x<-2\right\}$ c) $(-4,-2)$

37. $6|2x-1|-3<3$

   $6|2x-1|<6$

   $|2x-1|<1$

   $-1<2x-1<1$

   $0<2x<2$

   $0<x<1$

a)

b) $\left\{x\,|\,0<x<1\right\}$ c) $(0,1)$

39. $5-|w+4| > 10$

$-|w+4| > 5$

$|w+4| < -5$

Since absolute values cannot be negative, this inequality has no solution.

a)

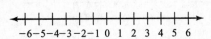

b) $\varnothing$ c) no interval notation

41. $\left|2-\dfrac{3}{2}k\right| \le 5$

$-5 \le 2-\dfrac{3}{2}k \le 5$

$-10 \le 4-3k \le 10$

$-14 \le -3k \le 6$

$\dfrac{14}{3} \ge k \ge -2$

a)

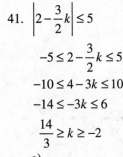

b) $\left\{k \middle| -2 \le k \le \dfrac{14}{3}\right\}$ c) $\left[-2, \dfrac{14}{3}\right]$

43. $|0.25x-3| + 2 > 4$

$|0.25x-3| > 2$

$0.25x-3 < -2$ or $0.25x-3 > 2$

$0.25x < 1$ $0.25x > 5$

$x < 4$ $x > 20$

a)

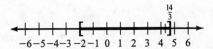

b) $\left\{x \middle| x < 4 \text{ or } x > 20\right\}$

c) $(-\infty, 4) \cup (20, \infty)$

45. $\left|2.4-\dfrac{3}{4}y\right| \le 7.2$

$-7.2 \le 2.4-\dfrac{3}{4}y \le 7.2$

$-9.6 \le -\dfrac{3}{4}y \le 4.8$

$12.8 \ge y \ge -6.4$

a)

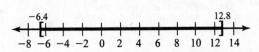

b) $\left\{y \middle| -6.4 \le y \le 12.8\right\}$

c) $[-6.4, 12.8]$

47. $|2p-8| + 5 > 1$

$|2p-8| > -4$

This inequality indicates that the absolute value is greater than a negative number. Because the absolute value of every real number is either positive or 0, the solution set is $\mathbb{R}$.

a)

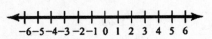

b) $\left\{p \middle| p \text{ is a real number}\right\}$

c) $(-\infty, \infty)$

49. $|x| < 3$

51. $|x| \ge 4$

53. $|x+1| > 1$

55. $|x-3| \le 2$

57. $|x| >$ any negative number

Review Exercises

1. $2x+3y = 3$

2. $2x+3 = -5$

$2(-4)+3 = -5$

$-8+3 = -5$

$-5 = -5$

Yes, $x = -4$ is a solution of $2x+3 = -5$.

3. $x^2 - 16 = 0$

$(x-4)(x+4) = 0$

$x-4 = 0$ or $x+4 = 0$

$x = 4$ $x = -4$

4. $|x+5| = -8$

Because the absolute value of every real number is a positive number or zero, this equation has no solution.

5. $y = 2x-3$

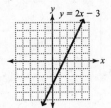

6. $2x - y = 4$

   $-y = -2x + 4$

   $y = 2x - 4$

Chapter 8 Review Exercises

1.   Intersection: $A \cap B = \{1, 5, 9\}$

   Union: $A \cup B = \{1, 5, 7, 9\}$

2.   Intersection: $A \cap B = \{ \ \}$ or $\varnothing$

   Union: $A \cup B = \{1, 2, 3, 4, 5, 6, 7\}$

3. $4 < -2x < 6$

   $-2 > x > -3$

   a)

   b) $\{x \mid -3 < x < -2\}$ c) $(-3, -2)$

4. $9 \le -3x \le 15$

   $-3 \ge x \ge -5$

   a)

   b) $\{x \mid -5 \le x \le -3\}$ c) $[-5, -3]$

5. $-3 < x + 4 < 7$

   $-7 < x < 3$

   a)

   b) $\{x \mid -7 < x < 3\}$ c) $(-7, 3)$

6. $0 < x - 1 \le 3$

   $1 < x \le 4$

   a)

   b) $\{x \mid 1 < x \le 4\}$ c) $(1, 4]$

7. $-1 \le 2x + 3 < 3$

   $-4 \le 2x < 0$

   $-2 \le x < 0$

   a)

b) $\{x \mid -2 \le x < 0\}$ c) $[-2, 0)$

8. $3x - 2 > 1$ and $3x - 2 < -8$

   $3x > 3$ $3x < -6$

   $x > 1$ $x < -2$

   a)

   b) $\{ \ \}$ or $\varnothing$ c)   no interval notation

9.   Let x represent the fourth test score.

   $80 \le \dfrac{80 + 89 + 83 + x}{4} < 90$

   $80 \le \dfrac{252 + x}{4} < 90$

   $320 \le 252 + x < 360$

   $68 \le x < 108$

   a)

   b) $\{x \mid 68 \le x < 108\}$

   c) $[68, 108)$

10.   Let x represent the length of the rectangular building.

   $12,000 \le 80x \le 16,000$

   $150 \le x \le 200$

   a)

   b) $\{x \mid 150 \le x \le 200\}$

   c) $[150, 200]$

11. $w + 4 \le -2$ or $w + 4 \ge 2$

   $w \le -6$ $w \ge -2$

   a)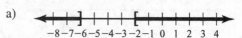

   b) $\{w \mid w \le -6 \text{ or } w \ge -2\}$

   c) $(-\infty, -6] \cup [-2, \infty)$

12. $4w - 3 < 1$ or $4w - 3 > 0$

   $4w < 4$ $4w > 3$

   $w < 1$ $w > \dfrac{3}{4}$

   a)

   b) $\{w \mid w \text{ is a real number}\}$

   c) $(-\infty, \infty)$

13. $2m - 5 < 0$ or $2m - 5 > 5$

$\qquad 2m < 5 \qquad\qquad 2m > 10$

$\qquad m < \dfrac{5}{2} \qquad\qquad m > 5$

a)

b) $\left\{ m \middle| m < \dfrac{5}{2} \text{ or } m > 5 \right\}$

c) $\left(-\infty, \dfrac{5}{2} \right) \cup (5, \infty)$

14. $3x + 2 \le -2$ or $3x + 2 \ge 8$

$\qquad 3x \le -4 \qquad\qquad 3x \ge 6$

$\qquad x \le -\dfrac{4}{3} \qquad\qquad x \ge 2$

a)

b) $\left\{ x \middle| x \le -\dfrac{4}{3} \text{ or } x \ge 2 \right\}$

c) $\left(-\infty, -\dfrac{4}{3} \right] \cup [2, \infty)$

15. $-x - 6 \le -2$ or $-x - 6 \ge 3$

$\qquad -x \le 4 \qquad\qquad -x \ge 9$

$\qquad x \ge -4 \qquad\qquad x \le -9$

a)

b) $\left\{ x \middle| x \le -9 \text{ or } x \ge -4 \right\}$

c) $(-\infty, -9] \cup [-4, \infty)$

16. $-4w + 1 \le -3$ or $-4w + 1 \ge 5$

$\qquad -4w \le -4 \qquad\qquad -4w \ge 4$

$\qquad w \ge 1 \qquad\qquad w \le -1$

a)

b) $\left\{ w \middle| w \le -1 \text{ or } w \ge 1 \right\}$

c) $(-\infty, -1] \cup [1, \infty)$

17. $-4, 4$

18. $x - 4 = -7$ or $x - 4 = 7$

$\qquad x = -3 \qquad\qquad x = 11$

19. $2w - 1 = -3$ or $2w - 1 = 3$

$\qquad 2w = -2 \qquad\qquad 2w = 4$

$\qquad w = -1 \qquad\qquad w = 2$

20. $|5r + 8| = -3$

Because the absolute value of every real number is a positive number or zero, this equation has no solution.

21. $|q - 4| - 3 = 8$

$\qquad |q - 4| = 11$

$\qquad q - 4 = -11$ or $q - 4 = 11$

$\qquad\quad q = -7 \qquad\qquad q = 15$

22. $|5w| - 2 = 13$

$\qquad |5w| = 15$

$\qquad 5w = -15$ or $5w = 15$

$\qquad\quad w = -3 \qquad\qquad w = 3$

23. $2|3x - 4| = 8$

$\qquad |3x - 4| = 4$

$\qquad 3x - 4 = -4$ or $3x - 4 = 4$

$\qquad\quad 3x = 0 \qquad\qquad 3x = 8$

$\qquad\quad x = 0 \qquad\qquad x = \dfrac{8}{3}$

24. $4 - 2|r - 5| = -8$

$\qquad -2|r - 5| = -12$

$\qquad |r - 5| = 6$

$\qquad r - 5 = -6$ or $r - 5 = 6$

$\qquad\quad r = -1 \qquad\qquad r = 11$

25. $3x - 2 = x + 2$ or $3x - 2 = -x - 2$

$\qquad 2x = 4 \qquad\qquad 4x = 0$

$\qquad x = 2 \qquad\qquad x = 0$

26. $2x - 1 = 3x + 2$ or $2x - 1 = -3x - 2$

$\qquad -x = 3 \qquad\qquad 5x = -1$

$\qquad x = -3 \qquad\qquad x = -\dfrac{1}{5}$

27. $-3x + 1 = 3 - 2x$ or $-3x + 1 = -3 + 2x$

$\qquad -x = 2 \qquad\qquad -5x = -4$

$\qquad x = -2 \qquad\qquad x = \dfrac{4}{5}$

28. $9-4x=7-2x$ or $9-4x=-7+2x$
    $-2x=-2$ $-6x=-16$
    $x=1$ $x=\dfrac{16}{6}$
                      $x=\dfrac{8}{3}$

29. $f(x)=|x-2|$

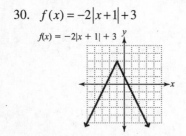

30. $f(x)=-2|x+1|+3$

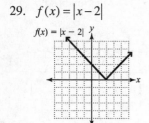

31. $|x|<5$
    a)

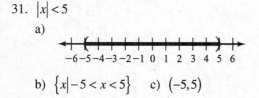

    b) $\{x|-5<x<5\}$ c) $(-5,5)$

32. $-4<2m+6<4$
    $-10<2m<-2$
    $-5<m<-1$
    a)

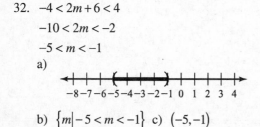

    b) $\{m|-5<m<-1\}$ c) $(-5,-1)$

33. $|3s-1|\le-2$
    Because absolute values cannot be negative, this inequality has no solution.
    a)

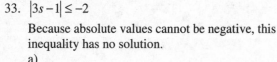

    b) $\{\ \}$ or $\varnothing$ c) no interval notation

34. $7|m+3|\le21$
    $|m+3|\le3$
    $-3\le m+3\le3$
    $-6\le m\le0$

a)

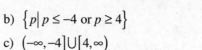

b) $\{m|-6\le m\le0\}$ c) $[-6,0]$

35. $|p|\ge4$
a)

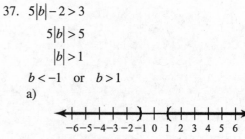

b) $\{p|p\le-4$ or $p\ge4\}$
c) $(-\infty,-4]\cup[4,\infty)$

36. $x-3<-7$ or $x-3>7$
    $x<-4$ $x>10$
a)

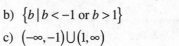

b) $\{x|x<-4$ or $x>10\}$
c) $(-\infty,-4)\cup(10,\infty)$

37. $5|b|-2>3$
    $5|b|>5$
    $|b|>1$
    $b<-1$ or $b>1$
a)

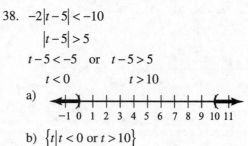

b) $\{b|b<-1$ or $b>1\}$
c) $(-\infty,-1)\cup(1,\infty)$

38. $-2|t-5|<-10$
    $|t-5|>5$
    $t-5<-5$ or $t-5>5$
    $t<0$ $t>10$
a)

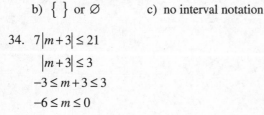

b) $\{t|t<0$ or $t>10\}$
c) $(-\infty,0)\cup(10,\infty)$

39. $5 - 2|2k - 3| \le -15$

$-2|2k - 3| \le -20$

$|2k - 3| \ge 10$

$2k - 3 \le -10$ or $2k - 3 \ge 10$

$2k \le -7$ $\qquad 2k \ge 13$

$k \le -\dfrac{7}{2}$ $\qquad k \ge \dfrac{13}{2}$

a)

b) $\left\{ k \mid k \le -\dfrac{7}{2} \text{ or } k \ge \dfrac{13}{2} \right\}$

c) $\left(-\infty, -\dfrac{7}{2} \right] \cup \left[\dfrac{13}{2}, \infty \right)$

40. $3 - 7|2p + 4| \le 24$

$-7|2p + 4| \le 21$

$|2p + 4| \ge -3$

a)

b) $\left\{ p \mid p \text{ is a real number} \right\}$ c) $(-\infty, \infty)$

Chapter 8 Practice Test

1. Let $A = \{h, o, m, e\}$ and $B = \{h, o, u, s, e\}$

$A \cap B = \{e, h, o\}$ because e, h, and o are the only elements in both A and B.

$A \cup B = \{e, h, m, o, s, u\}$ because these are the elements that are either in A or B or both.

2. $-3 < x + 4 \le 7$

$-7 < x \le 3$

a)

b) $\left\{ x \mid -7 < x \le 3 \right\}$ c) $(-7, 3]$

3. $4 < -2x \le 6$

$-2 > x \ge -3$

a)

b) $\left\{ x \mid -3 \le x < -2 \right\}$ c) $[-3, -2)$

4. $5x + 2 \le -3$ or $5x + 2 \ge 12$

$5x \le -5$ $\qquad 5x \ge 10$

$x \le -1$ $\qquad x \ge 2$

a)

b) $\left\{ x \mid x \le -1 \text{ or } x \ge 2 \right\}$

c) $(-\infty, -1] \cup [2, \infty)$

5. $6 - 2n < 2$ or $6 - 2n < 4$

$-2n < -4$ $\qquad -2n < -2$

$n > 2$ $\qquad n > 1$

a)

b) $\left\{ n \mid n > 1 \right\}$

c) $(1, \infty)$

6. Let x represent the number of books the person orders.

$40 \le 5x \le 50$

$8 \le x \le 10$

a)

b) $\left\{ x \mid 8 \le x \le 10 \right\}$ c) $[8, 10]$

7. $|x + 3| = 5$

$x + 3 = -5$ or $x + 3 = 5$

$x = -8$ $\qquad x = 2$

8. $3 - |2x - 3| = -6$

$-|2x - 3| = -9$

$|2x - 3| = 9$

$2x - 3 = -9$ or $2x - 3 = 9$

$2x = -6$ $\qquad 2x = 12$

$x = -3$ $\qquad x = 6$

9. $|2x + 3| = |x - 5|$

$2x + 3 = x - 5$ or $2x + 3 = -(x - 5)$

$x + 3 = -5$ $\qquad 2x + 3 = -x + 5$

$x = -8$ $\qquad 3x + 3 = 5$

$\qquad\qquad 3x = 2$

$\qquad\qquad x = \dfrac{2}{3}$

10. $|5x-4| = -3$

This equation has the absolute value equal to a negative number. Because the absolute value of every real number is a positive number or zero, this equation has no solution.

11. $f(x) = |x+1| - 2$

x	$f(x)$
–4	1
–3	0
–2	–1
–1	–2
0	–1
1	0
2	1

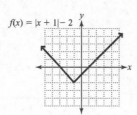

$f(x) = |x+1| - 2$

12. $f(x) = -2|x-3| + 1$

x	$f(x)$
0	–5
1	–3
2	–1
3	1
4	–1
5	–3
6	–5

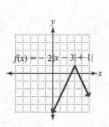

$f(x) = -2|x-3| + 1$

13. $|x+4| < 9$

$-9 < x+4 < 9$

$-13 < x < 5$

a)

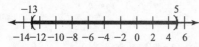

b) $\{x|-13 < x < 5\}$ c) $(-13,5)$

14. $2|x-1| > 4$

$|x-1| > 2$

$x-1 < -2$ or $x-1 > 2$

$x < -1$ $x > 3$

a)

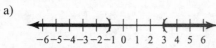

b) $\{x|x < -1 \text{ or } x > 3\}$

c) $(-\infty,-1) \cup (3,\infty)$

15. $3 - 2|x+4| > -3$

$-2|x+4| > -6$

$|x+4| < 3$

$-3 < x+4 < 3$

$-7 < x < -1$

a)

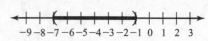

b) $\{x|-7 < x < -1\}$ c) $(-7,-1)$

16. $|3y-2| < -2$

Because the absolute value of every real number is a positive number or zero, this inequality has no solution.

a)

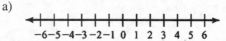

b) $\{\ \}$ or $\emptyset$ c) no interval notation

17. $|8t+4| \geq -12$

Because the absolute value of every real number is greater than or equal to -12, this inequality is true for all real numbers.

a)

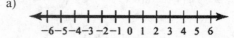

b) $\{t|t \text{ is a real number}\}$ c) $(-\infty,\infty)$

18. $2|3x-4| \leq 10$

$|3x-4| \leq 5$

$-5 \leq 3x-4 \leq 5$

$-1 \leq 3x \leq 9$

$-\dfrac{1}{3} \leq x \leq 3$

a)

b) $\left\{x \middle| -\dfrac{1}{3} \leq x \leq 3\right\}$ c) $\left[-\dfrac{1}{3},3\right]$

Chapters 1–8 Cumulative Review

1. False

2. True

3. True

4. True

5. least common multiple (LCM)

6. $2(x-8) = -4$

7. vertical line

8. $10 - 2(6 - 3^2) \div 2 \cdot 3 = 10 - 2(6-9) \div 2 \cdot 3$
$$= 10 - 2(-3) \div 2 \cdot 3$$
$$= 10 + 6 \div 2 \cdot 3$$
$$= 10 + 3 \cdot 3$$
$$= 10 + 9$$
$$= 19$$

9. $(3x - y)^2 = (3x)^2 - 2(3x)(y) + y^2$
$$= 9x^2 - 6xy + y^2$$

10. $\left(4xyz^{-3}\right)^2 \left(-2x^3 y^{-4}\right)^3$
$$= \left(4^2 x^2 y^2 z^{-6}\right)(-2)^3 \left(x^9 y^{-12}\right)$$
$$= \left(16x^2 y^2 z^{-6}\right)(-8)\left(x^9 y^{-12}\right)$$
$$= -128 x^{11} y^{-10} z^{-6}$$
$$= -\frac{128 x^{11}}{y^{10} z^6}$$

11. $\dfrac{x^3 - 5x^2 + 10x}{5x^2} = \dfrac{x^3}{5x^2} - \dfrac{5x^2}{5x^2} + \dfrac{10x}{5x^2}$
$$= \frac{1}{5}x - 1 + \frac{2}{x}$$

12. $\dfrac{a^2 - b^2}{x^2 - y^2} \div \dfrac{a+b}{x-y} = \dfrac{a^2 - b^2}{x^2 - y^2} \cdot \dfrac{x-y}{a+b}$
$$= \frac{(a+b)(a-b)}{(x+y)(x-y)} \cdot \frac{x-y}{a+b}$$
$$= \frac{a-b}{x+y}$$

13. $\dfrac{\dfrac{10x}{2x^2 + 7x - 4}}{\dfrac{5}{2x-1}} = \dfrac{10x}{2x^2 + 7x - 4} \cdot \dfrac{2x-1}{5}$
$$= \frac{2 \cdot 5 \cdot x}{(2x-1)(x+4)} \cdot \frac{2x-1}{5}$$
$$= \frac{2x}{x+4}$$

14. $\dfrac{x^2 + 4x}{x^2 + 8x + 16} + \dfrac{3}{x+4}$
$$= \frac{x^2 + 4x}{(x+4)(x+4)} + \frac{3(x+4)}{(x+4)(x+4)}$$
$$= \frac{x^2 + 4x + 3x + 12}{(x+4)(x+4)}$$
$$= \frac{x^2 + 7x + 12}{(x+4)(x+4)}$$
$$= \frac{(x+3)(x+4)}{(x+4)(x+4)}$$
$$= \frac{x+3}{x+4}$$

15. $12bc - 9b + 20c - 15$
$$= (12bc - 9b) + (20c - 15)$$
$$= 3b(4c-3) + 5(4c-3)$$
$$= (4c-3)(3b+5)$$

16. $2x^3 - 14x^2 + 20x = 2x\left(x^2 - 7x + 10\right)$
$$= 2x(x-5)(x-2)$$

17. $2x - 3(2x-4) = 4 + 4x$
$$2x - 6x + 12 = 4 + 4x$$
$$-4x + 12 = 4 + 4x$$
$$-8x + 12 = 4$$
$$-8x = -8$$
$$x = 1$$

18. $\dfrac{4}{3} = \dfrac{2x+2}{2x-1}$
$$4(2x-1) = 3(2x+2)$$
$$8x - 4 = 6x + 6$$
$$2x - 4 = 6$$
$$2x = 10$$
$$x = 5$$

19. $3x^2 - 7x - 20 = 0$
$$(3x+5)(x-4) = 0$$
$$3x + 5 = 0 \quad \text{or} \quad x - 4 = 0$$
$$3x = -5 \qquad\qquad x = 4$$
$$x = -\frac{5}{3}$$

20. $\dfrac{x}{x-2} - \dfrac{4}{x-1} = \dfrac{2}{x^2 - 3x + 2}$

$\dfrac{x}{x-2} - \dfrac{4}{x-1} = \dfrac{2}{(x-2)(x-1)}$

$(x-2)(x-1)\left(\dfrac{x}{x-2} - \dfrac{4}{x-1}\right)$

$= (x-2)(x-1)\left(\dfrac{2}{(x-2)(x-1)}\right)$

$x(x-1) - 4(x-2) = 2$

$x^2 - x - 4x + 8 = 2$

$x^2 - 5x + 8 = 2$

$x^2 - 5x + 6 = 0$

$(x-3)(x-2) = 0$

$x - 3 = 0 \quad \text{or} \quad x - 2 = 0$

$x = 3 \qquad\qquad x = 2$

In the original equation, note that two rational expressions are undefined when $x = 2$; so 2 is an extraneous solution. Consequently, 3 is the only solution.

21. $0 \le -1.5x - 3 \le 4.5$

$3 \le -1.5x \le 7.5$

$\dfrac{3}{-1.5} \ge \dfrac{-1.5x}{-1.5} \ge \dfrac{7.5}{-1.5}$

$-2 \ge x \ge -5$

$-5 \le x \le -2$

22. $|5x - 2| \ge 8$

$5x - 2 \le -8 \quad \text{or} \quad 5x - 2 \ge 8$

$5x \le -6 \qquad\qquad 5x \ge 10$

$x \le -\dfrac{6}{5} \qquad\qquad x \ge 2$

23. $f(x) = -\dfrac{2}{3}x + 4$

$m = -\dfrac{2}{3},\ (0, 4)$

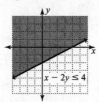

24. $x - 2y \le 4$

$-2y \le -x + 4$

$y \ge \dfrac{1}{2}x - 2$

Begin by graphing the related equation $x - 2y = 4$ with a solid line. Now choose $(0, 0)$ as a test point.

$x - 2y \le 4$

$0 - 2(0) \overset{?}{\le} 4$

$0 - 0 \overset{?}{\le} 4$

$0 \le 4$

Because $(0, 0)$ satisfies the inequality, shade the region which includes $(0, 0)$.

25. Translate "what percent of 90 is 32.4?" word for word. Let p represent the unknown percent.

$p \cdot 90 = 32.4$

$90p = 32.4$

$\dfrac{90p}{90} = \dfrac{32.4}{90}$

$p = 0.36$

To write 0.36 as a percent, multiply by 100%. The percent is $0.36 \cdot 100\% = 36\%$.

26. $f(x) = \dfrac{x+4}{x-2}$

a) $f(0) = \dfrac{0+4}{0-2} = \dfrac{4}{-2} = -2$

b) $f(5) = \dfrac{5+4}{5-2} = \dfrac{9}{3} = 3$

c) $f(2) = \dfrac{2+4}{2-2} = \dfrac{6}{0}$ is undefined

27. Create a table.

	Concentrate	Vol. of solution	Vol. of saline
40%	0.40	x	$0.40x$
15%	0.15	50	$0.15(50)$
30%	0.30	$x+50$	$0.30(x+50)$

Translate the information in the table to an equation and solve.

$$0.40x + 0.15(50) = 0.30(x+50)$$
$$100\big(0.40x + 0.15(50)\big) = 100\big(0.30(x+50)\big)$$
$$40x + 750 = 30(x+50)$$
$$40x + 750 = 30x + 1500$$
$$10x + 750 = 1500$$
$$10x = 750$$
$$x = 75$$

75 ml of the 40% solution should be added.

28. Let t represent the amount of time that the ship spends traveling. The time that the speedboat is traveling is $t-2$ hours. Using the relationship $\text{rate} \times \text{time} = \text{distance}$, complete a table.

Categories	Rate	Time	Distance
ship	15	t	$15t$
speedboat	40	$t-2$	$40(t-2)$

Translate to an equation.

$$15t = 40(t-2)$$
$$15t = 40t - 80$$
$$80 + 15t = 40t$$
$$80 = 25t$$
$$\frac{80}{25} = \frac{25t}{25}$$
$$3.2 = t$$

The speedboat will overtake the ship in 3.2 hours after the ship departs. Therefore, the speedboat will overtake the ship in $3.2 - 2 = 1.2$ or $1\frac{1}{5}$ hours.

29. Translate "y varies jointly as x and the square of z."

$$y = kxz^2 \qquad \text{Now, } y = 3xz^2$$
$$48 = k(4)(2)^2 \qquad y = 3 \cdot 3 \cdot 1^2$$
$$48 = 16k \qquad x = 9$$
$$3 = k$$

30.
$$h = 96t - 16t^2$$
$$80 = 96t - 16t^2$$
$$16t^2 - 96t + 80 = 0$$
$$16\big(t^2 - 6t + 5\big) = 0$$
$$16(t-1)(t-5) = 0$$
$$t - 1 = 0 \quad \text{or} \quad t - 5 = 0$$
$$t = 1 \qquad t = 5$$

The rock will be 80 feet above the ground at 1 second and 5 seconds.

Chapter 9

Rational Exponents, Radicals, and Complex Numbers

Exercise Set 9.1

1. The square roots of 36 are ± 6.

3. The square roots of 121 are ± 11.

5. The square roots of 196 are ± 14.

7. The square roots of 225 are ± 15.

9. $\sqrt{25} = 5$

11. $\sqrt{-64}$ is not a real number.

13. $-\sqrt{25} = -5$

15. $\pm\sqrt{25} = \pm 5$

17. $\sqrt{1.44} = 1.2$

19. $\sqrt{-10.64}$ is not a real number.

21. $-\sqrt{0.0121} = -0.11$

23. $\sqrt{\dfrac{49}{81}} = \dfrac{7}{9}$

25. $-\sqrt{\dfrac{144}{169}} = -\dfrac{12}{13}$

27. $\sqrt[3]{27} = 3$

29. $\sqrt[3]{-64} = -4$

31. $-\sqrt[3]{-216} = -(-6) = 6$

33. $\sqrt[4]{256} = 4$

35. $\sqrt[4]{-625}$ is not a real number.

37. $-\sqrt[4]{16} = -2$

39. $\sqrt[5]{32} = 2$

41. $\sqrt[5]{-243} = -3$

43. $-\sqrt[5]{-32} = -(-2) = 2$

45. $\sqrt[6]{64} = 2$

47. $\sqrt[3]{-\dfrac{8}{27}} = -\dfrac{2}{3}$

49. $\sqrt[4]{\dfrac{16}{81}} = \dfrac{2}{3}$

51. $\sqrt{7} \approx 2.646$

53. $-\sqrt{11} \approx -3.317$

55. $\sqrt[3]{50} \approx 3.684$

57. $\sqrt[3]{-53} \approx -3.756$

59. $\sqrt[4]{189} \approx 3.708$

61. $-\sqrt[4]{85} \approx -3.036$

63. $\sqrt[5]{89} \approx 2.454$

65. $\sqrt[6]{146} \approx 2.295$

67. $\sqrt{b^4} = b^2$

69. $\sqrt{16x^2} = 4x$

71. $\sqrt{100r^8 s^6} = 10r^4 s^3$

73. $\sqrt{0.25a^6 b^{12}} = 0.5a^3 b^6$

75. $\sqrt[3]{m^3} = m$

77. $\sqrt[3]{27a^9 b^6} = 3a^3 b^2$

79. $\sqrt[3]{-64a^3 b^{12}} = -4ab^4$

81. $\sqrt[3]{0.008x^{18}} = 0.2x^6$

83. $\sqrt[4]{a^4} = a$

85. $\sqrt[4]{16x^{16}} = 2x^4$

87. $\sqrt[5]{32x^{10}} = 2x^2$

89. $\sqrt[6]{x^{12} y^6} = x^2 y$

91. $\sqrt{36m^2} = 6|m|$

93. $\sqrt{(r-1)^2} = |r-1|$

95. $\sqrt[4]{256y^{12}} = 4|y^3|$

97. $\sqrt[3]{27y^3} = 3y$

99. $\sqrt{(y-3)^4} = (y-3)^2$

101. $\sqrt[3]{(y-4)^6} = (y-4)^2$

103. $f(x) = \sqrt{2x+4}$

$f(0) = \sqrt{2\cdot 0 + 4}$

$= \sqrt{0+4}$

$= \sqrt{4}$

$= 2$

105. $f(x) = \sqrt{4x+3}$

$f(3) = \sqrt{4\cdot 3 + 3}$

$= \sqrt{12+3}$

$= \sqrt{15}$

107. Because the index is even, the radicand must be nonnegative.

$2x - 8 \geq 0$

$2x \geq 8$

$x \geq 4$

Domain: $\{x \mid x \geq 4\}$, or $[4, \infty)$

109. Because the index is even, the radicand must be nonnegative.

$-4x + 16 \geq 0$

$-4x \geq -16$

$x \leq 4$

Domain: $\{x \mid x \leq 4\}$, or $(-\infty, 4]$

111. a)

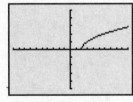

b) $\{x \mid x \geq 2\}$, or $[2, \infty)$

113. a)

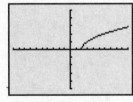

b) $\mathbb{R}$, or $(-\infty, \infty)$

115. $v = -\sqrt{19.6h}$

$v = -\sqrt{19.6(16)}$

$= -\sqrt{313.6}$

≈ -17.709 m/sec.

117. $T = 2\pi\sqrt{\dfrac{L}{9.8}}$

$T = 2\pi\sqrt{\dfrac{3}{9.8}}$

$\approx 2\pi\sqrt{0.306}$

≈ 3.476 sec.

119. $S = \dfrac{7}{2}\sqrt{2D}$

$S = \dfrac{7}{2}\sqrt{2\cdot 15}$

$= \dfrac{7}{2}\sqrt{30}$

$\approx \dfrac{7}{2}\cdot 5.477$

≈ 19.170 mph

121. $c = \sqrt{a^2 + b^2}$

$c = \sqrt{5^2 + 12^2}$

$= \sqrt{25 + 144}$

$= \sqrt{169}$

$= 13$ ft.

123. $R = \sqrt{F_1^2 + F_2^2}$

$R = \sqrt{9^2 + 12^2}$

$= \sqrt{81 + 144}$

$= \sqrt{225}$

$= 15$ N

125. a) $f(x) = 10.5\sqrt{x} + 21$

$f(2) = 10.5\sqrt{2} + 21$

≈ 35.849

≈ 36

Approximately 36 earthquakes with that magnitude occurred in 2010.

b) $f(x) = 10.5\sqrt{x} + 21$

$f(3) = 10.5\sqrt{3} + 21$

≈ 39.187

≈ 39

Approximately 39 earthquakes with that magnitude occurred in 2011.

Review Exercises

1. a) $\sqrt{16\cdot 9} = \sqrt{144}$
    $= 12$
    b) $\sqrt{16}\cdot\sqrt{9} = 4\cdot 3$
    $= 12$

2. $x^5 \cdot x^3 = x^{5+3} = x^8$

3. $\left(-9m^3 n\right)\left(5mn^2\right) = -9\cdot 5m^{3+1}n^{1+2} = -45m^4 n^3$

4. $4x^2\left(3x^2 - 5x + 1\right)$
    $= 4x^2 \cdot 3x^2 + 4x^2 \cdot\left(-5x\right) + 4x^2 \cdot 1$
    $= 12x^4 - 20x^3 + 4x^2$

5. $\left(7y - 4\right)\left(3y + 5\right)$
    $= 7y\cdot 3y + 7y\cdot 5 - 4\cdot 3y - 4\cdot 5$
    $= 21y^2 + 35y - 12y - 20$
    $= 21y^2 + 23y - 20$

6. $A = \left(x - 9\right)\left(2x + 1\right)$
    $= x\cdot 2x + x\cdot 1 - 9\cdot 2x - 9\cdot 1$
    $= 2x^2 + x - 18x - 9$
    $= 2x^2 - 17x - 9$

Exercise Set 9.2

1. $25^{1/2} = \sqrt{25} = 5$

3. $-100^{1/2} = -\sqrt{100} = -10$

5. $27^{1/3} = \sqrt[3]{27} = 3$

7. $\left(-64\right)^{1/3} = \sqrt[3]{-64} = -4$

9. $y^{1/4} = \sqrt[4]{y}$

11. $\left(144x^8\right)^{1/2} = \sqrt{144x^8} = 12x^4$

13. $18r^{1/2} = 18\sqrt{r}$

15. $\left(\dfrac{x^4}{81}\right)^{1/2} = \sqrt{\dfrac{x^4}{81}} = \dfrac{x^2}{9}$

17. $64^{2/3} = \left(\sqrt[3]{64}\right)^2 = \left(4\right)^2 = 16$

19. $-81^{3/4} = -\left(\sqrt[4]{81}\right)^3 = -\left(3\right)^3 = -27$

21. $\left(-8\right)^{4/3} = \left(\sqrt[3]{-8}\right)^4 = \left(-2\right)^4 = 16$

23. $16^{-3/4} = \dfrac{1}{16^{3/4}} = \dfrac{1}{\left(\sqrt[4]{16}\right)^3} = \dfrac{1}{2^3} = \dfrac{1}{8}$

25. $x^{4/5} = \sqrt[5]{x^4}$

27. $8n^{2/3} = 8\sqrt[3]{n^2}$

29. $\left(-32\right)^{-2/5} = \dfrac{1}{\left(-32\right)^{2/5}}$
    $= \dfrac{1}{\left(\sqrt[5]{-32}\right)^2}$
    $= \dfrac{1}{\left(-2\right)^2}$
    $= \dfrac{1}{4}$

31. $\left(\dfrac{1}{25}\right)^{3/2} = \left(\sqrt{\dfrac{1}{25}}\right)^3 = \left(\dfrac{1}{5}\right)^3 = \dfrac{1}{125}$

33. $\left(2a + 4\right)^{5/6} = \sqrt[6]{\left(2a + 4\right)^5}$

35. $\sqrt[4]{25} = 25^{1/4}$

37. $\sqrt[6]{z^5} = z^{5/6}$

39. $\dfrac{1}{\sqrt[6]{5^5}} = \dfrac{1}{5^{5/6}} = 5^{-5/6}$

41. $\dfrac{5}{\sqrt[5]{x^4}} = \dfrac{5}{x^{4/5}} = 5x^{-4/5}$

43. $\left(\sqrt[3]{5}\right)^7 = 5^{7/3}$

45. $\left(\sqrt[7]{x}\right)^2 = x^{2/7}$

47. $\sqrt[4]{\left(4a - 5\right)^7} = \left(4a - 5\right)^{7/4}$

49. $\left(\sqrt[5]{2r - 5}\right)^8 = \left(2r - 5\right)^{8/5}$

51. $x^{1/5} \cdot x^{3/5} = x^{1/5 + 3/5} = x^{4/5}$

53. $x^{3/2} \cdot x^{-1/3} = x^{3/2 + \left(-1/3\right)} = x^{9/6 - 2/6} = x^{7/6}$

55. $a^{2/3} \cdot a^{3/4} = a^{2/3 + 3/4} = a^{8/12 + 9/12} = a^{17/12}$

57. $\left(3w^{1/7}\right)\left(7w^{3/7}\right) = 21w^{1/7 + 3/7} = 21w^{4/7}$

59. $\left(-3a^{2/3}\right)\left(4a^{3/4}\right) = -12a^{2/3+3/4}$
$$= -12a^{8/12+9/12}$$
$$= -12a^{17/12}$$

61. $\dfrac{7^{7/3}}{7^{2/3}} = 7^{7/3-2/3} = 7^{5/3}$

63. $\dfrac{x^{1/6}}{x^{5/6}} = x^{1/6-5/6} = x^{-4/6} = x^{-2/3} = \dfrac{1}{x^{2/3}}$

65. $\dfrac{x^{3/4}}{x^{1/2}} = x^{3/4-1/2} = x^{3/4-2/4} = x^{1/4}$

67. $\dfrac{r^{3/4}}{r^{2/3}} = r^{3/4-2/3} = r^{9/12-8/12} = r^{1/12}$

69. $\dfrac{x^{-3/7}}{x^{2/7}} = x^{-3/7-2/7} = x^{-5/7} = \dfrac{1}{x^{5/7}}$

71. $\dfrac{a^{3/4}}{a^{-3/2}} = a^{3/4-(-3/2)} = a^{3/4+6/4} = a^{9/4}$

73. $\left(5s^{-2/7}\right)\left(4s^{5/7}\right) = 20s^{-2/7+5/7} = 20s^{3/7}$

75. $\left(-6b^{-5/4}\right)\left(4b^{3/2}\right) = -24b^{-5/4+3/2}$
$$= -24b^{-5/4+6/4}$$
$$= -24b^{1/4}$$

77. $\left(x^{2/3}\right)^3 = x^{(2/3)\cdot3} = x^2$

79. $\left(a^{5/6}\right)^2 = a^{(5/6)\cdot2} = a^{10/6} = a^{5/3}$

81. $\left(b^{2/3}\right)^{3/5} = b^{(2/3)\cdot(3/5)} = b^{6/15} = b^{2/5}$

83. $\left(2x^{2/3}y^{1/2}\right)^6 = 2^6 x^{(2/3)\cdot6} y^{(1/2)\cdot6}$
$$= 64x^{12/3} y^{6/2}$$
$$= 64x^4 y^3$$

85. $\left(8q^{3/2}t^{3/4}\right)^{1/3} = 8^{1/3} q^{(3/2)\cdot(1/3)} t^{(3/4)\cdot(1/3)}$
$$= 2q^{3/6} t^{3/12}$$
$$= 2q^{1/2} t^{1/4}$$

87. $\dfrac{\left(3a^{5/4}\right)^4}{a^2} = \dfrac{3^4 a^{(5/4)\cdot4}}{a^2}$
$$= \dfrac{81a^{20/4}}{a^2}$$
$$= \dfrac{81a^5}{a^2}$$
$$= 81a^{5-2}$$
$$= 81a^3$$

89. $\dfrac{\left(9z^{7/3}\right)^{1/2}}{z^{5/6}} = \dfrac{9^{1/2} z^{(7/3)\cdot(1/2)}}{z^{5/6}}$
$$= \dfrac{3z^{7/6}}{z^{5/6}}$$
$$= 3z^{7/6-5/6}$$
$$= 3z^{2/6}$$
$$= 3z^{1/3}$$

91. $\sqrt[4]{4} = 4^{1/4} = \left(2^2\right)^{1/4} = 2^{2\cdot(1/4)} = 2^{1/2} = \sqrt{2}$

93. $\sqrt[6]{49} = \left(7^2\right)^{1/6} = 7^{2\cdot(1/6)} = 7^{1/3} = \sqrt[3]{7}$

95. $\sqrt[4]{x^2} = \left(x^2\right)^{1/4} = x^{2\cdot(1/4)} = x^{1/2} = \sqrt{x}$

97. $\sqrt[8]{r^6} = \left(r^6\right)^{1/8} = r^{3/4} = \sqrt[4]{r^3}$

99. $\sqrt[8]{x^6 y^2} = \left(x^6 y^2\right)^{1/8}$
$$= x^{6\cdot(1/8)} y^{2\cdot(1/8)}$$
$$= x^{3/4} y^{1/4}$$
$$= \left(x^3 y\right)^{1/4}$$
$$= \sqrt[4]{x^3 y}$$

101. $\sqrt[10]{m^4 n^6} = \left(m^4 n^6\right)^{1/10}$
$$= m^{4\cdot(1/10)} n^{6\cdot(1/10)}$$
$$= m^{2/5} n^{3/5}$$
$$= \left(m^2 m^3\right)^{1/5}$$
$$= \sqrt[5]{m^2 n^3}$$

103. $\sqrt[3]{x} \cdot \sqrt{x} = x^{1/3} \cdot x^{1/2}$
$$= x^{1/3+1/2}$$
$$= x^{2/6+3/6}$$
$$= x^{5/6}$$
$$= \sqrt[6]{x^5}$$

105. $\sqrt[4]{y^2} \cdot \sqrt[3]{y^2} = y^{2/4} \cdot y^{2/3}$

$\qquad = y^{1/2+2/3}$

$\qquad = y^{3/6+4/6}$

$\qquad = y^{7/6}$

$\qquad = \sqrt[6]{y^7}$

107. $\dfrac{\sqrt[3]{x^4}}{\sqrt[4]{x^2}} = \dfrac{x^{4/3}}{x^{1/2}} = x^{4/3-1/2} = x^{8/6-3/6} = x^{5/6} = \sqrt[6]{x^5}$

109. $\dfrac{\sqrt[5]{n^4}}{\sqrt[3]{n^2}} = \dfrac{n^{4/5}}{n^{2/3}} = n^{4/5-2/3} = n^{12/15-10/15} = n^{2/15} = \sqrt[15]{n^2}$

111. $\sqrt{5} \cdot \sqrt[3]{3} = 5^{1/2} \cdot 3^{1/3}$

$\qquad = 5^{3/6} \cdot 3^{2/6}$

$\qquad = \left(5^3 \cdot 3^2\right)^{1/6}$

$\qquad = (125 \cdot 9)^{1/6}$

$\qquad = (1125)^{1/6}$

$\qquad = \sqrt[6]{1125}$

113. $\sqrt[4]{6} \cdot \sqrt[3]{2} = 6^{1/4} \cdot 2^{1/3}$

$\qquad = 6^{3/12} \cdot 2^{4/12}$

$\qquad = \left(6^3 \cdot 2^4\right)^{1/12}$

$\qquad = (216 \cdot 16)^{1/12}$

$\qquad = (3456)^{1/12}$

$\qquad = \sqrt[12]{3456}$

115. $\sqrt[3]{\sqrt[3]{x}} = \left(x^{1/3}\right)^{1/3} = x^{(1/3)\cdot(1/3)} = x^{1/9} = \sqrt[9]{x}$

117. $\sqrt{\sqrt[3]{n}} = \left(n^{1/3}\right)^{1/2} = n^{(1/3)\cdot(1/2)} = n^{1/6} = \sqrt[6]{n}$

Review Exercises

1. $2 \cdot 2 \cdot 2 \cdot 2 \cdot x \cdot x \cdot x \cdot y \cdot y = 2^4 x^3 y^2$

2. $\sqrt{16} \cdot \sqrt{9} = 4 \cdot 3 = 12$

3. $\sqrt[3]{27} \cdot \sqrt[3]{125} = 3 \cdot 5 = 15$

4. $(2.5 \times 10^6)(3.2 \times 10^5) = 8 \times 10^{6+5} = 8 \times 10^{11}$

5. $\left(\dfrac{3}{4}x^3 y\right)\left(-\dfrac{5}{6}xyz^2\right) = -\dfrac{5}{8}x^{3+1}y^{1+1}z^2$

$\qquad = -\dfrac{5}{8}x^4 y^2 z^2$

6. $\qquad \begin{array}{r} 2x^2 - 8x + 5 \\ x+3\overline{)2x^3 - 2x^2 - 19x + 18} \end{array}$

$\qquad \underline{2x^3 + 6x^2}$

$\qquad\qquad -8x^2 - 19x$

$\qquad\qquad \underline{-8x^2 - 24x}$

$\qquad\qquad\qquad 5x + 18$

$\qquad\qquad\qquad \underline{5x + 15}$

$\qquad\qquad\qquad\qquad 3$

Answer: $2x^2 - 8x + 5 + \dfrac{3}{x+3}$

Exercise Set 9.3

1. $\sqrt{2} \cdot \sqrt{32} = \sqrt{64} = 8$

3. $\sqrt{3x} \cdot \sqrt{27x^5} = \sqrt{3x \cdot 27x^5} = \sqrt{81x^6} = 9x^3$

5. $\sqrt{6xy^3} \cdot \sqrt{24xy} = \sqrt{6xy^3 \cdot 24xy}$

$\qquad = \sqrt{144x^2 y^4}$

$\qquad = 12xy^2$

7. $\sqrt{5} \cdot \sqrt{13} = \sqrt{5 \cdot 13} = \sqrt{65}$

9. $\sqrt{15} \cdot \sqrt{x} = \sqrt{15 \cdot x} = \sqrt{15x}$

11. $\sqrt[3]{3} \cdot \sqrt[3]{9} = \sqrt[3]{3 \cdot 9} = \sqrt[3]{27} = 3$

13. $\sqrt[3]{5y} \cdot \sqrt[3]{2y} = \sqrt[3]{5y \cdot 2y} = \sqrt[3]{10y^2}$

15. $\sqrt[4]{3} \cdot \sqrt[4]{7} = \sqrt[4]{3 \cdot 7} = \sqrt[4]{21}$

17. $\sqrt[4]{12w^3} \cdot \sqrt[4]{6w} = \sqrt[4]{12w^3 \cdot 6w}$

$\qquad = \sqrt[4]{72w^4}$

$\qquad = \sqrt[4]{w^4 \cdot 72}$

$\qquad = \sqrt[4]{w^4} \cdot \sqrt[4]{72}$

$\qquad = w\sqrt[4]{72}$

19. $\sqrt[4]{3x^2 y} \cdot \sqrt[4]{5xy^2} = \sqrt[4]{3x^2 y \cdot 5xy^2} = \sqrt[4]{15x^3 y^3}$

21. $\sqrt[5]{6x^3} \cdot \sqrt[5]{5x^4} = \sqrt[5]{6x^3 \cdot 5x^4}$

$\qquad = \sqrt[5]{30x^7}$

$\qquad = \sqrt[5]{x^5 \cdot 30x^2}$

$\qquad = \sqrt[5]{x^5} \cdot \sqrt[5]{30x^2}$

$\qquad = x\sqrt[5]{30x^2}$

23. $\sqrt[6]{4x^2 y^3} \cdot \sqrt[6]{2x^3 y} = \sqrt[6]{4x^2 y^3 \cdot 2x^3 y} = \sqrt[6]{8x^5 y^4}$

25. $\sqrt{\dfrac{7}{2}} \cdot \sqrt{\dfrac{3}{5}} = \sqrt{\dfrac{7}{2} \cdot \dfrac{3}{5}} = \sqrt{\dfrac{21}{10}}$

27. $\sqrt{\dfrac{6}{x}} \cdot \sqrt{\dfrac{y}{5}} = \sqrt{\dfrac{6}{x} \cdot \dfrac{y}{5}} = \sqrt{\dfrac{6y}{5x}}$

29. $\sqrt{\dfrac{25}{36}} = \dfrac{\sqrt{25}}{\sqrt{36}} = \dfrac{5}{6}$

31. $\sqrt{\dfrac{10}{9}} = \dfrac{\sqrt{10}}{\sqrt{9}} = \dfrac{\sqrt{10}}{3}$

33. $\dfrac{\sqrt{180}}{\sqrt{5}} = \sqrt{\dfrac{180}{5}} = \sqrt{36} = 6$

35. $\dfrac{\sqrt{15}}{\sqrt{5}} = \sqrt{\dfrac{15}{5}} = \sqrt{3}$

37. $\sqrt[3]{\dfrac{4}{w^6}} = \dfrac{\sqrt[3]{4}}{\sqrt[3]{w^6}} = \dfrac{\sqrt[3]{4}}{w^2}$

39. $\sqrt[3]{\dfrac{5y^2}{27x^9}} = \dfrac{\sqrt[3]{5y^2}}{\sqrt[3]{27x^9}} = \dfrac{\sqrt[3]{5y^2}}{3x^3}$

41. $\dfrac{\sqrt[3]{320}}{\sqrt[3]{5}} = \sqrt[3]{\dfrac{320}{5}} = \sqrt[3]{64} = 4$

43. $\sqrt[4]{\dfrac{3u^3}{16x^8}} = \dfrac{\sqrt[4]{3u^3}}{\sqrt[4]{16x^8}} = \dfrac{\sqrt[4]{3u^3}}{2x^2}$

45. $\sqrt{98} = \sqrt{49 \cdot 2} = \sqrt{49} \cdot \sqrt{2} = 7\sqrt{2}$

47. $\sqrt{128} = \sqrt{64 \cdot 2} = \sqrt{64} \cdot \sqrt{2} = 8\sqrt{2}$

49. $6\sqrt{80} = 6\sqrt{16 \cdot 5} = 6\sqrt{16} \cdot \sqrt{5} = 6 \cdot 4\sqrt{5} = 24\sqrt{5}$

51. $5\sqrt{112} = 5\sqrt{16 \cdot 7} = 5\sqrt{16} \cdot \sqrt{7} = 5 \cdot 4\sqrt{7} = 20\sqrt{7}$

53. $\sqrt{a^7} = \sqrt{a^6 \cdot a} = \sqrt{a^6} \cdot \sqrt{a} = a^3 \sqrt{a}$

55. $\sqrt{x^2 y^4} = xy^2$

57. $\sqrt{x^6 y^8 z^{10}} = x^3 y^4 z^5$

59. $rs^2 \sqrt{r^9 s^5} = rs^2 \sqrt{r^8 s^4 \cdot rs}$
$\qquad = rs^2 \sqrt{r^8 s^4} \cdot \sqrt{rs}$
$\qquad = rs^2 \cdot r^4 s^2 \cdot \sqrt{rs}$
$\qquad = r^5 s^4 \sqrt{rs}$

61. $3\sqrt{72x^5} = 3\sqrt{36x^4 \cdot 2x}$
$\qquad = 3\sqrt{36x^4} \cdot \sqrt{2x}$
$\qquad = 3 \cdot 6x^2 \cdot \sqrt{2x}$
$\qquad = 18x^2 \sqrt{2x}$

63. $\sqrt[3]{32} = \sqrt[3]{8 \cdot 4} = \sqrt[3]{8} \cdot \sqrt[3]{4} = 2\sqrt[3]{4}$

65. $\sqrt[3]{x^7} = \sqrt[3]{x^6 \cdot x} = \sqrt[3]{x^6} \cdot \sqrt[3]{x} = x^2 \sqrt[3]{x}$

67. $\sqrt[3]{x^6 y^5} = \sqrt[3]{x^6 y^3 \cdot y^2} = \sqrt[3]{x^6 y^3} \cdot \sqrt[3]{y^2} = x^2 y \sqrt[3]{y^2}$

69. $\sqrt[3]{128z^8} = \sqrt[3]{64z^6 \cdot 2z^2}$
$\qquad = \sqrt[3]{64z^6} \cdot \sqrt[3]{2z^2}$
$\qquad = 4z^2 \sqrt[3]{2z^2}$

71. $2\sqrt[3]{40} = 2\sqrt[3]{8 \cdot 5} = 2\sqrt[3]{8} \cdot \sqrt[3]{5} = 2 \cdot 2 \cdot \sqrt[3]{5} = 4\sqrt[3]{5}$

73. $\sqrt[4]{80} = \sqrt[4]{16 \cdot 5} = \sqrt[4]{16} \cdot \sqrt[4]{5} = 2\sqrt[4]{5}$

75. $3x^2 \sqrt[4]{243x^9} = 3x^2 \sqrt[4]{81x^8 \cdot 3x}$
$\qquad = 3x^2 \sqrt[4]{81x^8} \cdot \sqrt[4]{3x}$
$\qquad = 3x^2 \cdot 3x^2 \sqrt[4]{3x}$
$\qquad = 9x^4 \sqrt[4]{3x}$

77. $\sqrt[5]{486x^{16}} = \sqrt[5]{243x^{15} \cdot 2x}$
$\qquad = \sqrt[5]{243x^{15}} \cdot \sqrt[5]{2x}$
$\qquad = 3x^3 \sqrt[5]{2x}$

79. $\sqrt[6]{x^8 y^{14} z^{11}} = \sqrt[6]{x^6 y^{12} z^6 \cdot x^2 y^2 z^5}$
$\qquad = \sqrt[6]{x^6 y^{12} z^6} \cdot \sqrt[6]{x^2 y^2 z^5}$
$\qquad = xy^2 z \sqrt[6]{x^2 y^2 z^5}$

81. $\sqrt{3} \cdot \sqrt{21} = \sqrt{63} = \sqrt{9 \cdot 7} = 3\sqrt{7}$

83. $5\sqrt{10} \cdot 3\sqrt{6} = 15\sqrt{60}$
$\qquad = 15\sqrt{4 \cdot 15}$
$\qquad = 15 \cdot 2\sqrt{15}$
$\qquad = 30\sqrt{15}$

85. $\sqrt{y^3} \cdot \sqrt{y^2} = \sqrt{y^5} = \sqrt{y^4 \cdot y} = y^2 \sqrt{y}$

87. $x\sqrt{x^2 y^3} \cdot y^2 \sqrt{x^4 y^4} = xy^2 \sqrt{x^6 y^7}$
$\qquad = xy^2 \sqrt{x^6 y^6 \cdot y}$
$\qquad = xy^2 \cdot x^3 y^3 \sqrt{y}$
$\qquad = x^4 y^5 \sqrt{y}$

89. $4\sqrt{6c^3} \cdot 3\sqrt{10c^5} = 12\sqrt{60c^8}$
$$= 12\sqrt{4c^8 \cdot 15}$$
$$= 12 \cdot 2c^4\sqrt{15}$$
$$= 24c^4\sqrt{15}$$

91. $4\sqrt{3} \cdot 5\sqrt{6} = 20\sqrt{18}$
$$= 20\sqrt{9 \cdot 2}$$
$$= 20 \cdot 3\sqrt{2}$$
$$= 60\sqrt{2}$$

93. $\dfrac{\sqrt{48}}{\sqrt{6}} = \sqrt{\dfrac{48}{6}} = \sqrt{8} = \sqrt{4 \cdot 2} = 2\sqrt{2}$

95. $\dfrac{9\sqrt{160}}{3\sqrt{8}} = 3\sqrt{\dfrac{160}{8}}$
$$= 3\sqrt{20}$$
$$= 3\sqrt{4 \cdot 5}$$
$$= 3 \cdot 2\sqrt{5}$$
$$= 6\sqrt{5}$$

97. $\dfrac{\sqrt{c^5d^6}}{\sqrt{cd^3}} = \sqrt{\dfrac{c^5d^6}{cd^3}} = \sqrt{c^4d^3} = \sqrt{c^4d^2 \cdot d} = c^2d\sqrt{d}$

99. $\dfrac{8\sqrt{45a^5}}{2\sqrt{5a}} = 4\sqrt{\dfrac{45a^5}{5a}}$
$$= 4\sqrt{9a^4}$$
$$= 4 \cdot 3a^2$$
$$= 12a^2$$

101. $\dfrac{12\sqrt{72c^5}}{4\sqrt{6c^2}} = 3\sqrt{\dfrac{72c^5}{6c^2}}$
$$= 3\sqrt{12c^3}$$
$$= 3\sqrt{4c^2 \cdot 3c}$$
$$= 3 \cdot 2c\sqrt{3c}$$
$$= 6c\sqrt{3c}$$

103. $\dfrac{36\sqrt{96x^6y^{11}}}{4\sqrt{3x^2y^4}} = 9\sqrt{\dfrac{96x^6y^{11}}{3x^2y^4}}$
$$= 9\sqrt{32x^4y^7}$$
$$= 9\sqrt{16x^4y^6 \cdot 2y}$$
$$= 9 \cdot 4x^2y^3\sqrt{2y}$$
$$= 36x^2y^3\sqrt{2y}$$

105. $\sqrt{\dfrac{3}{7}} \cdot \sqrt{\dfrac{8}{7}} = \sqrt{\dfrac{24}{49}} = \dfrac{\sqrt{24}}{7} = \dfrac{\sqrt{4 \cdot 6}}{7} = \dfrac{2\sqrt{6}}{7}$

107. $\sqrt{\dfrac{a^3}{2}} \cdot \sqrt{\dfrac{a^5}{2}} = \sqrt{\dfrac{a^8}{4}} = \dfrac{a^4}{2}$

109. $\sqrt{\dfrac{3x^5}{2}} \cdot \sqrt{\dfrac{15x^5}{8}} = \sqrt{\dfrac{45x^{10}}{16}}$
$$= \dfrac{\sqrt{45x^{10}}}{\sqrt{16}}$$
$$= \dfrac{\sqrt{9x^{10} \cdot 5}}{4}$$
$$= \dfrac{3x^5\sqrt{5}}{4}$$

111. $\dfrac{1}{2} \cdot \sqrt{\dfrac{3}{8}} \cdot \sqrt{\dfrac{15}{2}} = \dfrac{1}{2}\sqrt{\dfrac{45}{16}}$
$$= \dfrac{1}{2} \cdot \dfrac{\sqrt{45}}{\sqrt{16}}$$
$$= \dfrac{1}{2} \cdot \dfrac{\sqrt{9 \cdot 5}}{4}$$
$$= \dfrac{1}{2} \cdot \dfrac{3\sqrt{5}}{4}$$
$$= \dfrac{3}{8}\sqrt{5}$$

Review Exercises

1. No   2.   Yes

3. $6x^2 - 4x - 3x + 2x^2 = 8x^2 - 7x$

4. $(2a - 3b)(4a + 3b)$
$$= 2a \cdot 4a + 2a \cdot 3b - 3b \cdot 4a - 3b \cdot 3b$$
$$= 8a^2 + 6ab - 12ab - 9b^2$$
$$= 8a^2 - 6ab - 9b^2$$

5. $(3m + 5n)(3m - 5n) = (3m)^2 - (5n)^2$
$$= 9m^2 - 25n^2$$

6. $(2x - 3y)^2 = (2x)^2 - 2(2x)(3y) + (3y)^2$
$$= 4x^2 - 12xy + 9y^2$$

Exercise Set 9.4

1. $9\sqrt{6} - 15\sqrt{6} = (9 - 15)\sqrt{6} = -6\sqrt{6}$

3. $7\sqrt{a} + 2\sqrt{a} = (7 + 2)\sqrt{a} = 9\sqrt{a}$

5. $4\sqrt{5} - 2\sqrt{6} + 8\sqrt{5} - 6\sqrt{6}$
    $= (4+8)\sqrt{5} + (-2-6)\sqrt{6}$
    $= 12\sqrt{5} - 8\sqrt{6}$

7. $3a\sqrt{5a} - 4b\sqrt{7b} + 8a\sqrt{5a} + 2b\sqrt{7b}$
    $= (3a+8a)\sqrt{5a} + (-4b+2b)\sqrt{7b}$
    $= 11a\sqrt{5a} - 2b\sqrt{7b}$

9. $6x\sqrt[3]{9} - 3x\sqrt[3]{9} = (6x-3x)\sqrt[3]{9} = 3x\sqrt[3]{9}$

11. $6x^2\sqrt[4]{5x} - 12x^2\sqrt[4]{5x} = (6x^2 - 12x^2)\sqrt[4]{5x}$
    $= -6x^2\sqrt[4]{5x}$

13. $3x\sqrt{5x} + 4x\sqrt[3]{5x}$
    Cannot combine because the radicals are not like.

15. $\sqrt{48} - \sqrt{75} = \sqrt{16 \cdot 3} - \sqrt{25 \cdot 3}$
    $= 4\sqrt{3} - 5\sqrt{3}$
    $= -\sqrt{3}$

17. $\sqrt{80y} - \sqrt{125y} = \sqrt{16 \cdot 5y} - \sqrt{25 \cdot 5y}$
    $= 4\sqrt{5y} - 5\sqrt{5y}$
    $= -\sqrt{5y}$

19. $\sqrt{80} - 4\sqrt{45} = \sqrt{16 \cdot 5} - 4\sqrt{9 \cdot 5}$
    $= 4\sqrt{5} - 4 \cdot 3\sqrt{5}$
    $= 4\sqrt{5} - 12\sqrt{5}$
    $= -8\sqrt{5}$

21. $3\sqrt{96} - 2\sqrt{54} = 3\sqrt{16 \cdot 6} - 2\sqrt{9 \cdot 6}$
    $= 3 \cdot 4\sqrt{6} - 2 \cdot 3\sqrt{6}$
    $= 12\sqrt{6} - 6\sqrt{6}$
    $= 6\sqrt{6}$

23. $6\sqrt{48a^3} - 2\sqrt{75a^3}$
    $= 6\sqrt{16a^2 \cdot 3a} - 2\sqrt{25a^2 \cdot 3a}$
    $= 6 \cdot 4a\sqrt{3a} - 2 \cdot 5a\sqrt{3a}$
    $= 24a\sqrt{3a} - 10a\sqrt{3a}$
    $= 14a\sqrt{3a}$

25. $\sqrt{150} - \sqrt{54} + \sqrt{24} = \sqrt{25 \cdot 6} - \sqrt{9 \cdot 6} + \sqrt{4 \cdot 6}$
    $= 5\sqrt{6} - 3\sqrt{6} + 2\sqrt{6}$
    $= 4\sqrt{6}$

27. $2\sqrt{8} - 3\sqrt{48} + 2\sqrt{98} - \sqrt{75}$
    $= 2\sqrt{4 \cdot 2} - 3\sqrt{16 \cdot 3} + 2\sqrt{49 \cdot 2} - \sqrt{25 \cdot 3}$
    $= 2 \cdot 2\sqrt{2} - 3 \cdot 4\sqrt{3} + 2 \cdot 7\sqrt{2} - 5\sqrt{3}$
    $= 4\sqrt{2} - 12\sqrt{3} + 14\sqrt{2} - 5\sqrt{3}$
    $= 18\sqrt{2} - 17\sqrt{3}$

29. $\sqrt[3]{128} + \sqrt[3]{54} = \sqrt[3]{64 \cdot 2} + \sqrt[3]{27 \cdot 2}$
    $= 4\sqrt[3]{2} + 3\sqrt[3]{2}$
    $= 7\sqrt[3]{2}$

31. $4\sqrt[3]{135x^5} - 6x\sqrt[3]{320x^2}$
    $= 4\sqrt[3]{27x^3 \cdot 5x^2} - 6x\sqrt[3]{64 \cdot 5x^2}$
    $= 4 \cdot 3x\sqrt[3]{5x^2} - 6x \cdot 4\sqrt[3]{5x^2}$
    $= 12x\sqrt[3]{5x^2} - 24x\sqrt[3]{5x^2}$
    $= -12x\sqrt[3]{5x^2}$

33. $-4\sqrt[4]{32x^9} + 2x\sqrt[4]{162x^5}$
    $= -4\sqrt[4]{16x^8 \cdot 2x} + 2x\sqrt[4]{81x^4 \cdot 2x}$
    $= -4 \cdot 2x^2\sqrt[4]{2x} + 2x \cdot 3x\sqrt[4]{2x}$
    $= -8x^2\sqrt[4]{2x} + 6x^2\sqrt[4]{2x}$
    $= -2x^2\sqrt[4]{2x}$

35. $\sqrt{2}(3+\sqrt{2}) = \sqrt{2} \cdot 3 + \sqrt{2} \cdot \sqrt{2}$
    $= 3\sqrt{2} + \sqrt{4}$
    $= 3\sqrt{2} + 2$

37. $\sqrt{3}(\sqrt{3} - \sqrt{15}) = \sqrt{3} \cdot \sqrt{3} - \sqrt{3} \cdot \sqrt{15}$
    $= \sqrt{9} - \sqrt{45}$
    $= 3 - \sqrt{9 \cdot 5}$
    $= 3 - 3\sqrt{5}$

39. $\sqrt{5}(\sqrt{6} + 2\sqrt{10}) = \sqrt{5} \cdot \sqrt{6} + \sqrt{5} \cdot 2\sqrt{10}$
    $= \sqrt{30} + 2\sqrt{50}$
    $= \sqrt{30} + 2\sqrt{25 \cdot 2}$
    $= \sqrt{30} + 2 \cdot 5\sqrt{2}$
    $= \sqrt{30} + 10\sqrt{2}$

41. $4\sqrt{3x}\left(2\sqrt{3x}-4\sqrt{6x}\right)$

$= 4\sqrt{3x}\cdot 2\sqrt{3x}-4\sqrt{3x}\cdot 4\sqrt{6x}$

$= 8\sqrt{9x^2}-16\sqrt{18x^2}$

$= 8\cdot 3x-16\sqrt{9x^2\cdot 2}$

$= 24x-16\cdot 3x\sqrt{2}$

$= 24x-48x\sqrt{2}$

43. $\left(3+\sqrt{5}\right)\left(4-\sqrt{2}\right)$

$= 3\cdot 4-3\cdot\sqrt{2}+\sqrt{5}\cdot 4-\sqrt{5}\cdot\sqrt{2}$

$= 12-3\sqrt{2}+4\sqrt{5}-\sqrt{10}$

45. $\left(3+\sqrt{x}\right)\left(2+\sqrt{x}\right)$

$= 3\cdot 2+3\cdot\sqrt{x}+\sqrt{x}\cdot 2+\sqrt{x}\cdot\sqrt{x}$

$= 6+3\sqrt{x}+2\sqrt{x}+\sqrt{x^2}$

$= 6+5\sqrt{x}+x$

47. $\left(2+3\sqrt{3}\right)\left(3+5\sqrt{2}\right)$

$= 2\cdot 3+2\cdot 5\sqrt{2}+3\sqrt{3}\cdot 3+3\sqrt{3}\cdot 5\sqrt{2}$

$= 6+10\sqrt{2}+9\sqrt{3}+15\sqrt{6}$

49. $\left(\sqrt{3}+\sqrt{5}\right)\left(\sqrt{5}+\sqrt{7}\right)$

$= \sqrt{3}\cdot\sqrt{5}+\sqrt{3}\cdot\sqrt{7}+\sqrt{5}\cdot\sqrt{5}+\sqrt{5}\cdot\sqrt{7}$

$= \sqrt{15}+\sqrt{21}+\sqrt{25}+\sqrt{35}$

$= \sqrt{15}+\sqrt{21}+5+\sqrt{35}$

51. $\left(\sqrt{x}+3\sqrt{y}\right)\left(\sqrt{x}-2\sqrt{y}\right)$

$= \sqrt{x}\cdot\sqrt{x}-\sqrt{x}\cdot 2\sqrt{y}+3\sqrt{y}\cdot\sqrt{x}-3\sqrt{y}\cdot 2\sqrt{y}$

$= \sqrt{x^2}-2\sqrt{xy}+3\sqrt{xy}-6\sqrt{y^2}$

$= x+\sqrt{xy}-6y$

53. $\left(4\sqrt{2}+2\sqrt{5}\right)\left(3\sqrt{7}-3\sqrt{3}\right)$

$= 4\sqrt{2}\cdot 3\sqrt{7}-4\sqrt{2}\cdot 3\sqrt{3}+2\sqrt{5}\cdot 3\sqrt{7}-2\sqrt{5}\cdot 3\sqrt{3}$

$= 12\sqrt{14}-12\sqrt{6}+6\sqrt{35}-6\sqrt{15}$

55. $\left(2\sqrt{a}+3\sqrt{b}\right)\left(4\sqrt{a}-\sqrt{b}\right)$

$= 2\sqrt{a}\cdot 4\sqrt{a}-2\sqrt{a}\cdot\sqrt{b}+3\sqrt{b}\cdot 4\sqrt{a}-3\sqrt{b}\cdot\sqrt{b}$

$= 8\sqrt{a^2}-2\sqrt{ab}+12\sqrt{ab}-3\sqrt{b^2}$

$= 8a+10\sqrt{ab}-3b$

57. $\left(\sqrt[3]{4}+5\right)\left(\sqrt[3]{4}-8\right)$

$= \sqrt[3]{4}\cdot\sqrt[3]{4}-\sqrt[3]{4}\cdot 8+5\sqrt[3]{4}-5\cdot 8$

$= \sqrt[3]{16}-8\sqrt[3]{4}+5\sqrt[3]{4}-40$

$= \sqrt[3]{8\cdot 2}-3\sqrt[3]{4}-40$

$= 2\sqrt[3]{2}-3\sqrt[3]{4}-40$

59. $\left(\sqrt[3]{9}+\sqrt[3]{4}\right)\left(\sqrt[3]{3}-\sqrt[3]{2}\right)$

$= \sqrt[3]{9}\cdot\sqrt[3]{3}-\sqrt[3]{9}\cdot\sqrt[3]{2}+\sqrt[3]{4}\cdot\sqrt[3]{3}-\sqrt[3]{4}\cdot\sqrt[3]{2}$

$= \sqrt[3]{27}-\sqrt[3]{18}+\sqrt[3]{12}-\sqrt[3]{8}$

$= 3-\sqrt[3]{18}+\sqrt[3]{12}-2$

$= 1-\sqrt[3]{18}+\sqrt[3]{12}$

61. $\left(\sqrt[3]{x}+2\right)\left(\sqrt[3]{x^2}-2\sqrt[3]{x}+4\right)$

$= \sqrt[3]{x}\cdot\sqrt[3]{x^2}-\sqrt[3]{x}\cdot 2\sqrt[3]{x}+\sqrt[3]{x}\cdot 4+2\sqrt[3]{x^2}$

$\quad -2\cdot 2\sqrt[3]{x}+2\cdot 4$

$= \sqrt[3]{x^3}-2\sqrt[3]{x^2}+4\sqrt[3]{x}+2\sqrt[3]{x^2}$

$\quad -4\sqrt[3]{x}+8$

$= x-2\sqrt[3]{x^2}+4\sqrt[3]{x}+2\sqrt[3]{x^2}-4\sqrt[3]{x}+8$

$= x+8$

63. $\left(4+\sqrt{6}\right)^2 = 4^2+2\cdot 4\sqrt{6}+\left(\sqrt{6}\right)^2$

$= 16+8\sqrt{6}+6$

$= 22+8\sqrt{6}$

65. $\left(4-\sqrt{2}\right)^2 = 4^2-2\cdot 4\sqrt{2}+\left(\sqrt{2}\right)^2$

$= 16-8\sqrt{2}+2$

$= 18-8\sqrt{2}$

67. $\left(2+2\sqrt{3}\right)^2 = 2^2+2\cdot 4\sqrt{3}+\left(2\sqrt{3}\right)^2$

$= 4+8\sqrt{3}+4\cdot 3$

$= 4+8\sqrt{3}+12$

$= 16+8\sqrt{3}$

69. $\left(2\sqrt{3}+3\sqrt{2}\right)^2 = \left(2\sqrt{3}\right)^2+2\cdot 6\sqrt{6}+\left(3\sqrt{2}\right)^2$

$= 4\cdot 3+12\sqrt{6}+9\cdot 2$

$= 12+12\sqrt{6}+18$

$= 30+12\sqrt{6}$

71. $\left(4+\sqrt{3}\right)\left(4-\sqrt{3}\right) = 4^2-\left(\sqrt{3}\right)^2 = 16-3 = 13$

73. $\left(\sqrt{2}+4\right)\left(\sqrt{2}-4\right)=\left(\sqrt{2}\right)^{2}-4^{2}=2-16=-14$

75. $\left(6+\sqrt{x}\right)\left(6-\sqrt{x}\right)=6^{2}-\left(\sqrt{x}\right)^{2}=36-x$

77. $\left(\sqrt{3}+\sqrt{2}\right)\left(\sqrt{3}-\sqrt{2}\right)=\left(\sqrt{3}\right)^{2}-\left(\sqrt{2}\right)^{2}=3-2=1$

79. $\left(\sqrt{x}+\sqrt{y}\right)\left(\sqrt{x}-\sqrt{y}\right)=\left(\sqrt{x}\right)^{2}-\left(\sqrt{y}\right)^{2}=x-y$

81. $\left(4+2\sqrt{3}\right)\left(4-2\sqrt{3}\right)=4^{2}-\left(2\sqrt{3}\right)^{2}$
$$=16-4\cdot3$$
$$=16-12$$
$$=4$$

83. $\left(3\sqrt{7}+\sqrt{13}\right)\left(3\sqrt{7}-\sqrt{13}\right)=\left(3\sqrt{7}\right)^{2}-\left(\sqrt{13}\right)^{2}$
$$=9\cdot7-13$$
$$=63-13$$
$$=50$$

85. $\sqrt{3}\cdot\sqrt{15}+\sqrt{8}\cdot\sqrt{10}=\sqrt{3\cdot15}+\sqrt{8\cdot10}$
$$=\sqrt{45}+\sqrt{80}$$
$$=\sqrt{9\cdot5}+\sqrt{16\cdot5}$$
$$=3\sqrt{5}+4\sqrt{5}$$
$$=7\sqrt{5}$$

87. $3\sqrt{3}\cdot\sqrt{18}-4\sqrt{18}\cdot\sqrt{12}=3\sqrt{3\cdot18}-4\sqrt{18\cdot12}$
$$=3\sqrt{54}-4\sqrt{216}$$
$$=3\sqrt{9\cdot6}-4\sqrt{36\cdot6}$$
$$=3\cdot3\sqrt{6}-4\cdot6\sqrt{6}$$
$$=9\sqrt{6}-24\sqrt{6}$$
$$=-15\sqrt{6}$$

89. $\dfrac{\sqrt{54}}{\sqrt{3}}+\sqrt{72}=\sqrt{\dfrac{54}{3}}+\sqrt{72}$
$$=\sqrt{18}+\sqrt{36\cdot2}$$
$$=\sqrt{9\cdot2}+6\sqrt{2}$$
$$=3\sqrt{2}+6\sqrt{2}$$
$$=9\sqrt{2}$$

91. $\dfrac{\sqrt{540}}{\sqrt{3}}-4\sqrt{125}=\sqrt{\dfrac{540}{3}}-4\sqrt{125}$
$$=\sqrt{180}-4\sqrt{125}$$
$$=\sqrt{36\cdot5}-4\sqrt{25\cdot5}$$
$$=6\sqrt{5}-4\cdot5\sqrt{5}$$
$$=6\sqrt{5}-20\sqrt{5}$$
$$=-14\sqrt{5}$$

93. $5\sqrt{3}+4\sqrt{12}+4\sqrt{27}$
$$=5\sqrt{3}+4\sqrt{4\cdot3}+4\sqrt{9\cdot3}$$
$$=5\sqrt{3}+4\cdot2\sqrt{3}+4\cdot3\sqrt{3}$$
$$=5\sqrt{3}+8\sqrt{3}+12\sqrt{3}$$
$$=25\sqrt{3}$$

95. a) $13+10+\sqrt{5}+9+\sqrt{5}+10=\left(42+2\sqrt{5}\right)$ ft.

b) 46.5 ft. c) $46.5(\$1.89)=\$87.89$

Review Exercises

1. $2x+5$

2. $(4x+3)(4x-3)=(4x)^{2}-3^{2}=16x^{2}-9$

3. $\sqrt{2}$ because $\sqrt{8}\cdot\sqrt{2}=\sqrt{8\cdot2}=\sqrt{16}$

4. $\sqrt[3]{4}$ because $\sqrt[3]{2}\cdot\sqrt[3]{4}=\sqrt[3]{2\cdot4}=\sqrt[3]{8}$

5. $5y-2x=10$
$$5y=2x+10$$
$$\dfrac{5y}{5}=\dfrac{2x+10}{5}$$
$$y=\dfrac{2}{5}x+2$$

slope is $\dfrac{2}{5}$; y-intercept is $(0,2)$

6. $y=-\dfrac{1}{3}x+4$

Exercise Set 9.5

1. $\dfrac{1}{\sqrt{3}} = \dfrac{1}{\sqrt{3}} \cdot \dfrac{\sqrt{3}}{\sqrt{3}} = \dfrac{\sqrt{3}}{\sqrt{9}} = \dfrac{\sqrt{3}}{3}$

3. $\dfrac{3}{\sqrt{8}} = \dfrac{3}{\sqrt{8}} \cdot \dfrac{\sqrt{2}}{\sqrt{2}} = \dfrac{3\sqrt{2}}{\sqrt{16}} = \dfrac{3\sqrt{2}}{4}$

5. $\sqrt{\dfrac{36}{7}} = \dfrac{\sqrt{36}}{\sqrt{7}} = \dfrac{6}{\sqrt{7}} \cdot \dfrac{\sqrt{7}}{\sqrt{7}} = \dfrac{6\sqrt{7}}{\sqrt{49}} = \dfrac{6\sqrt{7}}{7}$

7. $\sqrt{\dfrac{5}{12}} = \dfrac{\sqrt{5}}{\sqrt{12}} = \dfrac{\sqrt{5}}{\sqrt{12}} \cdot \dfrac{\sqrt{3}}{\sqrt{3}} = \dfrac{\sqrt{15}}{\sqrt{36}} = \dfrac{\sqrt{15}}{6}$

9. $\dfrac{\sqrt{7x^2}}{\sqrt{50}} = \dfrac{\sqrt{x^2 \cdot 7}}{\sqrt{25 \cdot 2}} = \dfrac{x\sqrt{7}}{5\sqrt{2}} \cdot \dfrac{\sqrt{2}}{\sqrt{2}}$

$= \dfrac{x\sqrt{14}}{5\sqrt{4}} = \dfrac{x\sqrt{14}}{5 \cdot 2} = \dfrac{x\sqrt{14}}{10}$

11. $\dfrac{\sqrt{8}}{\sqrt{56}} = \sqrt{\dfrac{8}{56}} = \sqrt{\dfrac{1}{7}} = \dfrac{\sqrt{1}}{\sqrt{7}} = \dfrac{1}{\sqrt{7}} \cdot \dfrac{\sqrt{7}}{\sqrt{7}}$

$= \dfrac{\sqrt{7}}{\sqrt{49}} = \dfrac{\sqrt{7}}{7}$

13. $\dfrac{5}{\sqrt{3a}} = \dfrac{5}{\sqrt{3a}} \cdot \dfrac{\sqrt{3a}}{\sqrt{3a}} = \dfrac{5\sqrt{3a}}{\sqrt{9a^2}} = \dfrac{5\sqrt{3a}}{3a}$

15. $\sqrt{\dfrac{3m}{11n}} = \dfrac{\sqrt{3m}}{\sqrt{11n}} = \dfrac{\sqrt{3m}}{\sqrt{11n}} \cdot \dfrac{\sqrt{11n}}{\sqrt{11n}} = \dfrac{\sqrt{33mn}}{\sqrt{121n^2}} = \dfrac{\sqrt{33mn}}{11n}$

17. $\dfrac{10}{\sqrt{5x}} = \dfrac{10}{\sqrt{5x}} \cdot \dfrac{\sqrt{5x}}{\sqrt{5x}} = \dfrac{10\sqrt{5x}}{\sqrt{25x^2}} = \dfrac{10\sqrt{5x}}{5x} = \dfrac{2\sqrt{5x}}{x}$

19. $\dfrac{\sqrt{6x}}{\sqrt{32x}} = \sqrt{\dfrac{6x}{32x}} = \sqrt{\dfrac{3}{16}} = \dfrac{\sqrt{3}}{\sqrt{16}} = \dfrac{\sqrt{3}}{4}$

21. $\dfrac{3}{\sqrt{x^3}} = \dfrac{3}{\sqrt{x^2 \cdot x}}$

$= \dfrac{3}{x\sqrt{x}} \cdot \dfrac{\sqrt{x}}{\sqrt{x}}$

$= \dfrac{3\sqrt{x}}{x\sqrt{x^2}}$

$= \dfrac{3\sqrt{x}}{x \cdot x}$

$= \dfrac{3\sqrt{x}}{x^2}$

23. $\dfrac{8x^2}{\sqrt{2x}} = \dfrac{8x^2}{\sqrt{2x}} \cdot \dfrac{\sqrt{2x}}{\sqrt{2x}}$

$= \dfrac{8x^2\sqrt{2x}}{\sqrt{4x^2}}$

$= \dfrac{8x^2\sqrt{2x}}{2x}$

$= 4x\sqrt{2x}$

25. Mistake: The product of $\sqrt{2}$ and 2 is not 2.

Correct: $\dfrac{\sqrt{3}}{\sqrt{2}} \cdot \dfrac{\sqrt{2}}{\sqrt{2}} = \dfrac{\sqrt{6}}{\sqrt{4}} = \dfrac{\sqrt{6}}{2}$

27. $\dfrac{5}{\sqrt[3]{3}} = \dfrac{5}{\sqrt[3]{3}} \cdot \dfrac{\sqrt[3]{9}}{\sqrt[3]{9}} = \dfrac{5\sqrt[3]{9}}{\sqrt[3]{27}} = \dfrac{5\sqrt[3]{9}}{3}$

29. $\sqrt[3]{\dfrac{5}{2}} = \dfrac{\sqrt[3]{5}}{\sqrt[3]{2}} \cdot \dfrac{\sqrt[3]{4}}{\sqrt[3]{4}} = \dfrac{\sqrt[3]{20}}{\sqrt[3]{8}} = \dfrac{\sqrt[3]{20}}{2}$

31. $\dfrac{6}{\sqrt[3]{4}} = \dfrac{6}{\sqrt[3]{4}} \cdot \dfrac{\sqrt[3]{2}}{\sqrt[3]{2}} = \dfrac{6\sqrt[3]{2}}{\sqrt[3]{8}} = \dfrac{6\sqrt[3]{2}}{2} = 3\sqrt[3]{2}$

33. $\dfrac{m}{\sqrt[3]{n}} = \dfrac{m}{\sqrt[3]{n}} \cdot \dfrac{\sqrt[3]{n^2}}{\sqrt[3]{n^2}} = \dfrac{m\sqrt[3]{n^2}}{\sqrt[3]{n^3}} = \dfrac{m\sqrt[3]{n^2}}{n}$

35. $\sqrt[3]{\dfrac{a}{b^2}} = \dfrac{\sqrt[3]{a}}{\sqrt[3]{b^2}} \cdot \dfrac{\sqrt[3]{b}}{\sqrt[3]{b}} = \dfrac{\sqrt[3]{ab}}{\sqrt[3]{b^3}} = \dfrac{\sqrt[3]{ab}}{b}$

37. $\dfrac{4}{\sqrt[3]{2x}} = \dfrac{4}{\sqrt[3]{2x}} \cdot \dfrac{\sqrt[3]{4x^2}}{\sqrt[3]{4x^2}}$

$= \dfrac{4\sqrt[3]{4x^2}}{\sqrt[3]{8x^3}}$

$= \dfrac{4\sqrt[3]{4x^2}}{2x}$

$= \dfrac{2\sqrt[3]{4x^2}}{x}$

39. $\sqrt[3]{\dfrac{6}{25a^2}} = \dfrac{\sqrt[3]{6}}{\sqrt[3]{25a^2}} \cdot \dfrac{\sqrt[3]{5a}}{\sqrt[3]{5a}} = \dfrac{\sqrt[3]{30a}}{\sqrt[3]{125a^3}} = \dfrac{\sqrt[3]{30a}}{5a}$

41. $\dfrac{5}{\sqrt[4]{4}} = \dfrac{5}{\sqrt[4]{4}} \cdot \dfrac{\sqrt[4]{4}}{\sqrt[4]{4}} = \dfrac{5\sqrt[4]{4}}{\sqrt[4]{16}} = \dfrac{5\sqrt[4]{4}}{2} = \dfrac{5\sqrt[4]{2^2}}{2} = \dfrac{5\sqrt{2}}{2}$

43. $\sqrt[4]{\dfrac{3}{x^2}} = \dfrac{\sqrt[4]{3}}{\sqrt[4]{x^2}} \cdot \dfrac{\sqrt[4]{x^2}}{\sqrt[4]{x^2}} = \dfrac{\sqrt[4]{3x^2}}{\sqrt[4]{x^4}} = \dfrac{\sqrt[4]{3x^2}}{x}$

45. $\dfrac{9}{\sqrt[4]{3x^3}} = \dfrac{9}{\sqrt[4]{3x^3}} \cdot \dfrac{\sqrt[4]{27x}}{\sqrt[4]{27x}}$

$\qquad = \dfrac{9\sqrt[4]{27x}}{\sqrt[4]{81x^4}}$

$\qquad = \dfrac{9\sqrt[4]{27x}}{3x}$

$\qquad = \dfrac{3\sqrt[4]{27x}}{x}$

47. $\dfrac{3}{\sqrt{2}+1} \cdot \dfrac{\sqrt{2}-1}{\sqrt{2}-1} = \dfrac{3\left(\sqrt{2}-1\right)}{\left(\sqrt{2}\right)^2 - 1^2}$

$\qquad = \dfrac{3\sqrt{2}-3}{2-1}$

$\qquad = \dfrac{3\sqrt{2}-3}{1}$

$\qquad = 3\sqrt{2}-3$

49. $\dfrac{4}{2-\sqrt{3}} \cdot \dfrac{2+\sqrt{3}}{2+\sqrt{3}} = \dfrac{4\left(2+\sqrt{3}\right)}{2^2 - \left(\sqrt{3}\right)^2}$

$\qquad = \dfrac{8+4\sqrt{3}}{4-3}$

$\qquad = \dfrac{8+4\sqrt{3}}{1}$

$\qquad = 8+4\sqrt{3}$

51. $\dfrac{5}{\sqrt{2}+\sqrt{3}} \cdot \dfrac{\sqrt{2}-\sqrt{3}}{\sqrt{2}-\sqrt{3}} = \dfrac{5\left(\sqrt{2}-\sqrt{3}\right)}{\left(\sqrt{2}\right)^2 - \left(\sqrt{3}\right)^2}$

$\qquad = \dfrac{5\sqrt{2}-5\sqrt{3}}{2-3}$

$\qquad = \dfrac{5\sqrt{2}-5\sqrt{3}}{-1}$

$\qquad = 5\sqrt{3}-5\sqrt{2}$

53. $\dfrac{4}{1-\sqrt{5}} \cdot \dfrac{1+\sqrt{5}}{1+\sqrt{5}} = \dfrac{4\left(1+\sqrt{5}\right)}{1^2 - \left(\sqrt{5}\right)^2}$

$\qquad = \dfrac{4\left(1+\sqrt{5}\right)}{1-5}$

$\qquad = \dfrac{4\left(1+\sqrt{5}\right)}{-4}$

$\qquad = -1\left(1+\sqrt{5}\right)$

$\qquad = -1-\sqrt{5}$

55. $\dfrac{\sqrt{3}}{\sqrt{3}-1} \cdot \dfrac{\sqrt{3}+1}{\sqrt{3}+1} = \dfrac{\sqrt{3}\left(\sqrt{3}+1\right)}{\left(\sqrt{3}\right)^2 - 1^2}$

$\qquad = \dfrac{3+\sqrt{3}}{3-1}$

$\qquad = \dfrac{3+\sqrt{3}}{2}$

57. $\dfrac{2\sqrt{3}}{\sqrt{3}-4} \cdot \dfrac{\sqrt{3}+4}{\sqrt{3}+4} = \dfrac{2\sqrt{3}\left(\sqrt{3}+4\right)}{\left(\sqrt{3}\right)^2 - 4^2}$

$\qquad = \dfrac{2\cdot3+8\sqrt{3}}{3-16}$

$\qquad = \dfrac{6+8\sqrt{3}}{-13}$

$\qquad = \dfrac{-6-8\sqrt{3}}{13}$

59. $\dfrac{4\sqrt{3}}{\sqrt{7}+\sqrt{2}} \cdot \dfrac{\sqrt{7}-\sqrt{2}}{\sqrt{7}-\sqrt{2}} = \dfrac{4\sqrt{3}\left(\sqrt{7}-\sqrt{2}\right)}{\left(\sqrt{7}\right)^2 - \left(\sqrt{2}\right)^2}$

$\qquad = \dfrac{4\sqrt{21}-4\sqrt{6}}{7-2}$

$\qquad = \dfrac{4\sqrt{21}-4\sqrt{6}}{5}$

61. $\dfrac{8\sqrt{2}}{4\sqrt{2}-\sqrt{6}} \cdot \dfrac{4\sqrt{2}+\sqrt{6}}{4\sqrt{2}+\sqrt{6}} = \dfrac{32\sqrt{4}+8\sqrt{12}}{16\cdot2-6}$

$\qquad = \dfrac{32\cdot2+8\sqrt{4\cdot3}}{32-6}$

$\qquad = \dfrac{64+8\cdot2\sqrt{3}}{26}$

$\qquad = \dfrac{64+16\sqrt{3}}{26}$

$\qquad = \dfrac{\cancel{2}\left(32+8\sqrt{3}\right)}{\cancel{2}\cdot13}$

$\qquad = \dfrac{32+8\sqrt{3}}{13}$

63. $\dfrac{6\sqrt{y}}{\sqrt{y}+1} \cdot \dfrac{\sqrt{y}-1}{\sqrt{y}-1} = \dfrac{6\sqrt{y}\left(\sqrt{y}-1\right)}{\left(\sqrt{y}\right)^2 - 1^2} = \dfrac{6y-6\sqrt{y}}{y-1}$

65. $\dfrac{3\sqrt{t}}{\sqrt{t}+2\sqrt{u}} \cdot \dfrac{\sqrt{t}-2\sqrt{u}}{\sqrt{t}-2\sqrt{u}} = \dfrac{3\sqrt{t}\left(\sqrt{t}-2\sqrt{u}\right)}{\left(\sqrt{t}\right)^2-\left(2\sqrt{u}\right)^2}$

$\qquad = \dfrac{3t-6\sqrt{tu}}{t-4u}$

67. $\dfrac{\sqrt{2y}}{\sqrt{x}-\sqrt{6y}} \cdot \dfrac{\sqrt{x}+\sqrt{6y}}{\sqrt{x}+\sqrt{6y}} = \dfrac{\sqrt{2y}\left(\sqrt{x}+\sqrt{6y}\right)}{\left(\sqrt{x}\right)^2-\left(\sqrt{6y}\right)^2}$

$\qquad = \dfrac{\sqrt{2xy}+\sqrt{12y^2}}{x-6y}$

$\qquad = \dfrac{\sqrt{2xy}+\sqrt{4y^2 \cdot 3}}{x-6y}$

$\qquad = \dfrac{\sqrt{2xy}+2y\sqrt{3}}{x-6y}$

69. $\dfrac{\sqrt{3}}{2} \cdot \dfrac{\sqrt{3}}{\sqrt{3}} = \dfrac{\sqrt{9}}{2\sqrt{3}} = \dfrac{3}{2\sqrt{3}}$

71. $\dfrac{\sqrt{2x}}{5} \cdot \dfrac{\sqrt{2x}}{\sqrt{2x}} = \dfrac{\sqrt{4x^2}}{5\sqrt{2x}} = \dfrac{2x}{5\sqrt{2x}}$

73. $\dfrac{\sqrt{8n}}{6} = \dfrac{\sqrt{4 \cdot 2n}}{6} = \dfrac{2\sqrt{2n}}{6}$

$\qquad = \dfrac{\sqrt{2n}}{3} \cdot \dfrac{\sqrt{2n}}{\sqrt{2n}}$

$\qquad = \dfrac{\sqrt{4n^2}}{3\sqrt{2n}}$

$\qquad = \dfrac{2n}{3\sqrt{2n}}$

75. $\dfrac{2+\sqrt{3}}{5} \cdot \dfrac{2-\sqrt{3}}{2-\sqrt{3}} = \dfrac{2^2-\left(\sqrt{3}\right)^2}{5\left(2-\sqrt{3}\right)}$

$\qquad = \dfrac{4-3}{10-5\sqrt{3}}$

$\qquad = \dfrac{1}{10-5\sqrt{3}}$

77. $\dfrac{\sqrt{5x}-6}{9} \cdot \dfrac{\sqrt{5x}+6}{\sqrt{5x}+6} = \dfrac{\left(\sqrt{5x}\right)^2-6^2}{9\left(\sqrt{5x}+6\right)}$

$\qquad = \dfrac{5x-36}{9\sqrt{5x}+54}$

79. $\dfrac{5\sqrt{n}+\sqrt{6n}}{2n} \cdot \dfrac{5\sqrt{n}-\sqrt{6n}}{5\sqrt{n}-\sqrt{6n}} = \dfrac{\left(5\sqrt{n}\right)^2-\left(\sqrt{6n}\right)^2}{2n\left(5\sqrt{n}-\sqrt{6n}\right)}$

$\qquad = \dfrac{25n-6n}{10n\sqrt{n}-2n\sqrt{6n}}$

$\qquad = \dfrac{19n}{10n\sqrt{n}-2n\sqrt{6n}}$

$\qquad = \dfrac{19\cancel{n}}{\cancel{n}\left(10\sqrt{n}-2\sqrt{6n}\right)}$

$\qquad = \dfrac{19}{10\sqrt{n}-2\sqrt{6n}}$

81. $f(x) = \dfrac{5\sqrt{2}}{x}$

a) $f\left(\sqrt{6}\right) = \dfrac{5\sqrt{2}}{\sqrt{6}} = 5\sqrt{\dfrac{2}{6}} = 5\sqrt{\dfrac{1}{3}}$

$\qquad = \dfrac{5}{\sqrt{3}} \cdot \dfrac{\sqrt{3}}{\sqrt{3}} = \dfrac{5\sqrt{3}}{3}$

b) $f\left(\sqrt{10}\right) = \dfrac{5\sqrt{2}}{\sqrt{10}} = 5\sqrt{\dfrac{2}{10}} = 5\sqrt{\dfrac{1}{5}}$

$\qquad = \dfrac{5}{\sqrt{5}} \cdot \dfrac{\sqrt{5}}{\sqrt{5}} = \dfrac{5\sqrt{5}}{\sqrt{25}} = \dfrac{5\sqrt{5}}{5}$

$\qquad = \sqrt{5}$

c) $f\left(\sqrt{22}\right) = \dfrac{5\sqrt{2}}{\sqrt{22}} = 5\sqrt{\dfrac{2}{22}} = 5\sqrt{\dfrac{1}{11}}$

$\qquad = \dfrac{5}{\sqrt{11}} \cdot \dfrac{\sqrt{11}}{\sqrt{11}} = \dfrac{5\sqrt{11}}{\sqrt{121}}$

$\qquad = \dfrac{5\sqrt{11}}{11}$

83. a) The graphs are identical. The functions are identical.

b) $f(x) = g(x)$

85. a) $T = \dfrac{2\pi\sqrt{L}}{\sqrt{9.8}} \cdot \dfrac{\sqrt{9.8}}{\sqrt{9.8}}$

$\qquad = \dfrac{2\pi\sqrt{9.8L}}{9.8}$

$\qquad = \dfrac{\pi\sqrt{9.8L}}{4.9}$

b) $T = \dfrac{2\pi\sqrt{L}}{\sqrt{9.8}} \cdot \dfrac{\sqrt{L}}{\sqrt{L}}$

$\qquad = \dfrac{2\pi L}{\sqrt{9.8L}}$

87. a) $s = \dfrac{\sqrt{3V}}{\sqrt{h}} \cdot \dfrac{\sqrt{h}}{\sqrt{h}} = \dfrac{\sqrt{3Vh}}{\sqrt{h^2}} = \dfrac{\sqrt{3Vh}}{h}$

b) $s = \dfrac{\sqrt{3(83,068,742)449}}{449} \approx 745$ ft.

89. a) $V_{rms} = \dfrac{V_m}{\sqrt{2}} \cdot \dfrac{\sqrt{2}}{\sqrt{2}} = \dfrac{\sqrt{2}\,V_m}{\sqrt{4}} = \dfrac{\sqrt{2}\,V_m}{2}$

b) $V_{rms} = \dfrac{163\sqrt{2}}{2}$ **c)** $V_{rms} \approx 115.3$

91. $\dfrac{5\sqrt{2}}{3+\sqrt{6}} \cdot \dfrac{3-\sqrt{6}}{3-\sqrt{6}} = \dfrac{5\sqrt{2}\left(3-\sqrt{6}\right)}{3^2 - \left(\sqrt{6}\right)^2}$

$\qquad = \dfrac{15\sqrt{2} - 5\sqrt{12}}{9-6}$

$\qquad = \dfrac{15\sqrt{2} - 5\cdot 2\sqrt{3}}{3}$

$\qquad = \dfrac{15\sqrt{2} - 10\sqrt{3}}{3}\,\Omega$

Review Exercises

1. $\pm\sqrt{28} = \pm\sqrt{4\cdot 7} = \pm 2\sqrt{7}$

2. $x^2 - 6x + 9 = (x-3)(x-3) = (x-3)^2$

3. $2x - 3 = 5$ **4.** $2x - 3 = -5$
$\quad\; 2x = 8$ $\qquad\qquad\; 2x = -2$
$\qquad x = 4$ $\qquad\qquad\;\; x = -1$

5. $\qquad x^2 - 36 = 0$
$(x+6)(x-6) = 0$
$x + 6 = 0 \quad$ or $\quad x - 6 = 0$
$\quad x = -6 \qquad\qquad x = 6$

6. $\qquad x^2 - 5x + 6 = 0$
$(x-2)(x-3) = 0$
$\qquad x - 2 = 0 \quad$ or $\quad x - 3 = 0$
$\qquad\qquad x = 2 \qquad\qquad x = 3$

Exercise Set 9.6

1. $\qquad \sqrt{x} = 2$
$\quad \left(\sqrt{x}\right)^2 = 2^2$
$\qquad\quad x = 4$

3. $\sqrt{k} = -4$ has no real-number solution

5. $\qquad \sqrt[3]{y} = 3$
$\quad \left(\sqrt[3]{y}\right)^3 = 3^3$
$\qquad\quad y = 27$

7. $\qquad \sqrt[3]{z} = -2$
$\quad \left(\sqrt[3]{z}\right)^3 = (-2)^3$
$\qquad\quad z = -8$

9. $\qquad \sqrt{n-1} = 4$
$\quad \left(\sqrt{n-1}\right)^2 = 4^2$
$\qquad\quad n - 1 = 16$
$\qquad\qquad n = 17$

11. $\qquad \sqrt{t+5} = 4$
$\quad \left(\sqrt{t+5}\right)^2 = 4^2$
$\qquad\quad t + 5 = 16$
$\qquad\qquad t = 11$

13. $\qquad \sqrt{3x-2} = 4$
$\quad \left(\sqrt{3x-2}\right)^2 = 4^2$
$\qquad\quad 3x - 2 = 16$
$\qquad\qquad 3x = 18$
$\qquad\qquad x = 6$

15. $\qquad \sqrt{2x+24} = 4$
$\quad \left(\sqrt{2x+24}\right)^2 = 4^2$
$\qquad\quad 2x + 24 = 16$
$\qquad\qquad 2x = -8$
$\qquad\qquad x = -4$

17. $\sqrt{2n-8} = -3$ has no real-number solution.

19. $\qquad \sqrt[3]{x-3} = 2$
$\quad \left(\sqrt[3]{x-3}\right)^3 = 2^3$
$\qquad\quad x - 3 = 8$
$\qquad\qquad x = 11$

21. $\qquad \sqrt[3]{3y-2} = -2$
$\quad \left(\sqrt[3]{3y-2}\right)^3 = (-2)^3$
$\qquad\quad 3y - 2 = -8$
$\qquad\qquad 3y = -6$
$\qquad\qquad y = -2$

23. $\sqrt{u-3} - 10 = 1$
$\qquad \sqrt{u-3} = 11$
$\quad \left(\sqrt{u-3}\right)^2 = 11^2$
$\qquad\quad u - 3 = 121$
$\qquad\qquad u = 124$

25. $\sqrt{y-6} + 2 = 9$
$\qquad \sqrt{y-6} = 7$
$\quad \left(\sqrt{y-6}\right)^2 = 7^2$
$\qquad\quad y - 6 = 49$
$\qquad\qquad y = 55$

27. $\sqrt{6x-5} - 2 = 3$
$\qquad \sqrt{6x-5} = 5$
$\quad \left(\sqrt{6x-5}\right)^2 = 5^2$
$\qquad\quad 6x - 5 = 25$
$\qquad\qquad 6x = 30$
$\qquad\qquad x = 5$

29. $\sqrt[3]{n+3} - 2 = -4$
$\qquad \sqrt[3]{n+3} = -2$
$\quad \left(\sqrt[3]{n+3}\right)^3 = (-2)^3$
$\qquad\quad n + 3 = -8$
$\qquad\qquad n = -11$

31. $\sqrt[4]{x-2} - 2 = -4$
$\qquad \sqrt[4]{x-2} = -2$

This equation has no real-number solution.

33. $\sqrt{3x-2} = \sqrt{8-2x}$

$\left(\sqrt{3x-2}\right)^2 = \left(\sqrt{8-2x}\right)^2$

$3x-2 = 8-2x$

$5x = 10$

$x = 2$

Check:

$\sqrt{3(2)-2} \stackrel{?}{=} \sqrt{8-2(2)}$

$\sqrt{6-2} \stackrel{?}{=} \sqrt{8-4}$

$\sqrt{4} \stackrel{?}{=} \sqrt{4}$

$2 = 2$

35. $\sqrt{4x-5} = \sqrt{6x+5}$

$\left(\sqrt{4x-5}\right)^2 = \left(\sqrt{6x+5}\right)^2$

$4x-5 = 6x+5$

$-10 = 2x$

$-5 = x$

Check:

$\sqrt{4(-5)-5} \stackrel{?}{=} \sqrt{6(-5)+5}$

$\sqrt{-20-5} \stackrel{?}{=} \sqrt{-30+5}$

$\sqrt{-25} \stackrel{?}{=} \sqrt{-25}$

$\sqrt{-25}$ is not a real number.

No real-number solution. (-5 is an extraneous solution.)

37. $\sqrt[3]{2r+2} = \sqrt[3]{3r-1}$

$\left(\sqrt[3]{2r+2}\right)^3 = \left(\sqrt[3]{3r-1}\right)^3$

$2r+2 = 3r-1$

$3 = r$

Check:

$\sqrt[3]{2(3)+2} \stackrel{?}{=} \sqrt[3]{3(3)-1}$

$\sqrt[3]{6+2} \stackrel{?}{=} \sqrt[3]{9-1}$

$\sqrt[3]{8} \stackrel{?}{=} \sqrt[3]{8}$

$2 = 2$

39. $\sqrt[4]{4x+4} = \sqrt[4]{5x+1}$

$\left(\sqrt[4]{4x+4}\right)^4 = \left(\sqrt[4]{5x+1}\right)^4$

$4x+4 = 5x+1$

$3 = x$

Check:

$\sqrt[4]{4(3)+4} \stackrel{?}{=} \sqrt[4]{5(3)+1}$

$\sqrt[4]{12+4} \stackrel{?}{=} \sqrt[4]{15+1}$

$\sqrt[4]{16} \stackrel{?}{=} \sqrt[4]{16}$

$2 = 2$

41. $\sqrt{2x+24} = x+8$

$\left(\sqrt{2x+24}\right)^2 = (x+8)^2$

$2x+24 = x^2+16x+64$

$0 = x^2+14x+40$

$0 = (x+4)(x+10)$

$x+4 = 0$ or $x+10 = 0$

$x = -4$ $x = -10$

Check $x = -4$:

$\sqrt{2(-4)+24} \stackrel{?}{=} (-4)+8$

$\sqrt{-8+24} \stackrel{?}{=} 4$

$\sqrt{16} \stackrel{?}{=} 4$

$4 = 4$

Check $x = -10$:

$\sqrt{2(-10)+24} \stackrel{?}{=} (-10)+8$

$\sqrt{-20+24} \stackrel{?}{=} -2$

$\sqrt{4} \stackrel{?}{=} -2$

$2 \neq -2$

-4 is the only solution. (-10 is an extraneous solution.)

43.
$$y - 1 = \sqrt{2y - 2}$$
$$(y - 1)^2 = \left(\sqrt{2y - 2}\right)^2$$
$$y^2 - 2y + 1 = 2y - 2$$
$$y^2 - 4y + 3 = 0$$
$$(y - 3)(y - 1) = 0$$

$y - 3 = 0 \quad \text{or} \quad y - 1 = 0$
$\quad y = 3 \qquad\qquad y = 1$

Check $y = 3$:
$$3 - 1 \overset{?}{=} \sqrt{2(3) - 2}$$
$$2 \overset{?}{=} \sqrt{6 - 2}$$
$$2 \overset{?}{=} \sqrt{4}$$
$$2 = 2$$

Check $y = 1$:
$$1 - 1 \overset{?}{=} \sqrt{2(1) - 2}$$
$$0 \overset{?}{=} \sqrt{2 - 2}$$
$$0 \overset{?}{=} \sqrt{0}$$
$$0 = 0$$

45. $\sqrt{3x + 10} - 4 = x$
$$\sqrt{3x + 10} = x + 4$$
$$\left(\sqrt{3x + 10}\right)^2 = (x + 4)^2$$
$$3x + 10 = x^2 + 8x + 16$$
$$0 = x^2 + 5x + 6$$
$$0 = (x + 2)(x + 3)$$

$x + 2 = 0 \quad \text{or} \quad x + 3 = 0$
$\quad x = -2 \qquad\qquad x = -3$

Check $x = -2$:
$$\sqrt{3(-2) + 10} - 4 \overset{?}{=} -2$$
$$\sqrt{-6 + 10} - 4 \overset{?}{=} -2$$
$$\sqrt{4} - 4 \overset{?}{=} -2$$
$$2 - 4 \overset{?}{=} -2$$
$$-2 = -2$$

Check $x = -3$:
$$\sqrt{3(-3) + 10} - 4 \overset{?}{=} -3$$
$$\sqrt{-9 + 10} - 4 \overset{?}{=} -3$$
$$\sqrt{1} - 4 \overset{?}{=} -3$$
$$1 - 4 \overset{?}{=} -3$$
$$-3 = -3$$

47. $\sqrt{10n+4} - 3n = n+1$

$\sqrt{10n+4} = 4n+1$

$\left(\sqrt{10n+4}\right)^2 = (4n+1)^2$

$10n+4 = 16n^2 + 8n + 1$

$0 = 16n^2 - 2n - 3$

$0 = (2n-1)(8n+3)$

$2n-1 = 0 \quad \text{or} \quad 8n+3 = 0$

$2n = 1 \qquad\qquad 8n = -3$

$n = \dfrac{1}{2} \qquad\qquad n = -\dfrac{3}{8}$

Check $n = \dfrac{1}{2}$:

$\sqrt{10\left(\dfrac{1}{2}\right)+4} - 3\left(\dfrac{1}{2}\right) \overset{?}{=} \left(\dfrac{1}{2}\right)+1$

$\sqrt{5+4} - \dfrac{3}{2} \overset{?}{=} \left(\dfrac{1}{2}\right) + \dfrac{2}{2}$

$\sqrt{9} - \dfrac{3}{2} \overset{?}{=} \dfrac{3}{2}$

$3 - \dfrac{3}{2} \overset{?}{=} \dfrac{3}{2}$

$\dfrac{3}{2} = \dfrac{3}{2}$

Check $n = -\dfrac{3}{8}$:

$\sqrt{10\left(-\dfrac{3}{8}\right)+4} - 3\left(-\dfrac{3}{8}\right) \overset{?}{=} \left(-\dfrac{3}{8}\right)+1$

$\sqrt{-\dfrac{30}{8}+\dfrac{32}{8}} + \dfrac{9}{8} \overset{?}{=} \left(-\dfrac{3}{8}\right)+\dfrac{8}{8}$

$\sqrt{\dfrac{2}{8}} + \dfrac{9}{8} \overset{?}{=} \dfrac{5}{8}$

$\sqrt{\dfrac{1}{4}} + \dfrac{9}{8} \overset{?}{=} \dfrac{5}{8}$

$\dfrac{1}{2} + \dfrac{9}{8} \overset{?}{=} \dfrac{5}{8}$

$\dfrac{13}{8} \neq \dfrac{5}{8}$

$\dfrac{1}{2}$ is the only solution.

$\left(-\dfrac{3}{8} \text{ is an extraneous solution.}\right)$

49. $\sqrt[3]{5x+2} + 2 = 5$

$\sqrt[3]{5x+2} = 3$

$\left(\sqrt[3]{5x+2}\right)^3 = 3^3$

$5x+2 = 27$

$5x = 25$

$x = 5$

Check:

$\sqrt[3]{5(5)+2} + 2 \overset{?}{=} 5$

$\sqrt[3]{25+2} + 2 \overset{?}{=} 5$

$\sqrt[3]{27} + 2 \overset{?}{=} 5$

$3 + 2 \overset{?}{=} 5$

$5 = 5$

51. $\sqrt[3]{n^2 - 2n + 5} = 2$

$\left(\sqrt[3]{n^2 - 2n + 5}\right)^3 = (2)^3$

$n^2 - 2n + 5 = 8$

$n^2 - 2n - 3 = 0$

$(n-3)(n+1) = 0$

$n-3 = 0 \quad \text{or} \quad n+1 = 0$

$n = 3 \qquad\qquad n = -1$

Check $n = 3$:

$\sqrt[3]{3^2 - 2(3) + 5} \overset{?}{=} 2$

$\sqrt[3]{9 - 6 + 5} \overset{?}{=} 2$

$\sqrt[3]{8} \overset{?}{=} 2$

$2 = 2$

Check $n = -1$:

$\sqrt[3]{(-1)^2 - 2(-1) + 5} \overset{?}{=} 2$

$\sqrt[3]{1 + 2 + 5} \overset{?}{=} 2$

$\sqrt[3]{8} \overset{?}{=} 2$

$2 = 2$

53. $\quad 1+\sqrt{x}=\sqrt{2x+1}$

$\left(1+\sqrt{x}\right)^2=\left(\sqrt{2x+1}\right)^2$

$1+2\sqrt{x}+x=2x+1$

$2\sqrt{x}=x$

$\left(2\sqrt{x}\right)^2=x^2$

$4x=x^2$

$0=x^2-4x$

$0=x(x-4)$

$x=0\quad\text{or}\quad x-4=0$

$\hspace{3.5cm}x=4$

Check $x=0$:

$1+\sqrt{0}\overset{?}{=}\sqrt{2(0)+1}$

$1+0\overset{?}{=}\sqrt{0+1}$

$1\overset{?}{=}\sqrt{1}$

$1=1$

Check $x=4$:

$1+\sqrt{4}\overset{?}{=}\sqrt{2(4)+1}$

$1+2\overset{?}{=}\sqrt{8+1}$

$3\overset{?}{=}\sqrt{9}$

$3=3$

55. $\sqrt{3x+1}+\sqrt{3x}=2$

$\left(\sqrt{3x+1}\right)^2=\left(2-\sqrt{3x}\right)^2$

$3x+1=4-4\sqrt{3x}+3x$

$(-3)^2=\left(-4\sqrt{3x}\right)^2$

$9=16\cdot 3x$

$9=48x$

$\dfrac{3}{16}=x$

Check:

$\sqrt{3\left(\dfrac{3}{16}\right)+1}+\sqrt{3\left(\dfrac{3}{16}\right)}\overset{?}{=}2$

$\sqrt{\dfrac{9}{16}+1}+\sqrt{\dfrac{9}{16}}\overset{?}{=}2$

$\sqrt{\dfrac{9}{16}+\dfrac{16}{16}}+\sqrt{\dfrac{9}{16}}\overset{?}{=}2$

$\sqrt{\dfrac{25}{16}}+\sqrt{\dfrac{9}{16}}\overset{?}{=}2$

$\dfrac{5}{4}+\dfrac{3}{4}\overset{?}{=}2$

$\dfrac{8}{4}\overset{?}{=}2$

$2=2$

57. $\quad\sqrt{6x+7}-2=\sqrt{2x+3}$

$\left(\sqrt{6x+7}-2\right)^2=\left(\sqrt{2x+3}\right)^2$

$6x+7-4\sqrt{6x+7}+4=2x+3$

$6x+11-4\sqrt{6x+7}=2x+3$

$4x+8=4\sqrt{6x+7}$

$x+2=\sqrt{6x+7}$

$(x+2)^2=\left(\sqrt{6x+7}\right)^2$

$x^2+4x+4=6x+7$

$x^2-2x-3=0$

$(x+1)(x-3)=0$

$x+1=0\quad\text{or}\quad x-3=0$

$x=-1\hspace{2cm}x=3$

Check $x=-1$:

$\sqrt{6(-1)+7}-2\overset{?}{=}\sqrt{2(-1)+3}$

$\sqrt{-6+7}-2\overset{?}{=}\sqrt{-2+3}$

$\sqrt{1}-2\overset{?}{=}\sqrt{1}$

$1-2\overset{?}{=}1$

$-1\neq 1$

Check $x = 3$:

$$\sqrt{6(3)+7} - 2 \overset{?}{=} \sqrt{2(3)+3}$$

$$\sqrt{18+7} - 2 \overset{?}{=} \sqrt{6+3}$$

$$\sqrt{25} - 2 \overset{?}{=} \sqrt{9}$$

$$5 - 2 \overset{?}{=} 3$$

$$3 = 3$$

3 is the only solution (-1 is an extraneous solution).

59. **Mistake:** You cannot take the principal square root of a number and get a negative.
Correct: No real-number solution.

61. **Mistake:** The binomial $x - 3$ was not squared correctly.
Correct:

$$\sqrt{x+3} = x - 3$$

$$\left(\sqrt{x+3}\right)^2 = \left(x-3\right)^2$$

$$x + 3 = x^2 - 6x + 9$$

$$0 = x^2 - 7x + 6$$

$$0 = \left(x-6\right)\left(x-1\right)$$

$$x - 6 = 0 \quad \text{or} \quad x - 1 = 0$$

$$x = 6 \qquad\qquad x = 1 \text{ is extraneous}$$

63. Substitute and solve.

$$T = 2\pi\sqrt{\frac{L}{9.8}}$$

$$2\pi = 2\pi\sqrt{\frac{L}{9.8}}$$

$$1 = \sqrt{\frac{L}{9.8}}$$

$$1^2 = \left(\sqrt{\frac{L}{9.8}}\right)^2$$

$$1 = \frac{L}{9.8}$$

$$9.8 = L$$

The length of the pendulum is 9.8 m.

65. Substitute and solve.

$$T = 2\pi\sqrt{\frac{L}{9.8}}$$

$$\frac{\pi}{2} = 2\pi\sqrt{\frac{L}{9.8}}$$

$$\frac{1}{4} = \sqrt{\frac{L}{9.8}}$$

$$\left(\frac{1}{4}\right)^2 = \left(\sqrt{\frac{L}{9.8}}\right)^2$$

$$\frac{1}{16} = \frac{L}{9.8}$$

$$\frac{9.8}{16} = L$$

$$0.6125 = L$$

The length of the pendulum is 0.6125 m.

67.
$$0.3 = \sqrt{\frac{h}{16}}$$

$$\left(0.3\right)^2 = \left(\sqrt{\frac{h}{16}}\right)^2$$

$$0.09 = \frac{h}{16}$$

$$1.44 = h$$

The distance is 1.44 ft.

69.
$$\frac{1}{4} = \sqrt{\frac{h}{16}}$$

$$\left(\frac{1}{4}\right)^2 = \left(\sqrt{\frac{h}{16}}\right)^2$$

$$\frac{1}{16} = \frac{h}{16}$$

$$1 = h$$

The distance is 1 ft.

71.
$$30 = \frac{7}{2}\sqrt{2D}$$

$$60 = 7\sqrt{2D}$$

$$\frac{60}{7} = \sqrt{2D}$$

$$\left(\frac{60}{7}\right)^2 = \left(\sqrt{2D}\right)^2$$

$$73.469 \approx 2D$$

$$36.73 \approx D$$

The skid distance is approximately 36.73 ft.

73.
$$45 = \frac{7}{2}\sqrt{2D}$$
$$90 = 7\sqrt{2D}$$
$$\frac{90}{7} = \sqrt{2D}$$
$$\left(\frac{90}{7}\right)^2 = \left(\sqrt{2D}\right)^2$$
$$165.306 \approx 2D$$
$$82.65 \text{ ft.} \approx D$$
The skid distance is approximately 82.65 ft.

75.
$$5 = \sqrt{F_1^2 + 3^2}$$
$$5^2 = \left(\sqrt{F_1^2 + 9}\right)^2$$
$$25 = F_1^2 + 9$$
$$16 = F_1^2$$
$$\sqrt{16} = F_1$$
$$4 = F_1$$
Force 1 has a value of 4 N.

77.
$$3\sqrt{5} = \sqrt{3^2 + F_2^2}$$
$$\left(3\sqrt{5}\right)^2 = \left(\sqrt{9 + F_2^2}\right)^2$$
$$9 \cdot 5 = 9 + F_2^2$$
$$45 - 9 = F_2^2$$
$$36 = F_2^2$$
$$\sqrt{36} = F_2$$
$$6 = F_2$$
Force 2 has a value of 6 N.

79. a)

$x = 9$	$x = 16$	$y = 5$
$y = \sqrt{9}$	$y = \sqrt{16}$	$5 = \sqrt{x}$
$y = 3$	$y = 4$	$5^2 = \left(\sqrt{x}\right)^2$
		$25 = x$

b)

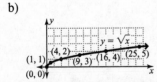

c) No. The x-values must be 0 or positive because real square roots exist only when $x \geq 0$. The y-values must be 0 or positive because, by definition, the principal square root is 0 or positive.

d) Yes, because it passes the vertical line test.

81. The graph becomes steeper from left to right.

83. The graph rises or lowers according to the value of the constant.

Review Exercises

1. $3^4 = 81$

2. $(-0.2)^3 = -0.008$

3. $\left(\frac{2}{5}\right)^{-4} = \left(\frac{5}{2}\right)^4 = \frac{625}{16}$ or 39.0625

4. $x^3 \cdot x^5 = x^{3+5} = x^8$

5. $\left(n^4\right)^6 = n^{4 \cdot 6} = n^{24}$

6. $\dfrac{y^7}{y^3} = y^{7-3} = y^4$

Exercise Set 9.7

1. $\sqrt{-36} = \sqrt{-1 \cdot 36} = \sqrt{-1} \cdot \sqrt{36} = i\sqrt{36} = 6i$

3. $\sqrt{-5} = \sqrt{-1 \cdot 5} = \sqrt{-1} \cdot \sqrt{5} = i\sqrt{5}$

5. $\sqrt{-8} = \sqrt{-1 \cdot 8} = \sqrt{-1} \cdot \sqrt{8} = i\sqrt{4 \cdot 2} = 2i\sqrt{2}$

7. $\sqrt{-18} = \sqrt{-1 \cdot 18} = \sqrt{-1} \cdot \sqrt{18} = i\sqrt{9 \cdot 2} = 3i\sqrt{2}$

9. $\sqrt{-27} = \sqrt{-1 \cdot 27} = \sqrt{-1} \cdot \sqrt{27} = i\sqrt{9 \cdot 3} = 3i\sqrt{3}$

11. $\sqrt{-125} = \sqrt{-1 \cdot 125}$
$$= \sqrt{-1} \cdot \sqrt{125}$$
$$= i\sqrt{25 \cdot 5}$$
$$= 5i\sqrt{5}$$

13. $\sqrt{-63} = \sqrt{-1 \cdot 63} = \sqrt{-1} \cdot \sqrt{63} = i\sqrt{9 \cdot 7} = 3i\sqrt{7}$

15. $\sqrt{-245} = \sqrt{-1 \cdot 245}$
$$= \sqrt{-1} \cdot \sqrt{245}$$
$$= i\sqrt{49 \cdot 5}$$
$$= 7i\sqrt{5}$$

17. $(9 + 3i) + (-3 + 4i) = 6 + 7i$

19. $(6 + 2i) + (5 - 8i) = 11 - 6i$

21. $(-4 + 6i) - (3 + 5i) = (-4 + 6i) + (-3 - 5i)$
$$= -7 + i$$

23. $(8 - 5i) - (-3i) = (8 - 5i) + (3i)$
$$= 8 - 2i$$

25. $(12+3i)+(-15-13i)=-3-10i$

27. $(-5-9i)-(-5-9i)=(-5-9i)+(5+9i)=0$

29. $(10+i)-(2-13i)+(6-5i)$
    $=(10+i)+(-2+13i)+(6-5i)$
    $=14+9i$

31. $(5-2i)-(9-14i)+(16i)$
    $=(5-2i)+(-9+14i)+(16i)$
    $=-4+28i$

33. $(8i)(3i)=24i^2=24(-1)=-24$

35. $(-8i)(5i)=-40i^2=-40(-1)=40$

37. $2i(6-7i)=12i-14i^2=12i-14(-1)=14+12i$

39. $-8i(4-9i)=-32i+72i^2$
    $=-32i+72(-1)$
    $=-72-32i$

41. $(6+i)(3-i)=18-6i+3i-i^2$
    $=18-3i-(-1)$
    $=18-3i+1$
    $=19-3i$

43. $(8+5i)(5-2i)=40-16i+25i-10i^2$
    $=40+9i-10(-1)$
    $=40+9i+10$
    $=50+9i$

45. $(8+i)(8-i)=64-i^2$
    $=64-(-1)$
    $=64+1$
    $=65$

47. $(3-4i)^2=3^2-2\cdot3\cdot4i+(4i)^2$
    $=9-24i+16i^2$
    $=9-24i+16(-1)$
    $=9-24i-16$
    $=-7-24i$

49. $\dfrac{2}{i}=\dfrac{2}{i}\cdot\dfrac{i}{i}=\dfrac{2i}{i^2}=\dfrac{2i}{-1}=-2i$

51. $\dfrac{4}{5i}=\dfrac{4}{5i}\cdot\dfrac{i}{i}=\dfrac{4i}{5i^2}=\dfrac{4i}{5(-1)}=-\dfrac{4i}{5}$

53. $\dfrac{6}{2i}=\dfrac{3}{i}\cdot\dfrac{i}{i}=\dfrac{3i}{i^2}=\dfrac{3i}{-1}=-3i$

55. $\dfrac{2+i}{2i}=\dfrac{2+i}{2i}\cdot\dfrac{i}{i}$
    $=\dfrac{2i+i^2}{2i^2}$
    $=\dfrac{2i+(-1)}{2(-1)}$
    $=\dfrac{-1+2i}{-2}$
    $=\dfrac{1}{2}-i$

57. $\dfrac{4+2i}{4i}=\dfrac{2(2+i)}{4i}$
    $=\dfrac{2+i}{2i}\cdot\dfrac{i}{i}$
    $=\dfrac{2i+i^2}{2i^2}$
    $=\dfrac{-1+2i}{-2}$
    $=\dfrac{1}{2}-i$

59. $\dfrac{7}{2+i}=\dfrac{7}{2+i}\cdot\dfrac{2-i}{2-i}$
    $=\dfrac{14-7i}{4-i^2}$
    $=\dfrac{14-7i}{4-(-1)}$
    $=\dfrac{14-7i}{4+1}$
    $=\dfrac{14-7i}{5}$
    $=\dfrac{14}{5}-\dfrac{7}{5}i$

61. $\dfrac{2i}{3-7i} = \dfrac{2i}{3-7i} \cdot \dfrac{3+7i}{3+7i}$

$= \dfrac{6i+14i^2}{9-49i^2}$

$= \dfrac{6i+14(-1)}{9-49(-1)}$

$= \dfrac{6i-14}{9+49}$

$= \dfrac{-14+6i}{58}$

$= -\dfrac{14}{58} + \dfrac{6}{58}i$

$= -\dfrac{7}{29} + \dfrac{3}{29}i$

63. $\dfrac{5-9i}{1-i} = \dfrac{5-9i}{1-i} \cdot \dfrac{1+i}{1+i}$

$= \dfrac{5+5i-9i-9i^2}{1-i^2}$

$= \dfrac{5-4i-9(-1)}{1-(-1)}$

$= \dfrac{5-4i+9}{1+1}$

$= \dfrac{14-4i}{2}$

$= 7-2i$

65. $\dfrac{3+i}{2+3i} = \dfrac{3+i}{2+3i} \cdot \dfrac{2-3i}{2-3i}$

$= \dfrac{6-9i+2i-3i^2}{4-9i^2}$

$= \dfrac{6-7i-3(-1)}{4-9(-1)}$

$= \dfrac{6-7i+3}{4+9}$

$= \dfrac{9-7i}{13}$

$= \dfrac{9}{13} - \dfrac{7}{13}i$

67. $\dfrac{1+6i}{4+5i} = \dfrac{1+6i}{4+5i} \cdot \dfrac{4-5i}{4-5i}$

$= \dfrac{4-5i+24i-30i^2}{16-25i^2}$

$= \dfrac{4+19i-30(-1)}{16-25(-1)}$

$= \dfrac{4+19i+30}{16+25}$

$= \dfrac{34+19i}{41}$

$= \dfrac{34}{41} + \dfrac{19}{41}i$

69. $i^{19} = i^{16} \cdot i^3 = \left(i^4\right)^4 \cdot i^3 = (1)^4 \cdot (-i) = 1 \cdot (-i) = -i$

71. $i^{42} = i^{40} \cdot i^2 = \left(i^4\right)^{10} \cdot i^2 = 1^{10} \cdot (-1) = 1(-1) = -1$

73. $i^{38} = i^{36} \cdot i^2 = \left(i^4\right)^9 \cdot i^2 = 1^9 \cdot (-1) = 1(-1) = -1$

75. $i^{60} = \left(i^4\right)^{15} = 1^{15} = 1$

77. $i^{-20} = \dfrac{1}{i^{20}} = \dfrac{1}{\left(i^4\right)^5} = \dfrac{1}{1^5} = \dfrac{1}{1} = 1$

79. $i^{-30} = \dfrac{1}{i^{30}} = \dfrac{1}{i^{28} \cdot i^2} = \dfrac{1}{\left(i^4\right)^7 \cdot i^2} = \dfrac{1}{1^7 \cdot (-1)}$

$= \dfrac{1}{1(-1)} = \dfrac{1}{-1} = -1$

81. $i^{-21} = \dfrac{1}{i^{21}} = \dfrac{1}{i^{20} \cdot i} = \dfrac{1}{\left(i^4\right)^5 \cdot i} = \dfrac{1}{1^5 \cdot i} = \dfrac{1}{1 \cdot i}$

$= \dfrac{1}{i} \cdot \dfrac{i}{i} = \dfrac{i}{i^2} = \dfrac{i}{-1} = -i$

83. $i^{-35} = \dfrac{1}{i^{35}} = \dfrac{1}{i^{32} \cdot i^3} = \dfrac{1}{\left(i^4\right)^8 \cdot i^3} = \dfrac{1}{1^8 \cdot (-i)} = \dfrac{1}{1(-i)}$

$= \dfrac{1}{-i} \cdot \dfrac{i}{i} = \dfrac{i}{-i^2} = \dfrac{i}{-(-1)} = \dfrac{i}{1} = i$

Review Exercises

1. $4x^2 - 12x + 9 = (2x-3)(2x-3) = (2x-3)^2$

2. $\pm\sqrt{48} = \pm\sqrt{16 \cdot 3} = \pm 4\sqrt{3}$

3. $3x - 2 = 4$

$\quad 3x = 6$

$\quad\ x = 2$

4. $2x^2 - x - 6 = 0$

$(2x+3)(x-2) = 0$

$2x+3 = 0$ or $x-2 = 0$

$2x = -3$ $x = 2$

$x = -\dfrac{3}{2}$

5. $2x + 3y = 6$

$y = 0: 2x + 3 \cdot 0 = 6$

$2x = 6$

$x = 3$

$(3, 0)$

$x = 0: 2 \cdot 0 + 3y = 6$

$3y = 6$

$y = 2$

$(0, 2)$

6. $y = x^2 - 3$

x	y	(x, y)
-2	1	$(-2, 1)$
-1	-2	$(-1, -2)$
0	-3	$(0, -3)$
1	-2	$(1, -2)$
2	1	$(2, 1)$

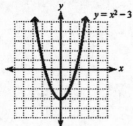

Chapter 9 Review Exercises

1.   The square roots of 121 are ± 11.

2.   The square roots of 49 are ± 7.

3. $\sqrt{169} = 13$

4. $-\sqrt{49} = -7$

5. $\sqrt{-36}$ is not a real number.

6. $\sqrt{\dfrac{1}{25}} = \dfrac{1}{5}$

7. $c = \sqrt{4^2 + 3^2}$

$c = \sqrt{16 + 9}$

$c = \sqrt{25}$

$c = 5$

The connecting piece should be 5 ft.

8. $T = 2\pi\sqrt{\dfrac{2.45}{9.8}}$

$= 2\pi\sqrt{0.25}$

$= 2\pi \cdot 0.5$

$= \pi$

The period is π sec.

9. $\sqrt{7} \approx 2.646$

10. $\sqrt{90} \approx 9.487$

11.   a) $S = 2\sqrt{2(40)} = 2\sqrt{80} = 2 \cdot 4\sqrt{5} = 8\sqrt{5}$

The exact speed is $8\sqrt{5}$ mph.

b) $8\sqrt{5} \approx 17.9$

The speed is approximately 17.9 mph.

12.   a) $t = \sqrt{\dfrac{40}{16}} = \dfrac{\sqrt{40}}{\sqrt{16}} = \dfrac{2\sqrt{10}}{4} = \dfrac{\sqrt{10}}{2}$

The exact time is $\dfrac{\sqrt{10}}{2}$ sec.

b) $\dfrac{\sqrt{10}}{2} \approx 1.58$

The time is approximately 1.58 sec.

13. $\sqrt{49x^8} = 7x^4$

14. $\sqrt{144a^6 b^{12}} = 12a^3 b^6$

15. $\sqrt{0.16m^2 n^{10}} = 0.4mn^5$

16. $\sqrt[3]{x^{15}} = x^5$

17. $\sqrt[3]{-64r^9 s^3} = -4r^3 s$

18. $\sqrt[4]{81x^{12}} = 3x^3$

19. $\sqrt[5]{32x^{15} y^{20}} = 2x^3 y^4$

20. $\sqrt[7]{x^{14} y^7} = x^2 y$

21. $\sqrt{81x^2} = 9|x|$

22. $\sqrt[4]{(x-1)^8} = (x-1)^2$

23. $f(x) = \sqrt{4x - 4}$.

$f(5) = \sqrt{4(5) - 4}$

$= \sqrt{20 - 4}$

$= \sqrt{16}$

$= 4$

24. $f(x) = \sqrt[3]{5x + 2}$

$f(-2) = \sqrt[3]{5(-2) + 2}$

$= \sqrt[3]{-10 + 2}$

$= \sqrt[3]{-8}$

$= -2$

25. The index is even, so the radicand must be nonnegative.

$2x + 10 \geq 0$

$2x \geq -10$

$x \geq -5$

Domain: $\{x \mid x \geq -5\}$, or $[-5, \infty)$.

26. The index is even, so the radicand must be nonnegative.

$9 - 3x \geq 0$

$-3x \geq -9$

$x \leq 3$

Domain: $\{x \mid x \leq 3\}$, or $(-\infty, 3]$.

27. $(-64)^{1/3} = \sqrt[3]{-64} = -4$

28. $(24a^4)^{1/2} = \sqrt{24a^4} = \sqrt{4a^4 \cdot 6} = 2a^2\sqrt{6}$

29. $\left(\dfrac{1}{32}\right)^{3/5} = \left(\sqrt[5]{\dfrac{1}{32}}\right)^3 = \left(\dfrac{1}{2}\right)^3 = \dfrac{1}{8}$

30. $(5r - 2)^{5/7} = \sqrt[7]{(5r - 2)^5}$

31. $121^{-1/2} = \dfrac{1}{121^{1/2}} = \dfrac{1}{\sqrt{121}} = \dfrac{1}{11}$

32. $81^{-3/4} = \dfrac{1}{81^{3/4}} = \dfrac{1}{\left(\sqrt[4]{81}\right)^3} = \dfrac{1}{3^3} = \dfrac{1}{27}$

33. $\left(\dfrac{16}{49}\right)^{3/2} = \left(\dfrac{16^{1/2}}{49^{1/2}}\right)^3 = \left(\dfrac{\sqrt{16}}{\sqrt{49}}\right)^3 = \left(\dfrac{4}{7}\right)^3 = \dfrac{64}{343}$

34. $81x^{3/4} = 81\sqrt[4]{x^3}$

35. $\sqrt[8]{33} = 33^{1/8}$

36. $\dfrac{8}{\sqrt[7]{n^3}} = \dfrac{8}{n^{3/7}} = 8n^{-3/7}$

37. $\left(\sqrt[5]{8}\right)^3 = 8^{3/5}$

38. $\left(\sqrt[3]{m}\right)^8 = m^{8/3}$

39. $\left(\sqrt[4]{3xw}\right)^3 = (3xw)^{3/4}$

40. $\sqrt[3]{(a + b)^4} = (a + b)^{4/3}$

41. $x^{2/3} \cdot x^{4/3} = x^{2/3 + 4/3} = x^{6/3} = x^2$

42. $\left(4m^{1/4}\right)\left(8m^{5/4}\right) = 32m^{1/4 + 5/4} = 32m^{6/4} = 32m^{3/2}$

43. $\dfrac{y^{3/5}}{y^{4/5}} = y^{3/5 - 4/5} = y^{-1/5} = \dfrac{1}{y^{1/5}}$

44. $\dfrac{b^{2/5}}{b^{-3/5}} = b^{2/5 - (-3/5)} = b^{2/5 + 3/5} = b^{5/5} = b$

45. $\left(k^{2/3}\right)^{3/4} = k^{(2/3) \cdot (3/4)} = k^{6/12} = k^{1/2}$

46. $\left(2xy^{1/5}\right)^{3/4} = 2^{3/4} x^{3/4} y^{(1/5) \cdot (3/4)} = 2^{3/4} x^{3/4} y^{3/20}$

47. $\sqrt[4]{\sqrt[3]{x^2}} = \sqrt[4]{x^{2/3}} = \left(x^{2/3}\right)^{1/4} = x^{(2/3) \cdot (1/4)}$

    $= x^{2/12} = x^{1/6} = \sqrt[6]{x}$

48. $\sqrt{3} \cdot \sqrt{27} = \sqrt{81} = 9$

49. $\sqrt{5x^5} \cdot \sqrt{20x^3} = \sqrt{100x^8} = 10x^4$

50. $\sqrt[3]{2} \cdot \sqrt[3]{4} = \sqrt[3]{8} = 2$

51. $\sqrt[4]{7} \cdot \sqrt[4]{6} = \sqrt[4]{42}$

52. $\sqrt[5]{3x^2y^3} \cdot \sqrt[5]{5x^2y} = \sqrt[5]{15x^4y^4}$

53. $4\sqrt{6} \cdot 7\sqrt{15} = 28\sqrt{90}$

    $= 28\sqrt{9 \cdot 10}$

    $= 28 \cdot 3\sqrt{10}$

    $= 84\sqrt{10}$

54. $\sqrt{x^9} \cdot \sqrt{x^6} = \sqrt{x^{15}} = \sqrt{x^{14} \cdot x} = x^7\sqrt{x}$

55. $4\sqrt{10c} \cdot 2\sqrt{6c^4} = 8\sqrt{60c^5}$

    $= 8\sqrt{4c^4 \cdot 15c}$

    $= 8 \cdot 2c^2\sqrt{15c}$

    $= 16c^2\sqrt{15c}$

56. $a\sqrt{a^3b^2} \cdot b^2\sqrt{a^5b^3} = ab^2\sqrt{a^8b^5}$
$$= ab^2\sqrt{a^8b^4 \cdot b}$$
$$= ab^2 \cdot a^4b^2\sqrt{b}$$
$$= a^5b^4\sqrt{b}$$

57. $\left(\sqrt[4]{6ab^2}\right)^4 = \left(\left(6ab^2\right)^{1/4}\right)^4 = 6ab^2$

58. $\sqrt{\dfrac{49}{121}} = \dfrac{7}{11}$

59. $\sqrt[3]{-\dfrac{27}{8}} = -\dfrac{3}{2}$

60. $\dfrac{9\sqrt{160}}{3\sqrt{8}} = 3\sqrt{\dfrac{160}{8}}$
$$= 3\sqrt{20}$$
$$= 3\sqrt{4 \cdot 5}$$
$$= 3 \cdot 2\sqrt{5}$$
$$= 6\sqrt{5}$$

61. $\dfrac{36\sqrt{96x^6y^{11}}}{4\sqrt{3x^2y^4}} = 9\sqrt{\dfrac{96x^6y^{11}}{3x^2y^4}}$
$$= 9\sqrt{32x^4y^7}$$
$$= 9\sqrt{16x^4y^6 \cdot 2y}$$
$$= 9 \cdot 4x^2y^3\sqrt{2y}$$
$$= 36x^2y^3\sqrt{2y}$$

62. $4b\sqrt{27b^7} = 4b\sqrt{9b^6 \cdot 3b}$
$$= 4b \cdot 3b^3\sqrt{3b}$$
$$= 12b^4\sqrt{3b}$$

63. $5\sqrt[3]{108} = 5\sqrt[3]{27 \cdot 4} = 5 \cdot 3\sqrt[3]{4} = 15\sqrt[3]{4}$

64. $2\sqrt[3]{40x^{10}} = 2\sqrt[3]{8x^9 \cdot 5x} = 2 \cdot 2x^3\sqrt[3]{5x} = 4x^3\sqrt[3]{5x}$

65. $2x^4\sqrt[4]{162x^7} = 2x^4\sqrt[4]{81x^4 \cdot 2x^3}$
$$= 2x^4 \cdot 3x\sqrt[4]{2x^3}$$
$$= 6x^5\sqrt[4]{2x^3}$$

66. $-5\sqrt{n} + 2\sqrt{n} = -3\sqrt{n}$

67. $3y^3\sqrt[4]{8y} - 9y^3\sqrt[4]{8y} = -6y^3\sqrt[4]{8y}$

68. $\sqrt{45} + \sqrt{20} = \sqrt{9 \cdot 5} + \sqrt{4 \cdot 5} = 3\sqrt{5} + 2\sqrt{5} = 5\sqrt{5}$

69. $4\sqrt{24} - 6\sqrt{54} = 4\sqrt{4 \cdot 6} - 6\sqrt{9 \cdot 6}$
$$= 4 \cdot 2\sqrt{6} - 6 \cdot 3\sqrt{6}$$
$$= 8\sqrt{6} - 18\sqrt{6}$$
$$= -10\sqrt{6}$$

70. $\sqrt{150} - \sqrt{54} + \sqrt{24} = \sqrt{25 \cdot 6} - \sqrt{9 \cdot 6} + \sqrt{4 \cdot 6}$
$$= 5\sqrt{6} - 3\sqrt{6} + 2\sqrt{6}$$
$$= 4\sqrt{6}$$

71. $4\sqrt{72x^2y} - 2x\sqrt{128y} + 5\sqrt{32x^2y}$
$$= 4\sqrt{36x^2 \cdot 2y} - 2x\sqrt{64 \cdot 2y} + 5\sqrt{16x^2 \cdot 2y}$$
$$= 24x\sqrt{2y} - 16x\sqrt{2y} + 20x\sqrt{2y}$$
$$= 28x\sqrt{2y}$$

72. $\sqrt[3]{250x^4y^5} + \sqrt[3]{128x^4y^5}$
$$= \sqrt[3]{125x^3y^3 \cdot 2xy^2} + \sqrt[3]{64x^3y^3 \cdot 2xy^2}$$
$$= 5xy\sqrt[3]{2xy^2} + 4xy\sqrt[3]{2xy^2}$$
$$= 9xy\sqrt[3]{2xy^2}$$

73. $\sqrt[4]{48} + \sqrt[4]{243} = \sqrt[4]{16 \cdot 3} + \sqrt[4]{81 \cdot 3}$
$$= 2\sqrt[4]{3} + 3\sqrt[4]{3}$$
$$= 5\sqrt[4]{3}$$

74. $\sqrt{5}\left(\sqrt{3} + \sqrt{2}\right) = \sqrt{5} \cdot \sqrt{3} + \sqrt{5} \cdot \sqrt{2} = \sqrt{15} + \sqrt{10}$

75. $\sqrt[3]{7}\left(\sqrt[3]{3} + 2\sqrt[3]{7}\right) = \sqrt[3]{7} \cdot \sqrt[3]{3} + \sqrt[3]{7} \cdot 2\sqrt[3]{7}$
$$= \sqrt[3]{21} + 2\sqrt[3]{49}$$

76. $3\sqrt{6}\left(2 - 3\sqrt{6}\right) = 3\sqrt{6} \cdot 2 + 3\sqrt{6} \cdot \left(-3\sqrt{6}\right)$
$$= 6\sqrt{6} - 9\sqrt{36}$$
$$= 6\sqrt{6} - 9 \cdot 6$$
$$= 6\sqrt{6} - 54$$

77. $\left(\sqrt{2} - \sqrt{3}\right)\left(\sqrt{5} + \sqrt{7}\right)$
$$= \sqrt{2} \cdot \sqrt{5} + \sqrt{2} \cdot \sqrt{7} - \sqrt{3} \cdot \sqrt{5} - \sqrt{3} \cdot \sqrt{7}$$
$$= \sqrt{10} + \sqrt{14} - \sqrt{15} - \sqrt{21}$$

78. $\left(\sqrt[3]{2} - 4\right)\left(\sqrt[3]{4} + 2\right)$
$$= \sqrt[3]{2} \cdot \sqrt[3]{4} + \sqrt[3]{2} \cdot 2 - 4 \cdot \sqrt[3]{4} - 4 \cdot 2$$
$$= \sqrt[3]{8} + 2\sqrt[3]{2} - 4\sqrt[3]{4} - 8$$
$$= 2 + 2\sqrt[3]{2} - 4\sqrt[3]{4} - 8$$
$$= -6 + 2\sqrt[3]{2} - 4\sqrt[3]{4}$$

79. $\left(\sqrt[4]{6x}+2\right)\left(\sqrt[4]{2x}-1\right)$

$=\sqrt[4]{6x}\cdot\sqrt[4]{2x}-\sqrt[4]{6x}\cdot1+2\cdot\sqrt[4]{2x}-2\cdot1$

$=\sqrt[4]{12x^2}-\sqrt[4]{6x}+2\sqrt[4]{2x}-2$

80. $\left(\sqrt{5a}+\sqrt{3b}\right)\left(\sqrt{5a}-\sqrt{3b}\right)=\left(\sqrt{5a}\right)^2-\left(\sqrt{3b}\right)^2$

$=5a-3b$

81. $\left(2\sqrt{3}-\sqrt{5}\right)\left(2\sqrt{3}+\sqrt{5}\right)=\left(2\sqrt{3}\right)^2-\left(\sqrt{5}\right)^2$

$=4\cdot3-5$

$=12-5$

$=7$

82. $\left(\sqrt{2}-\sqrt{5}\right)^2=\left(\sqrt{2}\right)^2-2\cdot\sqrt{2}\cdot\sqrt{5}+\left(\sqrt{5}\right)^2$

$=2-2\sqrt{10}+5$

$=7-2\sqrt{10}$

83. $\dfrac{1}{\sqrt{2}}=\dfrac{1}{\sqrt{2}}\cdot\dfrac{\sqrt{2}}{\sqrt{2}}=\dfrac{\sqrt{2}}{\sqrt{4}}=\dfrac{\sqrt{2}}{2}$

84. $\dfrac{3}{\sqrt[3]{3}}=\dfrac{3}{\sqrt[3]{3}}\cdot\dfrac{\sqrt[3]{9}}{\sqrt[3]{9}}=\dfrac{3\sqrt[3]{9}}{\sqrt[3]{27}}=\dfrac{3\sqrt[3]{9}}{3}=\sqrt[3]{9}$

85. $\sqrt{\dfrac{4}{7}}=\dfrac{\sqrt{4}}{\sqrt{7}}=\dfrac{2}{\sqrt{7}}\cdot\dfrac{\sqrt{7}}{\sqrt{7}}=\dfrac{2\sqrt{7}}{\sqrt{49}}=\dfrac{2\sqrt{7}}{7}$

86. $\dfrac{\sqrt[4]{5x^2}}{\sqrt[4]{2}}=\dfrac{\sqrt[4]{5x^2}}{\sqrt[4]{2}}\cdot\dfrac{\sqrt[4]{8}}{\sqrt[4]{8}}=\dfrac{\sqrt[4]{40x^2}}{\sqrt[4]{16}}=\dfrac{\sqrt[4]{40x^2}}{2}$

87. $\sqrt[3]{\dfrac{17}{3y^2}}=\dfrac{\sqrt[3]{17}}{\sqrt[3]{3y^2}}\cdot\dfrac{\sqrt[3]{9y}}{\sqrt[3]{9y}}=\dfrac{\sqrt[3]{153y}}{\sqrt[3]{27y^3}}=\dfrac{\sqrt[3]{153y}}{3y}$

88. $\dfrac{\sqrt[4]{9}}{\sqrt[4]{27}}=\dfrac{\sqrt[4]{9}}{\sqrt[4]{27}}\cdot\dfrac{\sqrt[4]{3}}{\sqrt[4]{3}}=\dfrac{\sqrt[4]{27}}{\sqrt[4]{81}}=\dfrac{\sqrt[4]{27}}{3}$

89. $\dfrac{4}{\sqrt{2}-\sqrt{3}}=\dfrac{4}{\sqrt{2}-\sqrt{3}}\cdot\dfrac{\sqrt{2}+\sqrt{3}}{\sqrt{2}+\sqrt{3}}$

$=\dfrac{4\sqrt{2}+4\sqrt{3}}{\left(\sqrt{2}\right)^2-\left(\sqrt{3}\right)^2}$

$=\dfrac{4\sqrt{2}+4\sqrt{3}}{2-3}$

$=\dfrac{4\sqrt{2}+4\sqrt{3}}{-1}$

$=-4\sqrt{2}-4\sqrt{3}$

90. $\dfrac{1}{4+\sqrt{3}}=\dfrac{1}{4+\sqrt{3}}\cdot\dfrac{4-\sqrt{3}}{4-\sqrt{3}}$

$=\dfrac{4-\sqrt{3}}{\left(4\right)^2-\left(\sqrt{3}\right)^2}$

$=\dfrac{4-\sqrt{3}}{16-3}$

$=\dfrac{4-\sqrt{3}}{13}$

91. $\dfrac{1}{2-\sqrt{n}}=\dfrac{1}{2-\sqrt{n}}\cdot\dfrac{2+\sqrt{n}}{2+\sqrt{n}}$

$=\dfrac{2+\sqrt{n}}{2^2-\left(\sqrt{n}\right)^2}$

$=\dfrac{2+\sqrt{n}}{4-n}$

92. $\dfrac{2\sqrt{3}}{3\sqrt{2}-2\sqrt{3}}=\dfrac{2\sqrt{3}}{3\sqrt{2}-2\sqrt{3}}\cdot\dfrac{3\sqrt{2}+2\sqrt{3}}{3\sqrt{2}+2\sqrt{3}}$

$=\dfrac{6\sqrt{6}+4\sqrt{9}}{\left(3\sqrt{2}\right)^2-\left(2\sqrt{3}\right)^2}$

$=\dfrac{6\sqrt{6}+4\cdot3}{9\cdot2-4\cdot3}$

$=\dfrac{6\sqrt{6}+12}{18-12}$

$=\dfrac{6\sqrt{6}+12}{6}$

$=\sqrt{6}+2$

93. $\dfrac{\sqrt{10}}{6}=\dfrac{\sqrt{10}}{6}\cdot\dfrac{\sqrt{10}}{\sqrt{10}}=\dfrac{\sqrt{100}}{6\sqrt{10}}=\dfrac{10}{6\sqrt{10}}=\dfrac{5}{3\sqrt{10}}$

94. $\dfrac{\sqrt{3x}}{5}=\dfrac{\sqrt{3x}}{5}\cdot\dfrac{\sqrt{3x}}{\sqrt{3x}}=\dfrac{\sqrt{9x^2}}{5\sqrt{3x}}=\dfrac{3x}{5\sqrt{3x}}$

95. $\dfrac{2-\sqrt{3}}{8}=\dfrac{2-\sqrt{3}}{8}\cdot\dfrac{2+\sqrt{3}}{2+\sqrt{3}}$

$=\dfrac{2^2-\left(\sqrt{3}\right)^2}{8\left(2+\sqrt{3}\right)}$

$=\dfrac{4-3}{16+8\sqrt{3}}$

$=\dfrac{1}{16+8\sqrt{3}}$

96. $\dfrac{2\sqrt{t}+\sqrt{3t}}{5t} = \dfrac{2\sqrt{t}+\sqrt{3t}}{5t} \cdot \dfrac{2\sqrt{t}-\sqrt{3t}}{2\sqrt{t}-\sqrt{3t}}$

$\qquad = \dfrac{\left(2\sqrt{t}\right)^2 - \left(\sqrt{3t}\right)^2}{5t\left(2\sqrt{t}-\sqrt{3t}\right)}$

$\qquad = \dfrac{4t-3t}{10t\sqrt{t}-5t\sqrt{3t}}$

$\qquad = \dfrac{t}{10t\sqrt{t}-5t\sqrt{3t}}$

$\qquad = \dfrac{\cancel{t}}{\cancel{t}\left(10\sqrt{t}-5\sqrt{3t}\right)}$

$\qquad = \dfrac{1}{10\sqrt{t}-5\sqrt{3t}}$

97. $\sqrt{x} = 9$

$\left(\sqrt{x}\right)^2 = 9^2$

$\qquad x = 81$

Check:

$\sqrt{81} \overset{?}{=} 9$

$\qquad 9 = 9$

98. $\sqrt{y} = -3$ has no real-number solution.

99. $\sqrt{w-1} = 3$

$\left(\sqrt{w-1}\right)^2 = 3^2$

$\qquad w-1 = 9$

$\qquad w = 10$

Check:

$\sqrt{10-1} \overset{?}{=} 3$

$\sqrt{9} \overset{?}{=} 3$

$\qquad 3 = 3$

100. $\sqrt[3]{3x-2} = -2$

$\left(\sqrt[3]{3x-2}\right)^3 = (-2)^3$

$\qquad 3x-2 = -8$

$\qquad 3x = -6$

$\qquad x = -2$

Check:

$\sqrt[3]{3(-2)-2} \overset{?}{=} -2$

$\sqrt[3]{-6-2} \overset{?}{=} -2$

$\sqrt[3]{-8} \overset{?}{=} -2$

$\qquad -2 = -2$

101. $\sqrt[4]{x-2} - 3 = -1$

$\sqrt[4]{x-2} = 2$

$\left(\sqrt[4]{x-2}\right)^4 = 2^4$

$\qquad x-2 = 16$

$\qquad x = 18$

Check:

$\sqrt[4]{18-2} - 3 \overset{?}{=} -1$

$\sqrt[4]{16} - 3 \overset{?}{=} -1$

$\qquad 2-3 \overset{?}{=} -1$

$\qquad -1 = -1$

102. $\sqrt{y+1} = \sqrt{2y-4}$

$\left(\sqrt{y+1}\right)^2 = \left(\sqrt{2y-4}\right)^2$

$\qquad y+1 = 2y-4$

$\qquad 5 = y$

Check:

$\sqrt{5+1} \overset{?}{=} \sqrt{2(5)-4}$

$\sqrt{6} \overset{?}{=} \sqrt{10-4}$

$\sqrt{6} = \sqrt{6}$

103. $\sqrt{x-6} = x+2$

$\left(\sqrt{x-6}\right)^2 = (x+2)^2$

$\qquad x-6 = x^2+4x+4$

$\qquad 0 = x^2+3x+10$

This polynomial cannot be factored. The equation has no real-number solution.

104. $\sqrt[3]{3x+10}-4=5$

$\sqrt[3]{3x+10}=9$

$\left(\sqrt[3]{3x+10}\right)^3=9^3$

$3x+10=729$

$3x=719$

$x=\dfrac{719}{3}$

Check:

$\sqrt[3]{3\left(\dfrac{719}{3}\right)+10}-4\overset{?}{=}5$

$\sqrt[3]{719+10}-4\overset{?}{=}5$

$\sqrt[3]{729}-4\overset{?}{=}5$

$9-4\overset{?}{=}5$

$5=5$

105. $\sqrt[4]{x+8}=\sqrt[4]{2x+1}$

$\left(\sqrt[4]{x+8}\right)^4=\left(\sqrt[4]{2x+1}\right)^4$

$x+8=2x+1$

$7=x$

Check:

$\sqrt[4]{7+8}\overset{?}{=}\sqrt[4]{2(7)+1}$

$\sqrt[4]{15}\overset{?}{=}\sqrt[4]{14+1}$

$\sqrt[4]{15}=\sqrt[4]{15}$

106. $\sqrt{5n-1}=4-2n$

$\left(\sqrt{5n-1}\right)^2=\left(4-2n\right)^2$

$5n-1=16-16n+4n^2$

$0=4n^2-21n+17$

$0=\left(4n-17\right)\left(n-1\right)$

$4n-17=0$ or $n-1=0$

$4n=17$ $\qquad n=1$

$n=\dfrac{17}{4}$

Check $n=\dfrac{17}{4}$:

$\sqrt{5\left(\dfrac{17}{4}\right)-1}\overset{?}{=}4-2\left(\dfrac{17}{4}\right)$

$\sqrt{\dfrac{85}{4}-\dfrac{4}{4}}\overset{?}{=}4-\dfrac{17}{2}$

$\sqrt{\dfrac{81}{4}}\overset{?}{=}\dfrac{8}{2}-\dfrac{17}{2}$

$\dfrac{9}{2}\neq-\dfrac{9}{2}$

Check $n=1$:

$\sqrt{5(1)-1}\overset{?}{=}4-2(1)$

$\sqrt{5-1}\overset{?}{=}4-2$

$\sqrt{4}\overset{?}{=}2$

$2=2$

The only solution is $n=1$ ($n=\dfrac{17}{4}$ is an extraneous solution).

107. $1+\sqrt{x}=\sqrt{2x+1}$

$\left(1+\sqrt{x}\right)^2=\left(\sqrt{2x+1}\right)^2$

$1+2\sqrt{x}+x=2x+1$

$2\sqrt{x}=x$

$\left(2\sqrt{x}\right)^2=x^2$

$4x=x^2$

$0=x^2-4x$

$0=x\left(x-4\right)$

$x=0$ or $x-4=0$

$\qquad\qquad x=4$

Check $x=0$:

$1+\sqrt{0}\overset{?}{=}\sqrt{2(0)+1}$

$1+0\overset{?}{=}\sqrt{0+1}$

$1\overset{?}{=}\sqrt{1}$

$1=1$

Check $x = 4$:

$$1 + \sqrt{4} \overset{?}{=} \sqrt{2(4)+1}$$

$$1 + 2 \overset{?}{=} \sqrt{8+1}$$

$$3 \overset{?}{=} \sqrt{9}$$

$$3 = 3$$

108. $$\sqrt{3x+1} = 2 - \sqrt{3x}$$

$$\left(\sqrt{3x+1}\right)^2 = \left(2 - \sqrt{3x}\right)^2$$

$$3x + 1 = 4 - 4\sqrt{3x} + 3x$$

$$4\sqrt{3x} = 3$$

$$\left(4\sqrt{3x}\right)^2 = 3^2$$

$$16 \cdot 3x = 9$$

$$48x = 9$$

$$x = \frac{9}{48}$$

$$x = \frac{3}{16}$$

Check:

$$\sqrt{3\left(\frac{3}{16}\right)+1} \overset{?}{=} 2 - \sqrt{3\left(\frac{3}{16}\right)}$$

$$\sqrt{\frac{9}{16} + \frac{16}{16}} \overset{?}{=} 2 - \sqrt{\frac{9}{16}}$$

$$\sqrt{\frac{25}{16}} \overset{?}{=} 2 - \frac{3}{4}$$

$$\frac{5}{4} \overset{?}{=} 2 - \frac{3}{4}$$

$$\frac{5}{4} \overset{?}{=} \frac{8}{4} - \frac{3}{4}$$

$$\frac{5}{4} = \frac{5}{4}$$

109. $$S = 2\sqrt{2L}$$

$$50 = 2\sqrt{2L}$$

$$25 = \sqrt{2L}$$

$$(25)^2 = \left(\sqrt{2L}\right)^2$$

$$625 = 2L$$

$$312.5 = L$$

The length of the skid marks is 312.5 ft.

110. $$T = 2\pi\sqrt{\frac{L}{9.8}}$$

$$\frac{\pi}{3} = 2\pi\sqrt{\frac{L}{9.8}}$$

$$\frac{1}{6} = \sqrt{\frac{L}{9.8}}$$

$$\left(\frac{1}{6}\right)^2 = \left(\sqrt{\frac{L}{9.8}}\right)^2$$

$$\frac{1}{36} = \frac{L}{9.8}$$

$$0.272 \approx L$$

The length of the pendulum is approximately 0.272 m.

111. $$t = \sqrt{\frac{h}{16}}$$

$$0.3 = \frac{\sqrt{h}}{\sqrt{16}}$$

$$0.3 = \frac{\sqrt{h}}{4}$$

$$1.2 = \sqrt{h}$$

$$(1.2)^2 = \left(\sqrt{h}\right)^2$$

$$1.44 = h$$

The distance the object falls is 1.44 ft.

112. $\sqrt{-9} = \sqrt{-1 \cdot 9} = \sqrt{-1} \cdot \sqrt{9} = 3i$

113. $\sqrt{-20} = \sqrt{-1 \cdot 20}$

$$= \sqrt{-1} \cdot \sqrt{20}$$

$$= i\sqrt{20}$$

$$= i\sqrt{4 \cdot 5}$$

$$= 2i\sqrt{5}$$

114. $(3 + 2i) + (5 - 8i) = 8 - 6i$

115. $(7 - 3i) - (-2 + 4i) = (7 - 3i) + (2 - 4i)$

$$= 9 - 7i$$

116. $(3i)(4i) = 12i^2 = 12(-1) = -12$

117. $2i(4 - i) = 8i - 2i^2 = 8i - 2(-1) = 2 + 8i$

118. $(6+2i)(4-i) = 6 \cdot 4 - 6 \cdot i + 2i \cdot 4 - 2i \cdot i$

$= 24 - 6i + 8i - 2i^2$

$= 24 + 2i - 2(-1)$

$= 24 + 2i + 2$

$= 26 + 2i$

119. $(5-i)^2 = 5^2 - 2(5)(i) + (-i)^2$

$= 25 - 10i + i^2$

$= 25 - 10i - 1$

$= 24 - 10i$

120. $\dfrac{5}{i} = \dfrac{5}{i} \cdot \dfrac{i}{i} = \dfrac{5i}{i^2} = \dfrac{5i}{-1} = -5i$

121. $\dfrac{3}{-i} = \dfrac{3}{-i} \cdot \dfrac{i}{i} = \dfrac{3i}{-i^2} = \dfrac{3i}{-(-1)} = \dfrac{3i}{1} = 3i$

122. $\dfrac{4}{3i} = \dfrac{4}{3i} \cdot \dfrac{i}{i} = \dfrac{4i}{3i^2} = \dfrac{4i}{3(-1)} = -\dfrac{4i}{3}$

123. $\dfrac{7+i}{5i} \cdot \dfrac{i}{i} = \dfrac{7i+i^2}{5i^2} = \dfrac{7i-1}{5(-1)} = \dfrac{-1+7i}{-5} = \dfrac{1}{5} - \dfrac{7}{5}i$

124. $\dfrac{3}{2+i} \cdot \dfrac{2-i}{2-i} = \dfrac{3(2-i)}{2^2 - i^2}$

$= \dfrac{6-3i}{4-(-1)}$

$= \dfrac{6-3i}{5}$

$= \dfrac{6}{5} - \dfrac{3}{5}i$

125. $\dfrac{5+i}{2-3i} = \dfrac{5+i}{2-3i} \cdot \dfrac{2+3i}{2+3i} = \dfrac{(5+i)(2+3i)}{2^2 - (3i)^2}$

$= \dfrac{10+15i+2i+3i^2}{4-9i^2} = \dfrac{10+17i+3(-1)}{4-9(-1)}$

$= \dfrac{10+17i-3}{4+9} = \dfrac{7+17i}{13}$

$= \dfrac{7}{13} + \dfrac{17}{13}i$

126. $i^{20} = (i^4)^5 = 1^5 = 1$

127. $i^{15} = i^{12} \cdot i^3 = (i^4)^3 \cdot i^3 = 1^3 \cdot (-i) = 1(-i) = -i$

Chapter 9 Practice Test

1. $\sqrt{36} = 6$

2. $\sqrt{-49} = \sqrt{-1 \cdot 49}$

$= \sqrt{-1} \cdot \sqrt{49}$

$= i \cdot 7$

$= 7i$

3. $\sqrt{81x^2 y^5} = \sqrt{81x^2 y^4 \cdot y}$

$= 9xy^2 \sqrt{y}$

4. $\sqrt[3]{54} = \sqrt[3]{27 \cdot 2} = 3\sqrt[3]{2}$

5. $\sqrt[4]{4x} \cdot \sqrt[4]{4x^5} = \sqrt[4]{16x^6}$

$= \sqrt[4]{16x^4 \cdot x^2}$

$= 2x\sqrt[4]{x^2} \text{ or } 2x\sqrt{x}$

6. $-\sqrt[3]{-27r^{15}} = -(-3r^5) = 3r^5$

7. $\dfrac{\sqrt{5}}{\sqrt{45}} = \sqrt{\dfrac{5}{45}} = \sqrt{\dfrac{1}{9}} = \dfrac{\sqrt{1}}{\sqrt{9}} = \dfrac{1}{3}$

8. $\dfrac{\sqrt[4]{1}}{\sqrt[4]{81}} = \dfrac{1}{3}$

9. $6\sqrt{7} - \sqrt{7} = (6-1)\sqrt{7}$

$= 5\sqrt{7}$

10. $\left(\sqrt{3} - 1\right)^2 = \left(\sqrt{3}\right)^2 - 2\left(\sqrt{3}\right)(1) + (-1)^2$

$= 3 - 2\sqrt{3} + 1$

$= 4 - 2\sqrt{3}$

11. $x^{2/3} \cdot x^{-4/3} = x^{2/3 + (-4/3)}$

$= x^{-2/3}$

$= \dfrac{1}{x^{2/3}}$

12. $\left(\sqrt[3]{2} - 4\right)\left(\sqrt[3]{4} + 2\right)$

$= \sqrt[3]{2} \cdot \sqrt[3]{4} + \sqrt[3]{2} \cdot 2 - 4 \cdot \sqrt[3]{4} - 4 \cdot 2$

$= \sqrt[3]{8} + 2\sqrt[3]{2} - 4\sqrt[3]{4} - 8$

$= 2 + 2\sqrt[3]{2} - 4\sqrt[3]{4} - 8$

$= -6 + 2\sqrt[3]{2} - 4\sqrt[3]{4}$

13. $\sqrt[5]{8x^3} = \left(8x^3\right)^{1/5}$

$\qquad = (8)^{1/5}\left(x^3\right)^{1/5}$

$\qquad = 8^{1/5}\,x^{3 \cdot 1/5}$

$\qquad = 8^{1/5}\,x^{3/5}$

14. $\sqrt[3]{\left(2x+5\right)^2} = \left(2x+5\right)^{2/3}$

15. $\dfrac{1}{\sqrt[3]{4}} = \dfrac{1}{\sqrt[3]{4}} \cdot \dfrac{\sqrt[3]{2}}{\sqrt[3]{2}} = \dfrac{\sqrt[3]{2}}{\sqrt[3]{8}} = \dfrac{\sqrt[3]{2}}{2}$

16. $\dfrac{\sqrt{x}}{\sqrt{x}+\sqrt{y}} = \dfrac{\sqrt{x}}{\sqrt{x}+\sqrt{y}} \cdot \dfrac{\sqrt{x}-\sqrt{y}}{\sqrt{x}-\sqrt{y}}$

$\qquad = \dfrac{\sqrt{x}\left(\sqrt{x}-\sqrt{y}\right)}{\left(\sqrt{x}\right)^2-\left(\sqrt{y}\right)^2}$

$\qquad = \dfrac{\sqrt{x^2}-\sqrt{xy}}{x-y}$

$\qquad = \dfrac{x-\sqrt{xy}}{x-y}$

17. $\sqrt{3x-2} = 8$

$\qquad \left(\sqrt{3x-2}\right)^2 = 8^2$

$\qquad\quad 3x-2 = 64$

$\qquad\qquad 3x = 66$

$\qquad\qquad\; x = 22$

18. $\sqrt[4]{x+8} = \sqrt[4]{2x+1}$

$\quad \left(\sqrt[4]{x+8}\right)^4 = \left(\sqrt[4]{2x+1}\right)^4$

$\qquad\quad x+8 = 2x+1$

$\qquad\qquad 8 = x+1$

$\qquad\qquad 7 = x$

19. $\sqrt{2x+1} - \sqrt{x+1} = 2$

$\qquad \sqrt{2x+1} = 2 + \sqrt{x+1}$

$\qquad \left(\sqrt{2x+1}\right)^2 = \left(2+\sqrt{x+1}\right)^2$

$\qquad 2x+1 = 4 + 4\sqrt{x+1} + x+1$

$\qquad 2x+1 = 5 + x + 4\sqrt{x+1}$

$\qquad x-4 = 4\sqrt{x+1}$

$\qquad \left(x-4\right)^2 = \left(4\sqrt{x+1}\right)^2$

$\qquad x^2-8x+16 = 16(x+1)$

$\qquad x^2-8x+16 = 16x+16$

$\qquad x^2-24x = 0$

$\qquad x(x-24) = 0$

$x = \cancel{0} \;\text{ or }\; x-24 = 0$

$\qquad\qquad x = 24$

0 is an extraneous solution.

20. $(2-i)-(4+3i) = (2-i)+(-4-3i)$

$\qquad\qquad\qquad = -2-4i$

21. $(4-i)(4+i) = 4^2 - i^2$

$\qquad\qquad\quad = 16-(-1)$

$\qquad\qquad\quad = 16+1$

$\qquad\qquad\quad = 17 \text{ or } 17+0i$

22. $\dfrac{2}{4-3i} = \dfrac{2}{4-3i} \cdot \dfrac{4+3i}{4+3i}$

$\qquad = \dfrac{2(4+3i)}{4^2-(3i)^2}$

$\qquad = \dfrac{8+6i}{16-9i^2}$

$\qquad = \dfrac{8+6i}{16-9(-1)}$

$\qquad = \dfrac{8+6i}{16+9}$

$\qquad = \dfrac{8+6i}{25}$

$\qquad = \dfrac{8}{25}+\dfrac{6}{25}i$

23. Area of a parallelogram is base $\times$ height.

$\quad$ Area $= 5\sqrt{12} \cdot 2\sqrt{3} = 10\sqrt{36} = 10 \cdot 6 = 60$ m^2

24. a) Let $h = 12$.

$$t = \sqrt{\frac{h}{16}} = \sqrt{\frac{12}{16}} = \sqrt{\frac{3}{4}} = \frac{\sqrt{3}}{\sqrt{4}} = \frac{\sqrt{3}}{2} \text{ seconds.}$$

b) Let $t = 2$.

$$2 = \sqrt{\frac{h}{16}}$$

$$2 = \frac{\sqrt{h}}{\sqrt{16}}$$

$$2 = \frac{\sqrt{h}}{4}$$

$$8 = \sqrt{h}$$

$$8^2 = \left(\sqrt{h}\right)^2$$

$$64 \text{ ft.} = h$$

25. a) Let $D = 40$ feet.

$$S = \frac{7}{4}\sqrt{D}$$

$$= \frac{7}{4}\sqrt{40}$$

$$= \frac{7}{4}\sqrt{4 \cdot 10}$$

$$= \frac{7}{4} \cdot 2\sqrt{10}$$

$$= \frac{7}{2}\sqrt{10}$$

$$= 3.5\sqrt{10}$$

The speed is $3.5\sqrt{10}$ mph.

b) Let $S = 30$ mph.

$$S = \frac{7}{4}\sqrt{D}$$

$$30 = \frac{7}{4}\sqrt{D}$$

$$\frac{120}{7} = \sqrt{D}$$

$$\left(\frac{120}{7}\right)^2 = \left(\sqrt{D}\right)^2$$

$$\frac{14,400}{49} = D$$

The length of the skid marks is approximately 293.88 ft.

Chapters 1–9 Cumulative Review

1. False; if $|x| > 3$, then $x > 3$ or $x < -3$.

2. False; $(2, -3)$ is not a solution because it does not satisfy both equations.

Equation 1	Equation 2
$4x + 3y = -1$	$2x - 5y = -11$
$4(2) + 3(-3) \overset{?}{=} -1$	$2(2) - 5(-3) \overset{?}{=} -11$
$8 - 9 \overset{?}{=} -1$	$4 + 15 \overset{?}{=} -11$
$-1 = -1$	$19 \neq -11$

3. True

4. True

5. x-intercept

6. inconsistent

7. $24x^5 y^2$

8. $4 - 6(5 - 3^2) + 12 \div 2 - 6$

$$= 4 - 6(5 - 9) + 12 \div 2 - 6$$

$$= 4 - 6(-4) + 12 \div 2 - 6$$

$$= 4 + 24 + 12 \div 2 - 6$$

$$= 4 + 24 + 6 - 6$$

$$= 28$$

9. $\dfrac{7.44 \times 10^{-2}}{3.1 \times 10^4} = 2.4 \times 10^{-2-4}$

$$= 2.4 \times 10^{-6}$$

$$= 0.0000024$$

10. $\dfrac{\left(x^{-2}\right)^3 \left(x^3\right)^4}{\left(x^{-3}\right)^3} = \dfrac{x^{-6} x^{12}}{x^{-9}}$

$$= \frac{x^{-6+12}}{x^{-9}}$$

$$= \frac{x^6}{x^{-9}}$$

$$= x^{6-(-9)}$$

$$= x^{6+9}$$

$$= x^{15}$$

11. $\left(3x^2 - 4y\right)^2$

$$= \left(3x^2\right)^2 - 2 \cdot 3x^2 \cdot 4y + \left(4y\right)^2$$

$$= 9x^4 - 24x^2 y + 16y^2$$

12. $2x-3\overline{\smash{)}8x^3+0x^2-22x+8}$ with quotient $4x^2+6x-2$

$$\underline{8x^3-12x^2}$$
$$12x^2-22x$$
$$\underline{12x^2-18x}$$
$$-4x+8$$
$$\underline{-4x+6}$$
$$2$$

Solution: $\dfrac{8x^3-22x+8}{2x-3}=4x^2+6x-2+\dfrac{2}{2x-3}$

13. $\dfrac{2x^2+7x+3}{x^2-9}\div\dfrac{2x^2+11x+5}{x^2-3x}$

$=\dfrac{2x^2+7x+3}{x^2-9}\cdot\dfrac{x^2-3x}{2x^2+11x+5}$

$=\dfrac{(2x+1)(x+3)}{(x+3)(x-3)}\cdot\dfrac{x(x-3)}{(2x+1)(x+5)}$

$=\dfrac{x}{x+5}$

14. $2\sqrt{5a^2}\cdot 3\sqrt{10a^5}=6\sqrt{50a^7}$

$\qquad\qquad\qquad =6\sqrt{25a^6\cdot 2a}$

$\qquad\qquad\qquad =6\cdot 5a^3\sqrt{2a}$

$\qquad\qquad\qquad =30a^3\sqrt{2a}$

15. $3x\sqrt[3]{24x^4}-4x^2\sqrt[3]{81x}$

$=3x\sqrt[3]{8x^3\cdot 3x}-4x^2\sqrt[3]{27\cdot 3x}$

$=3x\cdot 2x\sqrt[3]{3x}-4x^2\cdot 3\sqrt[3]{3x}$

$=6x^2\sqrt[3]{3x}-12x^2\sqrt[3]{3x}$

$=-6x^2\sqrt[3]{3x}$

16. $32^{-3/5}=\dfrac{1}{32^{3/5}}=\dfrac{1}{\left(\sqrt[5]{32}\right)^3}=\dfrac{1}{(2)^3}=\dfrac{1}{8}$

17. $(4+5i)(2-3i)=8-12i+10i-15i^2$

$\qquad\qquad\qquad =8-2i-15(-1)$

$\qquad\qquad\qquad =8-2i+15$

$\qquad\qquad\qquad =23-2i$

18. $\dfrac{2+\sqrt{2}}{4-\sqrt{2}}=\dfrac{2+\sqrt{2}}{4-\sqrt{2}}\cdot\dfrac{4+\sqrt{2}}{4+\sqrt{2}}$

$=\dfrac{8+2\sqrt{2}+4\sqrt{2}+\left(\sqrt{2}\right)^2}{4^2-\left(\sqrt{2}\right)^2}$

$=\dfrac{8+6\sqrt{2}+2}{16-2}$

$=\dfrac{10+6\sqrt{2}}{14}$

$=\dfrac{2\left(5+3\sqrt{2}\right)}{2\cdot 7}$

$=\dfrac{5+3\sqrt{2}}{7}$

19. $A=\dfrac{1}{2}h(B+b)$

$2A=h(B+b)$

$\dfrac{2A}{h}=B+b$

$\dfrac{2A}{h}-B=b$

$\dfrac{2A}{h}-\dfrac{Bh}{h}=b$

$\dfrac{2A-Bh}{h}=b$

20. $(x+2)(x+4)=63$

$x^2+6x+8=63$

$x^2+6x-55=0$

$(x-5)(x+11)=0$

$x-5=0\quad$ or $\quad x+11=0$

$x=5\qquad\qquad x=-11$

21. $\dfrac{3y}{y-4}-\dfrac{12}{y-4}=6$

$(y-4)\left(\dfrac{3y}{y-4}-\dfrac{12}{y-4}\right)=(y-4)(6)$

$3y-12=6y-24$

$-12=3y-24$

$12=3y$

$4=y$

Notice that if $y=4$, then $\dfrac{3y}{y-4}$ and $\dfrac{12}{y-4}$ are undefined. Therefore, $y=4$ is extraneous and the equation has no solution.

22. $\sqrt{3x+10} = x+2$

$\left(\sqrt{3x+10}\right)^2 = \left(x+2\right)^2$

$3x+10 = x^2 + 4x + 4$

$0 = x^2 + x - 6$

$0 = (x+3)(x-2)$

$x+3 = 0$ or $x-2 = 0$

$x = -3$ $x = 2$

Check $x = -3$:

$\sqrt{3(-3)+10} \overset{?}{=} -3+2$

$\sqrt{-9+10} \overset{?}{=} -1$

$\sqrt{1} \overset{?}{=} -1$

$1 \neq -1$

Check $x = 2$:

$\sqrt{3(2)+10} \overset{?}{=} 2+2$

$\sqrt{6+10} \overset{?}{=} 4$

$\sqrt{16} \overset{?}{=} 4$

$4 = 4$

The only solution is $x = 2$ ($x = -3$ is an extraneous solution).

23. $5x - 2y = -15$

$-2y = -5x - 15$

$y = \dfrac{5}{2}x + \dfrac{15}{2}$

The slope is $m = \dfrac{5}{2}$ and the y-intercept is $\dfrac{15}{2}$.

24. The slope of the given line is $m = 3$. Use $m = 3$ as the slope.

$y - y_1 = m\left(x - x_1\right)$

$y - 2 = 3\left(x - (-4)\right)$

$y - 2 = 3\left(x + 4\right)$

$y - 2 = 3x + 12$

$y = 3x + 14$

25. $y = -\dfrac{5}{3}x + 2$

$m = -\dfrac{5}{3}$; y-intercept: $\left(0, 2\right)$

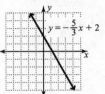

26. $\begin{cases} 2x - y > -4 \\ y > -\dfrac{2}{3}x - 2 \end{cases}$

27. Create a table.

Categories	Hourly wage	Hours worked	Pay
0 – 40 hours worked	w	40	$40w$
hours worked in excess of 40	$1.25w$	6	$6\left(1.25w\right)$

Translate the information in the table to an equation and solve.

$40w + 6\left(1.25w\right) = 782.80$

$40w + 7.5w = 782.80$

$47.5w = 782.80$

$w = 16.48$

Hernando's normal hourly wage is \$16.48 per hour.

28. Complete a table.

Categories	Rate	Time	Distance
against current	$r-5$	$\dfrac{20}{r-5}$	20
with current	$r+5$	$\dfrac{30}{r+5}$	30

$$\frac{20}{r-5} = \frac{30}{r+5}$$

$$30(r-5) = 20(r+5)$$

$$30r - 150 = 20r + 100$$

$$10r = 250$$

$$r = 25$$

The speed of the boat in still water is 25 mph.

29. Translate "y varies jointly as x and the square of z."

$$y = k \cdot x \cdot z^2 \qquad \text{Now,} \quad y = 2xz^2$$

$$72 = k \cdot 4 \cdot (3)^2 \qquad\qquad y = 2(3)(4)^2$$

$$72 = k \cdot 4 \cdot 9 \qquad\qquad\quad y = 2(3)(16)$$

$$72 = 36k \qquad\qquad\qquad\; y = 96$$

$$2 = k$$

30. Substitute and solve.

$$T = 2\pi\sqrt{\frac{L}{9.8}}$$

$$4\pi = 2\pi\sqrt{\frac{L}{9.8}}$$

$$2 = \sqrt{\frac{L}{9.8}}$$

$$2^2 = \left(\sqrt{\frac{L}{9.8}}\right)^2$$

$$4 = \frac{L}{9.8}$$

$$39.2 = L$$

The length of the pendulum is 39.2 m.

Chapter 10
Quadratic Equations and Functions

Exercise Set 10.1

1. $x^2 = 49$

   $x = \pm\sqrt{49}$

   $x = \pm 7$

3. $y^2 = \dfrac{4}{25}$

   $y = \pm\sqrt{\dfrac{4}{25}}$

   $y = \pm\dfrac{2}{5}$

5. $n^2 = 1.44$

   $n = \pm\sqrt{1.44}$

   $n = \pm 1.2$

7. $z^2 = 45$

   $z = \pm\sqrt{45}$

   $z = \pm\sqrt{9 \cdot 5}$

   $z = \pm 3\sqrt{5}$

9. $w^2 = -25$

   $w = \pm\sqrt{-25}$

   $w = \pm 5i$

11. $n^2 - 7 = 42$

    $n^2 = 49$

    $n = \pm\sqrt{49}$

    $n = \pm 7$

13. $y^2 - 16 = 65$

    $y^2 = 81$

    $y = \pm\sqrt{81}$

    $y = \pm 9$

15. $4n^2 = 36$

    $n^2 = 9$

    $n = \pm\sqrt{9}$

    $n = \pm 3$

17. $25t^2 = 9$

    $t^2 = \dfrac{9}{25}$

    $t = \pm\sqrt{\dfrac{9}{25}}$

    $t = \pm\dfrac{3}{5}$

19. $4h^2 = -16$

    $h^2 = -4$

    $h = \pm\sqrt{-4}$

    $h = \pm 2i$

21. $\dfrac{5}{6}x^2 = \dfrac{24}{5}$

    $x^2 = \dfrac{6}{5} \cdot \dfrac{24}{5}$

    $x^2 = \dfrac{144}{25}$

    $x = \pm\sqrt{\dfrac{144}{25}}$

    $x = \pm\dfrac{12}{5}$

23. $2x^2 + 5 = 21$

    $2x^2 = 16$

    $x^2 = 8$

    $x = \pm\sqrt{8}$

    $x = \pm\sqrt{4 \cdot 2}$

    $x = \pm 2\sqrt{2}$

25. $5y^2 - 7 = -97$

    $5y^2 = -90$

    $y^2 = -18$

    $y = \pm\sqrt{-18}$

    $y = \pm\sqrt{-9 \cdot 2}$

    $y = \pm 3i\sqrt{2}$

27. $\dfrac{3}{4}y^2 - 5 = 3$

$\dfrac{3}{4}y^2 = 8$

$y^2 = 8 \cdot \dfrac{4}{3}$

$y^2 = \dfrac{32}{3}$

$y = \pm\sqrt{\dfrac{32}{3}} = \pm\dfrac{\sqrt{32}}{\sqrt{3}} \cdot \dfrac{\sqrt{3}}{\sqrt{3}}$

$= \pm\dfrac{\sqrt{96}}{3} = \pm\dfrac{\sqrt{16 \cdot 6}}{3} = \pm\dfrac{4\sqrt{6}}{3}$

29. $0.2t^2 - 0.5 = 0.012$

$0.2t^2 = 0.512$

$t^2 = 2.56$

$t = \pm\sqrt{2.56}$

$t = \pm 1.6$

31. $(x+8)^2 = 49$

$x + 8 = \pm\sqrt{49}$

$x + 8 = \pm 7$

$x = -8 \pm 7$

$x = -8 + 7 = -1$

$x = -8 - 7 = -15$

33. $(5n-3)^2 = 16$

$5n - 3 = \pm\sqrt{16}$

$5n - 3 = \pm 4$

$5n = 3 \pm 4$

$n = \dfrac{3 \pm 4}{5}$

$n = \dfrac{3+4}{5} = \dfrac{7}{5}$

$n = \dfrac{3-4}{5} = -\dfrac{1}{5}$

35. $(m-8)^2 = -16$

$m - 8 = \pm\sqrt{-16}$

$m - 8 = \pm 4i$

$m = 8 \pm 4i$

37. $(4k-1)^2 = 40$

$4k - 1 = \pm\sqrt{40}$

$4k - 1 = \pm\sqrt{4 \cdot 10}$

$4k - 1 = \pm 2\sqrt{10}$

$4k = 1 \pm 2\sqrt{10}$

$k = \dfrac{1 \pm 2\sqrt{10}}{4}$

39. $(m-7)^2 = -12$

$m - 7 = \pm\sqrt{-12}$

$m - 7 = \pm\sqrt{-4 \cdot 3}$

$m - 7 = \pm 2i\sqrt{3}$

$m = 7 \pm 2i\sqrt{3}$

41. $\left(y - \dfrac{3}{4}\right)^2 = \dfrac{9}{16}$

$y - \dfrac{3}{4} = \pm\sqrt{\dfrac{9}{16}}$

$y - \dfrac{3}{4} = \pm\dfrac{3}{4}$

$y = \dfrac{3}{4} \pm \dfrac{3}{4}$

$y = \dfrac{3}{4} + \dfrac{3}{4} = \dfrac{3}{2}$

$y = \dfrac{3}{4} - \dfrac{3}{4} = 0$

43. $\left(\dfrac{5}{9}d - \dfrac{1}{2}\right)^2 = \dfrac{1}{36}$

$\dfrac{5}{9}d - \dfrac{1}{2} = \pm\sqrt{\dfrac{1}{36}}$

$\dfrac{5}{9}d - \dfrac{1}{2} = \pm\dfrac{1}{6}$

$\dfrac{5}{9}d = \dfrac{1}{2} \pm \dfrac{1}{6}$

$d = \dfrac{9}{5} \cdot \left(\dfrac{1}{2} \pm \dfrac{1}{6}\right)$

$d = \dfrac{9}{5} \cdot \left(\dfrac{1}{2} + \dfrac{1}{6}\right) = \dfrac{6}{5}$

$d = \dfrac{9}{5} \cdot \left(\dfrac{1}{2} - \dfrac{1}{6}\right) = \dfrac{3}{5}$

45. $(0.4x + 3.8)^2 = 2.56$

$\qquad 0.4x + 3.8 = \pm\sqrt{2.56}$

$\qquad 0.4x + 3.8 = \pm 1.6$

$\qquad\qquad 0.4x = -3.8 \pm 1.6$

$\qquad\qquad\quad x = \dfrac{-3.8 \pm 1.6}{0.4}$

$\qquad\qquad\quad x = \dfrac{-3.8 + 1.6}{0.4} = -5.5$

$\qquad\qquad\quad x = \dfrac{-3.8 - 1.6}{0.4} = -13.5$

47. Mistake: Gave only the positive solution.
    Correct: ± 7

49. Mistake: Changed -6 to 6.
    Correct: $5 \pm i\sqrt{6}$

51. $x^2 = 96$

$\qquad x = \pm\sqrt{96}$

$\qquad x = \pm\sqrt{16 \cdot 6}$

$\qquad x = \pm 4\sqrt{6} \approx \pm 9.798$

53. $y^2 - 15 = 5$

$\qquad y^2 = 20$

$\qquad y = \pm\sqrt{20}$

$\qquad y = \pm\sqrt{4 \cdot 5}$

$\qquad y = \pm 2\sqrt{5} \approx \pm 4.472$

55. $(n - 6)^2 = 15$

$\qquad n - 6 = \pm\sqrt{15}$

$\qquad\qquad n = 6 \pm \sqrt{15} \approx 9.873,\ 2.127$

57. a) $\left(\dfrac{14}{2}\right)^2 = 7^2 = 49$

$\qquad x^2 + 14x + 49$

   b) $(x + 7)^2$

59. a) $\left(\dfrac{-10}{2}\right)^2 = (-5)^2 = 25$

$\qquad n^2 - 10n + 25$

   b) $(n - 5)^2$

61. a) $\left(\dfrac{-9}{2}\right)^2 = \dfrac{81}{4}$

$\qquad y^2 - 9y + \dfrac{81}{4}$

   b) $\left(y - \dfrac{9}{2}\right)^2$

63. a) $\left(\dfrac{-\frac{2}{3}}{2}\right)^2 = \left(-\dfrac{1}{3}\right)^2 = \dfrac{1}{9}$

$\qquad s^2 - \dfrac{2}{3}s + \dfrac{1}{9}$

   b) $\left(s - \dfrac{1}{3}\right)^2$

65. $\qquad \left(\dfrac{2}{2}\right)^2 = 1^2 = 1$

$\quad w^2 + 2w + 1 = 15 + 1$

$\qquad (w + 1)^2 = 16$

$\qquad\qquad w + 1 = \pm 4$

$\qquad\qquad\qquad w = -1 \pm 4$

$\qquad\qquad\qquad w = -1 + 4 = 3$

$\qquad\qquad\qquad\quad = -1 - 4 = -5$

67. $\qquad \left(\dfrac{-2}{2}\right)^2 = (-1)^2 = 1$

$\quad r^2 - 2r + 1 = -50 + 1$

$\qquad (r - 1)^2 = -49$

$\qquad\qquad r - 1 = \pm 7i$

$\qquad\qquad\qquad r = 1 \pm 7i$

69. $$\left(\frac{-9}{2}\right)^2 = \frac{81}{4}$$

$$k^2 - 9k + \frac{81}{4} = -18 + \frac{81}{4}$$

$$\left(k - \frac{9}{2}\right)^2 = \frac{9}{4}$$

$$k - \frac{9}{2} = \pm\frac{3}{2}$$

$$k = \frac{9}{2} \pm \frac{3}{2}$$

$$k = \frac{9}{2} + \frac{3}{2} = \frac{12}{2} = 6$$

$$k = \frac{9}{2} - \frac{3}{2} = \frac{6}{2} = 3$$

71. $$\left(\frac{-2}{2}\right)^2 = (-1)^2 = 1$$

$$b^2 - 2b + 1 = 16 + 1$$

$$(b - 1)^2 = 17$$

$$b - 1 = \pm\sqrt{17}$$

$$b = 1 \pm \sqrt{17}$$

73. $$h^2 - 6h = -29$$

$$\left(-\frac{6}{2}\right)^2 = (-3)^2 = 9$$

$$h^2 - 6h + 9 = -29 + 9$$

$$(h - 3)^2 = -20$$

$$h - 3 = \pm\sqrt{-20}$$

$$h - 3 = \pm 2i\sqrt{5}$$

$$h = 3 \pm 2i\sqrt{5}$$

75. $$\left(\frac{\frac{1}{2}}{2}\right)^2 = \left(\frac{1}{4}\right)^2 = \frac{1}{16}$$

$$u^2 + \frac{1}{2}u + \frac{1}{16} = \frac{3}{2} + \frac{1}{16}$$

$$\left(u + \frac{1}{4}\right)^2 = \frac{25}{16}$$

$$u + \frac{1}{4} = \pm\frac{5}{4}$$

$$u = -\frac{1}{4} \pm \frac{5}{4}$$

$$u = -\frac{1}{4} + \frac{5}{4} = 1$$

$$u = -\frac{1}{4} - \frac{5}{4} = -\frac{3}{2}$$

77. $$\frac{4x^2}{4} + \frac{16x}{4} = \frac{9}{4}$$

$$x^2 + 4x = \frac{9}{4}$$

$$\left(\frac{b}{2}\right)^2 = \left(\frac{4}{2}\right)^2 = 2^2 = 4$$

$$x^2 + 4x + 4 = \frac{9}{4} + 4$$

$$(x + 2)^2 = \frac{25}{4}$$

$$x + 2 = \pm\sqrt{\frac{25}{4}}$$

$$x + 2 = \pm\frac{5}{2}$$

$$x = -2 + \frac{5}{2} = \frac{1}{2}$$

$$= -2 - \frac{5}{2} = -\frac{9}{2}$$

79. $$\frac{9x^2}{9} - \frac{18x}{9} = -\frac{5}{9}$$

$$x^2 - 2x = -\frac{5}{9}$$

$$x^2 - 2x + 1 = -\frac{5}{9} + 1$$

$$(x - 1)^2 = \frac{4}{9}$$

$$x - 1 = \pm\sqrt{\frac{4}{9}}$$

$$x = 1 \pm \frac{2}{3}$$

$$x = 1 + \frac{2}{3} = \frac{5}{3}$$

$$x = 1 - \frac{2}{3} = \frac{1}{3}$$

81.
$$n^2 - \frac{1}{2}n = \frac{3}{2}$$
$$\left(\frac{-\frac{1}{2}}{2}\right)^2 = \left(-\frac{1}{4}\right)^2 = \frac{1}{16}$$
$$n^2 - \frac{1}{2}n + \frac{1}{16} = \frac{3}{2} + \frac{1}{16}$$
$$\left(n - \frac{1}{4}\right)^2 = \frac{25}{16}$$
$$n - \frac{1}{4} = \pm\frac{5}{4}$$
$$n = \frac{1}{4} \pm \frac{5}{4}$$
$$n = \frac{1}{4} + \frac{5}{4} = \frac{3}{2}$$
$$n = \frac{1}{4} - \frac{5}{4} = -1$$

83.
$$\frac{4x^2}{4} + \frac{16x}{4} = \frac{7}{4}$$
$$x^2 + 4x = \frac{7}{4}$$
$$x^2 + 4x + 4 = \frac{7}{4} + 4$$
$$(x + 2)^2 = \frac{23}{4}$$
$$x + 2 = \pm\sqrt{\frac{23}{4}}$$
$$x + 2 = \pm\frac{\sqrt{23}}{2}$$
$$x = -2 \pm \frac{\sqrt{23}}{2}$$

85.
$$k^2 + \frac{1}{5}k = \frac{2}{5}$$
$$\left(\frac{\frac{1}{5}}{2}\right)^2 = \left(\frac{1}{10}\right)^2 = \frac{1}{100}$$
$$k^2 + \frac{1}{5}k + \frac{1}{100} = \frac{2}{5} + \frac{1}{100}$$
$$\left(k + \frac{1}{10}\right)^2 = \frac{41}{100}$$

$$k + \frac{1}{10} = \pm\sqrt{\frac{41}{100}}$$
$$k + \frac{1}{10} = \pm\frac{\sqrt{41}}{10}$$
$$k = \frac{-1 \pm \sqrt{41}}{10}$$

87.
$$3a^2 + 4a = 8$$
$$a^2 + \frac{4}{3}a = \frac{8}{3}$$
$$\left(\frac{4}{3} \cdot \frac{1}{2}\right)^2 = \left(\frac{2}{3}\right)^2 = \frac{4}{9}$$
$$a^2 + \frac{4}{3}a + \frac{4}{9} = \frac{8}{3} + \frac{4}{9}$$
$$\left(a + \frac{2}{3}\right)^2 = \frac{28}{9}$$
$$a + \frac{2}{3} = \pm\sqrt{\frac{28}{9}}$$
$$a + \frac{2}{3} = \pm\frac{2\sqrt{7}}{3}$$
$$a = -\frac{2}{3} \pm \frac{2\sqrt{7}}{3}$$
$$a = \frac{-2 \pm 2\sqrt{7}}{3}$$

89. Mistake: Did not divide by 3 so that x^2 has a coefficient of 1; then wrote an incorrect factored form.
Correct:
$$3x^2 + 4x = 2$$
$$x^2 + \frac{4}{3}x = \frac{2}{3}$$
$$x^2 + \frac{4}{3}x + \frac{4}{9} = \frac{2}{3} + \frac{4}{9}$$
$$\left(x + \frac{2}{3}\right)^2 = \frac{10}{9}$$
$$x + \frac{2}{3} = \pm\sqrt{\frac{10}{9}}$$
$$x = -\frac{2}{3} \pm \frac{\sqrt{10}}{3}$$
$$x = \frac{-2 \pm \sqrt{10}}{3}$$

91. $A = s^2$

$\sqrt{A} = \sqrt{s^2}$

$\sqrt{A} = s$

$\sqrt{196} = s$

$14 = s$

The length of each side is 14 inches.

93. $V = lwh$

$144 = 8 \cdot h \cdot 2h$

$144 = 16h^2$

$9 = h^2$

$\sqrt{9} = \sqrt{h^2}$

$3 = h$

height: 3 ft.

width: $2(3) = 6$ ft.

95. $A = \pi r^2$

$23,256.25\pi = \pi r^2$

$23,256.25 = r^2$

$\sqrt{23,256.25} = \sqrt{r^2}$

$152.5 = r$

So, the diameter is $2(152.5) = 305$ m

97.   Translate to the equation $(x-2)^2 = 256$ and solve.

$(x-2)^2 = 256$

$\sqrt{(x-2)^2} = \sqrt{256}$

$x - 2 = 16$

$x = 18$

The length x is 18 inches.

99.   Let the length be x and the width be $x - 4$.

$x^2 + (x-4)^2 = 20^2$

$x^2 + x^2 - 8x + 16 = 400$

$2x^2 - 8x = 384$

$x^2 - 4x = 192$

$x^2 - 4x + 4 = 192 + 4$

$(x-2)^2 = 196$

$x - 2 = \pm\sqrt{196}$

$x - 2 = \pm 14$

$x = 2 \pm 14$

$x = 2 - 14 = -12$

$x = 2 + 14 = 16$

Length must be positive, so disregard the negative solution. The length is 16 ft. and the width is $16 - 4 = 12$ ft.

101. a) Subtract the area of the interior rectangle from the area of the exterior rectangle.

$3l \cdot 2l - 8l = 1230$

$6l^2 - 8l = 1230$

$l^2 - \dfrac{4}{3}l = 205$

$l^2 - \dfrac{4}{3}l + \dfrac{4}{9} = 205 + \dfrac{4}{9}$

$\left(l - \dfrac{2}{3}\right)^2 = \dfrac{1849}{9}$

$l - \dfrac{2}{3} = \pm\sqrt{\dfrac{1849}{9}}$

$l - \dfrac{2}{3} = \pm\dfrac{43}{3}$

$l = \dfrac{2}{3} \pm \dfrac{43}{3}$

$l = \dfrac{2}{3} \pm \dfrac{43}{3} = \dfrac{45}{3} = 15$

$l = \dfrac{2}{3} \pm \dfrac{43}{3} = -\dfrac{41}{3}$

Because lengths cannot be negative, disregard the negative solution. The value of l is 15 cm.

b) length: $3 \cdot 15 = 45$ cm

width: $2 \cdot 15 = 30$ cm

103. $9 = 16t^2$

$$\frac{9}{16} = t^2$$

$$\sqrt{\frac{9}{16}} = t$$

$$\frac{3}{4} = t$$

The cover takes $\frac{3}{4}$ sec. to hit the floor.

105. $400 = \frac{1}{2} \cdot 50 \cdot v^2$

$$400 = 25v^2$$

$$16 = v^2$$

$$\sqrt{16} = \sqrt{v^2}$$

$$4 = v$$

The velocity is 4 m/sec.

Review Exercises

1. $x^2 - 10x + 25 = (x - 5)^2$

2. $x^2 + 6x + 9 = (x + 3)^2$

3. $\sqrt{36x^2} = 6x$

4. $\sqrt{(2x + 3)^2} = 2x + 3$

5. $\dfrac{-4 + \sqrt{8^2 - 4(2)(5)}}{2(2)} = \dfrac{-4 + \sqrt{64 - 40}}{4}$

$$= \frac{-4 + \sqrt{24}}{4}$$

$$= \frac{-4 + \sqrt{4 \cdot 6}}{4}$$

$$= \frac{-4 + 2\sqrt{6}}{4}$$

$$= \frac{-2 + \sqrt{6}}{2} \text{ or } -1 + \frac{\sqrt{6}}{2}$$

6. $\dfrac{-6 - \sqrt{6^2 - 4(5)(2)}}{2(5)} = \dfrac{-6 - \sqrt{36 - 40}}{10}$

$$= \frac{-6 - \sqrt{-4}}{10}$$

$$= \frac{-6 - 2i}{10}$$

$$= \frac{-3 - i}{5} \text{ or } -\frac{3}{5} - \frac{1}{5}i$$

Exercise Set 10.2

1. $x^2 - 3x + 7 = 0$

$a = 1$

$b = -3$

$c = 7$

3. $3x^2 - 9x - 4 = 0$

$a = 3$

$b = -9$

$c = -4$

5. $1.5x^2 - x + 0.2 = 0$

$a = 1.5$

$b = -1$

$c = 0.2$

7. $\dfrac{1}{2}x^2 + \dfrac{3}{4}x - 6 = 0$

$a = \dfrac{1}{2}$

$b = \dfrac{3}{4}$

$c = -6$

9. $x^2 + 9x + 20 = 0$

$a = 1, b = 9, c = 20$

$$x = \frac{-(9) \pm \sqrt{(9)^2 - 4(1)(20)}}{2(1)}$$

$$= \frac{-9 \pm \sqrt{81 - 80}}{2}$$

$$= \frac{-9 \pm \sqrt{1}}{2}$$

$$= \frac{-9 \pm 1}{2}$$

$$= -4, -5$$

11. $4x^2 + 5x - 6 = 0$
$a = 4, b = 5, c = -6$

$$x = \frac{-(5) \pm \sqrt{(5)^2 - 4(4)(-6)}}{2(4)}$$

$$= \frac{-5 \pm \sqrt{25 + 96}}{8}$$

$$= \frac{-5 \pm \sqrt{121}}{8}$$

$$= \frac{-5 \pm 11}{8}$$

$$= \frac{-5 + 11}{8} = \frac{3}{4}$$

$$= \frac{-5 - 11}{8} = -2$$

13. $x^2 - 9x = 0$
$a = 1, b = -9, c = 0$

$$x = \frac{-(-9) \pm \sqrt{(-9)^2 - 4(1)(0)}}{2(1)}$$

$$= \frac{9 \pm \sqrt{81}}{2}$$

$$= \frac{9 \pm 9}{2}$$

$$= \frac{9 + 9}{2} = 9$$

$$= \frac{9 - 9}{2} = 0$$

15. $x^2 - 8x + 16 = 0$
$a = 1, b = -8, c = 16$

$$x = \frac{-(-8) \pm \sqrt{(-8)^2 - 4(1)(16)}}{2(1)}$$

$$= \frac{8 \pm \sqrt{64 - 64}}{2}$$

$$= \frac{8 \pm \sqrt{0}}{2}$$

$$= \frac{8}{2}$$

$$= 4$$

17. $3x^2 + 4x - 4 = 0$
$a = 3, b = 4, c = -4$

$$x = \frac{-(4) \pm \sqrt{(4)^2 - 4(3)(-4)}}{2(3)}$$

$$= \frac{-4 \pm \sqrt{16 + 48}}{6}$$

$$= \frac{-4 \pm \sqrt{64}}{6}$$

$$= \frac{-4 \pm 8}{6}$$

$$= \frac{-4 + 8}{6} = \frac{2}{3}$$

$$= \frac{-4 - 8}{6} = -2$$

19. $x^2 - 2x + 2 = 0$
$a = 1, b = -2, c = 2$

$$x = \frac{-(-2) \pm \sqrt{(-2)^2 - 4(1)(2)}}{2(1)}$$

$$= \frac{2 \pm \sqrt{4 - 8}}{2}$$

$$= \frac{2 \pm \sqrt{-4}}{2}$$

$$= \frac{2 \pm 2i}{2}$$

$$= 1 \pm i$$

21. $x^2 - x - 1 = 0$
$a = 1, b = -1, c = -1$

$$x = \frac{-(-1) \pm \sqrt{(-1)^2 - 4(1)(-1)}}{2(1)}$$

$$= \frac{1 \pm \sqrt{1 + 4}}{2}$$

$$= \frac{1 \pm \sqrt{5}}{2}$$

23. $3x^2 + 10x + 5 = 0$
$a = 3, b = 10, c = 5$

$$x = \frac{-(10) \pm \sqrt{(10)^2 - 4(3)(5)}}{2(3)}$$

$$= \frac{-10 \pm \sqrt{100 - 60}}{6}$$

$$= \frac{-10 \pm \sqrt{40}}{6}$$

$$= \frac{-10 \pm 2\sqrt{10}}{6}$$

$$= \frac{-5 \pm \sqrt{10}}{3}$$

25. $-4x^2 + 6x - 5 = 0$
$a = -4, b = 6, c = -5$

$$x = \frac{-(6) \pm \sqrt{(6)^2 - 4(-4)(-5)}}{2(-4)}$$

$$= \frac{-6 \pm \sqrt{36 - 80}}{-8}$$

$$= \frac{-6 \pm \sqrt{-44}}{-8}$$

$$= \frac{-6 \pm 2i\sqrt{11}}{-8}$$

$$= \frac{3 \pm i\sqrt{11}}{4}$$

27. $18x^2 + 15x + 2 = 0$
$a = 18, b = 15, c = 2$

$$x = \frac{-(15) \pm \sqrt{(15)^2 - 4(18)(2)}}{2(18)}$$

$$= \frac{-15 \pm \sqrt{225 - 144}}{36}$$

$$= \frac{-15 \pm \sqrt{81}}{36}$$

$$= \frac{-15 \pm 9}{36}$$

$$= \frac{-15 + 9}{36} = -\frac{1}{6}$$

$$= \frac{-15 - 9}{36} = -\frac{2}{3}$$

29. $3x^2 - 4x + 3 = 0$
$a = 3, b = -4, c = 3$

$$x = \frac{-(-4) \pm \sqrt{(-4)^2 - 4(3)(3)}}{2(3)}$$

$$= \frac{4 \pm \sqrt{16 - 36}}{6}$$

$$= \frac{4 \pm \sqrt{-20}}{6}$$

$$= \frac{4 \pm 2i\sqrt{5}}{6}$$

$$= \frac{2 \pm i\sqrt{5}}{3}$$

31. $6x^2 - 3x - 4 = 0$
$a = 6, b = -3, c = -4$

$$x = \frac{-(-3) \pm \sqrt{(-3)^2 - 4(6)(-4)}}{2(6)}$$

$$= \frac{3 \pm \sqrt{9 + 96}}{12}$$

$$= \frac{3 \pm \sqrt{105}}{12}$$

33. $2x^2 + 0.1x - 0.03 = 0$
$a = 2, b = 0.1, c = -0.03$

$$x = \frac{-(0.1) \pm \sqrt{(0.1)^2 - 4(2)(-0.03)}}{2(2)}$$

$$= \frac{-0.1 \pm \sqrt{0.01 + 0.24}}{4}$$

$$= \frac{-0.1 \pm \sqrt{0.25}}{4}$$

$$= \frac{-0.1 \pm 0.5}{4}$$

$$= \frac{-0.1 + 0.5}{4} = 0.1$$

$$= \frac{-0.1 - 0.5}{4} = -0.15$$

35. $x^2 + \dfrac{1}{2}x - 3 = 0$

$a = 1, b = \dfrac{1}{2}, c = -3$

$x = \dfrac{-\left(\dfrac{1}{2}\right) \pm \sqrt{\left(\dfrac{1}{2}\right)^2 - 4(1)(-3)}}{2(1)}$

$= \dfrac{-\dfrac{1}{2} \pm \sqrt{\dfrac{1}{4} + 12}}{2}$

$= \dfrac{-\dfrac{1}{2} \pm \sqrt{\dfrac{49}{4}}}{2}$

$= \dfrac{-\dfrac{1}{2} \pm \dfrac{7}{2}}{2}$

$= \dfrac{-\dfrac{1}{2} + \dfrac{7}{2}}{2} = \dfrac{3}{2}$

$= \dfrac{-\dfrac{1}{2} - \dfrac{7}{2}}{2} = -2$

37. Use: $36x^2 - 49 = 0$

$a = 36, b = 0, c = -49$

$x = \dfrac{-(0) \pm \sqrt{(0)^2 - 4(36)(-49)}}{2(36)}$

$= \dfrac{0 \pm \sqrt{0 + 7056}}{72}$

$x = \dfrac{0 \pm 84}{72}$

$= \dfrac{\pm 84}{72}$

$= \pm \dfrac{7}{6}$

39. Use: $x^2 - 2x + 3 = 0$

$a = 1, b = -2, c = 3$

$x = \dfrac{-(-2) \pm \sqrt{(-2)^2 - 4(1)(3)}}{2(1)}$

$= \dfrac{2 \pm \sqrt{4 - 12}}{2}$

$= \dfrac{2 \pm \sqrt{-8}}{2}$

$= \dfrac{2 \pm 2i\sqrt{2}}{2} = 1 \pm i\sqrt{2}$

41. Use: $100x^2 + 6x - 50 = 0$

$a = 100, b = 6, c = -50$

$x = \dfrac{-(6) \pm \sqrt{(6)^2 - 4(100)(-50)}}{2(100)}$

$= \dfrac{-6 \pm \sqrt{36 + 20,000}}{200}$

$= \dfrac{-6 \pm \sqrt{20,036}}{200}$

$= \dfrac{-6 \pm 2\sqrt{5009}}{200}$

$= \dfrac{-3 \pm \sqrt{5009}}{100}$

$\approx -0.738, 0.678$

43. Use: $12x^2 - 6x + 5 = 0$

$a = 12, b = -6, c = 5$

$x = \dfrac{-(-6) \pm \sqrt{(-6)^2 - 4(12)(5)}}{2(12)}$

$= \dfrac{6 \pm \sqrt{36 - 240}}{24}$

$= \dfrac{6 \pm \sqrt{-204}}{24}$

$= \dfrac{6 \pm 2i\sqrt{51}}{24}$

$= \dfrac{3 \pm i\sqrt{51}}{12}$

45. Mistake: Did not evaluate $-b$.

Correct:

$\dfrac{-(-7) \pm \sqrt{(-7)^2 - (4)(3)(1)}}{2(3)} = \dfrac{7 \pm \sqrt{49 - 12}}{6}$

$= \dfrac{7 \pm \sqrt{37}}{6}$

47. Mistake: The result was not completely simplified.

Correct: $\dfrac{2 \pm \sqrt{-8}}{2} = \dfrac{2 \pm 2i\sqrt{2}}{2} = 1 + i\sqrt{2}$

49. $x^2 + 10x + 25 = 0$

$b^2 - 4ac = (10)^2 - 4(1)(25) = 100 - 100 = 0$

1 rational solution

51. $\dfrac{1}{4}x^2 - 4x + 4 = 0$

$b^2 - 4ac = (-4)^2 - 4\left(\dfrac{1}{4}\right)(4) = 16 - 4 = 12$

12 > 0 and 12 is not a perfect square
2 irrational solutions

53. $3x^2 + 10x - 8 = 0$

$b^2 - 4ac = (10)^2 - 4(3)(-8) = 100 + 96 = 196$

196 > 0 and 196 is a perfect square $\left(196 = 14^2\right)$

2 rational solutions

55. $x^2 - x + 3 = 0$

$b^2 - 4ac = (-1)^2 - 4(1)(3) = 1 - 12 = -11$

$-11 < 0$

2 nonreal complex solutions

57. $x^2 - 6x + 6 = 0$

$b^2 - 4ac = (-6)^2 - 4(1)(6) = 36 - 24 = 12$

12 > 0 and 12 is not a perfect square
2 irrational solutions

59. Square root principle or factoring

$x^2 - 81 = 0$

$\qquad x^2 = 81$

$\qquad x = \pm\sqrt{81}$

$\qquad x = \pm 9$

61. Quadratic formula

$2x^2 - 4x - 3 = 0$

$x = \dfrac{-(-4) \pm \sqrt{(-4)^2 - 4(2)(-3)}}{2(2)}$

$\quad = \dfrac{4 \pm \sqrt{16 + 24}}{4}$

$\quad = \dfrac{4 \pm \sqrt{40}}{4}$

$\quad = \dfrac{4 \pm 2\sqrt{10}}{4}$

$\quad = \dfrac{2 \pm \sqrt{10}}{2}$

63. Factoring

$x^2 + 6x = 0$

$x(x + 6) = 0$

$x = 0 \quad\text{or}\quad x + 6 = 0$

$\qquad\qquad\qquad x = -6$

65. Square root principle

$(x + 7)^2 = 40$

$x + 7 = \pm\sqrt{40}$

$x + 7 = \pm 2\sqrt{10}$

$\qquad x = -7 \pm 2\sqrt{10}$

67. Quadratic formula

$x^2 - 8x + 19 = 0$

$x = \dfrac{-(-8) \pm \sqrt{(-8)^2 - 4(1)(19)}}{2(1)}$

$x = \dfrac{8 \pm \sqrt{64 - 76}}{2}$

$x = \dfrac{8 \pm \sqrt{-12}}{2}$

$x = \dfrac{8 \pm 2i\sqrt{3}}{2}$

$x = 4 \pm i\sqrt{3}$

69. x-intercepts: $x^2 - x - 2 = 0$
$$(x-2)(x+1) = 0$$
$$x - 2 = 0 \ \text{ or } \ x + 1 = 0$$
$$x = 2 \qquad\quad x = -1$$
$$(2,0), (-1,0)$$

y-intercept: $y = 0^2 - 0 - 2$
$$y = -2$$
$$(0,-2)$$

71. x-intercept: $4x^2 - 12x + 9 = 0$
$$(2x-3)^2 = 0$$
$$2x - 3 = \pm\sqrt{0}$$
$$2x - 3 = 0$$
$$2x = 3$$
$$x = \frac{3}{2}$$
$$\left(\frac{3}{2}, 0\right)$$

y-intercept: $y = 4(0)^2 - 12(0) + 9$
$$y = 9$$
$$(0,9)$$

73. x-intercepts: $2x^2 + 15x - 8 = 0$
$$(2x-1)(x+8) = 0$$
$$2x - 1 = 0 \ \text{ or } \ x + 8 = 0$$
$$x = \frac{1}{2} \qquad\quad x = -8$$
$$\left(\frac{1}{2}, 0\right), (-8,0)$$

y-intercept: $y = 2 \cdot 0^2 + 15 \cdot 0 - 8$
$$y = -8$$
$$(0,-8)$$

75. x-intercepts: $-2x^2 + 3x - 6 = 0$
$$x = \frac{-3 \pm \sqrt{3^2 - 4(-2)(-6)}}{2(-2)}$$
$$= \frac{-3 \pm \sqrt{9 - 48}}{-4}$$
$$= \frac{-3 \pm \sqrt{-39}}{-4}$$
not real: no x-intercepts

y-intercept: $y = -2 \cdot 0^2 + 3 \cdot 0 - 6$
$$y = -6$$
$$(0,-6)$$

77. Let x represent the first positive integer and $x + 1$ represent the next consecutive integer.
$$x^2 + 5(x+1) = 71$$
$$x^2 + 5x + 5 = 71$$
$$x^2 + 5x - 66 = 0$$
$$(x+11)(x-6) = 0$$
$$x + 11 = 0 \qquad \text{or} \quad x - 6 = 0$$
$$x = \cancel{-11} \qquad\qquad x = 6$$
The integers are 6 and $6 + 1 = 7$.

79. Let x be the first integer, $x + 1$ be the next consecutive integer, and $x + 2$ be the third consecutive integer.
$$x^2 + (x+1)^2 = (x+2)^2$$
$$x^2 + x^2 + 2x + 1 = x^2 + 4x + 4$$
$$2x^2 + 2x + 1 = x^2 + 4x + 4$$
$$x^2 - 2x - 3 = 0$$
$$(x-3)(x+1) = 0$$
$$x - 3 = 0 \ \text{ or } \ x + 1 = 0$$
$$x = 3 \qquad\qquad x = \cancel{-1}$$
The sides are 3, $3 + 1 = 4$, and $3 + 2 = 5$.

81. If the width is w, then the length is $(3w - 3.5)$.
$$w(3w - 3.5) = 34$$
$$3w^2 - 3.5w - 34 = 0$$
$$w = \frac{-(-3.5) \pm \sqrt{(-3.5)^2 - 4(3)(-34)}}{2(3)}$$
$$w = \frac{3.5 \pm \sqrt{12.25 + 408}}{6}$$
$$w = \frac{3.5 \pm \sqrt{420.25}}{6}$$
$$w = \frac{3.5 \pm 20.5}{6}$$
$$w = \frac{3.5 + 20.5}{6} = 4$$
$$w = \frac{3.5 - 20.5}{6} = \cancel{-2.83}$$
The width is 4 ft. and the length is $3(4) - 3.5 = 8.5$ ft.

83. a) $22x = 1.5x \cdot x + 0.5x \cdot 8$

$22x = 1.5x^2 + 4x$

$0 = 1.5x^2 - 18x$

$0 = x(1.5x - 18)$

$0 = 1.5x - 18$ or $\cancel{0} = x$

$18 = 1.5x$

$12 \text{ ft.} = x$

b)

85. $460 = \frac{1}{2}(-32.2)t^2 + 48t + 485$

$0 = 16.1t^2 - 48t - 25$

$t = \dfrac{-(-48) \pm \sqrt{(-48)^2 - 4(16.1)(-25)}}{2(16.1)}$

$= \dfrac{48 \pm \sqrt{2304 + 1610}}{32.2}$

$= \dfrac{48 + \sqrt{3914}}{32.2} \approx 3.43$

$= \dfrac{48 - \sqrt{3914}}{32.2} \approx \cancel{-0.452}$

He was in flight for approximately 3.43 seconds.

87. a) $h = \frac{1}{2}(-9.8)t^2 + 0t + 10$

$h = -4.9t^2 + 10$

b) $5 = -4.9t^2 + 10$

$4.9t^2 = 5$

$t^2 = \dfrac{5}{4.9}$

$t \approx \pm\sqrt{1.02}$

$t \approx 1.01, \cancel{-1.01}$

It takes approximately 1.01 second.

c) $0 = -4.9t^2 + 10$

$4.9t^2 = 10$

$t^2 = \dfrac{10}{4.9}$

$t \approx \pm\sqrt{2.04}$

$t \approx 1.43, \cancel{-1.43}$

It takes approximately 1.43 seconds.

89. $0.5n^2 + 2.5n = 4.5n + 16$

$0.5n^2 - 2n - 16 = 0$

$5n^2 - 20n - 160 = 0$

$n^2 - 4n - 32 = 0$

$(n - 8)(n + 4) = 0$

$n = 8, \cancel{-4}$

The break-even point is 8000 units.

91. a) $b^2 - 4ac = 0$

$(-5)^2 - 4(2)c = 0$

$25 - 8c = 0$

$-8c = -25$

$c = \dfrac{25}{8}$

b) $b^2 - 4ac > 0$

$(-5)^2 - 4(2)c > 0$

$25 - 8c > 0$

$-8c > -25$

$c < \dfrac{25}{8}$

c) $b^2 - 4ac < 0$

$(-5)^2 - 4(2)c < 0$

$25 - 8c < 0$

$-8c < -25$

$c > \dfrac{25}{8}$

93. a) $b^2 - 4ac = 0$

$12^2 - 4a(8) = 0$

$144 - 32a = 0$

$-32a = -144$

$a = \dfrac{-144}{-32} = \dfrac{9}{2}$

b) $b^2 - 4ac > 0$

$12^2 - 4a(8) > 0$

$144 - 32a > 0$

$-32a > -144$

$a < \dfrac{-144}{-32}$

$a < \dfrac{9}{2}$

c) $b^2 - 4ac < 0$

$12^2 - 4a(8) < 0$

$144 - 32a < 0$

$-32a < -144$

$a > \dfrac{-144}{-32}$

$a > \dfrac{9}{2}$

Review Exercises

1. $u^2 - 9u + 14 = (u - 7)(u - 2)$

2. $3u^2 - 2u - 16 = (3u - 8)(u + 2)$

3. $\left(x^2\right)^2 = x^{2 \cdot 2} = x^4$

4. $\left(x^{1/3}\right)^2 = x^{(1/3) \cdot 2} = x^{2/3}$

5. $5u^2 + 13u - 6 = 0$

$(5u - 2)(u + 3) = 0$

$5u - 2 = 0 \quad \text{or} \quad u + 3 = 0$

$5u = 2 \qquad\qquad u = -3$

$u = \dfrac{2}{5}$

6. $\dfrac{7}{3u} = \dfrac{5}{u} - \dfrac{1}{u - 5}$

$3u(u - 5) \cdot \dfrac{7}{3u} = 3u(u - 5) \cdot \left[\dfrac{5}{u} - \dfrac{1}{u - 5}\right]$

$(u - 5)7 = 3(u - 5) \cdot 5 - 3u \cdot 1$

$7u - 35 = 15u - 75 - 3u$

$7u - 35 = 12u - 75$

$-5u = -40$

$u = 8$

Exercise Set 10.3

1. $\dfrac{60}{x + 2} = \dfrac{60}{x} - 5$

$x(x + 2)\left(\dfrac{60}{x + 2}\right) = x(x + 2)\left(\dfrac{60}{x} - 5\right)$

$60x = 60(x + 2) - 5x(x + 2)$

$60x = 60x + 120 - 5x^2 - 10x$

$5x^2 + 10x - 120 = 0$

$x^2 + 2x - 24 = 0$

$(x + 6)(x - 4) = 0$

$x + 6 = 0 \quad \text{or} \quad x - 4 = 0$

$x = -6 \qquad\qquad x = 4$

3. $\dfrac{1}{x} + \dfrac{1}{x + 2} = \dfrac{3}{4}$

$4x(x + 2)\left(\dfrac{1}{x} + \dfrac{1}{x + 2}\right) = 4x(x + 2)\left(\dfrac{3}{4}\right)$

$4(x + 2) + 4x = x(x + 2) \cdot 3$

$4x + 8 + 4x = 3x^2 + 6x$

$-3x^2 + 2x + 8 = 0$

$3x^2 - 2x - 8 = 0$

$(3x + 4)(x - 2) = 0$

$3x + 4 = 0 \quad \text{or} \quad x - 2 = 0$

$x = -\dfrac{4}{3} \qquad\qquad x = 2$

5. $\dfrac{1}{p-4}+\dfrac{1}{4}=\dfrac{8}{p^2-16}$

$4(p-4)(p+4)\left(\dfrac{1}{p-4}+\dfrac{1}{4}\right)$

$\qquad =4(p-4)(p+4)\left(\dfrac{8}{(p+4)(p-4)}\right)$

$4(p+4)+(p-4)(p+4)=4\cdot 8$

$4p+16+p^2-16=32$

$p^2+4p-32=0$

$(p+8)(p-4)=0$

$p+8=0 \quad$ or $\quad p-4=0$

$\quad p=-8 \qquad\qquad p=4$

Note that $p=4$ makes expressions in the original equation undefined. Therefore, $p=4$ is an extraneous solution.

7. $\dfrac{6}{2y+5}=\dfrac{2}{y+5}+\dfrac{1}{5}$

$5(2y+5)(y+5)\left(\dfrac{6}{2y+5}\right)$

$\qquad =5(2y+5)(y+5)\left(\dfrac{2}{y+5}+\dfrac{1}{5}\right)$

$5(y+5)\cdot 6=5(2y+5)\cdot 2+(2y+5)(y+5)$

$30y+150=20y+50+2y^2+15y+25$

$0=2y^2+5y-75$

$0=(y-5)(2y+15)$

$y-5=0 \quad$ or $\quad 2y+15=0$

$\quad y=5 \qquad\qquad y=-7.5$

9. $1+2x^{-1}-8x^{-2}=0$

$1+\dfrac{2}{x}-\dfrac{8}{x^2}=0$

$x^2\left(1+\dfrac{2}{x}-\dfrac{8}{x^2}\right)=x^2\cdot 0$

$x^2+2x-8=0$

$(x+4)(x-2)=0$

$x+4=0 \quad$ or $\quad x-2=0$

$\quad x=-4 \qquad\qquad x=2$

11. $3+13x^{-1}-10x^{-2}=0$

$3+\dfrac{13}{x}-\dfrac{10}{x^2}=0$

$x^2\left(3+\dfrac{13}{x}-\dfrac{10}{x^2}\right)=x^2\cdot 0$

$3x^2+13x-10=0$

$(3x-2)(x+5)=0$

$3x-2=0 \quad$ or $\quad x+5=0$

$x=\dfrac{2}{3} \qquad\qquad x=-5$

13. $x-8\sqrt{x}+15=0$

$-8\sqrt{x}=-x-15$

$8\sqrt{x}=x+15$

$\left(8\sqrt{x}\right)^2=(x+15)^2$

$64x=x^2+30x+225$

$0=x^2-34x+225$

$0=(x-9)(x-25)$

$x-9=0 \quad$ or $\quad x-25=0$

$x=9 \qquad\qquad x=25$

15. $2x-5\sqrt{x}-7=0$

$2x-7=5\sqrt{x}$

$(2x-7)^2=\left(5\sqrt{x}\right)^2$

$4x^2-28x+49=25x$

$4x^2-53x+49=0$

$x=\dfrac{53\pm\sqrt{53^2-4(4)(49)}}{2(4)}$

$x=\dfrac{53\pm\sqrt{2025}}{8}=\dfrac{53\pm 45}{8}=\dfrac{49}{4},\ \cancel{1}$

17. $\sqrt{2a+5}=3a-3$

$\left(\sqrt{2a+5}\right)^2=(3a-3)^2$

$2a+5=9a^2-18a+9$

$0=9a^2-20a+4$

$0=(a-2)(9a-2)$

$a-2=0 \quad$ or $\quad 9a-2=0$

$a=2 \qquad\qquad a=\cancel{\dfrac{2}{9}}$

19. $\sqrt{2m-8} - m - 1 = 0$

$\sqrt{2m-8} = m + 1$

$\left(\sqrt{2m-8}\right)^2 = (m+1)^2$

$2m - 8 = m^2 + 2m + 1$

$0 = m^2 + 9$

$-9 = m^2$

$\pm 3i = m$

21. $\sqrt{4x+1} = \sqrt{x+2} + 1$

$\left(\sqrt{4x+1}\right)^2 = \left(\sqrt{x+2}+1\right)^2$

$4x + 1 = x + 2 + 2\sqrt{x+2} + 1$

$3x - 2 = 2\sqrt{x+2}$

$(3x-2)^2 = \left(2\sqrt{x+2}\right)^2$

$9x^2 - 12x + 4 = 4(x+2)$

$9x^2 - 12x + 4 = 4x + 8$

$9x^2 - 16x - 4 = 0$

$(x-2)(9x+2) = 0$

$x - 2 = 0$ or $9x + 2 = 0$

$x = 2$ $x = \cancel{-\dfrac{2}{9}}$

23. $\sqrt{2x+1} - \sqrt{3x+4} = -1$

$\left(\sqrt{2x+1}\right)^2 = \left(\sqrt{3x+4}-1\right)^2$

$2x + 1 = 3x + 4 - 2\sqrt{3x+4} + 1$

$2\sqrt{3x+4} = x + 4$

$\left(2\sqrt{3x+4}\right)^2 = (x+4)^2$

$4(3x+4) = x^2 + 8x + 16$

$12x + 16 = x^2 + 8x + 16$

$0 = x^2 - 4x$

$0 = x(x-4)$

$x - 4 = 0$ or $x = 0$

$x = 4$

25. Let $u = x^2$.

$u^2 - 10u + 9 = 0$

$(u-9)(u-1) = 0$

$u - 9 = 0$ or $u - 1 = 0$

$u = 9$ $u = 1$

$x^2 = 9$ $x^2 = 1$

$x = \pm 3$ $x = \pm 1$

27. Let $u = x^2$.

$4u^2 - 13u + 9 = 0$

$(4u-9)(u-1) = 0$

$4u - 9 = 0$ or $u - 1 = 0$

$u = \dfrac{9}{4}$ $u = 1$

$x^2 = \dfrac{9}{4}$ $x^2 = 1$

$x = \pm\dfrac{3}{2}$ $x = \pm 1$

29. Let $u = x^2$.

$u^2 + 5u - 36 = 0$

$(u+9)(u-4) = 0$

$u + 9 = 0$ or $u - 4 = 0$

$u = -9$ $u = 4$

$x^2 = -9$ $x^2 = 4$

$x = \pm 3i$ $x = \pm 2$

31. Let $u = x + 2$.

$u^2 + 6u + 8 = 0$

$(u+4)(u+2) = 0$

$u + 4 = 0$ or $u + 2 = 0$

$u = -4$ $u = -2$

$x + 2 = -4$ $x + 2 = -2$

$x = -6$ $x = -4$

33. Let $u = x + 3$.

$$2u^2 - 9u - 5 = 0$$
$$(2u + 1)(u - 5) = 0$$
$$2u + 1 = 0 \quad \text{or} \quad u - 5 = 0$$
$$u = -\frac{1}{2} \qquad\qquad u = 5$$
$$\qquad\qquad\qquad x + 3 = 5$$
$$x + 3 = -\frac{1}{2} \qquad\quad x = 2$$
$$x = -3\frac{1}{2}$$

35. Let $u = \dfrac{x - 1}{2}$.

$$u^2 + 8u + 15 = 0$$
$$(u + 3)(u + 5) = 0$$
$$u + 3 = 0 \quad \text{or} \quad u + 5 = 0$$
$$u = -3 \qquad\qquad u = -5$$
$$\frac{x - 1}{2} = -3 \qquad \frac{x - 1}{2} = -5$$
$$x - 1 = -6 \qquad x - 1 = -10$$
$$x = -5 \qquad\qquad x = -9$$

37. Let $u = \dfrac{x + 2}{2}$.

$$2u^2 + u - 3 = 0$$
$$(2u + 3)(u - 1) = 0$$
$$2u + 3 = 0 \quad \text{or} \quad u - 1 = 0$$
$$u = \frac{-3}{2} \qquad\qquad u = 1$$
$$\frac{x + 2}{2} = \frac{-3}{2} \qquad \frac{x + 2}{2} = 1$$
$$\qquad\qquad\qquad x + 2 = 2$$
$$x + 2 = -3 \qquad\quad x = 0$$
$$x = -5$$

39. Let $u = x^{1/3}$.

$$u^2 - 5u + 6 = 0$$
$$(u - 2)(u - 3) = 0$$
$$u - 2 = 0 \quad \text{or} \quad u - 3 = 0$$
$$u = 2 \qquad\qquad u = 3$$
$$x^{1/3} = 2 \qquad\qquad x^{1/3} = 3$$
$$\left(x^{1/3}\right)^3 = 2^3 \qquad \left(x^{1/3}\right)^3 = 3^3$$
$$x = 8 \qquad\qquad x = 27$$

41. Let $u = x^{1/3}$.

$$2u^2 - 3u - 2 = 0$$
$$(2u + 1)(u - 2) = 0$$
$$2u + 1 = 0 \quad \text{or} \quad u - 2 = 0$$
$$u = -\frac{1}{2} \qquad\qquad u = 2$$
$$\qquad\qquad\qquad x^{1/3} = 2$$
$$x^{1/3} = -\frac{1}{2} \qquad \left(x^{1/3}\right)^3 = 2^3$$
$$\left(x^{1/3}\right)^3 = \left(-\frac{1}{2}\right)^3 \qquad x = 8$$
$$x = -\frac{1}{8}$$

43. Let $u = x^{1/4}$.

$$u^2 - 5u + 6 = 0$$
$$(u - 3)(u - 2) = 0$$
$$u - 3 = 0 \quad \text{or} \quad u - 2 = 0$$
$$u = 3 \qquad\qquad u = 2$$
$$x^{1/4} = 3 \qquad\qquad x^{1/4} = 2$$
$$\left(x^{1/4}\right)^4 = 3^4 \qquad \left(x^{1/4}\right)^4 = 2^4$$
$$x = 81 \qquad\qquad x = 16$$

45. Let $u = x^{1/4}$.

$$5u^2 + 8u - 4 = 0$$
$$(5u - 2)(u + 2) = 0$$
$$5u - 2 = 0 \quad \text{or} \quad u + 2 = 0$$
$$u = \frac{2}{5} \qquad\qquad u = -2$$
$$\qquad\qquad\qquad x^{1/4} = -2$$
$$x^{1/4} = \frac{2}{5} \qquad \left(x^{1/4}\right)^4 = (-2)^4$$
$$\left(x^{1/4}\right)^4 = \left(\frac{2}{5}\right)^4 \qquad x = \cancel{16}$$
$$x = \frac{16}{625}$$

47. Let x represent the time it takes the bus to travel 180 miles.

Vehicle	d	t	r
bus	180	x	$\dfrac{180}{x}$
truck	180	$x+1$	$\dfrac{180}{x+1}$

Rate of bus = Rate of truck + 15

$$\frac{180}{x} = \frac{180}{x+1} + 15$$

$$180(x+1) = 180x + 15x(x+1)$$

$$180x + 180 = 180x + 15x^2 + 15x$$

$$0 = 15x^2 + 15x - 180$$

$$0 = x^2 + x - 12$$

$$0 = (x+4)(x-3)$$

$$x = \cancel{-4}, 3$$

Because time cannot be negative, it takes the bus 3 hours to travel 180 miles.

49. Let x represent the time it takes the winner to run 26 miles.

Person	d	t	r
winner	26	x	$\dfrac{26}{x}$
other	26	$x+1$	$\dfrac{26}{x+1}$

Rate of winner = Rate of other + 3.66

$$\frac{26}{x} = \frac{26}{x+1} + 3.66$$

$$26(x+1) = 26x + 3.66x(x+1)$$

$$26x + 26 = 26x + 3.66x^2 + 3.66x$$

$$0 = 3.66x^2 + 3.66x - 26$$

$$x = \frac{-3.66 \pm \sqrt{(3.66)^2 - 4(3.66)(-26)}}{2(3.66)}$$

$$= \frac{-3.66 \pm \sqrt{394.0356}}{7.32}$$

$$\approx \frac{-3.66 \pm 19.850}{7.32} \approx 2.21, \cancel{-3.21}$$

Because time cannot be negative, it takes the winner approximately 2.21 hours to run 26 miles.

51. Let x represent the time it takes the motorcycle to travel 300 miles.

Vehicle	d	t	r
motorcycle	300	x	$\dfrac{300}{x}$
bus	360	$x+3$	$\dfrac{360}{x+3}$

Rate of motorcycle = Rate of bus + 15

$$\frac{300}{x} = \frac{360}{x+3} + 15$$

$$300(x+3) = 360x + 15x(x+3)$$

$$300x + 900 = 360x + 15x^2 + 45x$$

$$0 = 15x^2 + 105x - 900$$

$$0 = x^2 + 7x - 60$$

$$0 = (x+12)(x-5)$$

$$x = \cancel{-12}, 5$$

Because time cannot be negative, it takes the motorcycle approximately 5 hours to travel 300 miles. However, we are asked for the rate of the bus. Making the necessary substitution, we find that the bus is traveling at a rate of 45 mph.

$$\text{Rate of bus} = \frac{360}{5+3} = \frac{360}{8} = 45 \text{ mph}$$

53. Let x represent the time it took the cyclist to travel 36 miles a year ago.

	d	t	r
Old	36	x	$\dfrac{36}{x}$
New	36	$x-1$	$\dfrac{36}{x-1}$

New rate = Old rate + 6

$$\frac{36}{x-1} = \frac{36}{x} + 6$$

$$36x = 36(x-1) + 6x(x-1)$$

$$36x = 36x - 36 + 6x^2 - 6x$$

$$0 = 6x^2 - 6x - 36$$

$$0 = x^2 - x - 6$$

$$0 = (x-3)(x+2)$$

$$x = 3, \cancel{-2}$$

Because time cannot be negative, it used to take the cyclist approximately 3 hours to travel 36 miles.

a) Old rate $= \dfrac{36}{3} = 12$ mph

New rate $= \dfrac{36}{2} = 18$ mph

b) Old time is 3 hours, new time is 2 hours

55. Let x represent the number of hours for Jody to run the hoop nets.

Worker	Time to complete alone	Rate of work	Time at work	Portion of job completed
Billy	$x+2$	$\dfrac{1}{x+2}$	$\dfrac{12}{5}$	$\dfrac{12}{5(x+2)}$
Jody	x	$\dfrac{1}{x}$	$\dfrac{12}{5}$	$\dfrac{12}{5x}$

Billy's portion + Jody's portion = 1 (entire job)

$$\frac{12}{5(x+2)} + \frac{12}{5x} = 1$$

$$12x + 12(x+2) = 5x(x+2)$$

$$12x + 12x + 24 = 5x^2 + 10x$$

$$0 = 5x^2 - 14x - 24$$

$$0 = (5x+6)(x-4)$$

$$x = -\frac{6}{5}, 4$$

Since a negative time makes no sense in the context of this problem, it takes Jody 4 hours to run the hoop nets alone and Billy $4 + 2 = 6$ hours working alone.

57. Let x represent the number of hours for the new press to print the copies.

Press	Time to complete alone	Rate of work	Time at work	Portion of job completed
New	x	$\dfrac{1}{x}$	2	$\dfrac{2}{x}$
Old	$x+\dfrac{1}{2}$	$\dfrac{1}{x+\dfrac{1}{2}}$	2	$\dfrac{2}{x+\dfrac{1}{2}} = \dfrac{4}{2x+1}$

New press portion + old press portion = 1 (entire job.)

$$\frac{2}{x} + \frac{4}{2x+1} = 1$$

$$2(2x+1) + 4x = x(2x+1)$$

$$4x + 2 + 4x = 2x^2 + x$$

$$0 = 2x^2 - 7x - 2$$

$$x = \frac{-(-7) \pm \sqrt{(-7)^2 - 4(2)(-2)}}{2(2)}$$

$$= \frac{7 \pm \sqrt{49+16}}{4} = \frac{7 \pm \sqrt{65}}{4}$$

$$\approx 3.77, -0.27$$

Since a negative time makes no sense in the context of this problem, it takes the new press about 3.77 hours to run the copies alone and the old press about $3.77 + 0.5 = 4.27$ hours working alone.

59. String 1: tension = 50 lbs., frequency = x
String 2: tension = 60 lbs., frequency = $x + 40$ vps

$$\frac{x^2}{(40+x)^2} = \frac{50}{60}$$

$$60x^2 = 50(40+x)^2$$

$$60x^2 = 50(1600 + 80x + x^2)$$

$$60x^2 = 80{,}000 + 4000x + 50x^2$$

$$10x^2 - 4000x - 80{,}000 = 0$$

$$x^2 - 400x - 8000 = 0$$

$$x = \frac{-(-400) \pm \sqrt{(-400)^2 - 4(1)(-8000)}}{2(1)}$$

$$= \frac{400 \pm \sqrt{192{,}000}}{2} \approx 420, -19$$

Since string frequency must be positive, string 1's frequency is 420 vps, and string 2's frequency is $420 + 40 = 460$ vps.

Review Exercises

1. $-\dfrac{b}{2a} = -\dfrac{8}{2 \cdot 2} = -\dfrac{8}{4} = -2$

2. $\dfrac{4ac - b^2}{4a} = \dfrac{4(2)(5) - (-8)^2}{4(2)}$

$$= \frac{40 - 64}{8}$$

$$= \frac{-24}{8}$$

$$= -3$$

3. $2x + 3y = 6$

x-intercept: $2x + 3 \cdot 0 = 6$

$2x = 6$

$x = 3$

$(3, 0)$

y-intercept: $2 \cdot 0 + 3y = 6$

$3y = 6$

$y = 2$

$(0, 2)$

4. $f(x) = 2x^2 - 3x + 1$

$f(-1) = 2(-1)^2 - 3(-1) + 1$

$= 2 \cdot 1 + 3 + 1$

$= 2 + 3 + 1$

$= 6$

5. $f(x) = x^2 - 3$

x	$f(x)$	(x, y)
-2	1	$(-2, 1)$
-1	-2	$(-1, -2)$
0	-3	$(0, -3)$
1	-2	$(1, -2)$
2	1	$(2, 1)$

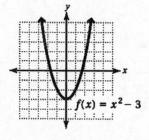

6. $f(x) = 3x^2$

x	$f(x)$	(x, y)
-2	12	$(-2, 12)$
-1	3	$(-1, 3)$
0	0	$(0, 0)$
1	3	$(1, 3)$
2	12	$(2, 12)$

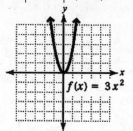

Exercise Set 10.4

1. $f(x) = 5(x - 2)^2 - 3$

Vertex: $(2, -3)$; axis of symmetry $x = 2$

3. $g(x) = -2(x + 1)^2 - 5$

Vertex: $(-1, -5)$; axis of symmetry $x = -1$

5. $k(x) = x^2 + 2$

Vertex: $(0, 2)$; axis of symmetry $x = 0$

7. Can be rewritten as $h(x) = 3(x + 5)^2 + 0$

Vertex: $(-5, 0)$; axis of symmetry $x = -5$

9. Can be rewritten as $f(x) = -0.5(x - 0)^2 + 0$

Vertex: $(0, 0)$; axis of symmetry $x = 0$

11. $f(x) = x^2 - 4x + 8$

x-coordinate: $\dfrac{-(-4)}{2(1)} = \dfrac{4}{2} = 2$

y-coordinate: $y = (2)^2 - 4(2) + 8$

$= 4 - 8 + 8$

$= 4$

vertex: $(2, 4)$

axis of symmetry: $x = 2$

13. $k(x) = 2x^2 + 16x + 27$

x-coordinate: $\dfrac{-(16)}{2(2)} = \dfrac{-16}{4} = -4$

y-coordinate: $y = 2(-4)^2 + 16(-4) + 27$

$\qquad = 32 - 64 + 27$

$\qquad = -5$

vertex: $(-4, -5)$

axis of symmetry: $x = -4$

15. $g(x) = -3x^2 + 2x + 1$

x-coordinate: $\dfrac{-(2)}{2(-3)} = \dfrac{-2}{-6} = \dfrac{1}{3}$

y-coordinate: $y = -3\left(\dfrac{1}{3}\right)^2 + 2\left(\dfrac{1}{3}\right) + 1$

$\qquad = -3 \cdot \dfrac{1}{9} + \dfrac{2}{3} + 1$

$\qquad = -\dfrac{1}{3} + \dfrac{2}{3} + 1$

$\qquad = \dfrac{4}{3}$

vertex: $\left(\dfrac{1}{3}, \dfrac{4}{3}\right)$

axis of symmetry: $x = \dfrac{1}{3}$

17. $h(x) = -3x^2$

a) downward

b) $(0, 0)$

c) $x = 0$

d)

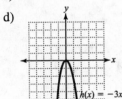

19. $k(x) = \dfrac{1}{4}x^2$

a) upward

b) $(0, 0)$

c) $x = 0$

d)

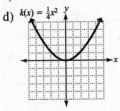

21. $f(x) = 4x^2 - 3$

a) upward

b) $(0, -3)$

c) $x = 0$

d)

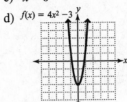

23. $g(x) = -0.5x^2 + 2$

a) downward

b) $(0, 2)$

c) $x = 0$

d)

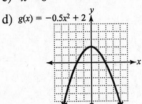

25. $f(x) = (x - 3)^2 + 2$

a) upward

b) $(3, 2)$

c) $x = 3$

d)

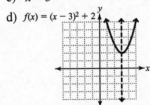

27. $k(x) = -2(x + 1)^2 - 3$

a) downward

b) $(-1, -3)$

c) $x = -1$

d)

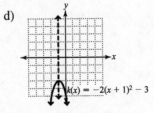

29. $h(x) = \frac{1}{3}(x-2)^2 - 1$

    a) upward

    b) $(2, -1)$

    c) $x = 2$

    d)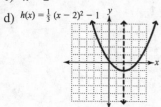

31. $h(x) = x^2 + 6x + 9$

    a) x-intercept: $x^2 + 6x + 9 = 0$

    $$(x+3)^2 = 0$$

    $$x + 3 = \pm\sqrt{0}$$

    $$x + 3 = 0$$

    $$x = -3$$

    $$(-3, 0)$$

    y-intercept: $y = 0^2 + 6(0) + 9 = 9$

    $(0, 9)$

    b) $y = x^2 + 6x + 9$

    $$y = (x+3)^2$$

    $$h(x) = (x+3)^2$$

    c) upward

    d) $(-3, 0)$

    e) $x = -3$

    f)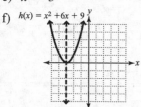

    g) Domain: $\{x \mid x \text{ is a real number}\}$ or $(-\infty, \infty)$

    Range: $\{y \mid y \ge 0\}$ or $[0, \infty)$

33. $g(x) = -3x^2 + 6x - 5$

    a) x-intercepts:

    $$-3x^2 + 6x - 5 = 0$$

    $$x = \frac{-6 \pm \sqrt{6^2 - 4(-3)(-5)}}{2(-3)}$$

    $$= \frac{-6 \pm \sqrt{36 - 60}}{-6}$$

    $$= \frac{-6 \pm \sqrt{-24}}{-6}$$

    Since the discriminant is negative, there are no x-intercepts.

    y-intercept: $y = -3(0)^2 + 6(0) - 5 = -5$

    $(0, -5)$

    b) $y = -3x^2 + 6x - 5$

    $$y + 5 = -3(x^2 - 2x)$$

    $$y + 5 - 3 = -3(x^2 - 2x + 1)$$

    $$y + 2 = -3(x-1)^2$$

    $$y = -3(x-1)^2 - 2$$

    $$g(x) = -3(x-1)^2 - 2$$

    c) downward

    d) $(1, -2)$

    e) $x = 1$

    f)

    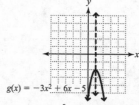

    g) Domain: $\{x \mid x \text{ is a real number}\}$ or $(-\infty, \infty)$

    Range: $\{y \mid y \le -2\}$ or $(-\infty, -2]$

35. $f(x) = 2x^2 + 6x + 3$

   a) x-intercepts:

$$2x^2 + 6x + 3 = 0$$

$$x = \frac{-6 \pm \sqrt{6^2 - 4(2)(3)}}{2(2)}$$

$$= \frac{-6 \pm \sqrt{36 - 24}}{4}$$

$$= \frac{-6 \pm \sqrt{12}}{4}$$

$$= \frac{-6 \pm 2\sqrt{3}}{4}$$

$$= \frac{-3 \pm \sqrt{3}}{2}$$

$$\left(\frac{-3 + \sqrt{3}}{2}, 0 \right), \left(\frac{-3 - \sqrt{3}}{2}, 0 \right)$$

y-intercept: $y = 2(0)^2 + 6(0) + 3 = 3$

$(0, 3)$

   b) $y = 2x^2 + 6x + 3$

$$y - 3 = 2x^2 + 6x$$

$$y - 3 = 2(x^2 + 3x)$$

$$y - 3 + \frac{9}{2} = 2\left(x^2 + 3x + \frac{9}{4} \right)$$

$$y + \frac{3}{2} = 2\left(x + \frac{3}{2} \right)^2$$

$$y = 2\left(x + \frac{3}{2} \right)^2 - \frac{3}{2}$$

$$f(x) = 2\left(x + \frac{3}{2} \right)^2 - \frac{3}{2}$$

   c) upward

   d) $\left(-\dfrac{3}{2}, -\dfrac{3}{2} \right)$

   e) $x = -\dfrac{3}{2}$

   f) $f(x) = 2x^2 + 6x + 3$

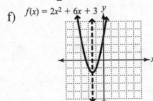

   g) Domain: $\{x \mid x \text{ is a real number}\}$ or $(-\infty, \infty)$

     Range: $\left\{ y \mid y \geq -\dfrac{3}{2} \right\}$ or $\left[-\dfrac{3}{2}, \infty \right)$

37. Finding the vertex will allow us to know the highest point and how many seconds it takes to reach the highest point.

   a) t-coordinate: $t = \dfrac{-45}{2(-4.9)} \approx 4.59$ sec is the time it takes to reach the highest point.

   b) h-coordinate:

$$h = -4.9(4.59)^2 + 45(4.59) \approx 103.32 \text{ meters}$$

     is the maximum height that the rocket reaches.

   c) The t-intercepts describe the times that the rocket is on the ground.

$$h = 0: 0 = -4.9t^2 + 45t$$

$$t = \frac{-(45) \pm \sqrt{(45)^2 - 4(-4.9)(0)}}{2(-4.9)}$$

$$= \frac{-45 \pm \sqrt{2025 + 0}}{-9.8} = \frac{-45 \pm 45}{-9.8}$$

$$= \frac{-45 + 45}{-9.8} = 0 \text{ (starting point)}$$

     or $= \dfrac{-45 - 45}{-9.8} = 9.18$ sec.

39. a) x-coordinate: $-\dfrac{3.2}{2(-0.8)} = 2$

     The maximum height will be the y-coordinate of the vertex.

     y-coordinate: $y = -0.8(2)^2 + 3.2(2) + 6 = 9.2$

     feet is the maximum height that the ball reaches.

   b) To find how far the ball travels, we need to find the x-value when it hits the ground, which is where $y = 0$.

$$y = 0: 0 = -0.8x^2 + 3.2x + 6$$

$$x = \frac{-(3.2) \pm \sqrt{(3.2)^2 - 4(-0.8)(6)}}{2(-0.8)}$$

$$\approx -1.39, 5.39$$

     The negative values does not make sense in this context, so the ball travels approximately 5.39 feet.

   c)

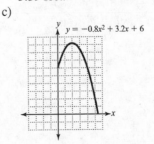

$y = -0.8x^2 + 3.2x + 6$

41. Finding the vertex will allow us to know the highest number of CDs sold and how many weeks it takes to reach that point.

$$n(t) = -200t^2 + 4000t$$

$$n(t) = -200\left(t^2 - 20t\right)$$

$$n(t) - 20,000 = -200\left(t^2 - 20t + 100\right)$$

$$n(t) = -200(t-10)^2 + 20,000$$

Vertex: $(10, 20,000)$

a) The tenth week   b) 20,000

43. Because the enclosed space is to be rectangular, the sum of the length and width is $\frac{1}{2}(400) = 200$ feet. Let x represent the width and $200 - x$ represent the length.

$$y = x(200 - x)$$

$$y = 200x - x^2$$

$$y = -x^2 + 200x$$

$$y - 10,000 = -\left(x^2 - 200x + 10,000\right)$$

$$y - 10,000 = -(x - 100)^2$$

$$y = -(x - 100)^2 + 10,000$$

The vertex, $(100, 10,000)$ shows the highest point on the graph, or the maximum area to be achieved.

The area is maximized if the width is 100 feet and the length is $200 - 100 = 100$ feet.

45. The vertex would be the lowest point on the graph.

$$C(n) = n^2 - 110n + 5000$$

$$m = n^2 - 110n + 5000$$

$$m - 5000 = n^2 - 110n$$

$$m - 5000 + 3025 = n^2 - 110n + 3025$$

$$m + 1075 = (n - 55)^2$$

$$m = (n - 55)^2 + 1075$$

$$c(n) = (n - 55)^2 + 1075$$

55 units would minimize the cost.

47. Let one of the integers be x. Then the other integer is $x + 12$.

$$y = x(x + 12)$$

$$y = x^2 + 12x$$

$$y + 36 = x^2 + 12x + 36$$

$$y + 36 = (x + 6)^2$$

$$y = (x + 6)^2 - 36$$

The vertex $(-6, -36)$ shows one of the integers is –6 the other is –6 + 12 = 6, and the product is –36.

Review Exercises

1. $-\left|-2 + 3 \cdot 4\right| - 4^0 = -\left|-2 + 12\right| - 1$
$$= -\left|10\right| - 1$$
$$= -10 - 1$$
$$= -11$$

2. $x^2 + 2x = 15$
$$x^2 + 2x - 15 = 0$$
$$(x + 5)(x - 3) = 0$$
$$x + 5 = 0 \quad \text{or} \quad x - 3 = 0$$
$$x = -5 \qquad\qquad x = 3$$

3. $x^2 = -20$
$$x = \pm\sqrt{-20}$$
$$x = \pm 2i\sqrt{5}$$

4. $4x - 8 \le 2x + 1$
$$2x \le 9$$
$$x \le 4.5$$

5. $\left|x - 3\right| \ge 8$
$$x - 3 \le -8 \quad \text{or} \quad x - 3 \ge 8$$
$$x \le -5 \qquad\qquad x \ge 11$$

6. $\begin{cases} x+y>2 \\ 2x-3y\le 6 \end{cases}$

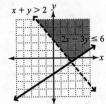

Exercise Set 10.5

1. a) $x=-5,-1$ b) $(-5,-1)$

   c) $(-\infty,-5)\cup(-1,\infty)$

3. a) $x=-1,3$ b) $(-\infty,-1]\cup[3,\infty)$

   c) $[-1,3]$

5. a) $x=-2$ b) $(-\infty,-2)\cup(-2,\infty)$

   c) $\varnothing$

7. a) $\varnothing$ b) $\varnothing$

   c) $\mathbb{R}$ or $(-\infty,\infty)$

9. $(x+4)(x+2)=0$

   $x+4=0$ or $x+2=0$

   $x=-4$ $x=-2$

   For $(-\infty,-4)$, we choose -5.

   $(-5+4)(-5+2)<0$ False

   $-1(-3)<0$

   $3<0$

   For $(-4,-2)$, we choose -3.

   $(-3+4)(-3+2)<0$ True

   $1(-1)<0$

   $-1<0$

   For $(-2,\infty)$, we choose 0.

   $(0+4)(0+2)<0$ False

   $4\cdot2<0$

   $8<0$

   Solution set: $(-4,-2)$

11. $(x-2)(x-5)=0$

    $x-2=0$ or $x-5=0$

    $x=2$ $x=5$

    For $(-\infty,2)$, we choose 0.

    $(0-2)(0-5)>0$ True

    $-2(-5)>0$

    $10>0$

    For $(2,5)$, we choose 3.

    $(3-2)(3-5)>0$ False

    $1(-2)>0$

    $-2>0$

    For $(5,\infty)$, we choose 6.

    $(6-2)(6-5)>0$ True

    $4(1)>0$

    $4>0$

    Solution set: $(-\infty,2)\cup(5,\infty)$

13. $x^2+5x+4=0$

    $(x+4)(x+1)=0$

    $x+4=0$ or $x+1=0$

    $x=-4$ $x=-1$

    For $(-\infty,-4)$, we choose -5

    $(-5)^2+5(-5)+4<0$ False

    $25-25+4<0$

    $4<0$

    For $(-4,-1)$, we choose -3.

    $(-3)^2+5(-3)+4<0$ True

    $9-15+4<0$

    $-2<0$

    For $(-1,\infty)$, we choose 0.

    $(0)^2+5(0)+4<0$ False

    $4<0$

    Solution set: $(-4,-1)$

15. $x^2 - 4x + 3 = 0$

$(x-3)(x-1) = 0$

$x - 3 = 0$ or $x - 1 = 0$

$x = 3$ $x = 1$

For $(-\infty, 1)$, we choose 0.

$(0)^2 - 4(0) + 3 > 0$ True

$3 > 0$

For $(1, 3)$, we choose 2.

$(2)^2 - 4(2) + 3 > 0$ False

$4 - 8 + 3 > 0$

$-1 > 0$

For $(3, \infty)$, we choose 4.

$(4)^2 - 4(4) + 3 > 0$ True

$16 - 16 + 3 > 0$

$3 > 0$

Solution set: $(-\infty, 1) \cup (3, \infty)$

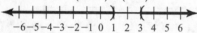

17. $b^2 - 6b + 8 = 0$

$(b-4)(b-2) = 0$

$b - 4 = 0$ or $b - 2 = 0$

$b = 4$ $b = 2$

For $(-\infty, 2)$, we choose 0.

$(0)^2 - 6(0) + 8 \le 0$ False

$8 \le 0$

For $(2, 4)$, we choose 3.

$(3)^2 - 6(3) + 8 \le 0$ True

$9 - 18 + 8 \le 0$

$-1 \le 0$

For $(4, \infty)$, we choose 5.

$(5)^2 - 6(5) + 8 \le 0$ False

$25 - 30 + 8 \le 0$

$3 \le 0$

Solution set: $[2, 4]$

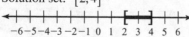

19. $y^2 - 4y - 5 = 0$

$(y+1)(y-5) = 0$

$y + 1 = 0$ or $y - 5 = 0$

$y = -1$ $y = 5$

For $(-\infty, -1)$, we choose -2.

$(-2)^2 - 5 \ge 4(-2)$ True

$4 - 5 \ge -8$

$-1 \ge -8$

For $(-1, 5)$, we choose 3.

$(3)^2 - 5 \ge 4(3)$ False

$9 - 5 \ge 12$

$4 \ge 12$

For $(5, \infty)$, we choose 6.

$(6)^2 - 5 \ge 4(6)$ True

$36 - 5 \ge 24$

$31 \ge 24$

Solution set: $(-\infty, -1] \cup [5, \infty)$

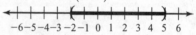

21. $a^2 - 3a - 10 = 0$

$(a-5)(a+2) = 0$

$a - 5 = 0$ or $a + 2 = 0$

$a = 5$ $a = -2$

For $(-\infty, -2)$, we choose -3.

$(-3)^2 - 3(-3) < 10$ False

$9 + 9 < 10$

$18 < 10$

For $(-2, 5)$, we choose 0.

$(0)^2 - 3(0) < 10$ True

$0 < 10$

For $(5, \infty)$, we choose 6.

$(6)^2 - 3(6) < 10$ False

$36 - 18 < 10$

$18 < 10$

Solution set: $(-2, 5)$

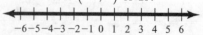

23. $y^2 + 6y + 9 = 0$

$(y+3)^2 = 0$

Since $(y+3)^2 \ge 0$ is always nonnegative, the

solution set is $(-\infty, \infty)$ or $\mathbb{R}$.

25. $2c^2 - 4c + 7 = 0$

$$c = \frac{-(-4) \pm \sqrt{(-4)^2 - 4(2)(7)}}{2 \cdot 2}$$

$$= \frac{4 \pm \sqrt{16 - 56}}{4}$$

$$= \frac{4 \pm \sqrt{-40}}{4}$$

$$= \frac{4 \pm 2i\sqrt{10}}{4} = \frac{2 \pm i\sqrt{10}}{2}$$

There are no c-intercepts, so the function is either always positive or always negative. Test the value zero.

$$2(0)^2 - 4(0) + 7 = 7 > 0$$

Therefore, the function is always positive, and the solution set is $\varnothing$.

$$\overset{\longleftrightarrow}{\underset{-6\,-5\,-4\,-3\,-2\,-1\ 0\ 1\ 2\ 3\ 4\ 5\ 6}{\rule{0pt}{0pt}}}$$

27. $4r^2 + 21r + 5 = 0$

$(4r + 1)(r + 5) = 0$

$4r + 1 = 0 \quad$ or $\quad r + 5 = 0$

$\quad\quad r = -0.25 \quad\quad\quad r = -5$

For $(-\infty, -5)$, we choose -6.

$4(-6)^2 + 21(-6) + 5 > 0 \quad\quad$ True

$\quad 144 - 126 + 5 > 0$

$\quad\quad\quad\quad\quad 23 > 0$

For $(-5, -0.25)$, we choose -1.

$4(-1)^2 + 21(-1) + 5 > 0 \quad\quad$ False

$\quad\quad 4 - 21 + 5 > 0$

$\quad\quad\quad\quad -12 > 0$

For $(-0.25, \infty)$, we choose 0.

$4(0)^2 + 21(0) + 5 > 0 \quad\quad$ True

$\quad\quad\quad\quad 5 > 0$

Solution set: $(-\infty, -5) \cup (-0.25, \infty)$

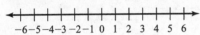

29. $x^2 - 5x = 0$

$x(x - 5) = 0$

$x = 0 \quad$ or $\quad x - 5 = 0$

$\quad\quad\quad\quad\quad\quad\quad x = 5$

For $(-\infty, 0)$, we choose -1.

$(-1)^2 - 5(-1) > 0 \quad\quad$ True

$\quad\quad 1 + 5 > 0$

$\quad\quad\quad\quad 6 > 0$

For $(0, 5)$, we choose 1.

$(1)^2 - 5(1) > 0 \quad\quad$ False

$\quad\quad 1 - 5 > 0$

$\quad\quad -4 > 0$

For $(5, \infty)$, we choose 6.

$(6)^2 - 5(6) > 0 \quad\quad$ True

$\quad\quad 36 - 30 > 0$

$\quad\quad\quad\quad 6 > 0$

Solution set: $(-\infty, 0) \cup (5, \infty)$

$$\overset{\longleftrightarrow}{\underset{-6\,-5\,-4\,-3\,-2\,-1\ 0\ 1\ 2\ 3\ 4\ 5\ 6}{\rule{0pt}{0pt}}}$$

31. $x^2 - 6x = 0$

$x(x - 6) = 0$

$x = 0 \quad$ or $\quad x - 6 = 0$

$\quad\quad\quad\quad\quad\quad\quad x = 6$

For $(-\infty, 0)$, we choose -1.

$(-1)^2 \leq 6(-1) \quad\quad$ False

$\quad 1 \leq -6$

For $(0, 6)$, we choose 1.

$(1)^2 \leq 6(1) \quad\quad$ True

$\quad 1 \leq 6$

For $(6, \infty)$, we choose 7.

$(7)^2 \leq 6(7) \quad\quad$ False

$\quad 49 \leq 42$

Solution set: $[0, 6]$

$$\overset{\longleftrightarrow}{\underset{-6\,-5\,-4\,-3\,-2\,-1\ 0\ 1\ 2\ 3\ 4\ 5\ 6}{\rule{0pt}{0pt}}}$$

33. Since $(x + 1)^2$ is always 0 or positive, every real number is a solution.

Solution set: $(-\infty, \infty)$ or $\mathbb{R}$

$$\overset{\longleftrightarrow}{\underset{-6\,-5\,-4\,-3\,-2\,-1\ 0\ 1\ 2\ 3\ 4\ 5\ 6}{\rule{0pt}{0pt}}}$$

35. $(x-4)(x+2)(x+4)=0$

$x-4=0$ or $x+2=0$ or $x+4=0$

$x=4$ $x=-2$ $x=-4$

Interval	$(-\infty,-4)$	$(-4,-2)$	$(-2,4)$	$(4,\infty)$
Test No.	-5	-3	0	5
Results	-27	7	-32	63
T/F	F	T	F	T

Solution set: $[-4,-2]\cup[4,\infty)$

37. $(x+2)(x+6)(x-1)=0$

$x+2=0$ or $x+6=0$ or $x-1=0$

$x=-2$ $x=-6$ $x=1$

Interval	$(-\infty,-6)$	$(-6,-2)$	$(-2,1)$	$(1,\infty)$
Test No.	-7	-3	0	2
Results	-40	12	-12	32
T/F	T	F	T	F

Solution set: $(-\infty,-6)\cup(-2,1)$

39. $3c^2+4c-1=0$

$$c=\frac{-4\pm\sqrt{4^2-4(3)(-1)}}{2(3)}=\frac{-2\pm\sqrt{7}}{3}$$

Interval	$\left(-\infty,\dfrac{-2-\sqrt{7}}{3}\right)$	$\left(\dfrac{-2-\sqrt{7}}{3},\dfrac{-2+\sqrt{7}}{3}\right)$
Test No.	-2	0
Results	3	-1
T/F	F	T

Interval	$\left(\dfrac{-2+\sqrt{7}}{3},\infty\right)$
Test No.	1
Results	6
T/F	F

Solution set: $\left(\dfrac{-2-\sqrt{7}}{3},\dfrac{-2+\sqrt{7}}{3}\right)$

41. $4r^2+8r-3=0$

$$r=\frac{-8\pm\sqrt{8^2-4(4)(-3)}}{2(4)}=\frac{-2\pm\sqrt{7}}{2}$$

Interval	$\left(-\infty,\dfrac{-2-\sqrt{7}}{2}\right)$	$\left(\dfrac{-2-\sqrt{7}}{2},\dfrac{-2+\sqrt{7}}{2}\right)$
Test No.	-3	0
Results	9	-3
T/F	T	F

Interval	$\left(\dfrac{-2+\sqrt{7}}{2},\infty\right)$
Test No.	1
Results	9
T/F	T

Solution set: $\left(-\infty,\dfrac{-2-\sqrt{7}}{2}\right]\cup\left[\dfrac{-2+\sqrt{7}}{2},\infty\right)$

43. $-0.2a^2-1.6a-2=0$

$$a=\frac{-(-1.6)\pm\sqrt{(-1.6)^2-4(-0.2)(-2)}}{2(-0.2)}$$

$$=-4\pm\sqrt{6}$$

Interval	$\left(-\infty,-4-\sqrt{6}\right)$	$\left(-4-\sqrt{6},-4+\sqrt{6}\right)$
Test No.	-7	-2
Results	-0.6	0.4
T/F	T	F

Interval	$\left(-4+\sqrt{6},\infty\right)$
Test No.	0
Results	-2
T/F	T

Solution set: $\left(-\infty,-4-\sqrt{6}\right]\cup\left[-4+\sqrt{6},\infty\right)$

45. $\dfrac{a+4}{a-1}=0$ $a-1=0$

                                  $a=1$

$(a-1)\dfrac{a+4}{a-1}=0(a-1)$

$a+4=0$

$a=-4$

Interval	$(-\infty,-4)$	$(-4,1)$	$(1,\infty)$
Test No.	-5	0	2
Results	$1/6$	-4	6
T/F	T	F	T

Solution set: $(-\infty,-4)\cup(1,\infty)$

47. $\dfrac{n+1}{n+5} = 0$ $\qquad n+5 = 0$

$\qquad\qquad\qquad\qquad\qquad n = -5$

$(n+5)\dfrac{n+1}{n+5} = 0(n+5)$

$\qquad\qquad n+1 = 0$

$\qquad\qquad\qquad n = -1$

Interval	$(-\infty,-5)$	$(-5,-1)$	$(-1,\infty)$
Test No.	-6	-2	0
Results	5	$-1/3$	$1/5$
T/F	F	T	F

Solution set: $(-5,-1]$

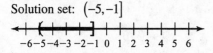

49. $\dfrac{6}{x+4} = 0$ $\qquad x+4 = 0$

$\qquad\qquad\qquad\qquad\qquad x = -4$

$(x+4)\dfrac{6}{x+4} = 0(x+4)$

$\qquad\qquad 6 \neq 0$

Interval	$(-\infty,-4)$	$(-4,\infty)$
Test No.	-6	0
Results	-3	$3/2$
T/F	F	T

Solution set: $(-4,\infty)$

51. $\dfrac{c}{c+3} = 3$ $\qquad c+3 = 0$

$\qquad\qquad\qquad\qquad\qquad c = -3$

$(c+3)\dfrac{c}{c+3} = 3(c+3)$

$\qquad\qquad c = 3c+9$

$\qquad\qquad -2c = 9$

$\qquad\qquad\qquad c = -\dfrac{9}{2}$

Interval	$(-\infty,-9/2)$	$(-9/2,-3)$
Test No.	-6	-4
Results	$2 < 3$	$4 < 3$
T/F	T	F

Interval	$(-3,\infty)$
Test No.	0
Results	$0 < 3$
T/F	T

Solution set: $\left(-\infty,-\dfrac{9}{2}\right)\cup(-3,\infty)$

53. $\dfrac{a+5}{a-4} = 4$ $\qquad a-4 = 0$

$\qquad\qquad\qquad\qquad\qquad a = 4$

$(a-4)\dfrac{a+5}{a-4} = 4(a-4)$

$\qquad\qquad a+5 = 4a-16$

$\qquad\qquad -3a = -21$

$\qquad\qquad\qquad a = 7$

Interval	$(-\infty,4)$	$(4,7)$
Test No.	0	5
Results	$-5/4 > 4$	$10 > 4$
T/F	F	T

Interval	$(7,\infty)$
Test No.	8
Results	$13/4 > 4$
T/F	F

Solution set: $(4,7)$

55. $\dfrac{p+3}{p-3} = 4$ $\qquad p-3 = 0$

$\qquad\qquad\qquad\qquad\qquad p = 3$

$(p-3)\dfrac{p+3}{p-3} = 4(p-3)$

$\qquad\qquad p+3 = 4p-12$

$\qquad\qquad -3p = -15$

$\qquad\qquad\qquad p = 5$

Interval	$(-\infty,3)$	$(3,5)$	$(5,\infty)$
Test No.	0	4	6
Results	$-1 \geq 4$	$7 \geq 4$	$3 \geq 4$
T/F	F	T	F

Solution set: $(3,5]$

57.
$$\frac{(k+3)(k-2)}{k-5}=0$$

$$(k-5)\frac{(k+3)(k-2)}{k-5}=0(k-5)$$

$$(k+3)(k-2)=0$$

$$k+3=0 \qquad k-2=0 \qquad k-5=0$$
$$k=-3 \qquad k=2 \qquad k=5$$

Interval	$(-\infty,-3)$	$(-3,2)$	$(2,5)$	$(5,\infty)$
Test No.	-4	0	3	6
Results	$-2/3$	$6/5$	-3	36
T/F	T	F	T	F

Solution set: $(-\infty,-3]\cup[2,5)$

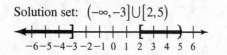

59. Because $(2x-1)^2$ is always nonnegative, we only set the denominator equal to 0; $x=0$.

Interval	$(-\infty,0)$	$(0,\infty)$
Test No.	-1	1
Results	-9	1
T/F	F	T

Solution set: $(0,\infty)$

61.
$$\frac{(x-5)(x-2)}{x+1}=0$$

$$(x+1)\frac{(x-5)(x-2)}{x+1}=0(x+1)$$

$$(x-5)(x-2)=0$$

$$x-5=0 \qquad x-2=0 \qquad x+1=0$$
$$x=5 \qquad x=2 \qquad x=-1$$

Interval	$(-\infty,-1)$	$(-1,2)$	$(2,5)$	$(5,\infty)$
Test No.	-2	0	3	6
Results	-28	10	-0.5	$4/7$
T/F	T	F	T	F

Solution set: $(-\infty,-1)\cup(2,5)$

63. When the ball is on the ground, $h=0$.

$$-16t^2+80t+96=0$$
$$-16(t^2-5t-6)=0$$
$$t^2-5t-6=0$$
$$(t-6)(t+1)=0$$
$$t=6,\cancel{-1}$$

a) 6 sec.

b) $-16t^2+80t+96=192$
$$-16t^2+80t-96=0$$
$$-16(t^2-5t+6)=0$$
$$t^2-5t+6=0$$
$$(t-2)(t-3)=0$$
$$t=2,3$$

The ball is 192 ft. above the ground at 2 sec. and then again at 3 sec.

c) $(2,3)$ sec.

d) $[0,2)\cup(3,6]$ sec.

65. Let $h=x-2$ and $b=x+6$
$$(x-2)(x+6)\ge20$$
$$x^2+4x-12\ge20$$
$$x^2+4x-32\ge0$$
$$(x+8)(x-4)\ge0$$
$$x\le-8 \text{ or } x\ge4$$

a) $x\ge4$ in.

b) $b\ge4+6$
$$b\ge10 \text{ in.}$$

$$h\ge4-2$$
$$h\ge2 \text{ in.}$$

67. $\dfrac{x_2-5}{x_2-2}\le\dfrac{1}{2}$ \qquad $x_2-2=0$
$$x_2=2$$

$$\frac{x_2-5}{x_2-2}=\frac{1}{2}$$
$$x_2-2=2x_2-10$$
$$-x_2=-8$$
$$x_2=8$$

Interval	$(-\infty,2)$	$(2,8)$	$(8,\infty)$
Test No.	0	4	10
Results	$\dfrac{5}{2}\le\dfrac{1}{2}$	$-\dfrac{1}{2}\le\dfrac{1}{2}$	$\dfrac{5}{8}\le\dfrac{1}{2}$
T/F	F	T	F

So, $2<x\le8$

Review Exercises

1. Add 9.

    $x^2 - 6x + 9 = (x - 3)^2$

2. Add $\dfrac{25}{4}$.

    $x^2 + 5x + \dfrac{25}{4} = \left(x + \dfrac{5}{2}\right)^2$

3. $\sqrt{x - 2} = 4$

    $\left(\sqrt{x - 2}\right)^2 = 4^2$

    $\qquad x - 2 = 16$

    $\qquad\quad x = 18$

4. Downward, because the coefficient of x^2 is negative.

5. $x^2 + 4x - 12 = 0$

    $(x + 6)(x - 2) = 0$

    $\qquad x + 6 = 0 \quad \text{or} \quad x - 2 = 0$

    $\qquad\quad x = -6 \qquad\qquad x = 2$

    $\qquad\quad (-6, 0), \ (2, 0)$

6. $\qquad\quad y = x^2 - 6x + 5$

    $\qquad\quad y - 5 = x^2 - 6x$

    $\quad y - 5 + 9 = x^2 - 6x + 9$

    $\qquad y + 4 = (x - 3)^2$

    $\qquad\quad y = (x - 3)^2 - 4$

    The vertex is $(3, -4)$

Exercises Set 10.6

1. $(f + g)(x) = f(x) + g(x)$

    $\qquad = (2x) + (3x - 1)$

    $\qquad = 5x - 1$

    $(f - g)(x) = f(x) - g(x)$

    $\qquad = (2x) - (3x - 1)$

    $\qquad = (2x) + (-3x + 1)$

    $\qquad = -x + 1$

3. $(f + g)(x) = f(x) + g(x)$

    $\qquad = (x - 5) + (2x - 3)$

    $\qquad = 3x - 8$

    $(f - g)(x) = f(x) - g(x)$

    $\qquad = (x - 5) - (2x - 3)$

    $\qquad = (x - 5) + (-2x + 3)$

    $\qquad = -x - 2$

5. $(f + g)(x) = f(x) + g(x)$

    $\qquad = (x^2 - 4x + 7) + (x - 3)$

    $\qquad = x^2 - 3x + 4$

    $(f - g)(x) = f(x) - g(x)$

    $\qquad = (x^2 - 4x + 7) - (x - 3)$

    $\qquad = (x^2 - 4x + 7) + (-x + 3)$

    $\qquad = x^2 - 5x + 10$

7. $(f + g)(x) = f(x) + g(x)$

    $\qquad = (2x^2 - 5x - 3) + (x^2 + 5)$

    $\qquad = 3x^2 - 5x + 2$

    $(f - g)(x) = f(x) - g(x)$

    $\qquad = (2x^2 - 5x - 3) - (x^2 + 5)$

    $\qquad = (2x^2 - 5x - 3) + (-x^2 - 5)$

    $\qquad = x^2 - 5x - 8$

9. $(f + g)(x) = f(x) + g(x)$

    $\qquad = (-5x^2 + 4x + 8) + (3x^2 - 2x - 1)$

    $\qquad = -2x^2 + 2x + 7$

    $(f - g)(x) = f(x) - g(x)$

    $\qquad = (-5x^2 + 4x + 8) - (3x^2 - 2x - 1)$

    $\qquad = (-5x^2 + 4x + 8) + (-3x^2 + 2x + 1)$

    $\qquad = -8x^2 + 6x + 9$

11. $(f \cdot g)(x) = f(x) g(x)$

    $\qquad = (2x)(x - 5)$

    $\qquad = 2x^2 - 10x$

13. $(f \cdot g)(x) = f(x) g(x)$

    $\qquad = (x + 1)(x - 2)$

    $\qquad = x^2 - 2x + x - 2$

    $\qquad = x^2 - x - 2$

15. $(f \cdot g)(x) = f(x)g(x)$
$$= (3x+2)(x+2)$$
$$= 3x^2 + 6x + 2x + 4$$
$$= 3x^2 + 8x + 4$$

17. $(f \cdot g)(x) = f(x)g(x)$
$$= (x^2 - 3x + 2)(x^2 + 2x - 7)$$
$$= x^4 + 2x^3 - 7x^2 - 3x^3 - 6x^2 + 21x$$
$$\quad + 2x^2 + 4x - 14$$
$$= x^4 - x^3 - 11x^2 + 25x - 14$$

19. $(f \cdot g)(x) = f(x)g(x)$
$$= (2x^2 - 3x + 1)(3x^2 - x - 1)$$
$$= 6x^4 - 2x^3 - 2x^2 - 9x^3 + 3x^2 + 3x$$
$$\quad + 3x^2 - x - 1$$
$$= 6x^4 - 11x^3 + 4x^2 + 2x - 1$$

21. $(f / g)(x) = \dfrac{f(x)}{g(x)} = \dfrac{2x^2 - 6x}{2x}$
$$= \dfrac{2x^2}{2x} - \dfrac{6x}{2x}$$
$$= x - 3, \ x \neq 0$$

23. $(f / g)(x) = \dfrac{f(x)}{g(x)} = \dfrac{6x^2 - 3x + 6}{3x}$
$$= \dfrac{6x^2}{3x} - \dfrac{3x}{3x} + \dfrac{6}{3x}$$
$$= 2x - 1 + \dfrac{2}{x}, \ x \neq 0$$

25. $(f / g)(x) = \dfrac{f(x)}{g(x)} = \dfrac{x^2 - 14x + 45}{x - 9}$

$$
\begin{array}{r}
x - 5 \\
x-9 \overline{\smash{)}\ x^2 - 14x + 45} \\
\underline{x^2 - 9x} \\
-5x + 45 \\
\underline{-5x + 45} \\
0
\end{array}
$$

$(f / g)(x) = x - 5, \ x \neq 9$

27. $(f / g)(x) = \dfrac{f(x)}{g(x)} = \dfrac{2x^2 - 3x - 5}{2x - 5}$

$$
\begin{array}{r}
x + 1 \\
2x-5 \overline{\smash{)}\ 2x^2 - 3x - 5} \\
\underline{2x^2 - 5x} \\
2x - 5 \\
\underline{2x - 5} \\
0
\end{array}
$$

$(f / g)(x) = x + 1, \ x \neq \dfrac{5}{2}$

29. $(f / g)(x) = \dfrac{f(x)}{g(x)} = \dfrac{2x^2 - 5x - 7}{x + 5}$

$$
\begin{array}{r}
2x - 15 \\
x+5 \overline{\smash{)}\ 2x^2 - 5x - 7} \\
\underline{2x^2 + 10x} \\
-15x - 7 \\
\underline{-15x - 75} \\
68
\end{array}
$$

$(f / g)(x) = 2x - 15 + \dfrac{68}{x + 5}, \ x \neq -5$

31. a) $(p + q)(x) = p(x) + q(x)$
$$= (5x - 1) + (2x + 3)$$
$$= 7x + 2$$

   b) $(p - q)(x) = p(x) - q(x)$
$$= (5x - 1) - (2x + 3)$$
$$= (5x - 1) + (-2x - 3)$$
$$= 3x - 4$$

   c) $(p \cdot q)(x) = p(x) \cdot q(x)$
$$= (5x - 1)(2x + 3)$$
$$= 10x^2 + 15x - 2x - 3$$
$$= 10x^2 + 13x - 3$$

   d) $(p / q)(x) = \dfrac{p(x)}{q(x)} = \dfrac{5x - 1}{2x + 3}, \ x \neq -\dfrac{3}{2}$

33. a) $(p+q)(x) = p(x) + q(x)$
$$= \left(2x^2 - 3x + 5\right) + (x - 1)$$
$$= 2x^2 - 2x + 4$$

b) $(p-q)(x) = p(x) - q(x)$
$$= \left(2x^2 - 3x + 5\right) - (x - 1)$$
$$= \left(2x^2 - 3x + 5\right) + (-x + 1)$$
$$= 2x^2 - 4x + 6$$

c) $(p \cdot q)(x) = p(x) \cdot q(x)$
$$= \left(2x^2 - 3x + 5\right) \cdot (x - 1)$$
$$= 2x^3 - 2x^2 - 3x^2 + 3x + 5x - 5$$
$$= 2x^3 - 5x^2 + 8x - 5$$

d) $(p/q)(x) = \dfrac{p(x)}{q(x)} = \dfrac{2x^2 - 3x + 5}{x - 1}$

$$\begin{array}{r} 2x - 1 \\ x-1 \overline{\smash{\big)}\, 2x^2 - 3x + 5} \\ \underline{2x^2 - 2x} \\ -x + 5 \\ \underline{-x + 1} \\ 4 \end{array}$$

$(p/q)(x) = 2x - 1 + \dfrac{4}{x-1}, \ x \neq 1$

35. a) $(p+q)(x) = 7x + 2$
$(p+q)(-2) = 7(-2) + 2$
$$= -14 + 2$$
$$= -12$$

b) $(p-q)(x) = 3x - 4$
$(p-q)(5) = 3(5) - 4$
$$= 15 - 4$$
$$= 11$$

c) $(p \cdot q)(x) = 10x^2 + 13x - 3$
$(p \cdot q)(-1) = 10(-1)^2 + 13(-1) - 3$
$$= 10(1) - 13 - 3$$
$$= 10 - 13 - 3$$
$$= -6$$

d) $(p/q)(x) = \dfrac{5x - 1}{2x + 3}$
$(p/q)(7) = \dfrac{5(7) - 1}{2(7) + 3}$
$$= \dfrac{35 - 1}{14 + 3}$$
$$= \dfrac{34}{17}$$
$$= 2$$

37. a) $(p+q)(x) = 2x^2 - 2x + 4$
$(p+q)(3) = 2(3)^2 - 2(3) + 4$
$$= 2(9) - 6 + 4$$
$$= 18 - 6 + 4$$
$$= 16$$

b) $(p-q)(x) = 2x^2 - 4x + 6$
$(p-q)(-4) = 2(-4)^2 - 4(-4) + 6$
$$= 2(16) + 16 + 6$$
$$= 32 + 16 + 6$$
$$= 54$$

c) $(p \cdot q)(x) = 2x^3 - 5x^2 + 8x - 5$
$(p \cdot q)(2) = 2(2)^3 - 5(2)^2 + 8(2) - 5$
$$= 2(8) - 5(4) + 16 - 5$$
$$= 16 - 20 + 16 - 5$$
$$= 7$$

d) $(p/q)(x) = 2x - 1 + \dfrac{4}{x-1}$
$(p/q)(-3) = 2(-3) - 1 + \dfrac{4}{(-3) - 1}$
$$= -6 - 1 + \dfrac{4}{-4}$$
$$= -6 - 1 - 1$$
$$= -8$$

39. a) $(f+g)(x) = f(x)+g(x)$
$$= (3x+1)+(x+5)$$
$$= 4x+6$$

b)

x	$f(x)$	$g(x)$	$(f+g)(x)$
−4	−11	1	−10
−2	−5	3	−2
0	1	5	6
2	7	7	14
4	13	9	22

c)

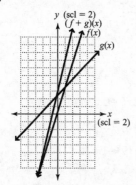

d) The output values for $f(x)$ added to the output values of $g(x)$ are equal to the output values of $(f+g)(x)$.

41. a) $(h-k)(x) = h(x)-k(x)$
$$= (2x-9)-(3x+5)$$
$$= (2x-9)+(-3x-5)$$
$$= -x-14$$

b)

x	$h(x)$	$k(x)$	$(h-k)(x)$
−4	−17	−7	−10
−2	−13	−1	−12
0	−9	5	−14
2	−5	11	−16
4	−1	17	−18

c)

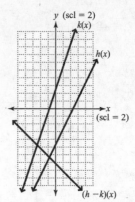

d) The output values for $k(x)$ subtracted from the output values of $h(x)$ are equal to the output values of $(h-k)(x)$.

43. a) $t(x) = w(x)+p(x)$
$$= (3x+2)+(x+2)$$
$$= 4x+4$$

b) $t(x) = 4x+4$
$$t(200) = 4(200)+4$$
$$= 800+4$$
$$= 804$$
The total cost is $804.

c)

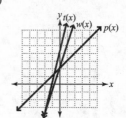

45. a) $p(x) = r(x) - c(x)$

$p(x) = (x^2 - 3x - 4) - (x + 8)$

$\quad = (x^2 - 3x - 4) + (-x - 8)$

$\quad = x^2 - 4x - 12$

b) $p(x) = x^2 - 4x - 12$

$p(100) = (100)^2 - 4(100) - 12$

$\quad = 10,000 - 400 - 12$

$\quad = 9588$

If $x = 100$, the net profit is \$9588.

c)

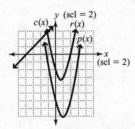

d) $\quad\quad p(x) = 0$

$x^2 - 4x - 12 = 0$

$(x - 6)(x + 2) = 0$

$x - 6 = 0 \quad$ or $\quad x + 2 = 0$

$x = 6 \quad\quad\quad x = -2$

A total of 6 units must be sold for the company to break even.

47. a) $A(x) = l(x) \cdot h(x)$

$\quad = (3x + 2)(2x)$

$\quad = 6x^2 + 4x$

b) $A(x) = 6x^2 + 4x$

$A(3) = 6(3)^2 + 4(3)$

$\quad = 6(9) + 12$

$\quad = 54 + 12$

$\quad = 66$

The area of the front face is 66 square feet.

49. a) $h(x) = \dfrac{V(x)}{l(x) \cdot w(x)}$

$h(x) = \dfrac{4x^3 - 7x^2 - 14x - 3}{(x - 3)(x + 1)}$

$\quad = \dfrac{4x^3 - 7x^2 - 14x - 3}{x^2 - 2x - 3}$

$$
\begin{array}{r}
4x + 1 \\
x^2 - 2x - 3 \overline{)\, 4x^3 - 7x^2 - 14x - 3} \\
\underline{4x^3 - 8x^2 - 12x} \\
x^2 - 2x - 3 \\
\underline{x^2 - 2x - 3} \\
0
\end{array}
$$

$h(x) = 4x + 1$

b) $h(x) = 4x + 1$

$h(4) = 4(4) + 1$

$\quad = 16 + 1$

$\quad = 17$

The height of the box is 17 cm.

Review Exercises

1. $f(x) = x^2 + 1$

$f(2) = (2)^2 + 1$

$\quad = 4 + 1$

$\quad = 5$

2. $f(x) = x^2 + 1$

$f(3a - 2) = (3a - 2)^2 + 1$

$\quad = (9a^2 - 12a + 4) + 1$

$\quad = 9a^2 - 12a + 5$

3. $f(x) = x^2 + 1$

$f\left(\sqrt{a - 1}\right) = \left(\sqrt{a - 1}\right)^2 + 1$

$\quad = (a - 1) + 1$

$\quad = a \quad\quad\quad$ for $a \geq 1$

4. $y = 2x + 4$

$y - 4 = 2x$

$\dfrac{y - 4}{2} = x$

5. $y = x^3 + 2$

$y - 2 = x^3$

$\sqrt[3]{y - 2} = x$

Chapter 10 Review Exercises

1. $x^2 = 16$

    $x = \pm\sqrt{16}$

    $x = \pm 4$

2. $y^2 = \dfrac{1}{36}$

    $y = \pm\sqrt{\dfrac{1}{36}}$

    $y = \pm\dfrac{1}{6}$

3. $k^2 + 2 = 30$

    $k^2 = 28$

    $k = \pm\sqrt{28}$

    $k = \pm 2\sqrt{7}$

4. $3x^2 = 42$

    $x^2 = \dfrac{42}{3}$

    $x^2 = 14$

    $x = \pm\sqrt{14}$

5. $5h^2 + 24 = 9$

    $5h^2 = -15$

    $h^2 = -3$

    $h = \pm\sqrt{-3}$

    $h = \pm i\sqrt{3}$

6. $(x+7)^2 = 25$

    $(x+7)^2 = \pm\sqrt{25}$

    $x + 7 = \pm 5$

    $x = -7 \pm 5$

      $= -7 + 5 = -2$

      $= -7 - 5 = -12$

7. $(x-9)^2 = -16$

    $x - 9 = \pm\sqrt{-16}$

    $x - 9 = \pm 4i$

    $x = 9 \pm 4i$

8. $\left(m + \dfrac{3}{5}\right)^2 = \dfrac{16}{25}$

    $m + \dfrac{3}{5} = \pm\sqrt{\dfrac{16}{25}}$

    $m + \dfrac{3}{5} = \pm\dfrac{4}{5}$

    $m = -\dfrac{3}{5} \pm \dfrac{4}{5}$

      $= -\dfrac{3}{5} + \dfrac{4}{5} = \dfrac{1}{5}$

      $= -\dfrac{3}{5} - \dfrac{4}{5} = -\dfrac{7}{5}$

9. $A = \pi r^2$

    $22{,}500\pi = \pi r^2$

      $0 = \pi r^2 - 22{,}500\pi$

      $0 = \pi\left(r^2 - 22{,}500\right)$

      $0 = r^2 - 22{,}500$

      $0 = (r+150)(r-150)$

      $r = \cancel{-150}, 150$

The radius of the circle was 150 feet.

10. $400 = \dfrac{1}{2}50v^2$

    $400 = 25v^2$

    $16 = v^2$

    $4 = v$

The velocity was 4 m/sec.

11. $\pi r^2(4) = \dfrac{4}{3}\pi\left(9^3\right)$

      $4\pi r^2 = 972\pi$

    $4\pi r^2 - 972\pi = 0$

    $4\pi\left(r^2 - 243\right) = 0$

      $r^2 - 243 = 0$

        $r^2 = 243$

        $r = \sqrt{243}$

        $r = 9\sqrt{3}$

The radius is $9\sqrt{3}$ inches.

12.　　　$m^2 + 8m = -7$

$$\left(\frac{8}{2}\right)^2 = (4)^2 = 16$$

$$m^2 + 8m + 16 = -7 + 16$$

$$(m+4)^2 = 9$$

$$m + 4 = \pm\sqrt{9}$$

$$m + 4 = \pm 3$$

$$m = -4 \pm 3$$

$$m = -4 - 3 = -7$$

$$m = -4 + 3 = -1$$

13.　$u^2 - 6u - 12 = 100$

$$u^2 - 6u = 112$$

$$\left(-\frac{6}{2}\right)^2 = (-3)^2 = 9$$

$$u^2 - 6u + 9 = 112 + 9$$

$$(u-3)^2 = 121$$

$$u - 3 = \pm\sqrt{121}$$

$$u - 3 = \pm 11$$

$$u = 3 \pm 11$$

$$u = 3 + 11 = 14$$

$$u = 3 - 11 = -8$$

14.　$2b^2 - 6b + 7 = 0$

$$b^2 - 3b = -\frac{7}{2}$$

$$\left(\frac{-3}{2}\right)^2 = \frac{9}{4}$$

$$b^2 - 3b + \frac{9}{4} = -\frac{7}{2} + \frac{9}{4}$$

$$\left(b - \frac{3}{2}\right)^2 = -\frac{5}{4}$$

$$b - \frac{3}{2} = \pm\sqrt{-\frac{5}{4}}$$

$$b - \frac{3}{2} = \pm\frac{i\sqrt{5}}{2}$$

$$b = \frac{3 \pm i\sqrt{5}}{2}$$

15.　　　$$\left(\frac{\frac{1}{4}}{2}\right)^2 = \left(\frac{1}{8}\right)^2 = \frac{1}{64}$$

$$u^2 + \frac{1}{4}u + \frac{1}{64} = \frac{3}{4} + \frac{1}{64}$$

$$\left(u + \frac{1}{8}\right)^2 = \frac{49}{64}$$

$$u + \frac{1}{8} = \pm\sqrt{\frac{49}{64}}$$

$$u + \frac{1}{8} = \pm\frac{7}{8}$$

$$u = -\frac{1}{8} \pm \frac{7}{8}$$

$$u = -\frac{1}{8} - \frac{7}{8} = -1$$

$$u = -\frac{1}{8} + \frac{7}{8} = \frac{3}{4}$$

16.　$p^2 + 2p - 5 = 0$

$$p = \frac{-(2) \pm \sqrt{(2)^2 - 4(1)(-5)}}{2(1)}$$

$$= \frac{-2 \pm \sqrt{4 + 20}}{2}$$

$$= \frac{-2 \pm \sqrt{24}}{2}$$

$$= \frac{-2 \pm 2\sqrt{6}}{2}$$

$$= -1 \pm \sqrt{6}$$

17.　$3x^2 - 2x + 1 = 0$

$$x = \frac{-(-2) \pm \sqrt{(-2)^2 - 4(3)(1)}}{2(3)}$$

$$= \frac{2 \pm \sqrt{4 - 12}}{6}$$

$$= \frac{2 \pm \sqrt{-8}}{6}$$

$$= \frac{2 \pm 2i\sqrt{2}}{6}$$

$$= \frac{1 \pm i\sqrt{2}}{3}$$

18. $2t^2 + t - 5 = 0$

$$t = \frac{-(1) \pm \sqrt{(1)^2 - 4(2)(-5)}}{2(2)}$$

$$= \frac{-1 \pm \sqrt{1 + 40}}{4}$$

$$= \frac{-1 \pm \sqrt{41}}{4}$$

19. Use $200x^2 + 10x - 3 = 0$.

$$x = \frac{-(10) \pm \sqrt{(10)^2 - 4(200)(-3)}}{2(200)}$$

$$= \frac{-10 \pm \sqrt{100 + 2400}}{400}$$

$$= \frac{-10 \pm \sqrt{2500}}{400}$$

$$= \frac{-10 \pm 50}{400}$$

$$= \frac{-10 - 50}{400} = -\frac{3}{20}$$

$$= \frac{-10 + 50}{400} = \frac{1}{10}$$

20.
$$x(x + 4) = 285$$

$$x^2 + 4x - 285 = 0$$

$$(x + 19)(x - 15) = 0$$

$$x = \cancel{-19}, 15$$

The width is 15 ft. and the length is
$15 + 4 = 19$ ft.

21.
$$x^2 + (x + 9)^2 = 17^2$$

$$x^2 + x^2 + 18x + 81 = 289$$

$$2x^2 + 18x - 208 = 0$$

$$x^2 + 9x - 104 = 0$$

$$x = \frac{-9 \pm \sqrt{9^2 - 4(1)(-104)}}{2(1)}$$

$$= \frac{-9 \pm \sqrt{81 + 416}}{2}$$

$$= \frac{-9 \pm \sqrt{497}}{2} \approx \cancel{-15.65}, 6.65$$

The height is 6.65 ft. and the base is $6.65 + 9 =$
15.65 ft.

22. $b^2 - 4b - 12 = 0$

$D = (-4)^2 - 4(1)(-12) = 64$; 2 rational

23. $6z^2 - 7z + 5 = 0$

$D = (-7)^2 - 4(6)(5) = -71$; 2 nonreal complex

24. $k^2 + 6k + 9 = 0$

$D = (6)^2 - 4(1)(9) = 0$; 1 rational

25. $0.8x^2 + 1.2x + 0.3 = 0$

$D = (1.2)^2 - 4(0.8)(0.3) = 0.48$; 2 irrational

26.
$$\frac{1}{y} + \frac{1}{y+3} = \frac{2}{3}$$

$$3y(y+3)\left(\frac{1}{y} + \frac{1}{y+3}\right) = 3y(y+3) \cdot \frac{2}{3}$$

$$3(y+3) + 3y = 2y(y+3)$$

$$3y + 9 + 3y = 2y^2 + 6y$$

$$0 = 2y^2 - 9$$

$$2y^2 = 9$$

$$y^2 = \frac{9}{2}$$

$$y = \pm\sqrt{\frac{9}{2}}$$

$$y = \pm\frac{3}{\sqrt{2}} \cdot \frac{\sqrt{2}}{\sqrt{2}}$$

$$y = \pm\frac{3\sqrt{2}}{2}$$

27.
$$\frac{1}{u} + \frac{1}{u-5} = \frac{10}{u^2 - 25}$$

$$u(u-5)(u+5)\left(\frac{1}{u} + \frac{1}{u-5} = \frac{10}{(u-5)(u+5)}\right)$$

$$(u-5)(u+5) + u(u+5) = 10u$$

$$u^2 - 25 + u^2 + 5u = 10u$$

$$2u^2 - 5u - 25 = 0$$

$$(2u+5)(u-5) = 0$$

$$2u + 5 = 0 \quad \text{or} \quad u - 5 = 0$$

$$u = -\frac{5}{2} \qquad\qquad u = \cancel{5}$$

28.
$$6 - 5x^{-1} + x^{-2} = 0$$
$$x^2\left(6 - 5x^{-1} + x^{-2}\right) = x^2 \cdot 0$$
$$6x^2 - 5x + 1 = 0$$
$$(3x - 1)(2x - 1) = 0$$
$$3x - 1 = 0 \quad \text{or} \quad 2x - 1 = 0$$
$$x = \frac{1}{3} \qquad x = \frac{1}{2}$$

29.
$$2 - 3x^{-1} - x^{-2} = 0$$
$$x^2\left(2 - 3x^{-1} - x^{-2}\right) = x^2 \cdot 0$$
$$2x^2 - 3x - 1 = 0$$
$$x = \frac{-(-3) \pm \sqrt{(-3)^2 - 4(2)(-1)}}{2(2)}$$
$$= \frac{3 \pm \sqrt{9 + 8}}{4}$$
$$= \frac{3 \pm \sqrt{17}}{4}$$

30. $14\sqrt{x} + 45 = 0$
$$14\sqrt{x} = -45$$
$$\sqrt{x} = -\frac{45}{14}$$
$$\text{no solution}$$

31.
$$\sqrt{4m} = 3m - 1$$
$$\left(\sqrt{4m}\right)^2 = (3m - 1)^2$$
$$4m = 9m^2 - 6m + 1$$
$$0 = 9m^2 - 10m + 1$$
$$0 = (9m - 1)(m - 1)$$
$$9m - 1 = 0 \quad \text{or} \quad m - 1 = 0$$
$$m = \cancel{\frac{1}{9}} \qquad m = 1$$

32.
$$\sqrt{6r + 13} = 2r + 1$$
$$\left(\sqrt{6r + 13}\right)^2 = (2r + 1)^2$$
$$6r + 13 = 4r^2 + 4r + 1$$
$$0 = 4r^2 - 2r - 12$$
$$0 = 2r^2 - r - 6$$
$$0 = (2r + 3)(r - 2)$$
$$2r + 3 = 0 \quad \text{or} \quad r - 2 = 0$$
$$r = \cancel{-\frac{3}{2}} \qquad r = 2$$

33.
$$\sqrt{21t + 2} + t = 2 + 4t$$
$$\sqrt{21t + 2} = 2 + 3t$$
$$\left(\sqrt{21t + 2}\right)^2 = (2 + 3t)^2$$
$$21t + 2 = 4 + 12t + 9t^2$$
$$0 = 2 - 9t + 9t^2$$
$$0 = (2 - 3t)(1 - 3t)$$
$$2 - 3t = 0 \quad \text{or} \quad 1 - 3t = 0$$
$$t = \frac{2}{3} \qquad t = \frac{1}{3}$$

34. Let $u = x^2$
$$u^2 - 5u + 6 = 0$$
$$(u - 2)(u - 3) = 0$$
$$u - 2 = 0 \quad \text{or} \quad u - 3 = 0$$
$$u = 2 \qquad u = 3$$
$$x^2 = 2 \qquad x^2 = 3$$
$$x = \pm\sqrt{2} \qquad x = \pm\sqrt{3}$$

35. Let $u = m^2$
$$2u^2 - 3u + 1 = 0$$
$$(2u - 1)(u - 1) = 0$$
$$2u - 1 = 0 \qquad u - 1 = 0$$
$$u = \frac{1}{2} \qquad u = 1$$
$$m^2 = \frac{1}{2} \qquad m^2 = 1$$
$$m = \pm 1$$
$$m = \pm\sqrt{\frac{1}{2} \cdot \frac{2}{2}} = \pm\frac{\sqrt{2}}{2}$$

36. Let $u = x + 5$
$$6u^2 - 5u + 1 = 0$$
$$(2u - 1)(3u - 1) = 0$$
$$2u - 1 = 0 \quad \text{or} \quad 3u - 1 = 0$$
$$u = \frac{1}{2} \qquad u = \frac{1}{3}$$
$$x + 5 = \frac{1}{2} \qquad x + 5 = \frac{1}{3}$$
$$x = -\frac{9}{2} \qquad x = -\frac{14}{3}$$

37. Let $u = \dfrac{x-1}{3}$

$u^2 + 10u + 9 = 0$

$(u+1)(u+9) = 0$

$u + 1 = 0 \quad$ or $\quad u + 9 = 0$

$u = -1 \qquad\qquad u = -9$

$\dfrac{x-1}{3} = -1 \qquad \dfrac{x-1}{3} = -9$

$x - 1 = -3 \qquad\quad x - 1 = -27$

$x = -2 \qquad\qquad x = -26$

38. Let $u = p^{1/3}$

$u^2 - 11u + 24 = 0$

$(u-8)(u-3) = 0$

$u - 8 = 0 \quad$ or $\quad u - 3 = 0$

$u = 8 \qquad\qquad u = 3$

$p^{1/3} = 8 \qquad\qquad p^{1/3} = 3$

$\left(p^{1/3}\right)^3 = 8^3 \qquad \left(p^{1/3}\right)^3 = 3^3$

$p = 512 \qquad\qquad p = 27$

39. Let $u = a^{1/4}$

$5u^2 + 13u - 6 = 0$

$(5u - 2)(u + 3) = 0$

$5u - 2 = 0 \qquad$ or $\quad u + 3 = 0$

$u = \dfrac{2}{5} \qquad\qquad\quad u = -3$

$a^{1/4} = \dfrac{2}{5} \qquad\quad a^{1/4} = -3$

$\left(a^{1/4}\right)^4 = \left(\dfrac{2}{5}\right)^4 \qquad \left(a^{1/4}\right)^4 = (-3)^4$

$\qquad\qquad\qquad\qquad\quad a = \cancel{81}$

$a = \dfrac{16}{625}$

40. $f(x) = -2x^2$

a) x-intercept: $-2x^2 = 0$

$x^2 = 0$

$x = \pm\sqrt{0}$

$x = 0$

$(0, 0)$

y-intercept: $y = -2(0)^2 = 0$

$(0, 0)$

b) downward

c) $(0, 0)$

d) $x = 0$

e)

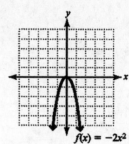

$f(x) = -2x^2$

41. $g(x) = \dfrac{1}{2}x^2 + 1$

a) x-intercept: $\dfrac{1}{2}x^2 + 1 = 0$

$\dfrac{1}{2}x^2 = -1$

$x^2 = -2$

$x = \pm\sqrt{-2}$

$x = \pm i\sqrt{2}$

Because these solutions are not real, there are no x-intercepts.

y-intercept: $y = \dfrac{1}{2}(0)^2 + 1 = 1$

$(0, 1)$

b) upward

c) $(0, 1)$

d) $x = 0$

e)

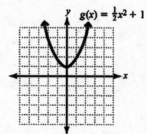

$g(x) = \frac{1}{2}x^2 + 1$

42. $h(x) = -\dfrac{1}{3}(x-2)^2$

a) x-intercept: $-\dfrac{1}{3}(x-2)^2 = 0$

$$(x-2)^2 = 0$$
$$x-2 = \pm\sqrt{0}$$
$$x = 2 \pm \sqrt{0}$$
$$x = 2$$
$$(2,0)$$

y-intercept: $y = -\dfrac{1}{3}(0-2)^2 = -\dfrac{1}{3}\cdot 4 = -\dfrac{4}{3}$

$$\left(0, -\dfrac{4}{3}\right)$$

b) downward
c) (2, 0)
d) $x = 2$
e)

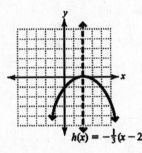

43. $k(x) = 4(x+3)^2 - 2$

a) x-intercepts: $\qquad 0 = 4(x+3)^2 - 2$

$$\dfrac{2}{4} = (x+3)^2$$
$$\pm\dfrac{\sqrt{2}}{2} = x+3$$
$$-3 \pm \dfrac{\sqrt{2}}{2} = x$$

$$\left(-3+\dfrac{\sqrt{2}}{2}, 0\right), \left(-3-\dfrac{\sqrt{2}}{2}, 0\right)$$

y-intercept: $y = 4(0+3)^2 - 2$

$$y = 4 \cdot 9 - 2$$
$$y = 34$$
$$(0, 34)$$

b) upward
c) $(-3, -2)$
d) $x = -3$

e)

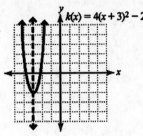

44. a) $\qquad y = x^2 + 2x - 1$

$$y + 1 = x^2 + 2x$$
$$y + 1 + 1 = x^2 + 2x + 1$$
$$y + 2 = (x+1)^2$$
$$y = (x+1)^2 - 2$$
$$m(x) = (x+1)^2 - 2$$

b) x-intercepts: $x = \dfrac{-2 \pm \sqrt{2^2 - 4(1)(-1)}}{2(1)}$

$$= \dfrac{-2 \pm 2\sqrt{2}}{2}$$
$$= -1 \pm \sqrt{2}$$
$$\left(-1+\sqrt{2}, 0\right), \left(-1-\sqrt{2}, 0\right)$$

y-intercept: $y = 0 + 2\cdot 0 - 1 = -1$

$$(0, -1)$$

c) upward
d) using part (a): $(-1, -2)$
e) $x = -1$
f)

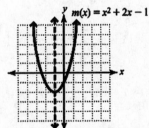

g) Domain: $\{x \mid x \text{ is a real number}\}$ or $(-\infty, \infty)$
    Range: $\{y \mid y \geq -2\}$ or $[-2, \infty)$

45. a) $y = -0.5x^2 + 4x - 6$

$y + 6 = -0.5x^2 + 4x$

$y + 6 = -0.5\left(x^2 - 8x\right)$

$y + 6 - 8 = -0.5\left(x^2 - 8x + 16\right)$

$y - 2 = -0.5\left(x - 4\right)^2$

$y = -0.5\left(x - 4\right)^2 + 2$

$p(x) = -0.5\left(x - 4\right)^2 + 2$

b) x-intercepts: $0 = -0.5x^2 + 4x - 6$

$0 = 5x^2 - 40x + 60$

$0 = x^2 - 8x + 12$

$0 = (x - 2)(x - 6)$

$x - 2 = 0 \quad\quad x - 6 = 0$

$x = 2 \quad\quad\quad x = 6$

$(2, 0), (6, 0)$

y-intercept: $y = -0.5 \cdot 0 + 4 \cdot 0 - 6 = -6$

$(0, -6)$

c) downward

d) from part (a): (4, 2)

e) $x = 4$

f)

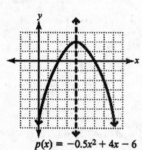

$p(x) = -0.5x^2 + 4x - 6$

g) Domain: $\{x \mid x \text{ is a real number}\}$ or $(-\infty, \infty)$

Range: $\{y \mid y \le 2\}$ or $(-\infty, 2]$

46. Finding the vertex will give the highest point that the acrobat will reach.

$t = \dfrac{-24}{2(-16)} = \dfrac{3}{4} = 0.75$

$h = -16(0.75)^2 + 24(0.75)$

$h = -9 + 18$

$h = 9$

The vertex is at $(0.75, 9)$

a) 0.75 sec.

b) 9 ft.

c) $-16t^2 + 24t = 0$

$-8t(2t - 3) = 0$

$t = \cancel{0}, 1.5 \text{ sec.}$

d)

$h = -16t^2 + 24t$

47. a) Find the vertex. $x = \dfrac{-2.16}{2(-0.03)} = 36$

$y = -0.03(36)^2 + 2.16(36)$

$y = 38.88$

The maximum height is 38.88 yd.

b) The distance of the punt is the x-coordinate of the vertex times 2, or 72 yd.

48. $(x + 5)(x - 3) = 0$

$x + 5 = 0 \quad$ or $\quad x - 3 = 0$

$x = -5 \quad\quad\quad x = 3$

Interval	$(-\infty, -5)$	$(-5, 3)$	$(3, \infty)$
Test No.	-6	0	4
Results	$9 > 0$	$-15 > 0$	$9 > 0$
T/F	T	F	T

Solution set: $(-\infty, -5) \cup (3, \infty)$

49. $n^2 - 6n + 8 = 0$

$(n - 4)(n - 2) = 0$

$n - 4 = 0 \quad$ or $\quad n - 2 = 0$

$n = 4 \quad\quad\quad n = 2$

Interval	$(-\infty, 2)$	$(2, 4)$	$(4, \infty)$
Test No.	0	3	5
Results	$0 \le -8$	$-9 \le -8$	$-5 \le -8$
T/F	F	T	F

Solution set: [2, 4]

50. $x^2 + 9x + 14 = 0$

$(x+7)(x+2) = 0$

$x + 7 = 0$ or $x + 2 = 0$

$x = -7$ $x = -2$

Interval	$(-\infty, -7)$	$(-7, -2)$	$(-2, \infty)$
Test No.	-8	-3	0
Results	$6 < 0$	$-4 < 0$	$14 < 0$
T/F	F	T	F

Solution set: $(-7, -2)$

51. $(x+3)(x-1)(x-2) = 0$

$x + 3 = 0$ or $x - 1 = 0$ or $x - 2 = 0$

$x = -3$ $x = 1$ $x = 2$

Interval	$(-\infty, -3)$	$(-3, 1)$	$(1, 2)$	$(2, \infty)$
Test No.	-4	0	1.5	3
Results	-30	6	-1.125	12
T/F	F	T	F	T

Solution set: $[-3, 1] \cup [2, \infty)$

52. a) $56 \le \dfrac{1}{2}(x+4)(x-2)$

$112 \le x^2 + 2x - 8$

$0 \le x^2 + 2x - 120$

$0 \le (x+12)(x-10)$

$x \le -12$ or $x \ge 10$

Since x must be at least 2, $x \ge 10$ in.

b) height ≥ 8 in., base ≥ 14 in.

53. $\dfrac{a+3}{a-1} = 0$ $a - 1 = 0$

$a = 1$

$(a-1)\dfrac{a+3}{a-1} = 0(a-1)$

$a + 3 = 0$

$a = -3$

Interval	$(-\infty, -3)$	$(-3, 1)$	$(1, \infty)$
Test No.	-5	0	2
Results	$1/3 \ge 0$	$-3 \ge 0$	$5 \ge 0$
T/F	T	F	T

Solution set: $(-\infty, -3] \cup (1, \infty)$

54. $\dfrac{r}{r+2} = 2$ $r + 2 = 0$

$r = -2$

$(r+2)\dfrac{r}{r+2} = 2(r+2)$

$r = 2r + 4$

$-4 = r$

Interval	$(-\infty, -4)$	$(-4, -2)$
Test No.	-5	-3
Results	$5/3 < 2$	$3 < 2$
T/F	T	F

Interval	$(-2, \infty)$
Test No.	0
Results	$0 < 2$
T/F	T

Solution set: $(-\infty, -4) \cup (-2, \infty)$

55. $\dfrac{n-3}{n-4} = 5$ $n - 4 = 0$

$n = 4$

$(n-4)\dfrac{n-3}{n-4} = 5(n-4)$

$n - 3 = 5n - 20$

$-4n = -17$

$n = \dfrac{17}{4}$

Interval	$(-\infty, 4)$	$\left(4, \dfrac{17}{4}\right)$
Test No.	0	4.1
Results	$3/4 \le 5$	$11 \le 5$
T/F	T	F

Interval	$\left(\dfrac{17}{4}, \infty\right)$
Test No.	5
Results	$2 \le 5$
T/F	T

Solution set: $(-\infty, 4) \cup \left[\dfrac{17}{4}, \infty\right)$

56. $\dfrac{(k+2)(k-3)}{k-5} = 0$

$(k-5)\dfrac{(k+2)(k-3)}{k-5} = 0(k-5)$

$(k+2)(k-3) = 0$

$k+2 = 0 \qquad k-3 = 0 \qquad k-5 = 0$

$k = -2 \qquad\; k = 3 \qquad\;\; k = 5$

Interval	$(-\infty,-2)$	$(-2,3)$	$(3,5)$	$(5,\infty)$
Test No.	-3	0	4	6
Results	$-3/4 < 0$	$\dfrac{6}{5} < 0$	$-6 < 0$	$24 < 0$
T/F	T	F	T	F

Solution set: $(-\infty,-2) \cup (3,5)$

57. a) $(f+g)(x) = f(x) + g(x)$
$= (-3x+9) + (4x+1)$
$= x + 10$

b) $(f-g)(x) = f(x) - g(x)$
$= (-3x+9) - (4x+1)$
$= (-3x+9) + (-4x-1)$
$= -7x + 8$

c) $(f \cdot g)(x) = f(x) \cdot g(x)$
$= (-3x+9)(4x+1)$
$= -12x^2 - 3x + 36x + 9$
$= -12x^2 + 33x + 9$

58. a) $(f+g)(x) = f(x) + g(x)$
$= (-x-2) + (8x+2)$
$= 7x$

b) $(f-g)(x) = f(x) - g(x)$
$= (-x-2) - (8x+2)$
$= (-x-2) + (-8x-2)$
$= -9x - 4$

c) $(f+g)(x) = f(x) \cdot g(x)$
$= (-x-2)(8x+2)$
$= -8x^2 - 2x - 16x - 4$
$= -8x^2 - 18x - 4$

59. a) $(f+g)(x) = f(x) + g(x)$
$= (3x^2 - x + 5) + (x+2)$
$= 3x^2 + 7$

b) $(f-g)(x) = f(x) - g(x)$
$= (3x^2 - x + 5) - (x+2)$
$= (3x^2 - x + 5) + (-x-2)$
$= 3x^2 - 2x + 3$

c) $(f \cdot g)(x) = f(x) \cdot g(x)$
$= (3x^2 - x + 5)(x+2)$
$= 3x^3 + 6x^2 - x^2 - 2x + 5x + 10$
$= 3x^3 + 5x^2 + 3x + 10$

60. a) $(f+g)(x) = f(x) + g(x)$
$= (x^2 - 2x - 1) + (3x+1)$
$= x^2 + x$

b) $(f-g)(x) = f(x) - g(x)$
$= (x^2 - 2x - 1) - (3x+1)$
$= (x^2 - 2x - 1) + (-3x-1)$
$= x^2 - 5x - 2$

c) $(f \cdot g)(x) = f(x) \cdot g(x)$
$= (x^2 - 2x - 1)(3x+1)$
$= 3x^3 + x^2 - 6x^2 - 2x - 3x - 1$
$= 3x^3 - 5x^2 - 5x - 1$

61. $(f/g)(x) = \dfrac{f(x)}{g(x)} = \dfrac{6x^4 - 15x^3 + 12x^2}{3x^2}$

$= \dfrac{6x^4}{3x^2} - \dfrac{15x^3}{3x^2} + \dfrac{12x^2}{3x^2}$

$= 2x^2 - 5x + 4, \; x \neq 0$

62. $(f \,/\, g)(x) = \dfrac{f(x)}{g(x)}$

$$= \frac{2x^3 - 17x^2 + 36x - 17}{2x - 5}, \; x \ne \frac{5}{2}$$

$$
\begin{array}{r}
x^2 - 6x + 3 \\
2x-5\overline{\smash{)}2x^3 - 17x^2 + 36x - 17} \\
\underline{2x^3 - 5x^2} \\
-12x^2 + 36x \\
\underline{-12x^2 + 30x} \\
6x - 17 \\
\underline{6x - 15} \\
-2
\end{array}
$$

$(f \,/\, g)(x) = x^2 - 6x + 3 + \dfrac{-2}{2x-5}, \; x \ne \dfrac{5}{2}$

63. a) $c(x) = w(x) + p(x)$

$= (3x + 5) + (x - 1)$

$= 4x + 4$

b) $c(x) = 4x + 4$

$c(225) = 4(225) + 4$

$= 900 + 4$

$= 904$

The total cost is $904.

c)

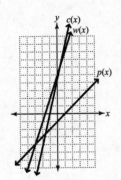

64. a) $h(x) = \dfrac{V(x)}{l(x) \cdot w(x)}$

$h(x) = \dfrac{6x^3 + 41x^2 + 26x - 24}{(3x+4)(x+6)}$

$= \dfrac{6x^3 + 41x^2 + 26x - 24}{3x^2 + 22x + 24}$

$$
\begin{array}{r}
2x - 1 \\
3x^2 + 22x + 24\overline{\smash{)}6x^3 + 41x^2 + 26x - 24} \\
\underline{6x^3 + 44x^2 + 48x} \\
-3x^2 - 22x - 24 \\
\underline{-3x^2 - 22x - 24} \\
0
\end{array}
$$

$h(x) = 2x - 1$

b) $h(x) = 2x - 1$

$h(9) = 2(9) - 1$

$= 18 - 1$

$= 17$

The height of the box is 17 in.

Chapter 10 Practice Test

1. $x^2 = 81$

$x = \pm\sqrt{81}$

$x = \pm 9$

2. $(x - 3)^2 = 20$

$x - 3 = \pm\sqrt{20}$

$x - 3 = \pm\sqrt{4 \cdot 5}$

$x - 3 = \pm 2\sqrt{5}$

$x = 3 \pm 2\sqrt{5}$

3. $x^2 - 8x = -4$

$x^2 - 8x + 16 = -4 + 16$

$(x - 4)^2 = 12$

$x - 4 = \pm\sqrt{12}$

$x - 4 = \pm\sqrt{4 \cdot 3}$

$x - 4 = \pm 2\sqrt{3}$

$x = 4 \pm 2\sqrt{3}$

4. $3m^2 - 6m = 5$

$\dfrac{3m^2 - 6m}{3} = \dfrac{5}{3}$

$m^2 - 2m = \dfrac{5}{3}$

$m^2 - 2m + 1 = \dfrac{5}{3} + 1$

$(m-1)^2 = \dfrac{8}{3}$

$m - 1 = \pm\sqrt{\dfrac{8}{3}}$

$m - 1 = \pm\dfrac{2\sqrt{2}}{\sqrt{3}}$

$m - 1 = \pm\dfrac{2\sqrt{2}}{\sqrt{3}} \cdot \dfrac{\sqrt{3}}{\sqrt{3}}$

$m - 1 = \dfrac{2\sqrt{6}}{3}$

$m = 1 \pm \dfrac{2\sqrt{6}}{3}$

$m = \dfrac{3 \pm 2\sqrt{6}}{3}$

5. $2x^2 + x - 6 = 0$
    Let $a = 2$, $b = 1$, and $c = -6$.

$x = \dfrac{-(1) \pm \sqrt{(1)^2 - 4(2)(-6)}}{2(2)}$

$= \dfrac{-1 \pm \sqrt{1 + 48}}{4}$

$= \dfrac{-1 \pm \sqrt{49}}{4}$

$= \dfrac{-1 \pm 7}{4}$

$= -2, \dfrac{3}{2}$

6. $x^2 - 8x + 15 = 0$
    Let $a = 1$, $b = -8$, and $c = 15$.

$x = \dfrac{-(-8) \pm \sqrt{(-8)^2 - 4(1)(15)}}{2(1)}$

$= \dfrac{8 \pm \sqrt{64 - 60}}{2}$

$= \dfrac{8 \pm \sqrt{4}}{2}$

$= \dfrac{8 \pm 2}{2}$

$= 3, 5$

7. $u^2 - 16 = -6u$

$u^2 + 6u - 16 = 0$

$(u + 8)(u - 2) = 0$

$u + 8 = 0 \qquad \text{or } u - 2 = 0$

$u = -8 \qquad\qquad u = 2$

8. $4w^2 + 6w + 3 = 0$
    Let $a = 4$, $b = 6$, and $c = 3$.

$x = \dfrac{-(6) \pm \sqrt{(6)^2 - 4(4)(3)}}{2(4)}$

$= \dfrac{-6 \pm \sqrt{36 - 48}}{8}$

$= \dfrac{-6 \pm \sqrt{-12}}{8}$

$= \dfrac{-6 \pm 2i\sqrt{3}}{8} = \dfrac{-3 \pm i\sqrt{3}}{4}$

9. $x^2 + 16 = 0$

$x^2 = -16$

$x = \pm\sqrt{-16}$

$x = \pm 4i$

10. $3k^2 = -5k$

$3k^2 + 5k = 0$

$k(3k + 5) = 0$

$k = 0 \quad \text{or} \quad 3k + 5 = 0$

$k = -\dfrac{5}{3}$

11.
$$\frac{1}{x+2}+\frac{1}{x}=\frac{5}{12}$$

$$12x(x+2)\left(\frac{1}{x+2}+\frac{1}{x}\right)=12x(x+2)\cdot\frac{5}{12}$$

$$12x+12(x+2)=5x(x+2)$$

$$12x+12x+24=5x^2+10x$$

$$0=5x^2-14x-24$$

$$0=(5x+6)(x-4)$$

$$5x+6=0 \quad\text{or}\quad x-4=0$$

$$x=-\frac{6}{5} \qquad\qquad x=4$$

12.
$$3-x^{-1}-2x^{-2}=0$$

$$x^2\left(3-x^{-1}-2x^{-2}\right)=x^2\cdot 0$$

$$3x^2-x-2=0$$

$$(3x+2)(x-1)=0$$

$$3x+2=0 \quad\text{or}\quad x-1=0$$

$$x=-\frac{2}{3} \qquad\qquad x=1$$

13. $9\sqrt{x}+8=0$ \qquad No solution

$$9\sqrt{x}=-8$$

$$\sqrt{x}=-\frac{8}{9}$$

14. $\sqrt{x+8}-x=2$

$$\sqrt{x+8}=x+2$$

$$\left(\sqrt{x+8}\right)^2=(x+2)^2$$

$$x+8=x^2+4x+4$$

$$0=x^2+3x-4$$

$$0=(x+4)(x-1)$$

$$x+4=0 \quad\text{or}\quad x-1=0$$

$$x=\cancel{-4} \qquad\qquad x=1$$

The solution −4 is extraneous. The only solution is 1.

15. Let $u=a^2$.

$$9u^2+26u-3=0$$

$$(9u-1)(u+3)=0$$

$$9u-1=0 \quad\text{or}\quad u+3=0$$

$$u=\frac{1}{9} \qquad\qquad u=-3$$

$$x^2=\frac{1}{9} \qquad\qquad x^2=-3$$

$$x=\pm\frac{1}{3} \qquad\qquad x=\pm i\sqrt{3}$$

16. Let $u=x+1$.

$$u^2+3u-4=0$$

$$(u+4)(u-1)=0$$

$$u+4=0 \quad\text{or}\quad u-1=0$$

$$u=-4 \qquad\qquad u=1$$

$$x+1=-4 \qquad\qquad x+1=1$$

$$x=-5 \qquad\qquad x=0$$

17. a) x-intercepts:

$$-x^2+6x-4=0$$

$$x^2-6x+4=0$$

$$x=\frac{-(-6)\pm\sqrt{(-6)^2-4(1)(4)}}{2(1)}$$

$$=\frac{6\pm\sqrt{36-16}}{2}$$

$$=\frac{6\pm\sqrt{20}}{2}$$

$$=\frac{6\pm 2\sqrt{5}}{2}$$

$$=3\pm\sqrt{5}$$

The x-intercepts are $\left(3+\sqrt{5},0\right),\left(3-\sqrt{5},0\right)$.

y-intercept: Let $x=0$.

$$y=-x^2+6x-4$$

$$y=-(0)^2+6(0)-4$$

$$y=-4$$

The y-intercept is (0, −4).

b)
$$y=-x^2+6x-4$$

$$y+4=-\left(x^2-6x\right)$$

$$y+4-9=-\left(x^2-6x+9\right)$$

$$y-5=-(x-3)^2$$

$$f(x)=-(x-3)^2+5$$

c) Because $a < 0$, the parabola opens downward.

d) From part (b) we see that $h = 3$ and $k = 5$. Therefore, the vertex is $(h, k) = (3, 5)$.

e) The axis of symmetry is given by $x = h$. Therefore, the axis of symmetry for this parabola is $x = 3$.

f)

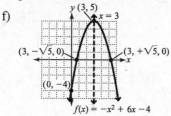

g) The domain is $\{x \mid x \text{ is a real number}\}$ or $(-\infty, \infty)$. The range is $\{y \mid y \le 5\}$ or $(-\infty, 5]$.

18. $(x+1)(x-4) = 0$

$x + 1 = 0 \quad$ or $\quad x - 4 = 0$

$x = -1 \qquad\qquad x = 4$

Interval	$(-\infty, -1]$	$[-1, 4]$	$[4, \infty)$
Test No.	-2	0	5
Results	$6 \le 0$	$-4 \le 0$	$6 \le 0$
T/F	F	T	F

Solution set: $[-1, 4]$

```
←+++++[+++++]+++→
-6 -5 -4 -3 -2 -1 0 1 2 3 4 5 6
```

19. $\dfrac{x+2}{x-1} = 0 \qquad\qquad x - 1 = 0$

$\qquad\qquad\qquad\qquad\qquad x = 1$

$(x-1)\dfrac{x+2}{x-1} = 0(x-1)$

$x + 2 = 0$

$x = -2$

Interval	$(-\infty, -2)$	$(-2, 1)$	$(1, \infty)$
Test No.	-5	0	2
Results	$1/2 > 0$	$-2 > 0$	$4 > 0$
T/F	T	F	T

Solution set: $(-\infty, -2) \cup (1, \infty)$

```
←+++)+++(++++→
-6 -5 -4 -3 -2 -1 0 1 2 3 4 5 6
```

20. $f(x) = x^2$, $g(x) = 3x^2 - 2$

a) $(f + g)(x) = f(x) + g(x)$

$\qquad\qquad = (x^2) + (3x^2 - 2)$

$\qquad\qquad = 4x^2 - 2$

b) $(f - g)(x) = f(x) - g(x)$

$\qquad\qquad = (x^2) - (3x^2 - 2)$

$\qquad\qquad = (x^2) + (-3x^2 + 2)$

$\qquad\qquad = -2x^2 + 2$

c) $(f \cdot g)(x) = f(x)g(x)$

$\qquad\qquad = (x^2)(3x^2 - 2)$

$\qquad\qquad = 3x^4 - 2x^2$

21. $f(x) = 20x^4 + 36x^3 - 28x^2$ and $g(x) = 4x^2$

$(f / g)(x) = \dfrac{f(x)}{g(x)} = \dfrac{20x^4 + 36x^3 - 28x^2}{4x^2}$

$\qquad\qquad = \dfrac{20x^4}{4x^2} + \dfrac{36x^3}{4x^2} - \dfrac{28x^2}{4x^2}$

$\qquad\qquad = 5x^2 + 9x - 7$

22. Let $d = 180$.

$16t^2 + 32t = d$

$16t^2 + 32t = 180$

$16t^2 + 32t - 180 = 0$

$4(4t^2 + 8t - 45) = 0$

$4t^2 + 8t - 45 = 0$

$(2t - 5)(2t + 9) = 0$

$2t - 5 = 0 \quad$ or $\quad 2t + 9 = 0$

$2t = 5 \qquad\qquad 2t = -9$

$t = \dfrac{5}{2} \qquad\qquad t = \cancel{-\dfrac{9}{2}}$

The solution $-\dfrac{9}{2}$ does not make sense in the context of the problem. Therefore, the ball takes 2.5 seconds to fall 180 feet.

23. Finding the vertex will show the maximum height that the arrow will reach.

$$x = \frac{-1.3}{2(-0.02)} = 32.5$$

$$y = -0.02(32.5)^2 + 1.3(32.5) + 8 = 29.125$$

The vertex is (32.5, 29.125).

a) The maximum height is 29.125 m.

b) $-0.02x^2 + 1.3x + 8 = 0$

$$x = \frac{-1.3 \pm \sqrt{1.3^2 - 4(-0.02)(8)}}{2(-0.02)}$$

$$x = \frac{-1.3 \pm \sqrt{2.33}}{-0.04}$$

$$x \approx 70.66, \cancel{-5.66}$$

The distance can't be negative, so the arrow travels 70.66 m.

c)

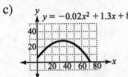

$$y = -0.02x^2 + 1.3x + 8$$

24. a)
$$12w(w + 15) \le 12{,}000$$
$$12w^2 + 180w - 12{,}000 \le 0$$
$$w^2 + 15w - 1000 \le 0$$
$$(w + 40)(w - 25) \le 0$$
$$-40 \le w \le 25$$

Since $w > 0$, $0 < w \le 25$ ft.

b) $15 < l \le 40$ ft.

25. a)
$$A(x) = l(x) \cdot h(x)$$
$$\frac{A(x)}{l(x)} = h(x)$$
$$\frac{36x^2 - x - 2}{4x - 1} = h(x)$$
$$\frac{\cancel{(4x - 1)}(9x + 2)}{\cancel{4x - 1}} = h(x)$$
$$9x + 2 = h(x)$$

b) Find $h(15)$.
$$h(x) = 9x + 2$$
$$h(15) = 9(15) + 2$$
$$= 135 + 2$$
$$= 137$$

The height of the box is 137 cm

Chapters 1–10 Cumulative Review

1. False; $(45, -12)$ is in quadrant IV.

2. False; a system of equations that has no solution is said to be inconsistent.

3. True

4. parabola, $a > 0$

5. $\sqrt[n]{ab}$

6. $\sqrt{a}, -\sqrt{a}$

7. $(-3x - 4) - (3x + 2) = (-3x - 4) + (-3x - 2)$
$$= -6x - 6$$

8. $(8n^2)(-7mn^3) = -56mn^5$

9. $(x - 3)(4x^2 - 2x + 1)$
$$= 4x^3 - 2x^2 + x - 12x^2 + 6x - 3$$
$$= 4x^3 - 14x^2 + 7x - 3$$

10. $m^3 + 8 = (m + 2)(m^2 - 2m + 4)$

11. $x^2 + 10x + 25 = (x + 5)^2$

12. $\dfrac{4u^2 + 4u + 1}{u + 2u^2} \cdot \dfrac{u}{2u^2 - u - 1}$
$$= \frac{\cancel{(2u + 1)}(2u + 1)}{\cancel{u}(1 + 2u)} \cdot \frac{\cancel{u}}{\cancel{(2u + 1)}(u - 1)}$$
$$= \frac{1}{u - 1}$$

13. $\dfrac{3}{x - 2} - \dfrac{2}{x + 2}$
$$= \frac{3(x + 2)}{(x - 2)(x + 2)} - \frac{2(x - 2)}{(x + 2)(x - 2)}$$
$$= \frac{3x + 6 - 2x + 4}{(x - 2)(x + 2)}$$
$$= \frac{x + 10}{(x - 2)(x + 2)}$$

14. $x^{3/4} \cdot x^{-1/4} = x^{3/4 + (-1/4)} = x^{1/2}$

15.
$$\frac{\sqrt{n}}{\sqrt{m}-\sqrt{n}} = \frac{\sqrt{n}}{\sqrt{m}-\sqrt{n}} \cdot \frac{\sqrt{m}+\sqrt{n}}{\sqrt{m}+\sqrt{n}}$$

$$= \frac{\sqrt{mn}+\sqrt{n^2}}{\left(\sqrt{m}\right)^2 - \left(\sqrt{n}\right)^2}$$

$$= \frac{\sqrt{mn}+n}{m-n}$$

16.
$$3n-5 = 7n+9$$
$$-4n-5 = 9$$
$$-4n = 14$$
$$n = -\frac{7}{2}$$

17.
$$|3x+4|+3 = 7$$
$$|3x+4| = 4$$
$$3x+4 = -4 \quad \text{or} \quad 3x+4 = 4$$
$$3x = -8 \qquad\qquad 3x = 0$$
$$x = -\frac{8}{3} \qquad\qquad x = 0$$

18.
$$2x^2+7x = 15$$
$$2x^2+7x-15 = 0$$
$$(2x-3)(x+5) = 0$$
$$2x-3 = 0 \quad \text{or} \quad x+5 = 0$$
$$x = \frac{3}{2} \qquad\qquad x = -5$$

19.
$$\frac{5}{x-2}-\frac{3}{x} = \frac{11}{3x}$$
$$3x(x-2)\left(\frac{5}{x-2}-\frac{3}{x}\right) = 3x(x-2)\cdot\frac{11}{3x}$$
$$3x\cdot 5 - 3(x-2)\cdot 3 = 11(x-2)$$
$$15x-9x+18 = 11x-22$$
$$6x+18 = 11x-22$$
$$40 = 5x$$
$$8 = x$$

20.
$$\sqrt{5x-4} = 9$$
$$\left(\sqrt{5x-4}\right)^2 = 9^2$$
$$5x-4 = 81$$
$$5x = 85$$
$$x = 17$$

21. $x^2-6x+11 = 0$

Let $a = 1$, $b = -6$, and $c = 11$.

$$x = \frac{-(-6)\pm\sqrt{(-6)^2-4(1)(11)}}{2(1)}$$

$$= \frac{6\pm\sqrt{36-44}}{2}$$

$$= \frac{6\pm\sqrt{-8}}{2}$$

$$= \frac{6\pm 2i\sqrt{2}}{2}$$

$$= 3\pm i\sqrt{2}$$

22. a) $-2 < x+4 < 5$
$$-6 < x < 1$$
$$(-6,1)$$

b) $\{x|-6 < x < 1\}$ c) $(-6,1)$

23.
$$x^2+x = 12$$
$$x^2+x-12 = 0$$
$$(x+4)(x-3) = 0$$
$$x+4 = 0 \quad \text{or} \quad x-3 = 0$$
$$x = -4 \qquad\qquad x = 3$$

Interval	$(-\infty,-4)$	$(-4,3)$
Test No.	-5	0
Results	$20 \le 12$	$0 \le 12$
T/F	F	T

Interval	$(3,\infty)$
Test No.	4
Results	$20 \le 12$
T/F	F

a)

b) $\{x|-4 \le x \le 3\}$ c) $[-4,3]$

24. $f(x) = -2x$

x	$f(x)$	(x, y)
-1	2	$(-1, 2)$
0	0	$(0, 0)$
1	-2	$(1, -2)$

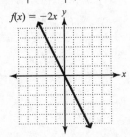

$f(x) = -2x$

25. Solve for y. $3x - 2y = 6$

$$-2y = -3x + 6$$

$$y = \frac{3}{2}x - 3$$

So $m = \frac{3}{2}$. The perpendicular slope is $-\frac{2}{3}$.

Use $m = -\frac{2}{3}$ and $(-2, 4)$.

$$y - 4 = -\frac{2}{3}\left(x - (-2)\right)$$

$$y - 4 = -\frac{2}{3}x - \frac{4}{3}$$

$$3y - 12 = -2x - 4$$

$$2x + 3y = 8$$

26. $\begin{cases} x + y + z = 5 & \text{Eqtn. 1} \\ 2x + y - 2z = -5 & \text{Eqtn. 2} \\ x - 2y + z = 8 & \text{Eqtn. 3} \end{cases}$

Multiply equation 1 by -1 and add to equation 3. This makes equation 4.

$$-x - y - z = -5$$
$$\underline{x - 2y + z = 8}$$
$$-3y = 3$$
$$y = -1 \quad \text{Eqtn. 4}$$

Multiply equation 1 by -2 and add to equation 2. This makes equation 5.

$$-2x - 2y - 2z = -10$$
$$\underline{2x + y - 2z = -5}$$
$$-y - 4z = -15 \quad \text{Eqtn. 5}$$

Substitute equation 4 into equation 5 and solve for z.

$$-(-1) - 4z = -15$$
$$1 - 4z = -15$$
$$-4z = -16$$
$$z = 4$$

Substitute the values for y and z to solve for x.

$$x - 1 + 4 = 5$$
$$x + 3 = 5$$
$$x = 2$$

Solution: $(2, -1, 4)$

27. Complete a table.

	Rate	Time	Distance
not experienced	$\dfrac{300}{t} - 10$	$t + 1$	300
experienced	$\dfrac{300}{t}$	t	300

$$\left(\frac{300}{t} - 10\right)(t + 1) = 300$$

$$300 + \frac{300}{t} - 10t - 10 = 300$$

$$\frac{300}{t} - 10t - 10 = 0$$

$$300 - 10t^2 - 10t = 0$$

$$t^2 + t - 30 = 0$$

$$(t + 6)(t - 5) = 0$$

$$t + 6 = 0 \qquad t - 5 = 0$$

$$t = \cancel{-6} \qquad t = 5$$

It takes the experienced driver 5 hours to drive 300 miles.

28. Create a table.

	Concentrate	Vol. of solution	Vol. of saline
15%	0.15	50	$0.15(50)$
40%	0.40	x	$0.40x$
30%	0.30	$x + 50$	$0.30(x + 50)$

Translate the information in the table to an equation and solve.

$$0.15(50) + 0.40x = 0.3(x + 50)$$

$$7.5 + 0.4x = 0.3x + 15$$

$$0.1x = 7.5$$

$$x = 75$$

75 ml of 40% solution must be added.

29. $v = \dfrac{k}{T}$ So, $v = \dfrac{48}{T}$

    $6 = \dfrac{k}{8}$ $v = \dfrac{48}{15}$

    $48 = k$ $v = 3.2$ cm/sec.

30. a) $t = \sqrt{\dfrac{24}{16}} = \dfrac{\sqrt{24}}{4} = \dfrac{\sqrt{4 \cdot 6}}{4} = \dfrac{2\sqrt{6}}{4} = \dfrac{\sqrt{6}}{2}$ sec.

    b) $t = \sqrt{\dfrac{h}{16}}$

       $3 = \dfrac{\sqrt{h}}{4}$

       $12 = \sqrt{h}$

       $12^2 = \left(\sqrt{h}\right)^2$

       144 ft. $= h$

Chapter 11

Exponential and Logarithmic Functions

---

1. $(f \circ g)(0) = f[g(0)]$
$$= f(3)$$
$$= 3 \cdot 3 + 5$$
$$= 14$$

3. $(h \circ f)(1) = h[f(1)]$
$$= h(8)$$
$$= \sqrt{8+1}$$
$$= 3$$

5. $(f \circ g)(-2) = f[g(-2)]$
$$= f(7)$$
$$= 3 \cdot 7 + 5$$
$$= 26$$

7. $(f \circ g)(x) = f[g(x)]$
$$= f(x^2 + 3)$$
$$= 3(x^2 + 3) + 5$$
$$= 3x^2 + 9 + 5$$
$$= 3x^2 + 14$$

9. $(f \circ h)(x) = f[h(x)]$
$$= f(\sqrt{x+1})$$
$$= 3\sqrt{x+1} + 5$$

11. $(h \circ f)(0) = h[f(0)]$
$$= h(5)$$
$$= \sqrt{5+1}$$
$$= \sqrt{6}$$

13. $(f \circ g)(x) = f[g(x)]$
$$= f(3x + 4)$$
$$= 2(3x + 4) - 2$$
$$= 6x + 8 - 2$$
$$= 6x + 6$$

$(g \circ f)(x) = g[f(x)]$
$$= g(2x - 2)$$
$$= 3(2x - 2) + 4$$
$$= 6x - 6 + 4$$
$$= 6x - 2$$

15. $(f \circ g)(x) = f[g(x)]$
$$= f(x^2 + 1)$$
$$= x^2 + 1 + 2$$
$$= x^2 + 3$$

asdf

$(g \circ f)(x) = g[f(x)]$
$$= g(x + 2)$$
$$= (x + 2)^2 + 1$$
$$= x^2 + 4x + 4 + 1$$
$$= x^2 + 4x + 5$$

17. $(f \circ g)(x) = f[g(x)]$
$$= f(3x)$$
$$= (3x)^2 + 3(3x) - 4$$
$$= 9x^2 + 9x - 4$$

$(g \circ f)(x) = g[f(x)]$
$$= g(x^2 + 3x - 4)$$
$$= 3(x^2 + 3x - 4)$$
$$= 3x^2 + 9x - 12$$

19. $(f \circ g)(x) = f[g(x)]$
$$= f(2x - 5)$$
$$= \sqrt{2x - 5 + 2}$$
$$= \sqrt{2x - 3}$$

$(g \circ f)(x) = g[f(x)]$
$$= g(\sqrt{x+2})$$
$$= 2\sqrt{x+2} - 5$$

21. $(f \circ g)(x) = f\big[g(x)\big]$

$$= f\left(\frac{x-3}{x}\right)$$

$$= \frac{\dfrac{x-3}{x}+1}{\dfrac{x-3}{x}}$$

$$= \frac{2x-3}{x-3}$$

$(g \circ f)(x) = g\big[f(x)\big]$

$$= g\left(\frac{x+1}{x}\right)$$

$$= \frac{\dfrac{x+1}{x}-3}{\dfrac{x+1}{x}}$$

$$= \frac{1-2x}{x+1}$$

23. The domain is $[0,\infty)$ and the range is $[3,\infty)$

25. Yes

27. No

29. $(f \circ g)(x) = f\big[g(x)\big]$

$$= f(x-5)$$

$$= x-5+5$$

$$= x$$

$(g \circ f)(x) = g\big[f(x)\big]$

$$= g(x+5)$$

$$= x+5-5$$

$$= x$$

Yes

31. $(f \circ g)(x) = f\big[g(x)\big]$

$$= f\left(\frac{x}{6}\right)$$

$$= 6 \cdot \frac{x}{6}$$

$$= x$$

$(g \circ f)(x) = g\big[f(x)\big]$

$$= g(6x)$$

$$= \frac{6x}{6}$$

$$= x$$

Yes

33. $(f \circ g)(x) = f\big[g(x)\big]$

$$= f\left(\frac{x+3}{2}\right)$$

$$= 2 \cdot \frac{x+3}{2} - 3$$

$$= x+3-3$$

$$= x$$

$(g \circ f)(x) = g\big[f(x)\big]$

$$= g(2x-3)$$

$$= \frac{2x-3+3}{2}$$

$$= \frac{2x}{2}$$

$$= x$$

Yes

35. $(f \circ g)(x) = f\big[g(x)\big]$

$$= f\left(\sqrt[3]{x+4}\right)$$

$$= \left(\sqrt[3]{x+4}\right)^3 - 4$$

$$= x+4-4$$

$$= x$$

$(g \circ f)(x) = g\big[f(x)\big]$

$$= g\left(x^3 - 4\right)$$

$$= \sqrt[3]{x^3 - 4 + 4}$$

$$= \sqrt[3]{x^3}$$

$$= x$$

Yes

37. $(f \circ g)(x) = f\big[g(x)\big]$

$$= f\left(\sqrt{x}\right)$$

$$= \left(\sqrt{x}\right)^2$$

$$= x$$

$(g \circ f)(x) = g\big[f(x)\big]$

$$= g\left(x^2\right)$$

$$= \sqrt{x^2}$$

$$= |x|$$

No

39. $(f \circ g)(x) = f\left[g(x)\right]$

$\qquad = f\left(\sqrt{x}\right)$

$\qquad = \left(\sqrt{x}\right)^2$

$\qquad = x$

$(g \circ f)(x) = g\left[f(x)\right]$

$\qquad = g\left(x^2\right)$

$\qquad = \sqrt{x^2}$

$\qquad = |x|$

$\qquad = x, \text{ since } x \ge 0$

Yes

41. $(f \circ g)(x) = f\left[g(x)\right]$

$\qquad = f\left(\dfrac{3-5x}{x}\right)$

$\qquad = \dfrac{3}{\dfrac{3-5x}{x}+5} = \dfrac{3x}{3-5x+5x}$

$\qquad = \dfrac{3x}{3} = x$

$(g \circ f)(x) = g\left[f(x)\right]$

$\qquad = g\left(\dfrac{3}{x+5}\right)$

$\qquad = \dfrac{3-5\left(\dfrac{3}{x+5}\right)}{\dfrac{3}{x+5}} = \dfrac{3-\dfrac{15}{x+5}}{\dfrac{3}{x+5}}$

$\qquad = \dfrac{3(x+5)-15}{3} = \dfrac{3x+15-15}{3}$

$\qquad = \dfrac{3x}{3} = x$

Yes

43. Yes, because $(f \circ g)(x) = (g \circ f)(x) = x$:

$(f \circ f)(x) = f\left[f(x)\right] = f\left(\dfrac{1}{x}\right) = \dfrac{1}{\dfrac{1}{x}} = x$

45. Yes. Every horizontal line that can intersect this graph does so at one and only one point, so the function is one to one.

47. No. A horizontal line can intersect this graph in more than one point, so the function is not one to one.

49. $f^{-1} = \{(2,-3),(-3,-1),(4,0),(6,4)\}$

51. $f^{-1} = \{(-2,7),(2,9),(1,-4),(3,3)\}$

53. $\qquad y = x+6$

$\qquad x = y+6$

$\qquad x-6 = y$

$\qquad f^{-1}(x) = x-6$

55. $\qquad y = 2x+3$

$\qquad x = 2y+3$

$\qquad x-3 = 2y$

$\qquad \dfrac{x-3}{2} = y$

$\qquad f^{-1}(x) = \dfrac{x-3}{2}$

57. $\qquad y = x^3-1$

$\qquad x = y^3-1$

$\qquad x+1 = y^3$

$\qquad \sqrt[3]{x+1} = y$

$\qquad f^{-1}(x) = \sqrt[3]{x+1}$

59. $\qquad y = \dfrac{2}{x+2}$

$\qquad x = \dfrac{2}{y+2}$

$\qquad x(y+2) = 2$

$\qquad xy+2x = 2$

$\qquad xy = 2-2x$

$\qquad y = \dfrac{2-2x}{x}$

$\qquad f^{-1}(x) = \dfrac{2-2x}{x}$

61. $\qquad y = \dfrac{x+2}{x-3}$

$\qquad x = \dfrac{y+2}{y-3}$

$\qquad x(y-3) = y+2$

$\qquad xy-3x = y+2$

$\qquad xy-y = 3x+2$

$\qquad y(x-1) = 3x+2$

$\qquad y = \dfrac{3x+2}{x-1}$

$\qquad f^{-1}(x) = \dfrac{3x+2}{x-1}$

63. $$y = \sqrt{x-2}$$
$$x = \sqrt{y-2}$$
$$x^2 = y - 2$$
$$x^2 + 2 = y$$
$$f^{-1}(x) = x^2 + 2, x \geq 0$$

65. $$y = 2x^3 + 4$$
$$x = 2y^3 + 4$$
$$x - 4 = 2y^3$$
$$\frac{x-4}{2} = y^3$$
$$\sqrt[3]{\frac{x-4}{2}} = y$$
$$f^{-1}(x) = \sqrt[3]{\frac{x-4}{2}}$$

67. $$y = \sqrt[3]{x+2}$$
$$x = \sqrt[3]{y+2}$$
$$x^3 = y + 2$$
$$x^3 - 2 = y$$
$$f^{-1}(x) = x^3 - 2$$

69. $$y = 2\sqrt[3]{2x+4}$$
$$x = 2\sqrt[3]{2y+4}$$
$$\frac{x}{2} = \sqrt[3]{2y+4}$$
$$\frac{x^3}{8} = 2y + 4$$
$$\frac{x^3}{8} - 4 = 2y$$
$$\frac{1}{2}\left(\frac{x^3}{8} - 4\right) = \frac{1}{2} \cdot 2y$$
$$\frac{x^3}{16} - 2 = y$$
$$f^{-1}(x) = \frac{x^3}{16} - 2$$

71.

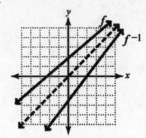

73.

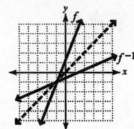

75.

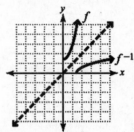

77.

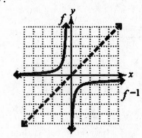

79.

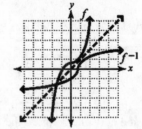

81.

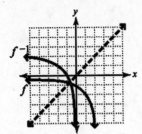

83. a) $y = 0.05x + 100$

$x = 0.05y + 100$

$x - 100 = 0.05y$

$\dfrac{x - 100}{0.05} = y$

b) x represents the salary, y represents the sales

c) $y = \dfrac{x - 100}{0.05}$

$= \dfrac{350 - 100}{0.05}$

$= \dfrac{250}{0.05}$

$= 5000$

If the salary was $350, sales were $5000.

85. a) $y = 45x + 65(105 - x)$

$y = 45x + 6825 - 65x$

$y = -20x + 6825$

$x = -20y + 6825$

$x - 6825 = -20y$

$\dfrac{x - 6825}{-20} = y$

$y = \dfrac{6825 - x}{20}$

b) x represents the cost, y represents the number of 36-in. fans.

c) $f^{-1}(x) = \dfrac{6825 - x}{20}$

$f^{-1}(5225) = \dfrac{6825 - 5225}{20}$

$= 80$

The number of 36-inch fans was 80.

87. $(5, -4)$

89. $y = ax + b$

$x = ay + b$

$x - b = ay$

$\dfrac{x - b}{a} = y$

$f^{-1}(x) = \dfrac{x - b}{a}$

Review Exercises

1. $32 = 2^5$

2. $2^3 = 8$

3. $\left(-\dfrac{1}{3}\right)^3 = -\dfrac{1}{27}$

4. $4^{-2} = \left(\dfrac{1}{4}\right)^2 = \dfrac{1}{16}$

5. $4^{3/2} = \left(\sqrt{4}\right)^3 = 2^3 = 8$

6. Vertex: $(3, 1)$

Exercise Set 11.2

1. $f(x) = 3^x$

x	$y = f(x)$
-1	$\dfrac{1}{3}$
0	1
1	3
2	9

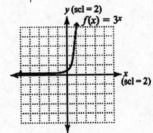

3. $f(x) = 4^x - 3$

x	$y = f(x)$
-1	$4^{-1} - 3 = -\dfrac{11}{4}$
0	$4^0 - 3 = -2$
1	$4^1 - 3 = 1$
2	$4^2 - 3 = 13$

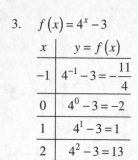

5. $f(x) = \left(\dfrac{1}{3}\right)^x$

x	$y = f(x)$
-1	$\left(\dfrac{1}{3}\right)^{-1} = 3$
0	$\left(\dfrac{1}{3}\right)^0 = 1$
1	$\left(\dfrac{1}{3}\right)^1 = \dfrac{1}{3}$
2	$\left(\dfrac{1}{3}\right)^2 = \dfrac{1}{9}$

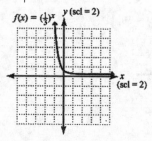

7. $f(x) = \left(\dfrac{2}{3}\right)^x + 2$

x	$y = f(x)$
-1	$\left(\dfrac{2}{3}\right)^{-1} + 2 = \dfrac{7}{2}$
0	$\left(\dfrac{2}{3}\right)^0 + 2 = 3$
1	$\left(\dfrac{2}{3}\right)^1 + 2 = \dfrac{8}{3}$
2	$\left(\dfrac{2}{3}\right)^2 + 2 = \dfrac{22}{9}$

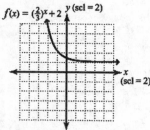

9. $f(x) = -3^x$

x	$y = f(x)$
-1	$-\dfrac{1}{3}$
0	-1
1	-3
2	-9

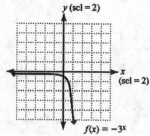

11. $f(x) = 2^{x-2}$

x	$y = f(x)$
-1	$2^{-1-2} = \dfrac{1}{8}$
0	$2^{0-2} = \dfrac{1}{4}$
1	$2^{1-2} = \dfrac{1}{2}$
2	$2^{2-2} = 1$

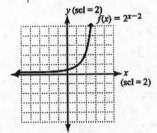

13. $f(x) = 2^{-x}$

x	$y = f(x)$
-1	$2^{-(-1)} = 2$
0	$2^{-(0)} = 1$
1	$2^{-(1)} = \dfrac{1}{2}$
2	$2^{-(2)} = \dfrac{1}{4}$

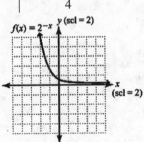

15. $f(x) = 2^{2x-3}$

x	$y = f(x)$
-1	$2^{2(-1)-3} = \dfrac{1}{32}$
0	$2^{2(0)-3} = \dfrac{1}{8}$
1	$2^{2(1)-3} = \dfrac{1}{2}$
2	$2^{2(2)-3} = 2$

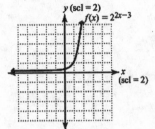

17. $f(x) = 3^{-x+2}$

x	$y = f(x)$
-1	$3^{-(-1)+2} = 27$
0	$3^{-(0)+2} = 9$
1	$3^{-(1)+2} = 3$
2	$3^{-(2)+2} = 1$

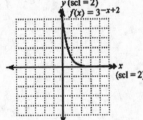

19. $2^x = 8$

$2^x = 2^3$

$x = 3$

21. $8^x = 32$

$\left(2^3\right)^x = 2^5$

$2^{3x} = 2^5$

$3x = 5$

$x = \dfrac{5}{3}$

23. $16^x = 4$

     $\left(2^4\right)^x = 2^2$

     $2^{4x} = 2^2$

     $4x = 2$

     $x = \dfrac{1}{2}$

25. $6^x = \dfrac{1}{36}$

     $6^x = 6^{-2}$

     $x = -2$

27. $\left(\dfrac{1}{3}\right)^x = 9$

     $\left(3^{-1}\right)^x = 3^2$

     $3^{-x} = 3^2$

     $-x = 2$

     $x = -2$

29. $\left(\dfrac{2}{3}\right)^x = \dfrac{8}{27}$

     $\left(\dfrac{2}{3}\right)^x = \left(\dfrac{2}{3}\right)^3$

     $x = 3$

31. $\left(\dfrac{1}{2}\right)^x = 16$

     $\left(2^{-1}\right)^x = 2^4$

     $2^{-x} = 4$

     $-x = 4$

     $x = -4$

33. $25^{x+1} = 125$

     $\left(5^2\right)^{x+1} = 5^3$

     $5^{2x+2} = 5^3$

     $2x + 2 = 3$

     $2x = 1$

     $x = \dfrac{1}{2}$

35. $8^{2x-1} = 32^{x-3}$

     $\left(2^3\right)^{2x-1} = \left(2^5\right)^{x-3}$

     $2^{6x-3} = 2^{5x-15}$

     $6x - 3 = 5x - 15$

     $x = -12$

37.   a)

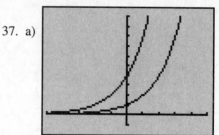

     b) The graph of g is the graph of f shifted 2 units to the left.

39.   a)

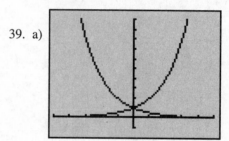

     b) The graph of g is the graph of f reflected about the y-axis.

41. $A = A_0 \left(2\right)^{t/20}$

     $A = 100\left(2\right)^{120/20}$

     $A = 6400$

     After 120 minutes, there will be 6400 cells present.

43. $A = A_0 \left(2\right)^{t/50}$

     $A = 6\left(2\right)^{500/50}$

     $A = 6144$

     There will be 6144 billion people in 2500 if their growth is uncontrolled.

45. $A = P\left(1 + \dfrac{r}{n}\right)^{nt}$

     $A = 10{,}000\left(1 + \dfrac{0.08}{4}\right)^{4 \cdot 12}$

     $A = 25{,}870.70$

     The account value will be \$25,870.70 after 12 years.

47. $A = A_0 \left(\frac{1}{2}\right)^{t/h}$

$A = 5\left(\frac{1}{2}\right)^{2160/270}$

$A = 0.020$

After 2160 days, 0.020 gram would remain.

49. $y = 2632.31(1.033)^x$

$y = 2632.31(1.033)^{59}$

$y \approx 17,875.41$

51. $f(x) = 2.5(0.7)^x$

$f(2) = 2.5(0.7)^2$

$\quad\quad = 1.225$

The chlorine level after 2 days is 1.225 parts per million.

53. $T(x) = 1.97(1.503)^x$

$T(27) = 1.97(1.503)^{27}$

$\quad\quad \approx 118,130$

If the trend continues, a total of about 118,130 million transistors could be put on a single chip in 2020.

Review Exercises

1. $5^3 = 125$

2. $\left(\frac{1}{3}\right)^{-2} = \left(\frac{3}{1}\right)^2 = 9$

3. $4^{-3} = \frac{1}{4^3} = \frac{1}{64}$

4. $5^{2/3} = \sqrt[3]{5^2} = \left(\sqrt[3]{5}\right)^2$

5. $\quad f(x) = 3x + 4$

$\quad\quad y = 3x + 4$

$\quad\quad x = 3y + 4$

$\quad x - 4 = 3y$

$\quad \dfrac{x-4}{3} = y$

$f^{-1}(x) = \dfrac{x-4}{3}$

6.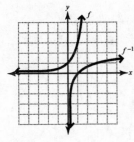

Exercise Set 11.3

1. $\log_2 32 = 5$

3. $\log_{10} 1000 = 3$

5. $\log_e x = 4$

7. $\log_5 \dfrac{1}{125} = -3$

9. $\log_{10} \dfrac{1}{100} = -2$

11. $\log_{625} 5 = \dfrac{1}{4}$

13. $\log_{1/4} \dfrac{1}{16} = 2$

15. $\log_7 \sqrt{7} = \dfrac{1}{2}$

17. $3^4 = 81$

19. $4^{-2} = \dfrac{1}{16}$

21. $10^2 = 100$

23. $e^5 = a$

25. $e^{-4} = \dfrac{1}{e^4}$

27. $\left(\dfrac{1}{8}\right)^2 = \dfrac{1}{64}$

29. $\left(\dfrac{1}{5}\right)^{-2} = 25$

31. $7^{1/2} = \sqrt{7}$

33. $2^5 = x$

$\quad 32 = x$

35. $5^{-2} = x$

$\quad \dfrac{1}{5^2} = x$

$\quad \dfrac{1}{25} = x$

37. $3^y = 81$

$\quad 3^y = 3^4$

$\quad\quad y = 4$

39. $5^y = \dfrac{1}{25}$

$\quad 5^y = 5^{-2}$

$\quad\quad y = -2$

41. $b^3 = 1000$

$\quad b^3 = 10^3$

$\quad\quad b = 10$

43. $m^{-4} = \dfrac{1}{16}$

$\quad m^{-4} = 16^{-1}$

$\quad m^{-4} = \left(2^4\right)^{-1}$

$\quad m^{-4} = 2^{-4}$

$\quad\quad m = 2$

45. $\left(\dfrac{1}{2}\right)^2 = x$

$\quad \dfrac{1}{4} = x$

47. $\left(\dfrac{1}{3}\right)^{-5} = h$

$\quad 3^5 = h$

$\quad 243 = h$

49. $\left(\dfrac{1}{3}\right)^y = \dfrac{1}{9}$

$\left(\dfrac{1}{3}\right)^y = \left(\dfrac{1}{3}\right)^2$

$\quad y = 2$

51. $\left(\dfrac{1}{2}\right)^t = 64$

$\left(2^{-1}\right)^t = 2^6$

$\quad 2^{-t} = 2^6$

$\quad -t = 6$

$\quad t = -6$

53. $\log_2 x = 4$

$\quad x = 2^4$

$\quad x = 16$

55. $\log_{1/4} h = 3$

$\quad h = \left(\dfrac{1}{4}\right)^3$

$\quad h = \dfrac{1}{64}$

57. $\log_b 16 = 4$

$\quad b^4 = 16$

$\quad b^4 = 2^4$

$\quad b = 2$

59. $\dfrac{1}{3}\log_5 c = 1$

$\quad \log_5 c = 3$

$\quad c = 5^3$

$\quad c = 125$

61. $3\log_t 9 = 6$

$\quad \log_t 9 = 2$

$\quad t^2 = 9$

$\quad t^2 = 3^2$

$\quad t = 3$

63. $\dfrac{1}{2}\log_4 m = -1$

$\quad \log_4 m = -2$

$\quad m = 4^{-2}$

$\quad m = \dfrac{1}{4^2}$

$\quad m = \dfrac{1}{16}$

65. $f(x) = \log_4 x$

$\quad y = \log_4 x$

$\quad 4^y = x$

y	-1	0	1	2
x	$\dfrac{1}{4}$	1	4	16

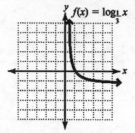

67. $f(x) = \log_{1/3} x$

$\quad y = \log_{1/3} x$

$\quad \left(\dfrac{1}{3}\right)^y = x$

y	-2	-1	0	1
x	9	3	1	$\dfrac{1}{3}$

69. $f(x) = 62 + 35\log_{10}(x-4)$

$f(14) = 62 + 35\log_{10}(14-4)$

$\quad = 62 + 35\log_{10} 10$

$\quad = 62 + 35(1)$

$\quad = 97$

At age 14, a girl has reached 97% of her adult height.

71. $\quad 35 = 95 - 30\log_2 x$

$\quad -60 = -30\log_2 x$

$\quad 2 = \log_2 x$

$\quad x = 2^2$

$\quad x = 4$

After 4 days, 35% of the students recall important features of the lecture.

73. $\log_b b = 1$ because $b^1 = b$.

75. $\log_b 1 = 0$ because $b^0 = 1$.

Review Exercises

1. x^{-6}

2. $x^4 \cdot x^2 = x^{4+2} = x^6$

3. $\left(x^3\right)^5 = x^{3\cdot5} = x^{15}$

4. $\dfrac{x^6}{x^3} = x^{6-3} = x^3$

5. $\dfrac{\left(x^3\right)^2 \cdot x^4}{x^5} = \dfrac{x^6 \cdot x^4}{x^5} = \dfrac{x^{10}}{x^5} = x^{10-5} = x^5$

6. $\sqrt[4]{x^3} = x^{3/4}$

Exercise Set 11.4

1. $8^{\log_8 2} = 2$

3. $a^{\log_a r} = r$

5. $a^{\log_a 4x} = 4x$

7. $\log_3 3^5 = 5$

9. $\log_e e^y = y$

11. $\log_a a^{7x} = 7x$

13. $\log_2 5y = \log_2 5 + \log_2 y$

15. $\log_a pq = \log_a p + \log_a q$

17. $\log_4 mnp = \log_4 m + \log_4 n + \log_4 p$

19. $\log_a x(x-5) = \log_a x + \log_a (x-5)$

21. $\log_3 5 + \log_3 8 = \log_3 (5 \cdot 8) = \log_3 40$

23. $\log_4 3 + \log_4 9 = \log_4 (3 \cdot 9) = \log_4 27$

25. $\log_a 7 + \log_a m = \log_a 7m$

27. $\log_4 a + \log_4 b = \log_4 ab$

29. $\log_a 2 + \log_a x + \log_a (x+5) = \log_a 2x(x+5)$
$$= \log_a (2x^2 + 10x)$$

31. $\log_4 (x+1) + \log_4 (x+3) = \log_4 (x+1)(x+3)$
$$= \log_4 (x^2 + 4x + 3)$$

33. $\log_2 \dfrac{7}{9} = \log_2 7 - \log_2 9$

35. $\log_a \dfrac{x}{5} = \log_a x - \log_a 5$

37. $\log_a \dfrac{a}{b} = \log_a a - \log_a b = 1 - \log_a b$

39. $\log_a \dfrac{x}{x-3} = \log_a x - \log_a (x-3)$

41. $\log_4 \dfrac{2x-3}{4x+5} = \log_4 (2x-3) - \log_4 (4x+5)$

43. $\log_6 24 - \log_6 3 = \log_6 \dfrac{24}{3} = \log_6 8$

45. $\log_2 24 - \log_2 12 = \log_2 \dfrac{24}{12} = \log_2 2 = 1$

47. $\log_a x - \log_a 3 = \log_a \dfrac{x}{3}$

49. $\log_4 p - \log_4 q = \log_4 \dfrac{p}{q}$

51. $\log_b x - \log_b (x-4) = \log_b \dfrac{x}{x-4}$

53. $\log_x (x^2 - x) - \log_x (x-1) = \log_x \dfrac{x^2 - x}{x-1}$
$$= \log_x \dfrac{x(x-1)}{x-1}$$
$$= \log_x x$$
$$= 1$$

55. $\log_4 3^6 = 6\log_4 3$

57. $\log_a x^7 = 7\log_a x$

59. $\log_a \sqrt{3} = \log_a 3^{1/2} = \dfrac{1}{2}\log_a 3$

61. $\log_3 \sqrt[3]{x^2} = \log_3 x^{2/3} = \dfrac{2}{3}\log_3 x$

63. $\log_a \dfrac{1}{6^2} = \log_a 6^{-2} = -2\log_a 6$

65. $\log_a \dfrac{1}{y^2} = \log_a y^{-2} = -2\log_a y$

67. $4\log_3 5 = \log_3 5^4$

69. $-3\log_2 x = \log_2 x^{-3} = \log_2 \dfrac{1}{x^3}$

71. $\dfrac{1}{2}\log_7 64 = \log_7 64^{1/2} = \log_7 \sqrt{64} = \log_7 8$

73. $\dfrac{3}{4}\log_a x = \log_a x^{3/4} = \log_a \sqrt[4]{x^3}$

75. $\dfrac{2}{3}\log_a 8 = \log_a 8^{2/3}$
$$= \log_a \sqrt[3]{8^2}$$
$$= \log_a \sqrt[3]{64}$$
$$= \log_a 4$$

77. $-\dfrac{1}{2}\log_3 x = \log_3 x^{-1/2} = \log_3 \dfrac{1}{x^{1/2}} = \log_3 \dfrac{1}{\sqrt{x}}$

79. $\log_a \dfrac{x^3}{y^4} = \log_a x^3 - \log_a y^4 = 3\log_a x - 4\log_a y$

81. $\log_3 a^4 b^2 = \log_3 a^4 + \log_3 b^2 = 4\log_3 a + 2\log_3 b$

83. $\log_a \dfrac{xy}{z} = \log_a xy - \log_a z$

    $\quad = \log_a x + \log_a y - \log_a z$

85. $\log_x \dfrac{a^2}{bc^3} = \log_x a^2 - \log_x bc^3$

    $\quad = \log_x a^2 - \left(\log_x b + \log_x c^3\right)$

    $\quad = 2\log_x a - \left(\log_x b + 3\log_x c\right)$

    $\quad = 2\log_x a - \log_x b - 3\log_x c$

87. $\log_4 \sqrt[4]{\dfrac{x^3}{y}} = \log_4 \left(\dfrac{x^3}{y}\right)^{1/4}$

    $\quad = \log_4 \left(\dfrac{x^{3/4}}{y^{1/4}}\right)$

    $\quad = \log_4 x^{3/4} - \log_4 y^{1/4}$

    $\quad = \dfrac{3}{4}\log_4 x - \dfrac{1}{4}\log_4 y$

89. $\log_a \sqrt[3]{\dfrac{x^2 y}{z^3}} = \log_a \left(\dfrac{x^2 y}{z^3}\right)^{1/3}$

    $\quad = \log_a \left(\dfrac{x^{2/3} y^{1/3}}{z^{3/3}}\right)$

    $\quad = \log_a x^{2/3} y^{1/3} - \log_a z$

    $\quad = \log_a x^{2/3} + \log_a y^{1/3} - \log_a z$

    $\quad = \dfrac{2}{3}\log_a x + \dfrac{1}{3}\log_a y - \log_a z$

91. $3\log_3 2 - 2\log_3 4 = \log_3 2^3 - \log_3 4^2$

    $\quad = \log_3 8 - \log_3 16$

    $\quad = \log_3 \dfrac{8}{16}$

    $\quad = \log_3 \dfrac{1}{2}$

93. $4\log_b x + 3\log_b y = \log_b x^4 + \log_b y^3$

    $\quad = \log_b x^4 y^3$

95. $\dfrac{1}{2}\left(\log_a 5 - \log_a 7\right) = \dfrac{1}{2}\log_a\left(\dfrac{5}{7}\right)$

    $\quad = \log_a \left(\dfrac{5}{7}\right)^{1/2}$

    $\quad = \log_a \sqrt{\dfrac{5}{7}}$

97. $\dfrac{2}{3}\left(\log_a x^2 + \log_a y^3\right) = \dfrac{2}{3}\log_a\left(x^2 y^3\right)$

    $\quad = \log_a \left(x^2 y^3\right)^{2/3}$

    $\quad = \log_a \sqrt[3]{\left(x^2 y^3\right)^2}$

99. $\log_b x + \log_b (3x-2) = \log_b x(3x-2)$

    $\quad = \log_b \left(3x^2 - 2x\right)$

101. $3\log_a (x-2) - 4\log_a (x+1)$

    $\quad = \log_a (x-2)^3 - \log_a (x+1)^4$

    $\quad = \log_a \dfrac{(x-2)^3}{(x+1)^4}$

103. $2\log_a x + 4\log_a z - 3\log_a w - 6\log_a u$

    $\quad = \log_a x^2 + \log_a z^4 - \log_a w^3 - \log_a u^6$

    $\quad = \log_a x^2 z^4 - \log_a w^3 - \log_a u^6$

    $\quad = \log_a \dfrac{x^2 z^4}{w^3} - \log_a u^6$

    $\quad = \log_a \dfrac{x^2 z^4}{w^3 u^6}$

Review Exercises

1. $\left(10^{9.5}\right)\left(10^{-12}\right) = 10^{9.5+(-12)} = 10^{-2.5}$

2. $\dfrac{10^{-3}}{10^{-12}} = 10^{-3-(-12)} = 10^9$

3. $A = 10,000\left(1+\dfrac{0.06}{4}\right)^{5(4)}$

    $A = \$13,468.55$

4. $\log_{10} 50 = 1.6990$

5. $10^{1.6532} = 45$

6. $e^{-1.3863} = 0.25$

Exercise Set 11.5

1. $\log 64 = 1.8062$

3. $\log 0.0067 = -2.1739$

5. $\log 435.6 = 2.6391$

7. $\log\left(1.5\times 10^4\right) = 4.1761$

9. $\log\left(1.6\times10^{-6}\right)=-5.7959$

11. $\ln 9.34 = 2.2343$

13. $\ln 79.2 = 4.3720$

15. $\ln 0.034 = -3.3814$

17. $\ln\left(5.4\times e^{4}\right)=5.6864$

19. $\log e = 0.4343$

21. Error results because the domain of $\log_a x$ is $(0,\infty)$; so $\log 0$ is undefined.

23. $\log 100 = \log_{10} 10^2 = 2$

25. $\log\dfrac{1}{100}=\log_{10}10^{-2}=-2$

27. $\log\sqrt[3]{10}=\log_{10}10^{1/3}=\dfrac{1}{3}$

29. $\log 0.001 = \log_{10} 10^{-3} = -3$

31. $\ln e^3 = \log_e e^3 = 3$

33. $\ln\sqrt{e}=\log_e e^{1/2}=\dfrac{1}{2}$

35. $d = 10\log\dfrac{I}{I_0}$

$d = 10\log\dfrac{10^{-3}}{10^{-12}}$

$d = 10\log 10^9$

$d = 10\cdot 9$

$d = 90$

The reading for a firecracker is 90dB.

37. $d = 10\log\dfrac{I}{I_0}$

$60 = 10\log\dfrac{I}{10^{-12}}$

$6 = \log\dfrac{I}{10^{-12}}$

$6 = \log I - \log 10^{-12}$

$6 = \log I - (-12)$

$6 = \log I + 12$

$-6 = \log_{10} I$

$I = 10^{-6}$

The sound intensity of a noisy office is 10^{-6} watts/m^2.

39. $\text{pH} = -\log\left[H_3O^+\right]$

$\text{pH} = -\log\left[1.6\times10^{-3}\right]$

$\text{pH} \approx 2.796$

41. $3.5 = -\log\left[H_3O^+\right]$

$\log\left[H_3O^+\right] = -3.5$

$10^{-3.5} = \left[H_3O^+\right]$

The hydronium ion concentration is $10^{-3.5}$ moles/L.

43. $t = \dfrac{1}{r}\ln\dfrac{A}{P}$

$t = \dfrac{1}{0.04}\ln\dfrac{5000}{2000}$

≈ 22.9

It will take approximately 23 years.

45. San Francisco: $7.8 = \log\dfrac{I}{I_0}$

$7.8 = \log I - \log I_0$

$7.8 + \log I_0 = \log I$

$10^{7.8 + \log I_0} = I$

Alaska: $8.4 = \log\dfrac{I}{I_0}$

$8.4 = \log I - \log I_0$

$8.4 + \log I_0 = \log I$

$10^{8.4 + \log I_0} = I$

$\dfrac{10^{8.4+\log I_0}}{10^{7.8+\log I_0}} = 10^{(8.4+\log I_0)-(7.8+\log I_0)}$

$= 10^{0.6}$

≈ 4

The 1964 Alaska earthquake was about four times as severe as the 1906 San Francisco earthquake.

47. $\log E_s = 11.8 + 1.5(7.3)$

$\log_{10} E_s = 22.75$

$E_s = 10^{22.75}$

The energy released was $10^{22.75}$ ergs.

49. $y = -67.89 + 22.56 \ln x$

$y = -67.89 + 22.56 \ln(41)$

$y \approx -67.89 + 83.78$

$y \approx 15.89$

The purchasing power of \$1.00 in 2010 was approximately 40 cents in 1986.

51. $y = 3\log 110$

$y \approx 6.124$

and

$6.124(1000) = 6124$ ft.

Review Exercises

1. $\log_3(2x+1) - \log_3(x-1) = \log_3 \dfrac{2x+1}{x-1}$

2. $3^2 = 2x + 5$

3. $3x - (7x+2) = 12 - 2(x-4)$

$3x - 7x - 2 = 12 - 2x + 8$

$-4x - 2 = 20 - 2x$

$-2x - 2 = 20$

$-2x = 22$

$x = -11$

4. $x^2 + 2x = 15$

$x^2 + 2x - 15 = 0$

$(x+5)(x-3) = 0$

$x + 5 = 0 \ \text{ or } \ x - 3 = 0$

$x = -5 \qquad\quad x = 3$

5. $\dfrac{5}{x} + \dfrac{3}{x+1} = \dfrac{23}{3x}$

$3x(x+1)\left[\dfrac{5}{x} + \dfrac{3}{x+1}\right] = 3x(x+1)\left[\dfrac{23}{3x}\right]$

$3(x+1)(5) + 3x(3) = (x+1)23$

$15x + 15 + 9x = 23x + 23$

$24x + 15 = 23x + 23$

$x = 8$

6. $\sqrt{5x-1} = 7$

$\left(\sqrt{5x-1}\right)^2 = 7^2$

$5x - 1 = 49$

$5x = 50$

$x = 10$

Exercise Set 11.6

1. $2^x = 9$

$\log 2^x = \log 9$

$x \log 2 = \log 9$

$x = \dfrac{\log 9}{\log 2} \approx 3.1699$

3. $5^{2x} = 32$

$\log 5^{2x} = \log 32$

$2x \log 5 = \log 32$

$2x = \dfrac{\log 32}{\log 5}$

$x = \dfrac{\log 32}{\log 5} \div 2 \approx 1.0767$

5. $5^{x+3} = 10$

$\log 5^{x+3} = \log 10$

$(x+3)\log 5 = 1$

$x \log 5 + 3 \log 5 = 1$

$x \log 5 = 1 - 3 \log 5$

$x = \dfrac{1 - 3\log 5}{\log 5} \approx -1.5693$

7. $8^{x-2} = 6$

$\log 8^{x-2} = \log 6$

$(x-2)\log 8 = \log 6$

$x \log 8 - 2 \log 8 = \log 6$

$x \log 8 = \log 6 + 2 \log 8$

$x = \dfrac{\log 6 + 2\log 8}{\log 8} \approx 2.8617$

9. $4^{x+2} = 5^x$

$\log 4^{x+2} = \log 5^x$

$(x+2)\log 4 = x \log 5$

$x \log 4 + 2 \log 4 = x \log 5$

$x \log 4 - x \log 5 = -2 \log 4$

$x(\log 4 - \log 5) = -2 \log 4$

$x = \dfrac{-2\log 4}{\log 4 - \log 5} \approx 12.4251$

11. $$2^{x+1} = 3^{x-2}$$
$$\log 2^{x+1} = \log 3^{x-2}$$
$$(x+1)\log 2 = (x-2)\log 3$$
$$x\log 2 + \log 2 = x\log 3 - 2\log 3$$
$$x\log 2 - x\log 3 = -2\log 3 - \log 2$$
$$x(\log 2 - \log 3) = -2\log 3 - \log 2$$
$$x = \frac{-2\log 3 - \log 2}{\log 2 - \log 3} \approx 7.1285$$

13. $$e^{3x} = 5$$
$$\ln e^{3x} = \ln 5$$
$$3x = \ln 5$$
$$x = \frac{\ln 5}{3} \approx 0.5365$$

15. $$e^{0.03x} = 25$$
$$\ln e^{0.03x} = \ln 25$$
$$0.03x = \ln 25$$
$$x = \frac{\ln 25}{0.03} \approx 107.2959$$

17. $$e^{-0.022x} = 5$$
$$\ln e^{-0.022x} = \ln 5$$
$$-0.022x = \ln 5$$
$$x = -\frac{\ln 5}{0.022} \approx -73.1563$$

19. $$\ln e^{4x} = 24$$
$$4x = 24$$
$$x = 6$$

21. $$\log_4(x+5) = 2$$
$$4^2 = x+5$$
$$16 = x+5$$
$$11 = x$$

23. $$\log_4(4x-8) = 2$$
$$4^2 = 4x-8$$
$$16 = 4x-8$$
$$24 = 4x$$
$$6 = x$$

25. $$\log_4 x^2 = 2$$
$$4^2 = x^2$$
$$16 = x^2$$
$$\pm\sqrt{16} = x$$
$$\pm 4 = x$$

27. $$\log_6\left(x^2+5x\right) = 2$$
$$6^2 = x^2+5x$$
$$36 = x^2+5x$$
$$0 = x^2+5x-36$$
$$0 = (x+9)(x-4)$$
$$x+9 = 0 \quad \text{or} \quad x-4 = 0$$
$$x = -9 \qquad\qquad x = 4$$

29. $$\log(4x-3) = \log(3x+4)$$
$$4x-3 = 3x+4$$
$$x = 7$$

31. $$\ln(3x+4) = \ln(x-6)$$
$$3x+4 = x-6$$
$$2x = -10$$
$$x = \cancel{-5}$$
No solution

33. $$\log_9\left(x^2+4x\right) = \log_9 12$$
$$x^2+4x = 12$$
$$x^2+4x-12 = 0$$
$$(x+6)(x-2) = 0$$
$$x+6 = 0 \quad \text{or} \quad x-2 = 0$$
$$x = -6 \qquad\qquad x = 2$$

35. $$\log_4 x + \log_4 8 = 2$$
$$\log_4 8x = 2$$
$$4^2 = 8x$$
$$16 = 8x$$
$$2 = x$$

37. $$\log_2 x - \log_2 5 = 1$$
$$\log_2 \frac{x}{5} = 1$$
$$2^1 = \frac{x}{5}$$
$$10 = x$$

39. $\log_3 x + \log_3 (x+6) = 3$

  $\log_3 x(x+6) = 3$

  $\log_3 (x^2 + 6x) = 3$

  $3^3 = x^2 + 6x$

  $27 = x^2 + 6x$

  $0 = x^2 + 6x - 27$

  $0 = (x+9)(x-3)$

  $x+9 = 0$ or $x - 3 = 0$

  $x = \cancel{-9}$ $x = 3$

41. $\log_3 (2x+15) + \log_3 x = 3$

  $\log_3 x(2x+15) = 3$

  $\log_3 (2x^2 + 15x) = 3$

  $3^3 = 2x^2 + 15x$

  $27 = 2x^2 + 15x$

  $0 = 2x^2 + 15x - 27$

  $0 = (2x-3)(x+9)$

  $2x - 3 = 0$ or $x + 9 = 0$

  $x = \dfrac{3}{2}$ $x = \cancel{-9}$

43. $\log_2 (7x+3) - \log_2 (2x-3) = 3$

  $\log_2 \dfrac{7x+3}{2x-3} = 3$

  $2^3 = \dfrac{7x+3}{2x-3}$

  $8(2x-3) = 7x+3$

  $16x - 24 = 7x + 3$

  $9x = 27$

  $x = 3$

45. $\log_2 (3x+8) - \log_2 (x+1) = 2$

  $\log_2 \dfrac{3x+8}{x+1} = 2$

  $2^2 = \dfrac{3x+8}{x+1}$

  $4 = \dfrac{3x+8}{x+1}$

  $4x + 4 = 3x + 8$

  $x = 4$

47. $\log_8 2x + \log_8 6 = \log_8 10$

  $\log_8 12x = \log_8 10$

  $12x = 10$

  $x = \dfrac{5}{6}$

49. $\ln x + \ln (2x-1) = \ln 10$

  $\ln x(2x-1) = \ln 10$

  $\ln (2x^2 - x) = \ln 10$

  $2x^2 - x = 10$

  $2x^2 - x - 10 = 0$

  $(2x-5)(x+2) = 0$

  $2x - 5 = 0$ or $x + 2 = 0$

  $x = \dfrac{5}{2}$ $x = \cancel{-2}$

51. $\log x - \log (x-5) = \log 6$

  $\log \dfrac{x}{x-5} = \log 6$

  $\dfrac{x}{x-5} = 6$

  $x = 6x - 30$

  $-5x = -30$

  $x = 6$

53. $\log_6 (3x+4) - \log_6 (x-2) = \log_6 8$

  $\log_6 \dfrac{3x+4}{x-2} = \log_6 8$

  $\dfrac{3x+4}{x-2} = 8$

  $3x + 4 = 8x - 16$

  $20 = 5x$

  $4 = x$

55. $8000 = 5000\left(1 + \dfrac{.05}{4}\right)^{4t}$

  $\dfrac{8}{5} = (1.0125)^{4t}$

  $\log \dfrac{8}{5} = (4t)\log 1.0125$

  $\dfrac{\log \dfrac{8}{5}}{4 \log 1.0125} = t$

  $9.5 \approx t$

  It will take about 9.5 years.

57. a) $A = 8000e^{0.06(15)}$

$A = 19,676.82$

The value of the account will be $19,676.82.

b) $14000 = 8000e^{0.06t}$

$\dfrac{14}{8} = e^{0.06t}$

$\ln\dfrac{7}{4} = \ln e^{0.06t}$

$\ln\dfrac{7}{4} = 0.06t$

$\dfrac{\ln\dfrac{7}{4}}{0.06} = t$

$9.3 \approx t$

It will take about 9.3 years.

59. a) $P(0.08) = e^{21.5(0.08)}$

$P(0.08) = 5.58\%$

b) $50 = e^{21.5b}$

$\ln 50 = \ln e^{21.5b}$

$\ln 50 = 21.5b$

$\dfrac{\ln 50}{21.5} = b$

$0.18 = b$

61. a) $A = 500e^{0.04(14)}$

$A \approx 875$

There will be 875 mosquitoes.

b) $10,000 = 500e^{0.04t}$

$20 = e^{0.04t}$

$\ln 20 = \ln e^{0.04t}$

$\ln 20 = 0.04t$

$\dfrac{\ln 20}{0.04} = t$

$75 \approx t$

It will take about 75 days.

63. a) $A = 6.8e^{0.011(10)}$

$A \approx 7.59$

The world population will be about 7.59 billion in 2020.

b) $7.0 = 6.8e^{0.011t}$

$\dfrac{7.0}{6.8} = e^{0.011t}$

$\ln\dfrac{7.0}{6.8} = \ln e^{0.011t}$

$\ln\dfrac{7.0}{6.8} = 0.011t$

$3 \approx t$

Year: $2010 + 3 = 2013$

65. a) $P = 0.48\ln(50+1)$

$P = 0.48\ln 51$

$P \approx 1.89$

The barometric pressure is 1.89 inches of mercury.

b) $1.5 = 0.48\ln(x+1)$

$3.125 = \ln(x+1)$

$e^{3.125} = x+1$

$e^{3.125} - 1 = x$

$21.8 \approx x$

The distance is about 21.8 miles.

67. a) $f(10) = 29 + 48.8\log(10+1)$

$f(10) = 29 + 48.8\log(11)$

$f(10) \approx 79.8$

At age 10, a boy has reached 79.8% of his adult height.

b) $75 = 29 + 48.8\log(x+1)$

$46 = 48.8\log(x+1)$

$0.9426 = \log(x+1)$

$10^{0.9426} = x+1$

$10^{0.9426} - 1 = x$

$7.76 = x$

A boy will attain 75% of his adult height at about age 8.

69. $y = 1.244(1.043)^x$

$y = 1.244(1.043)^{50}$

$y \approx 10.21$

In 2020, it will take $10.21 to have the same purchasing power as $1.00 in 1970.

71. $y = 78.62(1.092)^x$

 $y = 78.62(1.092)^{33}$

 $y \approx 1435.1$

Approximately \$1435.1 billion will be spent on recreation in 2018.

73. $\log_4 12 = \dfrac{\log 12}{\log 4} \approx 1.7925$

75. $\log_8 3 = \dfrac{\log 3}{\log 8} \approx 0.5283$

77. $\dfrac{\log 5}{\log 0.5} \approx -2.3219$

79. $\log_{1/4} \dfrac{3}{5} = \dfrac{\log \dfrac{3}{5}}{\log \dfrac{1}{4}} \approx 0.3685$

81. $P = 95 - 30\log_2 x$

 $P = 95 - 30\log_2 3$

 $P = 95 - 30\dfrac{\log 3}{\log 2}$

 $P \approx 47$

After 3 days, about 47% of the lecture is retained.

Review Exercises

1. $(3x+2) - (2x-1) = (3x+2) + (-2x+1)$
 $= x+3$

2. $(3\sqrt{5} + 2)(2\sqrt{5} - 1)$
 $= 3\sqrt{5} \cdot 2\sqrt{5} + 3\sqrt{5} \cdot (-1) + 2 \cdot 2\sqrt{5} + 2 \cdot (-1)$
 $= 30 - 3\sqrt{5} + 4\sqrt{5} - 2$
 $= 28 + \sqrt{5}$

3. $f(-5) = (-5)^2 + 4 = 25 + 4 = 29$

4. $(f+g)(x) = x^2 + 4 + 2x - 1 = x^2 + 2x + 3$

5. $f \circ g = f\big[g(x)\big]$
 $= (2x-1)^2 + 4$
 $= 4x^2 - 4x + 1 + 4$
 $= 4x^2 - 4x + 5$

6. Using the answer from #5:
 $(f \circ g)(2) = 4(2)^2 - 4(2) + 5$
 $= 16 - 8 + 5$
 $= 13$

Chapter 11 Review Exercises

1. $(f \circ g)(3) = f\big[g(3)\big]$
 $= f(7)$
 $= 3(7) + 4$
 $= 21 + 4$
 $= 25$

2. $(g \circ f)(3) = g\big[f(3)\big]$
 $= g(13)$
 $= 13^2 - 2$
 $= 169 - 2$
 $= 167$

3. $f\big[g(0)\big] = f(-2)$
 $= 3(-2) + 4$
 $= -6 + 4$
 $= -2$

4. $g\big[f(0)\big] = g(4)$
 $= 4^2 - 2$
 $= 16 - 2$
 $= 14$

5. $(f \circ g)(x) = f\big[g(x)\big]$
 $= f(2x+3)$
 $= 3(2x+3) - 6$
 $= 6x + 9 - 6$
 $= 6x + 3$
 $(g \circ f)(x) = g\big[f(x)\big]$
 $= g(3x-6)$
 $= 2(3x-6) + 3$
 $= 6x - 12 + 3$
 $= 6x - 9$

6. $(f \circ g)(x) = f[g(x)]$
$= f(3x-7)$
$= (3x-7)^2 + 4$
$= 9x^2 - 42x + 49 + 4$
$= 9x^2 - 42x + 53$

$(g \circ f)(x) = g[f(x)]$
$= g(x^2+4)$
$= 3(x^2+4) - 7$
$= 3x^2 + 12 - 7$
$= 3x^2 + 5$

7. $(f \circ g)(x) = f[g(x)]$
$= f(2x-1)$
$= \sqrt{2x-1-3}$
$= \sqrt{2x-4}$

$(g \circ f)(x) = g[f(x)]$
$= g(\sqrt{x-3})$
$= 2\sqrt{x-3} - 1$

8. $(f \circ g)(x) = f[g(x)]$
$= f\left(\dfrac{x-4}{x}\right)$
$= \dfrac{\dfrac{x-4}{x} + 3}{\dfrac{x-4}{x}}$
$= \dfrac{x-4+3x}{x-4} = \dfrac{4x-4}{x-4}$

$(g \circ f)(x) = g[f(x)]$
$= g\left(\dfrac{x+3}{x}\right)$
$= \dfrac{\dfrac{x+3}{x} - 4}{\dfrac{x+3}{x}}$
$= \dfrac{x+3-4x}{x+3}$
$= \dfrac{3-3x}{x+3}$

9. $(f \circ g)(x) = f[g(x)]$
$= f\left(\dfrac{x-2}{3}\right)$
$= 3\left(\dfrac{x-2}{3}\right) + 2$
$= x - 2 + 2$
$= x$

$(g \circ f)(x) = g[f(x)]$
$= g(3x+2)$
$= \dfrac{3x+2-2}{3}$
$= \dfrac{3x}{3}$
$= x$

Yes, f and g are inverse functions.

10. $(f \circ g)(x) = f[g(x)]$
$= f(\sqrt[3]{x-6})$
$= (\sqrt[3]{x-6})^3 + 6$
$= x - 6 + 6$
$= x$

$(g \circ f)(x) = g[f(x)]$
$= g(x^3+6)$
$= \sqrt[3]{x^3+6-6}$
$= x$

Yes, f and g are inverse functions.

11. $(f \circ g)(x) = f[g(x)]$
$= f(\sqrt{x+2})$
$= (\sqrt{x+2})^2 - 3$
$= x + 2 - 3$
$= x - 1$

No, f and g are not inverse functions.

12. $(f \circ g)(x) = f[g(x)]$

$$= f\left(\frac{-4x}{x+1}\right)$$

$$= \frac{\dfrac{-4x}{x+1}}{\dfrac{-4x}{x+1}+4}$$

$$= \frac{(x+1)\left(\dfrac{-4x}{x+1}\right)}{(x+1)\left(\dfrac{-4x}{x+1}+4\right)}$$

$$= \frac{-4x}{-4x+4x+4}$$

$$= \frac{-4x}{4}$$

$$= -x$$

No, f and g are not inverse functions.

13. No, not a one-to-one function

14. Yes, it is a one-to-one function.

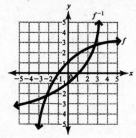

15. $y = 5x + 4$

$x = 5y + 4$

$x - 4 = 5y$

$\dfrac{x-4}{5} = y$

$f^{-1}(x) = \dfrac{x-4}{5}$

16. $y = x^3 + 6$

$x = y^3 + 6$

$x - 6 = y^3$

$\sqrt[3]{x-6} = y$

$f^{-1}(x) = \sqrt[3]{x-6}$

17. $y = \dfrac{4}{x+5}$

$x = \dfrac{4}{y+5}$

$xy + 5x = 4$

$xy = 4 - 5x$

$y = \dfrac{4-5x}{x}$

$f^{-1}(x) = \dfrac{4-5x}{x}$

18. $y = \sqrt[3]{3x+2}$

$x = \sqrt[3]{3y+2}$

$x^3 = 3y + 2$

$x^3 - 2 = 3y$

$\dfrac{x^3-2}{3} = y$

$f^{-1}(x) = \dfrac{x^3-2}{3}$

19. $(-6, 4)$

20. $f^{-1}(b) = a$

21. $f(x) = 3^x$

x	$y = f(x)$
-1	$\dfrac{1}{3}$
0	1
1	3
2	9

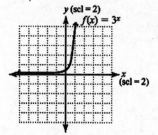

22. $f(x) = \left(\dfrac{1}{2}\right)^x$

x	$y = f(x)$
-1	$\left(\dfrac{1}{2}\right)^{(-1)} = 2$
0	$\left(\dfrac{1}{2}\right)^{(0)} = 1$
1	$\left(\dfrac{1}{2}\right)^{(1)} = \dfrac{1}{2}$
2	$\left(\dfrac{1}{2}\right)^{(2)} = \dfrac{1}{4}$

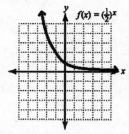

23. $f(x) = 2^{x-3}$

x	$y = f(x)$
-1	$2^{-1-3} = \dfrac{1}{16}$
0	$2^{0-3} = \dfrac{1}{8}$
1	$2^{1-3} = \dfrac{1}{4}$
2	$2^{2-3} = \dfrac{1}{2}$

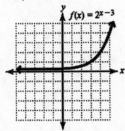

24. $f(x) = 3^{-x+2}$

x	$y = f(x)$
-1	$3^{-(-1)+2} = 27$
0	$3^{-(0)+2} = 9$
1	$3^{-(1)+2} = 3$
2	$3^{-(2)+2} = 1$

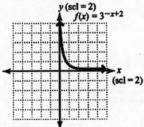

25. $5^x = 625$
$5^x = 5^4$
$x = 4$

26. $16^x = 64$
$(4^2)^x = 4^3$
$4^{2x} = 4^3$
$2x = 3$
$x = \dfrac{3}{2}$

27. $6^x = \dfrac{1}{36}$
$6^x = \dfrac{1}{6^2}$
$6^x = 6^{-2}$
$x = -2$

28. $\left(\dfrac{3}{4}\right)^x = \dfrac{16}{9}$
$\left(\dfrac{3}{4}\right)^x = \left(\dfrac{4}{3}\right)^2$
$\left(\dfrac{3}{4}\right)^x = \left(\dfrac{3}{4}\right)^{-2}$
$x = -2$

29. $4^{x-1} = 64$
$4^{x-1} = 4^3$
$x - 1 = 3$
$x = 4$

30. $5^{x+2} = 25^x$
$5^{x+2} = \left(5^2\right)^x$
$5^{x+2} = 5^{2x}$
$x + 2 = 2x$
$2 = x$

31. $\left(\dfrac{1}{3}\right)^{-x} = 27$
$\left(3^{-1}\right)^{-x} = 3^3$
$3^x = 3^3$
$x = 3$

32. $8^{3x-2} = 16^{4x}$
$\left(2^3\right)^{3x-2} = \left(2^4\right)^{4x}$
$2^{9x-6} = 2^{16x}$
$9x - 6 = 16x$
$-6 = 7x$
$-\dfrac{6}{7} = x$

33. $A = A_0 \cdot 2^{t/100}$
$A = 500 \cdot 2^{365/100}$
$A \approx 6277$
There are 6277 cells after one year.

34. $A = 25000\left(1 + \dfrac{0.06}{12}\right)^{12 \cdot 8}$
$A = 40{,}353.57$
The account is worth \$40,353.57 after eight years.

35. $A = A_0\left(\dfrac{1}{2}\right)^{t/h}$
$A = 50\left(\dfrac{1}{2}\right)^{300/107}$
$A \approx 7.16$
After 300 days, 7.16 grams remain.

36. $y = 4.0144(1.028)^x$

   $y = 4.0144(1.028)^{30}$

   $y \approx 9.19$

   In 2020, the minimum wage will be approximately \$9.19.

37. $\log_7 343 = 3$

38. $\log_4 \dfrac{1}{64} = -3$

39. $\log_{3/2} \dfrac{81}{16} = 4$

40. $\log_{11} \sqrt[3]{11} = \dfrac{1}{3}$

41. $9^2 = 81$

42. $\left(\dfrac{1}{5}\right)^{-3} = 125$

43. $a^4 = 16$

44. $e^b = c$

45. $3^{-4} = x$

   $\dfrac{1}{3^4} = x$

   $\dfrac{1}{81} = x$

46. $\left(\dfrac{1}{2}\right)^{-2} = x$

   $2^2 = x$

   $4 = x$

47. $2^x = 32$

   $2^x = 2^5$

   $x = 5$

48. $\left(\dfrac{1}{4}\right)^x = 16$

   $\left(4^{-1}\right)^x = 4^2$

   $4^{-x} = 4^2$

   $-x = 2$

   $x = -2$

49. $x^4 = 81$

   $x^4 = 3^4$

   $x = 3$

50. $x^3 = \dfrac{1}{1000}$

   $x = \sqrt[3]{\dfrac{1}{1000}}$

   $x = \dfrac{1}{10}$

51. $\left(\dfrac{3}{4}\right)^x = \dfrac{3}{4}$

   $\left(\dfrac{3}{4}\right)^x = \left(\dfrac{3}{4}\right)^1$

   $x = 1$

52. $81^x = 1$

   $81^x = 1^0$

   $x = 0$

53. $f(x) = \log_4 x$

   $y = \log_4 x$

   $4^y = x$

y	-1	0	1	2
x	$\dfrac{1}{4}$	1	4	16

   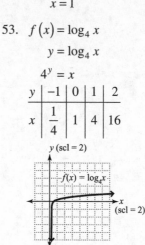

54. $f(x) = \log_{1/3} x$

   $y = \log_{1/3} x$

   $\left(\dfrac{1}{3}\right)^y = x$

y	-2	-1	0	1
x	9	3	1	$\dfrac{1}{3}$

   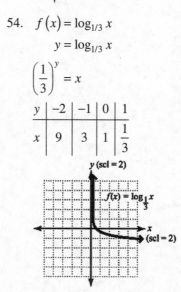

55. Domain is $(-\infty, \infty)$ and range is $(0, \infty)$ because $f(x) = \log_b x$ and $g(x) = b^x$ are inverses.

56. $d = 10 \log \dfrac{I}{I_0}$

$d = 10 \log \dfrac{1000 I_0}{I_0}$

$d = 10 \log 1000$

$d = 10 \log 10^3$

$d = 10 \cdot 3$

$d = 30$

The decibel reading is 30 dB.

57. $3^{\log_3 8} = 8$

58. $\log_4 4^6 = 6$

59. $\log_6 6x = \log_6 6 + \log_6 x = 1 + \log_6 x$

60. $\log_4 x(2x-5) = \log_4 x + \log_4 (2x-5)$

61. $\log_3 4 + \log_3 8 = \log_3 (4 \cdot 8) = \log_3 32$

62. $\log_5 3 + \log_5 x + \log_5 (x-2) = \log_5 3x(x-2)$
$$= \log_5 \left(3x^2 - 6x\right)$$

63. $\log_b \dfrac{x}{5} = \log_b x - \log_b 5$

64. $\log_a \dfrac{3x-2}{4x+3} = \log_a (3x-2) - \log_a (4x+3)$

65. $\log_8 32 - \log_8 16 = \log_8 \dfrac{32}{16} = \log_8 2$

66. $\log_2 (x+5) - \log_2 (2x-3) = \log_2 \dfrac{x+5}{2x-3}$

67. $\log_3 7^4 = 4 \log_3 7$

68. $\log_a \sqrt[3]{x} = \log_a x^{1/3} = \dfrac{1}{3} \log_a x$

69. $\log_4 \dfrac{1}{a^4} = \log_4 a^{-4} = -4 \log_4 a$

70. $\log_a \sqrt[5]{a^4} = \log_a a^{4/5} = \dfrac{4}{5} \log_a a = \dfrac{4}{5}$

71. $4 \log_a x = \log_a x^4$

72. $\dfrac{3}{5} \log_a y = \log_a y^{3/5} = \log_a \sqrt[5]{y^3}$

73. $\log_a x^2 y^3 = \log_a x^2 + \log_a y^3$
$$= 2 \log_a x + 3 \log_a y$$

74. $\log_a \dfrac{c^4}{d^3} = \log_a c^4 - \log_a d^3$
$$= 4 \log_a c - 3 \log_a d$$

75. $\log_a \dfrac{x^2 y^3}{z^4} = \log_a x^2 + \log_a y^3 - \log_a z^4$
$$= 2 \log_a x + 3 \log_a y - 4 \log_a z$$

76. $\log_a \sqrt{\dfrac{a^3}{b^4}} = \log_a \left(\dfrac{a^3}{b^4}\right)^{1/2}$
$$= \log_a \left(\dfrac{a^{3/2}}{b^{4/2}}\right)$$
$$= \log_a a^{3/2} - \log_a b^2$$
$$= \dfrac{3}{2} \log_a a - 2 \log_a b$$
$$= \dfrac{3}{2} - 2 \log_a b$$

77. $3 \log_x y + 5 \log_x z = \log_x y^3 + \log_x z^5$
$$= \log_x y^3 z^5$$

78. $3 \log_a 4 - 2 \log_a 3 = \log_a 4^3 - \log_a 3^2$
$$= \log_a \dfrac{4^3}{3^2}$$

79. $\dfrac{1}{4} \left(2 \log_a x + 3 \log_a y\right) = \dfrac{2}{4} \log_a x + \dfrac{3}{4} \log_a y$
$$= \log_a x^{2/4} + \log_a y^{3/4}$$
$$= \log_a x^{2/4} y^{3/4}$$
$$= \log_a \left(x^2 y^3\right)^{1/4}$$
$$= \log_a \sqrt[4]{x^2 y^3}$$

80. $4 \log_a (x+5) + 2 \log_a (x-3)$
$$= \log_a (x+5)^4 + \log_a (x-3)^2$$
$$= \log_a (x+5)^4 (x-3)^2$$

81. $\log 326 = 2.5132$

82. $\log 0.0035 = -2.4559$

83. $\ln 0.043 = -3.1466$

84. $\ln 92 = 4.5218$

85. $\log 0.00001 = \log_{10} 10^{-5} = -5$

86. $\ln \sqrt[4]{e} = \ln e^{1/4} = \dfrac{1}{4}$

87. $d = 10\log\dfrac{I}{I_0}$

$d = 10\log\dfrac{10^{-3.5}}{10^{-12}}$

$d = 10\log 10^{8.5}$

$d = 10(8.5)$

$d = 85$

The decibel reading of the thunder was 85 dB.

88. $\text{pH} = -\log\left[H_3O^+\right]$

$\text{pH} = -\log 0.001259$

$\text{pH} \approx 2.9$

89. $\qquad A = Pe^{rt}$

$10,000 = 6000e^{0.03t}$

$\dfrac{10}{6} = e^{0.03t}$

$\ln\dfrac{5}{3} = \ln e^{0.03t}$

$\ln\dfrac{5}{3} = 0.03t$

$\dfrac{\ln\dfrac{5}{3}}{0.03} = t$

$17 \approx t$

It will take approximately 17 years.

90. Earthquake 1: $\qquad 7.8 = \log\dfrac{I}{I_0}$

$7.8 = \log I - \log I_0$

$7.8 + \log I_0 = \log I$

$10^{7.8 + \log I_0} = I$

Earthquake 2: $\qquad 6.8 = \log\dfrac{I}{I_0}$

$6.8 = \log I - \log I_0$

$6.8 + \log I_0 = \log I$

$10^{6.8 + \log I_0} = I$

$\dfrac{10^{7.8 + \log I_0}}{10^{6.8 + \log + I_0}} = 10^{(7.8 + \log I_0) - (6.8 + \log + I_0)}$

$= 10^{7.8 - 6.8}$

$= 10^1$

$= 10$

Earthquake 1 was ten times as severe as earthquake 2.

91. $\qquad 9^x = 32$

$\ln 9^x = \ln 32$

$x\ln 9 = \ln 32$

$x = \dfrac{\ln 32}{\ln 9} \approx 1.5773$

92. $\qquad 3^{5x} = 19$

$\ln 3^{5x} = \ln 19$

$5x\ln 3 = \ln 19$

$5x = \dfrac{\ln 19}{\ln 3}$

$x = \dfrac{\dfrac{\ln 19}{\ln 3}}{5} \approx 0.5360$

93. $\qquad 6^{2x-1} = 22$

$\ln 6^{2x-1} = \ln 22$

$(2x-1)\ln 6 = \ln 22$

$2x\ln 6 - \ln 6 = \ln 22$

$2x\ln 6 = \ln 22 + \ln 6$

$x = \dfrac{\ln 22 + \ln 6}{2\ln 6} \approx 1.3626$

94. $\qquad 4^{2x-3} = 5^{x+1}$

$\ln 4^{2x-3} = \ln 5^{x+1}$

$(2x-3)\ln 4 = (x+1)\ln 5$

$2x\ln 4 - 3\ln 4 = x\ln 5 + \ln 5$

$2x\ln 4 - x\ln 5 = \ln 5 + 3\ln 4$

$x(2\ln 4 - \ln 5) = \ln 5 + 3\ln 4$

$x = \dfrac{\ln 5 + 3\ln 4}{2\ln 4 - \ln 5} \approx 4.9592$

95. $e^{4x} = 11$

$\ln e^{4x} = \ln 11$

$4x = \ln 11$

$x = \dfrac{\ln 11}{4} \approx 0.5995$

96. $e^{-0.003x} = 5$

$\ln e^{-0.003x} = \ln 5$

$-0.003x = \ln 5$

$x = -\dfrac{\ln 5}{0.003} \approx -536.4793$

97.
$$A = P\left(1 + \frac{r}{n}\right)^{nt}$$

$$18000 = 15000\left(1 + \frac{0.03}{12}\right)^{12 \cdot t}$$

$$1.2 = (1.0025)^{12t}$$

$$\ln 1.2 = \ln (1.0025)^{12t}$$

$$\ln 1.2 = 12t \ln (1.0025)$$

$$\frac{\ln 1.2}{12 \ln (1.0025)} = t$$

$$6.1 \approx t$$

It will take approximately 6.1 years.

98. a) $A = 7000e^{0.07(8)}$

$A = \$12,254.71$

b) $12,000 = 7000e^{0.07t}$

$$\frac{12}{7} = e^{0.07t}$$

$$\ln \frac{12}{7} = \ln e^{0.07t}$$

$$\ln \frac{12}{7} = 0.07t$$

$$\frac{\ln \frac{12}{7}}{0.07} = t \approx 7.7$$

It will take about 7.7 years.

99. a) $A = 400e^{0.03(12)}$

$A \approx 573$

The population of the colony will be about 573 ants after one year.

b) $800 = 400e^{0.03t}$

$$2 = e^{0.03t}$$

$$\ln 2 = \ln e^{0.03t}$$

$$\ln 2 = 0.03t$$

$$\frac{\ln 2}{0.03} = t$$

$$23 \approx t$$

The population will be 800 ants after about 23 months.

100. $G = \dfrac{t}{3.3 \log \dfrac{b}{B}}$

$$G = \frac{240}{3.3 \log \dfrac{1,000,000}{1000}}$$

$$G = \frac{240}{3.3 \log 1000}$$

$$G = \frac{240}{3.3 \log 10^3}$$

$$G = \frac{240}{3.3(3)} \approx 24$$

The generation time is about 24 minutes.

101. $\log_4 (x + 8) = 2$

$$4^2 = x + 8$$

$$16 = x + 8$$

$$8 = x$$

102. $\log_2 (x^2 + 2x) = 3$

$$2^3 = x^2 + 2x$$

$$0 = x^2 + 2x - 8$$

$$0 = (x + 4)(x - 2)$$

$$x + 4 = 0 \quad \text{or} \quad x - 2 = 0$$

$$x = -4 \qquad\quad x = 2$$

103. $\log (3x - 8) = \log (x - 2)$

$$3x - 8 = x - 2$$

$$2x = 6$$

$$x = 3$$

104. $\log 5 + \log x = 2$

$$\log 5x = 2$$

$$10^2 = 5x$$

$$100 = 5x$$

$$20 = x$$

105. $\log_3 x - \log_3 4 = 2$

$$\log_3 \frac{x}{4} = 2$$

$$3^2 = \frac{x}{4}$$

$$9 = \frac{x}{4}$$

$$36 = x$$

106. $\log_2 x + \log_2 (x-6) = 4$

$\log_2 x(x-6) = 4$

$\log_2 (x^2 - 6x) = 4$

$2^4 = x^2 - 6x$

$0 = x^2 - 6x - 16$

$0 = (x-8)(x+2)$

$x - 8 = 0$ or $x + 2 = 0$

$x = 8$ $x = \cancel{-2}$

107. $\log_4 x + \log_4 (x+2) = \log_4 8$

$\log_4 x(x+2) = \log_4 8$

$x(x+2) = 8$

$x^2 + 2x = 8$

$x^2 + 2x - 8 = 0$

$(x+4)(x-2) = 0$

$x + 4 = 0$ or $x - 2 = 0$

$x = \cancel{-4}$ $x = 2$

108. $\log_3 (5x+2) - \log_3 (x-2) = 2$

$\log_3 \dfrac{5x+2}{x-2} = 2$

$3^2 = \dfrac{5x+2}{x-2}$

$9(x-2) = 5x+2$

$9x - 18 = 5x + 2$

$4x = 20$

$x = 5$

109. $\log_4 15 = \dfrac{\log 15}{\log 4} \approx 1.9534$

110. $\log_{1/2} 6 = \dfrac{\log 6}{\log 0.5} \approx -2.5850$

Chapter 11 Practice Test

1. Let $f(x) = x^2 - 6$ and $g(x) = 3x - 5$.

$f[g(x)] = f(3x-5)$

$= (3x-5)^2 - 6$

$= 9x^2 - 30x + 25 - 6$

$= 9x^2 - 30x + 19$

2. $f(x) = 4x - 3$

a) $y = 4x - 3$

$x = 4y - 3$

$x + 3 = 4y$

$\dfrac{x+3}{4} = y$

$f^{-1}(x) = \dfrac{x+3}{4}$

b) $f\left[f^{-1}(x) \right] = f\left(\dfrac{x+3}{4} \right)$

$= 4\left(\dfrac{x+3}{4} \right) - 3$

$= x + 3 - 3$

$= x$

$f^{-1}\left[f(x) \right] = f^{-1}(4x-3)$

$= \dfrac{4x-3+3}{4}$

$= \dfrac{4x}{4}$

$= x$

c) The graphs are symmetric about the graph of $y = x$.

3.
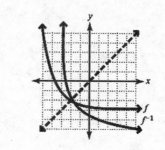

4. $f(x) = 2^{x-1}$

x	$y = f(x)$
-1	$2^{-1-1} = \dfrac{1}{4}$
0	$2^{0-1} = \dfrac{1}{2}$
1	$2^{1-1} = 1$
2	$2^{2-1} = 2$

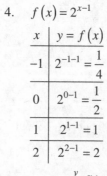

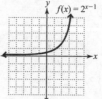

5. $32^{x-2} = 8^{2x}$

$\left(2^5\right)^{x-2} = \left(2^3\right)^{2x}$

$2^{5x-10} = 2^{6x}$

$5x - 10 = 6x$

$-10 = x$

6. a) $A = 20,000\left(1 + \dfrac{0.05}{4}\right)^{4(15)}$

$A = \$42,143.63$

b) $30,000 = 20,000\left(1 + \dfrac{0.05}{4}\right)^{4 \cdot t}$

$1.5 = (1.0125)^{4t}$

$\ln 1.5 = \ln (1.0125)^{4t}$

$\ln 1.5 = 4t \ln (1.0125)$

$\dfrac{\ln 1.5}{4 \ln (1.0125)} = t$

$8.2 \approx t$

It will take about 8.2 years.

7. a) $A = 100\left(\dfrac{1}{2}\right)^{216/30}$

$A = 0.68$

About 0.68 gram remains.

b) $20 = 100\left(\dfrac{1}{2}\right)^{t/30}$

$0.2 = (0.5)^{t/30}$

$\ln 0.2 = \ln (0.5)^{t/30}$

$\ln 0.2 = \dfrac{t}{30} \ln 0.5$

$\dfrac{30 \ln 0.2}{\ln 0.5} = t$

$69.66 \approx t$

It will take about 69.66 minutes, or 1.16 hours.

8. a) In 1960: $y = 3.17(1.026)^{60}$

$y \approx 14.8$

There were about 14.8 million people.

In 2018: $y = 3.17(1.026)^{118}$

$y \approx 65.5$

There were about 65.5 million people.

b) $31.4 = 3.17(1.026)^x$

$\dfrac{31.4}{3.17} = 1.026^x$

$\ln\left(\dfrac{31.4}{3.17}\right) = \ln 1.026^x$

$\ln\left(\dfrac{31.4}{3.17}\right) = x \ln 1.026$

$\dfrac{\ln\left(\dfrac{31.4}{3.17}\right)}{\ln 1.026} = x$

$89 \approx x$

89 years after 1900, in 1989.

9. $\left(\dfrac{1}{3}\right)^{-4} = 81$

10. $\log_6 \dfrac{1}{216} = x$

$6^x = \dfrac{1}{216}$

$6^x = 216^{-1}$

$6^x = \left(6^3\right)^{-1}$

$6^x = 6^{-3}$

$x = -3$

11. $\log_x 625 = 4$

$x^4 = 625$

$x^4 = 5^4$

$x = 5$

12. $f(x) = \log_2 x$

$y = \log_2 x$

$2^y = x$

y	-1	0	1	2
x	$\dfrac{1}{2}$	1	2	4

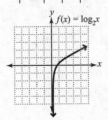

13. $\log_b \dfrac{x^4 y^2}{z} = \log_b x^4 y^2 - \log_b z$

$\qquad = \log_b x^4 + \log_b y^2 - \log_b z$

$\qquad = 4\log_b x + 2\log_b y - \log_b z$

14. $\log_b \sqrt[4]{\dfrac{x^5}{y^7}} = \log_b \left(\dfrac{x^5}{y^7}\right)^{1/4}$

$\qquad = \log_b \dfrac{x^{5/4}}{y^{7/4}}$

$\qquad = \log_b x^{5/4} - \log_b y^{7/4}$

$\qquad = \dfrac{5}{4}\log_b x - \dfrac{7}{4}\log_b y$

15. $\dfrac{3}{4}\left(2\log_b x + 3\log_b y\right) = \dfrac{6}{4}\log_b x + \dfrac{9}{4}\log_b y$

$\qquad = \log_b x^{6/4} + \log_b y^{9/4}$

$\qquad = \log_b x^{6/4} y^{9/4}$

$\qquad = \log_b \left(x^6 y^9\right)^{1/4}$

$\qquad = \log_b \sqrt[4]{x^6 y^9}$

16. a) $A = 300e^{-0.0028(500)}$

$\qquad A \approx 74$

About 74 grams remain.

b) $\qquad A = 300e^{-0.0028t}$

$\qquad 150 = 300e^{-0.0028t}$

$\qquad 0.5 = e^{-0.0028t}$

$\qquad \ln 0.5 = \ln e^{-0.0028t}$

$\qquad \ln 0.5 = -0.0028t$

$\qquad \dfrac{\ln 0.5}{-0.0028} = t$

$\qquad 247.6 \approx t$

The half-life is about 247.6 days.

17. $\qquad 6^{x-3} = 19$

$\qquad \ln 6^{x-3} = \ln 19$

$\qquad (x-3)\ln 6 = \ln 19$

$\qquad x\ln 6 - 3\ln 6 = \ln 19$

$\qquad x\ln 6 = \ln 19 + 3\ln 6$

$\qquad x = \dfrac{\ln 19 + 3\ln 6}{\ln 6}$

$\qquad x \approx 4.6433$

18. $\log_3 x + \log_3 (x+6) = 3$

$\qquad \log_3 x(x+6) = 3$

$\qquad \log_3 (x^2 + 6x) = 3$

$\qquad 3^3 = x^2 + 6x$

$\qquad 0 = x^2 + 6x - 27$

$\qquad 0 = (x+9)(x-3)$

$\qquad x+9 = 0 \quad$ or $\quad x-3 = 0$

$\qquad x = \cancel{-9} \qquad\qquad x = 3$

19. $\log(4x+2) - \log(3x-2) = \log 2$

$\qquad \log\dfrac{4x+2}{3x-2} = \log 2$

$\qquad \dfrac{4x+2}{3x-2} = 2$

$\qquad 4x+2 = 6x-4$

$\qquad 6 = 2x$

$\qquad 3 = x$

20. a) $n = 3.3\log\dfrac{b}{B}$

$\qquad n = 3.3\log\dfrac{10,000,000}{100}$

$\qquad n = 3.3\log(100,000)$

$\qquad n = 16.5$

There were 16.5 generations.

b) $n = 3.3\log\dfrac{b}{B}$

$\qquad 9.9 = 3.3\log\dfrac{b}{100}$

$\qquad 3 = \log\dfrac{b}{100}$

$\qquad 10^3 = \dfrac{b}{100}$

$\qquad 1000 = \dfrac{b}{100}$

$\qquad 100,000 = b$

There will be 100,000 bacteria.

Chapters 1–11 Cumulative Review

1.   True

2.   True

3.   False; the range of $f(x) = x^2 + 4$ is $[4, \infty)$.

4. -8

5. x-intercepts

6. $x - 3 = 4$; $x - 3 = -4$

7. $6 - 8 \div 2 \cdot 3^2 - 4\left(3 - 4 \cdot 2^3\right)$

$= 6 - 8 \div 2 \cdot 3^2 - 4\left(3 - 4 \cdot 8\right)$

$= 6 - 8 \div 2 \cdot 3^2 - 4\left(3 - 32\right)$

$= 6 - 8 \div 2 \cdot 3^2 - 4\left(-29\right)$

$= 6 - 8 \div 2 \cdot 9 - 4\left(-29\right)$

$= 6 - 4 \cdot 9 - \left(-116\right)$

$= 6 - 36 + 116$

$= 86$

8. $\dfrac{\left(4a^{-4}\right)^3 \left(2a^2\right)^{-3}}{\left(2a^{-2}\right)^3} = \dfrac{64a^{-12} \cdot 2^{-3} a^{-6}}{2^3 a^{-6}}$

$= \dfrac{2^6 a^{-12} \cdot 2^{-3} a^{-6}}{2^3 a^{-6}}$

$= \dfrac{2^3 a^{-18}}{2^3 a^{-6}}$

$= a^{-18 - (-6)}$

$= a^{-12}$

$= \dfrac{1}{a^{12}}$

9. $\dfrac{21p^2 q^4 - 14p^5 q^3 + 6p^3 q^7}{7p^3 q^5}$

$= \dfrac{21p^2 q^4}{7p^3 q^5} - \dfrac{14p^5 q^3}{7p^3 q^5} + \dfrac{6p^3 q^7}{7p^3 q^5}$

$= 3p^{-1} q^{-1} - 2p^2 q^{-2} + \dfrac{6}{7} q^2$

$= \dfrac{3}{pq} - \dfrac{2p^2}{q^2} + \dfrac{6q^2}{7}$

10. $\dfrac{2x+3}{x^2 + 6x + 9} - \dfrac{x+4}{x+3} = \dfrac{2x+3}{(x+3)^2} - \dfrac{(x+4)(x+3)}{(x+3)^2}$

$= \dfrac{2x+3}{(x+3)^2} - \dfrac{x^2 + 7x + 12}{(x+3)^2}$

$= \dfrac{2x + 3 - x^2 - 7x - 12}{(x+3)^2}$

$= \dfrac{-x^2 - 5x - 9}{(x+3)^2}$

11. $4\sqrt{6c^3} \cdot 3\sqrt{10c^5} = 12\sqrt{6c^3 \cdot 10c^5}$

$= 12\sqrt{60c^8}$

$= 12\sqrt{4c^8 \cdot 15}$

$= 12 \cdot 2c^4 \sqrt{15}$

$= 24c^4 \sqrt{15}$

12. $(4 + 3i)(2 - 4i) = 8 - 16i + 6i - 12i^2$

$= 8 - 10i - 12(-1)$

$= 8 - 10i + 12$

$= 20 - 10i$

13. $\log_b b^{2x} = 2x$

14. $\dfrac{d^2 - d - 12}{2d^2} \div \dfrac{3d^2 + 13d + 12}{d}$

$= \dfrac{d^2 - d - 12}{2d^2} \cdot \dfrac{d}{3d^2 + 13d + 12}$

$= \dfrac{(d-4)\cancel{(d+3)}}{2 \cdot d \cdot \cancel{d}} \cdot \dfrac{\cancel{d}}{(3d+4)\cancel{(d+3)}}$

$= \dfrac{d-4}{2d(3d+4)}$

15. $x^4 - 9x^2 - 4x^2 y^2 + 36y^2$

$= \left(x^4 - 9x^2\right) - \left(4x^2 y^2 - 36y^2\right)$

$= x^2\left(x^2 - 9\right) - 4y^2\left(x^2 - 9\right)$

$= \left(x^2 - 9\right)\left(x^2 - 4y^2\right)$

$= (x+3)(x-3)(x+2y)(x-2y)$

16. $24x^4 y - 30x^3 y^2 + 9x^2 y^3$

$= 3x^2 y\left(8x^2 - 10xy + 3y^2\right)$

$= 3x^2 y(4x - 3y)(2x - y)$

17. $|2x - 3| - 5 = -4$

$|2x - 3| = 1$

$2x - 3 = 1$ or $2x - 3 = -1$

$2x = 4 \qquad\qquad 2x = 2$

$x = 2 \qquad\qquad x = 1$

18. $2|6x - 10| > -8$

$|6x - 10| > -4$

Because an absolute value is always nonnegative, this statement is true for all real numbers.

19. $\begin{cases} 4x - 5y = -22 \\ 3x + 2y = -5 \end{cases}$

Solve the first equation for x: $x = \dfrac{5}{4}y - \dfrac{22}{4}$

Substitute $\dfrac{5}{4}y - \dfrac{22}{4}$ for x in the second equation and solve for y.

$$3x + 2y = -5$$

$$3\left(\dfrac{5}{4}y - \dfrac{22}{4}\right) + 2y = -5$$

$$\dfrac{15}{4}y - \dfrac{66}{4} + \dfrac{8}{4}y = -5$$

$$\dfrac{23}{4}y - \dfrac{66}{4} = -5$$

$$4\left(\dfrac{23}{4}y - \dfrac{66}{4}\right) = 4(-5)$$

$$23y - 66 = -20$$

$$23y = 46$$

$$y = 2$$

$$x = \dfrac{5}{4}y - \dfrac{22}{4}$$

$$x = \dfrac{5}{4}(2) - \dfrac{22}{4}$$

$$x = \dfrac{10}{4} - \dfrac{22}{4}$$

$$x = -\dfrac{12}{4}$$

$$x = -3$$

Solution: $(-3, 2)$

20. Let $u = 2x - 3$.

$$u^2 - 2u = 8$$

$$u^2 - 2u - 8 = 0$$

$$(u - 4)(u + 2) = 0$$

$u - 4 = 0$ or $u + 2 = 0$

$\quad u = 4 \qquad\qquad u = -2$

$2x - 3 = 4 \qquad 2x - 3 = -2$

$\quad 2x = 7 \qquad\qquad 2x = 1$

$\quad\; x = \dfrac{7}{2} \qquad\qquad\; x = \dfrac{1}{2}$

21. $\dfrac{x}{x-2} - \dfrac{4}{x-1} = \dfrac{2}{x^2 - 3x + 2}$

$$\dfrac{x}{x-2} - \dfrac{4}{x-1} = \dfrac{2}{(x-2)(x-1)}$$

$$(x-2)(x-1)\left(\dfrac{x}{x-2} - \dfrac{4}{x-1}\right)$$

$$= (x-2)(x-1)\left(\dfrac{2}{(x-2)(x-1)}\right)$$

$$x(x-1) - 4(x-2) = 2$$

$$x^2 - x - 4x + 8 = 2$$

$$x^2 - 5x + 6 = 0$$

$$(x-3)(x-2) = 0$$

$x - 3 = 0$ or $x - 2 = 0$

$\quad x = 3 \qquad\qquad x = 2$

Notice that if $x = 2$, then $\dfrac{x}{x-2}$ and $\dfrac{2}{x^2 - 3x + 2}$ are undefined. Therefore, $x = 2$ is extraneous. The only solution is $x = 3$.

22. $\sqrt{x+4} - \sqrt{x-4} = 4$

$$\sqrt{x+4} = \sqrt{x-4} + 4$$

$$\left(\sqrt{x+4}\right)^2 = \left(\sqrt{x-4} + 4\right)^2$$

$$x + 4 = x - 4 + 8\sqrt{x-4} + 16$$

$$x + 4 = x + 12 + 8\sqrt{x-4}$$

$$-8 = 8\sqrt{x-4}$$

$$-1 = \sqrt{x-4}$$

$$(-1)^2 = \left(\sqrt{x-4}\right)^2$$

$$1 = x - 4$$

$$5 = x$$

Check:

$$\sqrt{5+4} - \sqrt{5-4} \overset{?}{=} 4$$

$$\sqrt{9} - \sqrt{1} \overset{?}{=} 4$$

$$3 - 1 \overset{?}{=} 4$$

$$2 \neq 4$$

Because 5 is an extraneous solution, this equation has no solution.

23. $4^{x+2} = 8$

$\left(2^2\right)^{x+2} = 2^3$

$2^{2x+4} = 2^3$

$2x+4 = 3$

$2x = -1$

$x = -\dfrac{1}{2}$

24. $\dfrac{1}{2}\log_2 x = 2$

$\log_2 x^{1/2} = 2$

$2^2 = x^{1/2}$

$4 = \sqrt{x}$

$4^2 = \left(\sqrt{x}\right)^2$

$16 = x$

25. $\log(3x-5) + \log x = \log 12$

$\log\left[(3x-5)x\right] = \log 12$

$\log\left(3x^2 - 5x\right) = \log 12$

$3x^2 - 5x = 12$

$3x^2 - 5x - 12 = 0$

$(3x+4)(x-3) = 0$

$3x+4 = 0 \qquad \text{or} \quad x-3 = 0$

$x = \cancel{-\dfrac{4}{3}} \qquad\qquad x = 3$

26. $2x - 5y = 10$

$-5y = -2x + 10$

$y = \dfrac{2}{5}x - 2$

a) $m = \dfrac{2}{5}$

b) y-intercept: $(0, -2)$

   x-intercept:

   $2x - 5(0) = 10$

   $2x = 10$

   $x = 5$

   $(5, 0)$

c)

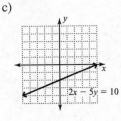

27. $a^2 + b^2 = c^2$

$n^2 + (n-1)^2 = (n+8)^2$

$n^2 + n^2 - 2n + 1 = n^2 + 16n + 64$

$n^2 - 18n - 63 = 0$

$(n-21)(n+3) = 0$

$n - 21 = 0 \quad \text{or} \quad n + 3 = 0$

$n = 21 \qquad\qquad n = \cancel{-3}$

Lengths cannot be negative, so disregard the negative solution. One leg has a length of 21. The other leg has length $21 - 1 = 20$, and the hypotenuse has length $21 + 8 = 29$.

28. $A = P\left(1 + \dfrac{r}{n}\right)^{nt}$

$A = 15{,}000\left(1 + \dfrac{0.04}{12}\right)^{12 \cdot 3}$

$A = 15{,}000\left(1.00\overline{3}\right)^{36}$

$A \approx 16{,}909.08$

After three years, there will be $16,909.08 in the account.

29. Complete a table.

Categories	Rate of Work	Time at Work	Amt of Task Completed
Ethan	$\dfrac{1}{2}$	t	$\dfrac{t}{2}$
Pam	$\dfrac{1}{1.5} = \dfrac{1}{\frac{3}{2}} = \dfrac{2}{3}$	t	$\dfrac{2t}{3}$

$\dfrac{t}{2} + \dfrac{2t}{3} = 1$

$6 \cdot \left(\dfrac{t}{2} + \dfrac{2t}{3}\right) = 6 \cdot (1)$

$3t + 4t = 6$

$7t = 6$

$t = \dfrac{6}{7} \approx 0.857$

Working together, it takes Ethan and Pam 0.857 hour, or about 51 minutes, to complete the job.

30. Create a table.

	rate	time	distance
car	$x+10$	$\dfrac{300}{x+10}$	300
bus	x	$\dfrac{300}{x}$	300

Translate the information in the table to an equation and solve.

$$\frac{300}{x+10}+1=\frac{300}{x}$$

$$x(x+10)\left(\frac{300}{x+10}+1\right)=x(x+10)\left(\frac{300}{x}\right)$$

$$300x+x(x+10)=300(x+10)$$

$$300x+x^2+10x=300x+3000$$

$$x^2+10x-3000=0$$

$$(x+60)(x-50)=0$$

$$x+60=0 \qquad \text{or} \quad x-50=0$$

$$x=\cancel{-60} \qquad\qquad x=50$$

Disregard the negative rate. If the bus travels at a rate of 50 miles per hour, then it takes

$$\frac{300}{50+10}=\frac{300}{60}=5 \text{ hours for the car to travel 300 miles.}$$

Chapter 12
Conic Sections

Exercise Set 12.1

1. $y = (x-1)^2 + 2$

   opens upward

   vertex is $(1, 2)$

   axis of symmetry is $x = 1$

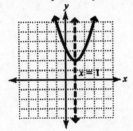

3. $y = -x^2 - 2x + 3$

   $y - 3 = -1(x^2 + 2x)$

   $y - 3 - 1 = -1(x^2 + 2x + 1)$

   $y - 4 = -1(x+1)^2$

   $y = -1(x+1)^2 + 4$

   opens downward

   vertex is $(-1, 4)$

   axis of symmetry is $x = -1$

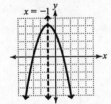

5. $x = (y+2)^2 - 2$

   opens right

   vertex is $(-2, -2)$

   axis of symmetry is $y = -2$

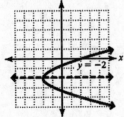

7. $x = -(y-1)^2 + 3$

   opens left

   vertex is $(3, 1)$

   axis of symmetry is $y = 1$

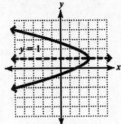

9. $x = 2(y+2)^2 - 4$

   opens right

   vertex is $(-4, -2)$

   axis of symmetry is $y = -2$

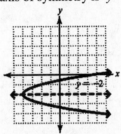

11. $x = -3(y+2)^2 - 5$

    opens left

    vertex is $(-5, -2)$

    axis of symmetry is $y = -2$

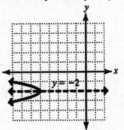

13. $$x = y^2 + 4y + 3$$
$$x - 3 = y^2 + 4y$$
$$x - 3 + 4 = y^2 + 4y + 4$$
$$x + 1 = (y + 2)^2$$
$$x = (y + 2)^2 - 1$$

opens right

vertex is $(-1, -2)$

axis of symmetry is $y = -2$

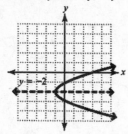

15. $$x = -y^2 + 6y - 5$$
$$x + 5 = -(y^2 - 6y)$$
$$x + 5 - 9 = -(y^2 - 6y + 9)$$
$$x - 4 = -(y - 3)^2$$
$$x = -(y - 3)^2 + 4$$

opens left

vertex is $(4, 3)$

axis of symmetry is $y = 3$

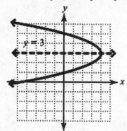

17. $$x = 2y^2 + 8y + 3$$
$$x - 3 = 2(y^2 + 4y)$$
$$x - 3 + 8 = 2(y^2 + 4y + 4)$$
$$x + 5 = 2(y + 2)^2$$
$$x = 2(y + 2)^2 - 5$$

opens right

vertex is $(-5, -2)$

axis of symmetry is $y = -2$

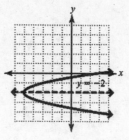

19. $$x = 3y^2 - 6y + 1$$
$$x - 1 = 3(y^2 - 2y)$$
$$x - 1 + 3 = 3(y^2 - 2y + 1)$$
$$x = (y - 1)^2 - 2$$

opens right

vertex is $(-2, 1)$

axis of symmetry is $y = 1$

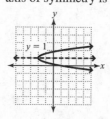

21. $$x = -2y^2 + 4y + 5$$
$$x - 5 = -2(y^2 - 2y)$$
$$x - 5 - 2 = -2(y^2 - 2y + 1)$$
$$x - 7 = -2(y - 1)^2$$
$$x = -2(y - 1)^2 + 7$$

opens left

vertex is $(7, 1)$

axis of symmetry is $y = 1$

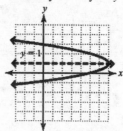

23. c   25. a

27. $(-4,2)$ and $(-1,6)$

$$d = \sqrt{(-4-(-1))^2 + (2-6)^2}$$
$$= \sqrt{(-3)^2 + (-4)^2}$$
$$= \sqrt{9+16}$$
$$= \sqrt{25}$$
$$= 5$$

$$\text{Midpoint} = \left(\frac{-4+(-1)}{2}, \frac{2+6}{2}\right)$$
$$= \left(\frac{-5}{2}, \frac{8}{2}\right)$$
$$= \left(-\frac{5}{2}, 4\right)$$

29. $(-8,-4)$ and $(-3,8)$

$$d = \sqrt{(-8-(-3))^2 + (-4-8)^2}$$
$$= \sqrt{(-5)^2 + (-12)^2}$$
$$= \sqrt{25+144}$$
$$= \sqrt{169}$$
$$= 13$$

$$\text{Midpoint} = \left(\frac{-8+(-3)}{2}, \frac{-4+8}{2}\right)$$
$$= \left(\frac{-11}{2}, \frac{4}{2}\right)$$
$$= \left(-\frac{11}{2}, 2\right)$$

31. $(-8,-10)$ and $(4,-5)$

$$d = \sqrt{(-8-4)^2 + (-10-(-5))^2}$$
$$= \sqrt{(-12)^2 + (-5)^2}$$
$$= \sqrt{144+25}$$
$$= \sqrt{169}$$
$$= 13$$

$$\text{Midpoint} = \left(\frac{-8+4}{2}, \frac{-10+(-5)}{2}\right)$$
$$= \left(\frac{-4}{2}, \frac{-15}{2}\right)$$
$$= \left(-2, -\frac{15}{2}\right)$$

33. $(2,4)$ and $(4,8)$

$$d = \sqrt{(2-4)^2 + (4-8)^2}$$
$$= \sqrt{(-2)^2 + (-4)^2}$$
$$= \sqrt{4+16}$$
$$= \sqrt{20}$$
$$= 2\sqrt{5}$$

$$\text{Midpoint} = \left(\frac{2+4}{2}, \frac{4+8}{2}\right)$$
$$= \left(\frac{6}{2}, \frac{12}{2}\right)$$
$$= (3,6)$$

35. $(-5,2)$ and $(3,-2)$

$$d = \sqrt{(-5-3)^2 + (2-(-2))^2}$$
$$= \sqrt{(-8)^2 + (4)^2}$$
$$= \sqrt{64+16}$$
$$= \sqrt{80}$$
$$= 4\sqrt{5}$$

$$\text{Midpoint} = \left(\frac{-5+3}{2}, \frac{2+(-2)}{2}\right)$$
$$= \left(\frac{-2}{2}, \frac{0}{2}\right)$$
$$= (-1,0)$$

37. $(6,-2)$ and $(1,-5)$

$$d = \sqrt{(6-1)^2 + (-2-(-5))^2}$$
$$= \sqrt{(5)^2 + (3)^2}$$
$$= \sqrt{25+9}$$
$$= \sqrt{34}$$

$$\text{Midpoint} = \left(\frac{6+1}{2}, \frac{-2+(-5)}{2}\right)$$
$$= \left(\frac{7}{2}, \frac{-7}{2}\right)$$
$$= \left(\frac{7}{2}, -\frac{7}{2}\right)$$

39. Find the distance from the center to a point on the circle. This is the radius.

$(4,2)$ and $(8,-1)$

$$r = \sqrt{(8-4)^2 + (-1-2)^2}$$
$$= \sqrt{(4)^2 + (-3)^2}$$
$$= \sqrt{16+9}$$
$$= \sqrt{25}$$
$$= 5$$

41. Find the distance from the center to a point on the circle. This is the radius.

$(2,-6)$ and $(10,-1)$

$$r = \sqrt{(10-2)^2 + (-1-(-6))^2}$$
$$= \sqrt{(8)^2 + (5)^2}$$
$$= \sqrt{64+25}$$
$$= \sqrt{89}$$

43. $(x-2)^2 + (y-1)^2 = 4$

Center $(2,1)$; radius 2

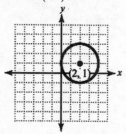

45. $(x+3)^2 + (y+2)^2 = 81$

Center $(-3,-2)$; radius 9

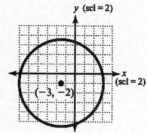

47. $(x-5)^2 + (y+3)^2 = 49$

Center $(5,-3)$; radius 7

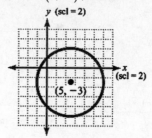

49. $(x-1)^2 + (y+1)^2 = 12$

Center $(1,-1)$; radius $\sqrt{12} = 2\sqrt{3}$

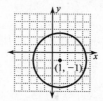

51. $(x+4)^2 + (y+2)^2 = 32$

Center $(-4,-2)$; radius $\sqrt{32} = 4\sqrt{2}$

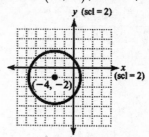

53.
$$x^2 + y^2 + 8x - 6y + 16 = 0$$
$$(x^2 + 8x) + (y^2 - 6y) = -16$$
$$(x^2 + 8x + 16) + (y^2 - 6y + 9) = -16 + 16 + 9$$
$$(x+4)^2 + (y-3)^2 = 9$$

Center $(-4,3)$; radius 3

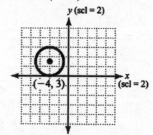

55.
$$x^2 + y^2 + 10x - 4y - 35 = 0$$
$$\left(x^2 + 10x\right) + \left(y^2 - 4y\right) = 35$$
$$\left(x^2 + 10x + 25\right) + \left(y^2 - 4y + 4\right) = 35 + 25 + 4$$
$$\left(x + 5\right)^2 + \left(y - 2\right)^2 = 64$$

Center $\left(-5, 2\right)$; radius 8

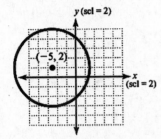

57.
$$x^2 + y^2 + 14x - 4y + 49 = 0$$
$$\left(x^2 + 14x\right) + \left(y^2 - 4y\right) = -49$$
$$\left(x^2 + 14x + 49\right) + \left(y^2 - 4y + 4\right) = -49 + 49 + 4$$
$$\left(x + 7\right)^2 + \left(y - 2\right)^2 = 4$$

Center $\left(-7, 2\right)$; radius 2

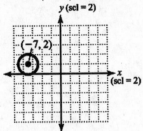

59. b   **61.** c

63. $x^2 + y^2 = 49$

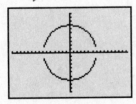

65. $\left(x - 2\right)^2 + \left(y + 3\right)^2 = 25$

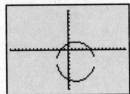

67. Center: $\left(4, 2\right) = \left(h, k\right)$; radius: 4
$$\left(x - h\right)^2 + \left(y - k\right)^2 = r^2$$
$$\left(x - 4\right)^2 + \left(y - 2\right)^2 = 4^2$$
$$\left(x - 4\right)^2 + \left(y - 2\right)^2 = 16$$

69. Center: $\left(-4, -3\right) = \left(h, k\right)$; radius: 5
$$\left(x - h\right)^2 + \left(y - k\right)^2 = r^2$$
$$\left(x - \left(-4\right)\right)^2 + \left(y - \left(-3\right)\right)^2 = 5^2$$
$$\left(x + 4\right)^2 + \left(y + 3\right)^2 = 25$$

71. Center: $\left(6, -2\right) = \left(h, k\right)$; radius: $\sqrt{14}$
$$\left(x - h\right)^2 + \left(y - k\right)^2 = r^2$$
$$\left(x - 6\right)^2 + \left(y - \left(-2\right)\right)^2 = \left(\sqrt{14}\right)^2$$
$$\left(x - 6\right)^2 + \left(y + 2\right)^2 = 14$$

73. Center: $\left(-5, 2\right) = \left(h, k\right)$; radius: $3\sqrt{5}$
$$\left(x - h\right)^2 + \left(y - k\right)^2 = r^2$$
$$\left(x - \left(-5\right)\right)^2 + \left(y - 2\right)^2 = \left(3\sqrt{5}\right)^2$$
$$\left(x + 5\right)^2 + \left(y - 2\right)^2 = 45$$

75. Find the distance from the center to a point on the circle. This is the radius.
$\left(2, 4\right)$ and $\left(5, 8\right)$
$$r = \sqrt{\left(2 - 5\right)^2 + \left(4 - 8\right)^2}$$
$$= \sqrt{\left(-3\right)^2 + \left(-4\right)^2}$$
$$= \sqrt{9 + 16}$$
$$= \sqrt{25}$$
$$= 5$$
The equation in standard form is:
$$\left(x - 2\right)^2 + \left(y - 4\right)^2 = 25$$

77. Find the distance from the center to a point on the circle. This is the radius.
$\left(2, 4\right)$ and $\left(7, 16\right)$
$$r = \sqrt{\left(2 - 7\right)^2 + \left(4 - 16\right)^2}$$
$$= \sqrt{\left(-5\right)^2 + \left(-12\right)^2}$$
$$= \sqrt{25 + 144}$$
$$= \sqrt{169}$$
$$= 13$$
The equation in standard form is
$$\left(x - 2\right)^2 + \left(y - 4\right)^2 = 169$$

79. The center is $(2,-5)$ since all points are an equal distance from that point. Also, since all points are 8 units from the center, the radius is 8.

The equation in standard form is:
$(x-2)^2 + (y+5)^2 = 64$

81. $(4,-2)$ and $(-2,6)$

$$\text{diameter} = \sqrt{(4-(-2))^2 + (-2-6)^2}$$
$$= \sqrt{(4+2)^2 + (-8)^2}$$
$$= \sqrt{36+64}$$
$$= \sqrt{100}$$
$$= 10$$

The radius is $10 \div 2 = 5$.

$$\text{Midpoint} = \left(\frac{4+(-2)}{2}, \frac{-2+6}{2}\right)$$
$$= \left(\frac{2}{2}, \frac{4}{2}\right)$$
$$= (1,2)$$

The center of the circle is $(1,2)$.
The equation in standard form is:
$(x-1)^2 + (y-2)^2 = 25$

83. $(6,-2)$ and $(-2,8)$

$$\text{diameter} = \sqrt{(6-(-2))^2 + (-2-8)^2}$$
$$= \sqrt{(6+2)^2 + (-10)^2}$$
$$= \sqrt{64+100}$$
$$= \sqrt{164}$$
$$= 2\sqrt{41}$$

The radius is $2\sqrt{41} \div 2 = \sqrt{41}$.

$$\text{Midpoint} = \left(\frac{6+(-2)}{2}, \frac{-2+8}{2}\right)$$
$$= \left(\frac{4}{2}, \frac{6}{2}\right)$$
$$= (2,3)$$

The center of the circle is $(2,3)$.
The equation in standard form is:
$(x-2)^2 + (y-3)^2 = 41$

85. a)
$$h = -16t^2 + 96t + 112$$
$$h - 112 = -16(t^2 - 6t)$$
$$h - 112 - 144 = -16(t^2 - 6t + 9)$$
$$h - 256 = -16(t-3)^2$$
$$h = -16(t-3)^2 + 256$$

The vertex for the parabola is (3, 256). The maximum height for the rock will be 256 ft.

b) It will take 3 seconds to reach the maximum height.

c)
$$0 = -16t^2 + 96t + 112$$
$$0 = -16(t^2 - 6t - 7)$$
$$0 = t^2 - 6t - 7$$
$$0 = (t-7)(t+1)$$
$$t - 7 = 0 \qquad t + 1 = 0$$
$$t = 7 \qquad\quad t = \cancel{-1}$$

The rock will hit the ground after 7 seconds.

87. We must find a to use the standard form $y = a(x-h)^2 + k$. The parabola has vertex $(0,18)$. When $y = 0$, $x = \frac{30}{2} = 15$. Then
$$y = a(x-h)^2 + k$$
$$0 = a(15-0)^2 + 18$$
$$-18 = a \cdot 225$$
$$\frac{-18}{225} = a$$
$$a = -\frac{2}{25} = -0.08$$

Substitute into the general equation.
$$y = a(x-h)^2 + k$$
$$y = -0.08(x-0)^2 + 18$$
$$y = -0.08x^2 + 18$$
$$\text{or}$$
$$y = -\frac{2}{25}x^2 + 18$$

89. $y = 0.0038x^2 - 0.3475x + 8.316$
$y = 0.0038(17)^2 - 0.3475(17) + 8.316$
$y \approx 3.5\%$

91. $x^2 + y^2 = 20^2$
$x^2 + y^2 = 400$

93. $x^2 + (y-110)^2 = 100^2$
$x^2 + (y-110)^2 = 10,000$

95. $x^2 + y^2 = (6370 + 230)^2$
$x^2 + y^2 = 6600^2$
$x^2 + y^2 = 43,560,000$

Review Exercises

1.

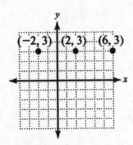

2. $25x^2 - 9y^2 = (5x + 3y)(5x - 3y)$

3.

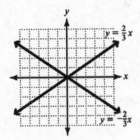

4. $\dfrac{x^2}{16} = 1$
$x^2 = 16$
$x = \pm\sqrt{16}$
$x = \pm 4$

5. $9x^2 + 16 \cdot 0 = 144$
$9x^2 = 144$
$x^2 = 16$
$x = \pm 4$
$(-4, 0), (4, 0)$

$9 \cdot 0 + 16y^2 = 144$
$16y^2 = 144$
$y^2 = \dfrac{144}{16}$
$y^2 = 9$
$y = \pm 3$
$(0, -3), (0, 3)$

6. $25x^2 + 9 \cdot 0 = 225$
$25x^2 = 225$
$x^2 = 9$
$x = \pm 3$
$(-3, 0), (3, 0)$

$25 \cdot 0 + 9y^2 = 225$
$9y^2 = 225$
$y^2 = \dfrac{225}{9}$
$y^2 = 25$
$y = \pm 5$
$(0, -5), (0, 5)$

Exercise Set 12.2

1. $\dfrac{x^2}{81} + \dfrac{y^2}{64} = 1$

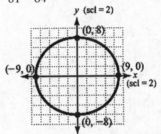

3. $\dfrac{x^2}{36} + y^2 = 1$

5. $4x^2 + 9y^2 = 36$
$\dfrac{4x^2}{36} + \dfrac{9y^2}{36} = \dfrac{36}{36}$
$\dfrac{x^2}{9} + \dfrac{y^2}{4} = 1$

7. $36x^2 + 4y^2 = 144$

$$\frac{36x^2}{144} + \frac{4y^2}{144} = \frac{144}{144}$$

$$\frac{x^2}{4} + \frac{y^2}{36} = 1$$

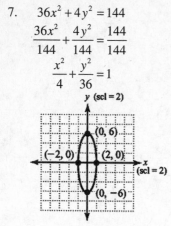

9. $\dfrac{(x-1)^2}{49} + \dfrac{(y+3)^2}{25} = 1$

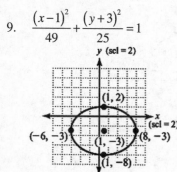

11. $\dfrac{(x-4)^2}{4} + \dfrac{(y+3)^2}{36} = 1$

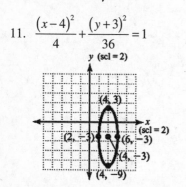

13. $\dfrac{x^2}{25} + \dfrac{y^2}{4} = 1$

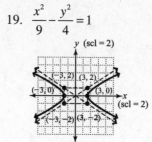

15. $\dfrac{(y-2)^2}{36} + \dfrac{(x+4)^2}{25} = 1$

17. $\dfrac{x^2}{9} + \dfrac{y^2}{16} = 1$

19. $\dfrac{x^2}{9} - \dfrac{y^2}{4} = 1$

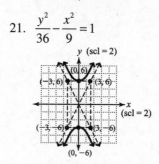

21. $\dfrac{y^2}{36} - \dfrac{x^2}{9} = 1$

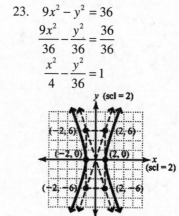

23. $9x^2 - y^2 = 36$

$$\frac{9x^2}{36} - \frac{y^2}{36} = \frac{36}{36}$$

$$\frac{x^2}{4} - \frac{y^2}{36} = 1$$

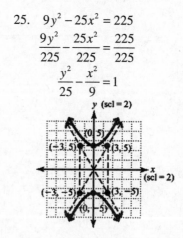

25. $9y^2 - 25x^2 = 225$

$$\frac{9y^2}{225} - \frac{25x^2}{225} = \frac{225}{225}$$

$$\frac{y^2}{25} - \frac{x^2}{9} = 1$$

27. $\dfrac{x^2}{36} - \dfrac{y^2}{4} = 1$

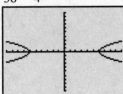

29. $\dfrac{x^2}{9} - \dfrac{y^2}{4} = 1$

31. c   33. b

35. $\quad 9x^2 + 16y^2 = 144$

$\quad \dfrac{9x^2}{144} + \dfrac{16y^2}{144} = \dfrac{144}{144}$

$\quad\quad \dfrac{x^2}{16} + \dfrac{y^2}{9} = 1$

ellipse

37. $\quad x^2 + y^2 - 6x + 8y - 75 = 0$

$\quad\quad x^2 - 6x + y^2 + 8y = 75$

$\quad x^2 - 6x + 9 + y^2 + 8y + 16 = 75 + 9 + 16$

$\quad\quad (x-3)^2 + (y+4)^2 = 100$

circle

39. $(x-2)^2 + (y+2)^2 = 49$

circle

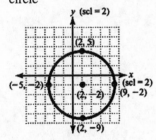

41. $y = 2(x+1)^2 + 3$

parabola

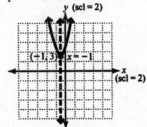

43. $\dfrac{x^2}{36} + \dfrac{y^2}{81} = 1$

ellipse

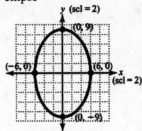

45. $\dfrac{x^2}{4} - \dfrac{y^2}{25} = 1$

hyperbola

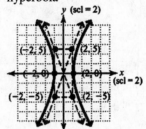

47. $\quad y = 2x^2 + 8x + 6$

$\quad y - 6 = 2(x^2 + 4x)$

$\quad y - 6 + 8 = 2(x^2 + 4x + 4)$

$\quad\quad y = 2(x+2)^2 - 2$

parabola

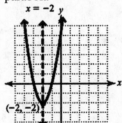

49. $\dfrac{x^2}{16} - \dfrac{y^2}{16} = 1$

hyperbola

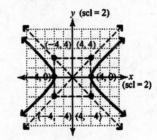

51. $\dfrac{(x+3)^2}{9}+\dfrac{(y+2)^2}{36}=1$

ellipse

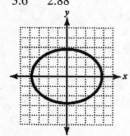

53. a) $400x^2+256y^2=102,400$

$\dfrac{400}{102,400}x^2+\dfrac{256}{102,400}y^2=\dfrac{102,400}{102,400}$

$\dfrac{x^2}{256}+\dfrac{y^2}{400}=1$

Yes, the sailboat will clear the bridge. The height of the bridge at the center is $\sqrt{400}=20$ ft., and the boat's mast is only 18 ft. above the water.

b) $\sqrt{256}=16$

The bridge is $2(16)=32$ ft. at the base of the arch.

55. $\dfrac{x^2}{3.6^2}+\dfrac{y^2}{2.88^2}=1$

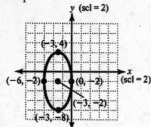

57. $\dfrac{x^2}{1.5^2}+\dfrac{y^2}{2^2}=1$

$\dfrac{x^2}{2.25}+\dfrac{y^2}{4}=1$

59. Rewrite as: $\dfrac{x^2}{5^2}+\dfrac{y^2}{4^2}=1$

Now, use substitution: $c^2=5^2-4^2$

$c^2=25-16$

$c^2=9$

$c=\sqrt{9}$

$c=3$ units

Review Exercises

1. $3x^2+y=6$

$y=-3x^2+6$

2. $\begin{cases} x+y=3 \\ 2x+y=4 \end{cases}$

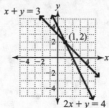

Solution: $(1,2)$

3. $\begin{cases} 2x+y=1 \\ 3x+4y=-6 \end{cases}$

Solve the first equation for y: $y=1-2x$

$3x+4(1-2x)=-6$ $y=1-2\cdot 2$

$3x+4-8x=-6$ $y=-3$

$-5x=-10$

$x=2$

Solution: $(2,-3)$

4. $4(2x+3y=6)$ $3(-3)-4y=-25$

$3(3x-4y=-25)$ $-9-4y=-25$

$\overline{8x+12y=24}$ $-4y=-16$

$\underline{9x-12y=-75}$ $y=4$

$17x=-51$

$x=-3$

Solution: $(-3,4)$

5. $3x^2+10x-8=0$

$(3x-2)(x+4)=0$

$3x-2=0$ or $x+4=0$

$x=\dfrac{2}{3}$ $x=-4$

6. $4x^2=36$

$x^2=9$

$x=\pm 3$

Exercise Set 12.3

1. $\begin{cases} x^2 + 2y = 1 \\ 2x + y = 2 \end{cases}$

   Solve the second equation for y: $y = -2x + 2$

   $$x^2 + 2y = 1$$
   $$x^2 + 2(-2x + 2) = 1$$
   $$x^2 - 4x + 4 = 1$$
   $$x^2 - 4x + 3 = 0$$
   $$(x-3)(x-1) = 0$$
   $$x - 3 = 0 \quad \text{or} \quad x - 1 = 0$$
   $$x = 3 \qquad x = 1$$

   $x = 3 \quad y = -2(3) + 2 = -6 + 2 = -4 \quad (3, -4)$

   $x = 1 \quad y = -2(1) + 2 = -2 + 2 = 0 \quad (1, 0)$

3. $\begin{cases} y = x^2 + 4x + 4 \\ 3x - y = -6 \end{cases}$

   The first equation is solved for y.

   $$3x - (x^2 + 4x + 4) = -6$$
   $$3x - x^2 - 4x - 4 = -6$$
   $$0 = x^2 + x - 2$$
   $$0 = (x+2)(x-1)$$
   $$x + 2 = 0 \quad \text{or} \quad x - 1 = 0$$
   $$x = -2 \qquad x = 1$$

   $x = -2: \quad y = (-2)^2 + 4(-2) + 4$
   $$y = 4 - 8 + 4$$
   $$y = 0$$
   $$(-2, 0)$$

   $x = 1: \quad y = (1)^2 + 4(1) + 4$
   $$y = 1 + 4 + 4$$
   $$y = 9$$
   $$(1, 9)$$

5. $\begin{cases} x^2 + y^2 = 25 \\ x - y = -1 \end{cases}$

   Solve the second equation for x. $x = y - 1$

   $$(y-1)^2 + y^2 = 25$$
   $$y^2 - 2y + 1 + y^2 = 25$$
   $$2y^2 - 2y - 24 = 0$$
   $$y^2 - y - 12 = 0$$
   $$(y-4)(y+3) = 0$$
   $$y - 4 = 0 \quad \text{or} \quad y + 3 = 0$$
   $$y = 4 \qquad y = -3$$

   $y = 4: \quad x = 4 - 1$
   $$x = 3$$
   $$(3, 4)$$

   $y = -3: \quad x = -3 - 1$
   $$x = -4$$
   $$(-4, -3)$$

7. $\begin{cases} x^2 + y^2 = 10 \\ 3x + y = 6 \end{cases}$

   Solve the second equation for y. $y = 6 - 3x$

   $$x^2 + y^2 = 10$$
   $$x^2 + (6 - 3x)^2 = 10$$
   $$x^2 + 36 - 36x + 9x^2 = 10$$
   $$10x^2 - 36x + 26 = 0$$
   $$2(5x^2 - 18x + 13) = 0$$
   $$2(5x - 13)(x - 1) = 0$$
   $$5x - 13 = 0 \qquad \text{or} \quad x - 1 = 0$$
   $$x = \frac{13}{5} = 2.6 \qquad \qquad x = 1$$

   $x = 2.6: \quad y = 6 - 3(2.6)$
   $$y = 6 - 7.8$$
   $$y = -1.8$$
   $$(2.6, -1.8)$$

   $x = 1: \quad y = 6 - 3(1)$
   $$y = 6 - 3$$
   $$y = 3$$
   $$(1, 3)$$

9. $\begin{cases} x^2 + 2y^2 = 4 \\ x + y = 5 \end{cases}$

   Solve the second equation for x. $x = 5 - y$

   $$(5 - y)^2 + 2y^2 = 4$$
   $$25 - 10y + y^2 + 2y^2 - 4 = 0$$
   $$3y^2 - 10y + 21 = 0$$
   $$y = \frac{10 \pm \sqrt{10^2 - 4(3)(21)}}{2(3)}$$
   $$y = \frac{10 \pm \sqrt{-152}}{6}$$

   Because the discriminant of the quadratic equation is negative, we know that the equation cannot be solved using real numbers. Therefore there is no real solution.

11. $\begin{cases} y = 2x^2 + 3 \\ 2x - y = -3 \end{cases}$

The first equation is solved for y.

$2x - (2x^2 + 3) = -3$

$2x - 2x^2 - 3 = -3$

$0 = 2x^2 - 2x$

$0 = 2x(x - 1)$

$2x = 0 \quad$ or $\quad x - 1 = 0$

$x = 0 \qquad\qquad x = 1$

$x = 0: \quad y = 2 \cdot 0 + 3$

$\qquad\qquad y = 3$

$\qquad\qquad (0, 3)$

$x = 1: \quad y = 2 \cdot 1 + 3$

$\qquad\qquad y = 2 + 3$

$\qquad\qquad y = 5$

$\qquad\qquad (1, 5)$

13. $\begin{cases} y = (x-3)^2 + 2 \\ y = -(x-2)^2 + 3 \end{cases}$

$(x-3)^2 + 2 = -(x-2)^2 + 3$

$x^2 - 6x + 9 + 2 = -x^2 + 4x - 4 + 3$

$2x^2 - 10x + 12 = 0$

$x^2 - 5x + 6 = 0$

$(x-2)(x-3) = 0$

$x - 2 = 0 \quad$ or $\quad x - 3 = 0$

$x = 2 \qquad\qquad x = 3$

$x = 2: \quad y = -(2-2)^2 + 3 = 0 + 3 = 3 \quad (2, 3)$

$x = 3: \quad y = (3-3)^2 + 2 = 0 + 2 = 2 \quad (3, 2)$

15. $\begin{cases} x^2 + y^2 = 20 \\ x^2 - y^2 = 12 \end{cases}$

$\begin{array}{r} x^2 + y^2 = 20 \\ x^2 - y^2 = 12 \\ \hline 2x^2 \quad\quad = 32 \end{array}$

$x^2 \quad = 16$

$x \quad = \pm 4$

$x = -4: \quad (-4)^2 + y^2 = 20$

$\qquad\qquad 16 + y^2 = 20$

$\qquad\qquad y^2 = 4$

$\qquad\qquad y = \pm 2$

$\qquad\qquad (-4, -2), (-4, 2)$

$x = 4: \quad (4)^2 + y^2 = 20$

$\qquad\qquad 16 + y^2 = 20$

$\qquad\qquad y^2 = 4$

$\qquad\qquad y = \pm 2$

$\qquad\qquad (4, -2), (4, 2)$

17. $\begin{cases} 4x^2 + 9y^2 = 72 \\ x^2 + y^2 = 13 \end{cases}$

$4x^2 + 9y^2 = 72$

$\underline{-4(x^2 + y^2 = 13)}$

$\begin{array}{r} 4x^2 + 9y^2 = 72 \\ -4x^2 - 4y^2 = -52 \\ \hline 5y^2 = 20 \end{array}$

$y^2 = 4$

$y = \pm 2$

$y = -2: \quad x^2 + (-2)^2 = 13$

$\qquad\qquad x^2 + 4 = 13$

$\qquad\qquad x^2 = 9$

$\qquad\qquad x = \pm 3$

$\qquad\qquad (-3, -2), (3, -2)$

$y = 2: \quad x^2 + (2)^2 = 13$

$\qquad\qquad x^2 + 4 = 13$

$\qquad\qquad x^2 = 9$

$\qquad\qquad x = \pm 3$

$\qquad\qquad (-3, 2), (3, 2)$

19. $\begin{cases} 4x^2 - y^2 = 15 \\ x^2 + y^2 = 5 \end{cases}$

$\begin{array}{r} 4x^2 - y^2 = 15 \\ x^2 + y^2 = 5 \\ \hline 5x^2 \quad\quad = 20 \end{array}$

$x^2 \quad = 4$

$x \quad = \pm 2$

$x = -2: \quad (-2)^2 + y^2 = 5$

$\qquad\qquad 4 + y^2 = 5$

$\qquad\qquad y^2 = 1$

$\qquad\qquad y = \pm 1$

$\qquad\qquad (-2, -1), (-2, 1)$

$x = 2: \quad (2)^2 + y^2 = 5$

$\qquad\qquad 4 + y^2 = 5$

$\qquad\qquad y^2 = 1$

$\qquad\qquad y = \pm 1$

$\qquad\qquad (2, -1), (2, 1)$

21. $\begin{cases} x^2 + 3y^2 = 36 \\ x = y^2 - 6 \end{cases}$

The second equation is solved for x.

$$\left(y^2 - 6\right)^2 + 3y^2 = 36$$
$$y^4 - 12y^2 + 36 + 3y^2 - 36 = 0$$
$$y^4 - 9y^2 = 0$$
$$y^2\left(y^2 - 9\right) = 0$$
$$y^2\left(y + 3\right)\left(y - 3\right) = 0$$
$$y = 0, -3, 3$$

$y = 0:\quad x = 0^2 - 6$
$\qquad\quad x = -6$
$\qquad\quad (-6, 0)$

$y = -3:\quad x = (-3)^2 - 6$
$\qquad\qquad x = 9 - 6$
$\qquad\qquad x = 3$
$\qquad\qquad (3, -3)$

$y = 3:\quad x = (3)^2 - 6$
$\qquad\quad x = 9 - 6$
$\qquad\quad x = 3$
$\qquad\quad (3, 3)$

23. $\begin{cases} 9x^2 + 4y^2 = 36 \\ 4x^2 - 9y^2 = 36 \end{cases}$

$$4\left(9x^2 + 4y^2 = 36\right)$$
$$-9\left(4x^2 - 9y^2 = 36\right)$$
$$\overline{\quad\quad\quad\quad\quad\quad\quad}$$
$$36x^2 + 16y^2 = 144$$
$$-36x^2 + 81y^2 = -324$$
$$\overline{\quad\quad\quad\quad\quad\quad\quad}$$
$$97y^2 = -180$$
$$y^2 = -1.856$$

y is imaginary, no solution

25. $\begin{cases} y = x^2 \\ x^2 + y^2 = 20 \end{cases}$

The first equation is already solved for y.

$$x^2 + \left(x^2\right)^2 = 20$$
$$x^2 + x^4 = 20$$
$$x^4 + x^2 - 20 = 0$$
$$\left(x^2 - 4\right)\cancel{\left(x^2 + 5\right)} = 0$$
$$x^2 = 4$$
$$x = \pm 2$$

$x^2 + 5 = 0$ would have yielded an imaginary answer.

$x = -2:\quad y = (-2)^2$
$\qquad\qquad y = 4$
$\qquad\qquad (-2, 4)$

$x = 2:\quad y = (2)^2$
$\qquad\quad y = 4$
$\qquad\quad (2, 4)$

27. $\begin{cases} 25x^2 - 16y^2 = 400 \\ x^2 + 4y^2 = 16 \end{cases}$

Use elimination.

$$25x^2 - 16y^2 = 400$$
$$\underline{4\left(x^2 + 4y^2 = 16\right)}$$
$$25x^2 - 16y^2 = 400$$
$$\underline{4x^2 + 16y^2 = 64}$$
$$29x^2 \qquad\quad = 464$$
$$x^2 \qquad\quad = 16$$
$$x \qquad\quad = \pm 4$$

$x = -4:\quad (-4)^2 + 4y^2 = 16$
$\qquad\qquad\quad 16 + 4y^2 = 16$
$\qquad\qquad\qquad\quad 4y^2 = 0$
$\qquad\qquad\qquad\quad\; y^2 = 0$
$\qquad\qquad\qquad\quad\;\; y = 0$
$\qquad\qquad\qquad\qquad (-4, 0)$

$x = 4:\quad 4^2 + 4y^2 = 16$
$\qquad\qquad 16 + 4y^2 = 16$
$\qquad\qquad\qquad 4y^2 = 0$
$\qquad\qquad\qquad\; y^2 = 0$
$\qquad\qquad\qquad\;\; y = 0$
$\qquad\qquad\qquad\quad (4, 0)$

29. $\begin{cases} xy = 4 \\ 2x^2 - y^2 = 4 \end{cases}$

Solve the first equation for y. $y = \dfrac{4}{x}$

$$2x^2 - \left(\dfrac{4}{x}\right)^2 = 4$$
$$2x^2 - \dfrac{16}{x^2} = 4$$
$$2x^4 - 16 = 4x^2$$
$$x^4 - 2x^2 - 8 = 0$$
$$(x^2 - 4)(x^2 + 2) = 0$$

$x^2 - 4 = 0$ or $x^2 + 2 = 0$
$x^2 = 4$ $\qquad\quad$ $x^2 = -2$
$x = \pm 2$ $\qquad$ Leads to imaginary solutions

$x = -2:$ $y = \dfrac{4}{-2}$
$\qquad\qquad y = -2$
$\qquad\qquad (-2, -2)$

$x = 2:$ $y = \dfrac{4}{2}$
$\qquad\qquad y = 2$
$\qquad\qquad (2, 2)$

31. $\begin{cases} y = x^2 - 2x - 3 \\ y = -x^2 + 6x + 7 \end{cases}$

They are both solved for y, so use substitution.
$$-x^2 + 6x + 7 = x^2 - 2x - 3$$
$$0 = 2x^2 - 8x - 10$$
$$0 = x^2 - 4x - 5$$
$$0 = (x - 5)(x + 1)$$
$$x = 5, -1$$

$x = 5:$ $y = (5)^2 - 2(5) - 3$
$\qquad\qquad = 25 - 10 - 3$
$\qquad\qquad = 12$
$\qquad\qquad (5, 12)$

$x = -1:$ $y = (-1)^2 - 2(-1) - 3$
$\qquad\qquad\quad = 1 + 2 - 3$
$\qquad\qquad\quad = 0$
$\qquad\qquad\quad (-1, 0)$

33. $\begin{cases} 4x^2 + 5y^2 = 36 \\ 4x^2 - 3y^2 = 4 \end{cases}$

$\qquad 4x^2 + 5y^2 = 36$
$\underline{-(4x^2 - 3y^2 = 4)}$
$\qquad 4x^2 + 5y^2 = 36$
$\underline{\,-4x^2 + 3y^2 = -4}$
$\qquad\qquad 8y^2 = 32$
$\qquad\qquad\ y^2 = 4$
$\qquad\qquad\ y = \pm 2$

$y = -2:$ $4x^2 + 5(-2)^2 = 36$
$\qquad\qquad 4x^2 + 5 \cdot 4 = 36$
$\qquad\qquad 4x^2 + 20 = 36$
$\qquad\qquad\quad 4x^2 = 16$
$\qquad\qquad\qquad x^2 = 4$
$\qquad\qquad\qquad x = \pm 2$
$\qquad\qquad\qquad (-2, -2), (2, -2)$

$y = 2:$ $4x^2 + 5(2)^2 = 36$
$\qquad\qquad 4x^2 + 5 \cdot 4 = 36$
$\qquad\qquad 4x^2 + 20 = 36$
$\qquad\qquad\quad 4x^2 = 16$
$\qquad\qquad\qquad x^2 = 4$
$\qquad\qquad\qquad x = \pm 2$
$\qquad\qquad\qquad (-2, 2), (2, 2)$

35. $\begin{cases} x = -y^2 + 2 \\ x^2 - 5y^2 = 4 \end{cases}$

The first equation is already solved for x.
$$\left(-y^2 + 2\right)^2 - 5y^2 = 4$$
$$y^4 - 4y^2 + 4 - 5y^2 = 4$$
$$y^4 - 9y^2 = 0$$
$$y^2\left(y^2 - 9\right) = 0$$
$$y = 0, \pm 3$$

$y = 0:$ $x = -0^2 + 2$
$\qquad\qquad x = 2$
$\qquad\qquad (2, 0)$

$y = -3:$ $x = -(-3)^2 + 2$
$\qquad\qquad x = -9 + 2$
$\qquad\qquad x = -7$
$\qquad\qquad (-7, -3)$

$y = 3:$ $x = -(3)^2 + 2$
$\qquad\qquad x = -9 + 2$
$\qquad\qquad x = -7$
$\qquad\qquad (-7, 3)$

37. Answers may vary, but one possible system is
$$\begin{cases} x + y = 4 \\ x^2 + y^2 = 1 \end{cases}$$

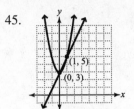

39. Let x be one of the integers and y be the other integer.
$$x^2 + y^2 = 34$$
$$\underline{x^2 - y^2 = 16}$$
$$2x^2 \quad = 50$$
$$x^2 \quad = 25$$
$$x \quad = \pm 5$$

$x = -5: \ (-5)^2 + y^2 = 34$
$$25 + y^2 = 34$$
$$y^2 = 9$$
$$y = \pm 3$$
$$(-5, -3), (-5, 3)$$

$x = 5: \ (5)^2 + y^2 = 34$
$$25 + y^2 = 34$$
$$y^2 = 9$$
$$y = \pm 3$$
$$(5, -3), (5, 3)$$

41. Let x be one dimension and let y be the other.
$$xy = 144$$
$$2x + 2y = 52$$
Solve the second equation for x. $x = 26 - y$
$$(26 - y)y = 144$$
$$26y - y^2 = 144$$
$$0 = y^2 - 26y + 144$$
$$0 = (y - 8)(y - 18)$$
$$y = 8, 18$$
If $y = 8$: $xy = 144$ and $x = 18$.
If $y = 18$: $xy = 144$ and $x = 8$.
The computer keyboard is 8 in. by 18 in.

43. $-3x^2 + 120 = 11x + 28$
$$0 = 3x^2 + 11x - 92$$
$$0 = \cancel{(3x + 23)}(x - 4)$$
$$x = 4$$
$3x + 23 = 0$ would have yielded a negative answer.
So, because $x = 4$ and x is in hundreds of units, at equilibrium there would be 400 chairs.

The price per chair would be $p = 11(4) + 28 = 44 + 28 = \$72$.

45.

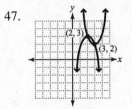

47.

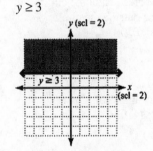

Review Exercises

1. $x + 3y \le 6$ Yes
$$0 + 3 \cdot 0 \le 6$$
$$0 \le 6$$

2. $x + 3y \le 6$ No
$$3 + 3 \cdot 4 \le 6$$
$$3 + 12 \le 6$$
$$15 \le 6$$

3. $y \ge 3$

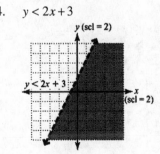

4. $y < 2x + 3$

5. $2x + 3y < -6$

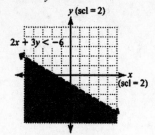

6. $x \geq -2$

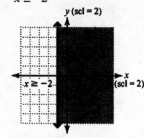

Exercise Set 12.4

1. $x^2 + y^2 \leq 9$

First graph the related equation $x^2 + y^2 = 9$ with a solid line.
Let region 1 be the area inside the circle, and let region 2 be the area outside the circle.
Region 1: Choose (0, 0).
$0^2 + 0^2 \leq 9$
$0 + 0 \leq 9$
$0 \leq 9$
True, so region 1 is in the solution set.
Region 2: Choose (4, 0).
$4^2 + 0^2 \leq 9$
$16 + 0 \leq 9$
$16 \leq 9$
False, so region 2 is not in the solution set.

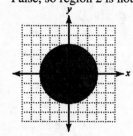

3. $y > x^2$

First graph the related equation $y = x^2$ with a dotted line.
Let region 1 be the area above the boundary, and let region 2 be the area below the boundary.
Region 1: Choose (0, 1).
$1 > 0^2$
$1 > 0$
True, so region 1 is in the solution set.
Region 2: Choose (2, 0).
$0 > 2^2$
$0 > 4$
False, so region 2 is not in the solution set.

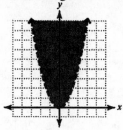

5. $y > 2(x+2)^2 - 3$

First graph the related equation $y = 2(x+2)^2 - 3$ with a dotted line.
Let region 1 be the area above the boundary, and let region 2 be the area below the boundary.
Region 1: Choose $(-2, 0)$.

$0 > 2(-2+2)^2 - 3$
$0 > 2(0)^2 - 3$
$0 > 0 - 3$
$0 > -3$
True, so region 1 is in the solution set.
Region 2: Choose (0, 0).
$0 > 2(0+2)^2 - 3$
$0 > 2(2)^2 - 3$
$0 > 2(4) - 3$
$0 > 8 - 3$
$0 > 5$
False, so region 2 is not in the solution set.

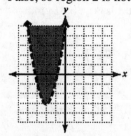

7. $\dfrac{x^2}{16} + \dfrac{y^2}{9} \leq 1$

First graph the related equation $\dfrac{x^2}{16} + \dfrac{y^2}{9} = 1$ with

a solid line.
Let region 1 be the area inside the ellipse, and let region 2 be the area outside the ellipse.

Region 1: Choose (0, 0).

$\dfrac{0^2}{16} + \dfrac{0^2}{9} \leq 1$

$0 + 0 \leq 1$

$0 \leq 1$

True, so region 1 is in the solution set.

Region 2: Choose (5, 0).

$\dfrac{5^2}{16} + \dfrac{0^2}{9} \leq 1$

$\dfrac{25}{16} + 0 \leq 1$

$\dfrac{25}{16} \leq 1$

False, so region 2 is not in the solution set.

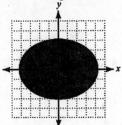

9. $\dfrac{x^2}{25} - \dfrac{y^2}{4} > 1$

First graph the related equation $\dfrac{x^2}{25} - \dfrac{y^2}{4} = 1$ with

a dotted line.
Let region 1 be the area to the left of the left branch of the hyperbola, let region 3 be the area to the right of the right branch of the hyperbola, and let region 2 be the area between the two branches.

Region 1: Choose $(-8, 0)$.

$\dfrac{(-8)^2}{25} - \dfrac{0^2}{4} > 1$

$\dfrac{64}{25} - \dfrac{0}{4} > 1$

$\dfrac{64}{25} > 1$

True, so region 1 is in the solution set.

Region 2: Choose $(0,0)$.

$\dfrac{0^2}{25} - \dfrac{0^2}{4} > 1$

$\dfrac{0}{25} - \dfrac{0}{4} > 1$

$0 > 1$

False, so region 2 is not in the solution set.

Region 3: Choose $(8,0)$.

$\dfrac{8^2}{25} - \dfrac{0^2}{4} > 1$

$\dfrac{64}{25} - \dfrac{0}{4} > 1$

$\dfrac{64}{25} > 1$

True, so region 3 is in the solution set.

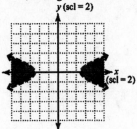

11. $x^2 + y^2 < 16$

First graph the related equation $x^2 + y^2 = 16$
with a dotted line.
Let region 1 be the area inside the circle, and let region 2 be the area outside the circle.

Region 1: Choose (0, 0).

$0^2 + 0^2 < 16$

$0 < 16$

True, so region 1 is in the solution set.

Region 2: Choose (5, 0).

$5^2 + 0^2 < 16$

$25 + 0 < 16$

$25 < 16$

False, so region 2 is not in the solution set.

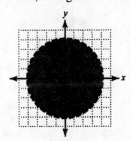

13. $y \geq -2(x-3)^2 + 3$

First graph the related equation

$y = -2(x-3)^2 + 3$ with a solid line.

Let region 1 be the area above the boundary, and let region 2 be the area below the boundary.

Region 1: Choose $(0,0)$.

$0 \geq -2(0-3)^2 + 3$

$0 \geq -2(-3)^2 + 3$

$0 \geq -2(9) + 3$

$0 \geq -18 + 3$

$0 \geq -15$

True, so region 1 is in the solution set.

Region 2: Choose $(4,-2)$.

$-2 \geq -2(4-3)^2 + 3$

$-2 \geq -2(1)^2 + 3$

$-2 \geq -2(1) + 3$

$-2 \geq -2 + 3$

$-2 \geq 1$

False, so region 2 is not in the solution set.

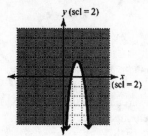

15. $\dfrac{x^2}{16} + \dfrac{y^2}{4} \leq 1$

First graph the related equation $\dfrac{x^2}{16} + \dfrac{y^2}{4} = 1$ with a solid line.

Let region 1 be the area inside the ellipse, and let region 2 be the area outside the ellipse.

Region 1: Choose $(0, 0)$.

$\dfrac{0^2}{16} + \dfrac{0^2}{4} \leq 1$

$0 + 0 \leq 1$

$0 \leq 1$

True, so region 1 is in the solution set.

Region 2: Choose $(5, 0)$.

$\dfrac{5^2}{16} + \dfrac{0^2}{4} \leq 1$

$\dfrac{25}{16} + 0 \leq 1$

$\dfrac{25}{16} \leq 1$

False, so region 2 is not in the solution set.

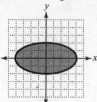

17. $\dfrac{y^2}{25} - \dfrac{x^2}{4} \leq 1$

First graph the related equation $\dfrac{y^2}{25} - \dfrac{x^2}{4} = 1$ with a solid line.

Let region 1 be the area above the top branch of the hyperbola, let region 3 be the area below the bottom branch of the hyperbola, and let region 2 be the area between the two branches.

Region 1: Choose $(0,8)$.

$\dfrac{8^2}{25} - \dfrac{0^2}{4} \leq 1$

$\dfrac{64}{25} - \dfrac{0}{4} \leq 1$

$\dfrac{64}{25} \leq 1$

False, so region 1 is not in the solution set.

Region 2: Choose $(0,0)$.

$\dfrac{0^2}{25} - \dfrac{0^2}{4} \leq 1$

$\dfrac{0}{25} - \dfrac{0}{4} \leq 1$

$0 - 0 \leq 1$

$0 \leq 1$

True, so region 2 is in the solution set.

Region 3: Choose $(0,-8)$.

$\dfrac{(-8)^2}{25} - \dfrac{0^2}{4} \leq 1$

$\dfrac{64}{25} - \dfrac{0}{4} \leq 1$

$\dfrac{64}{25} \leq 1$

False, so region 3 is not in the solution set.

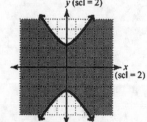

19. $y < x^2 + 4x - 5$

    First graph the related equation $y = x^2 + 4x - 5$ with a dotted line.
    Let region 1 be the area above the boundary, and let region 2 be the area below the boundary.
    Region 1: Choose $(0,0)$.

    $0 < 0^2 + 4(0) - 5$
    $0 < 0 + 0 - 5$
    $0 < -5$
    False, so region 1 is not in the solution set.
    Region 2: Choose $(4, 0)$.

    $0 < 4^2 + 4(4) - 5$
    $0 < 16 + 16 - 5$
    $0 < 27$
    True, so region 2 is in the solution set.

    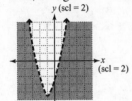

21. $\begin{cases} y < -x^2 \\ 2x - y < 4 \end{cases}$

    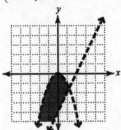

23. $\begin{cases} 3x + 2y \geq -6 \\ x^2 + y^2 \leq 25 \end{cases}$

    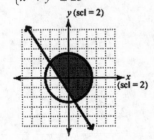

25. $\begin{cases} x^2 + y^2 > 4 \\ x^2 + y^2 > 9 \end{cases}$

    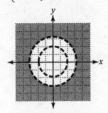

27. $\begin{cases} y > x^2 + 1 \\ 2x + y < 3 \end{cases}$

    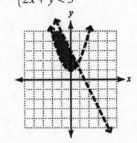

29. $\begin{cases} y < -x^2 + 3 \\ y > x^2 - 2 \end{cases}$

    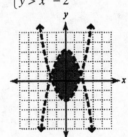

31. $\begin{cases} \dfrac{x^2}{25} + \dfrac{y^2}{9} \leq 1 \\ x^2 + y^2 \geq 4 \end{cases}$

    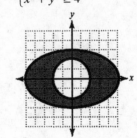

33. $\begin{cases} \dfrac{x^2}{25} + \dfrac{y^2}{9} < 1 \\ y > x^2 + 1 \end{cases}$

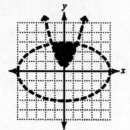

35. $\begin{cases} \dfrac{x^2}{9} - \dfrac{y^2}{4} \le 1 \\ \dfrac{x^2}{25} + \dfrac{y^2}{9} \le 1 \end{cases}$

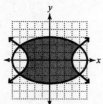

37. $\begin{cases} \dfrac{x^2}{4} - \dfrac{y^2}{4} > 1 \\ y > 2 \end{cases}$

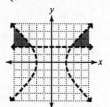

39. $\begin{cases} 3x + 2y \le 6 \\ x - y > -3 \\ x + 6y \ge 2 \end{cases}$

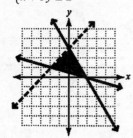

41. $\begin{cases} \dfrac{x^2}{16} + \dfrac{y^2}{4} \le 1 \\ x^2 + y^2 \le 9 \\ y \le x \end{cases}$

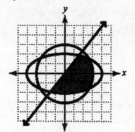

43. $\begin{cases} \dfrac{x^2}{49} - \dfrac{y^2}{16} \le 1 \\ \dfrac{x^2}{64} + \dfrac{y^2}{36} \le 1 \\ 2x - y \le -3 \end{cases}$

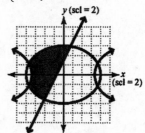

45. $\begin{cases} y < 2x^2 + 8 \\ 2x + y > 3 \\ x - y < 4 \end{cases}$

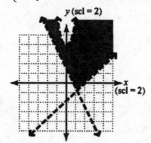

Review Exercises

1. $(3 \cdot 1 + 2) + (3 \cdot 2 + 2) + (3 \cdot 3 + 2) + (3 \cdot 4 + 2)$
   $= (3 + 2) + (6 + 2) + (9 + 2) + (12 + 2)$
   $= 5 + 8 + 11 + 14$
   $= 38$

2. $\dfrac{n}{2}(a_1 + a_n) = \dfrac{12}{2}(-8 + 60)$
   $= 6(52)$
   $= 312$

3. $a_1 + (n-1)d = -12 + (25-1)(-3)$
$$= -12 + (-72)$$
$$= -84$$

4. $n = 1 \quad 2(1^2) - 3 = 2 - 3 = -1$
$n = 2 \quad 2(2^2) - 3 = 8 - 3 = 5$
$n = 3 \quad 2(3^2) - 3 = 18 - 3 = 15$
$n = 4 \quad 2(4^2) - 3 = 32 - 3 = 29$

5. $n = 1 \quad \dfrac{(-1)^1}{3(1)-2} = \dfrac{-1}{3-2} = -1$

$n = 2 \quad \dfrac{(-1)^2}{3(2)-2} = \dfrac{1}{6-2} = \dfrac{1}{4}$

$n = 3 \quad \dfrac{(-1)^3}{3(3)-2} = \dfrac{-1}{9-2} = -\dfrac{1}{7}$

$n = 4 \quad \dfrac{(-1)^4}{3(4)-2} = \dfrac{1}{12-2} = \dfrac{1}{10}$

6. Let n, $(n + 2)$, and $(n + 4)$ be the three consecutive odd integers.
$$n + (n+2) + (n+4) = 207$$
$$3n + 6 = 207$$
$$3n = 201$$
$$n = 67$$
The three integers are 67, 69, and 71.

Chapter 12 Review Exercises

1. $y = 2(x-3)^2 - 5$
opens upward
vertex is $(3, -5)$
axis of symmetry is $x = 3$

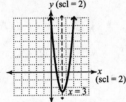

2. $y = -2(x+2)^2 + 3$
opens downward
vertex is $(-2, 3)$
axis of symmetry is $x = -2$

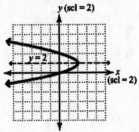

3. $x = -(y-2)^2 + 4$
opens left
vertex is $(4, 2)$
axis of symmetry is $y = 2$

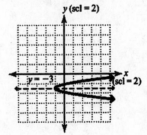

4. $x = 2(y+3)^2 - 2$
opens right
vertex is $(-2, -3)$
axis of symmetry is $y = -3$

5.
$$x = y^2 + 6y + 8$$
$$x - 8 = y^2 + 6y$$
$$x - 8 + 9 = y^2 + 6y + 9$$
$$x + 1 = (y + 3)^2$$
$$x = (y + 3)^2 - 1$$

opens right

vertex is $(-1, -3)$

axis of symmetry is $y = -3$

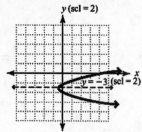

6.
$$x = 2y^2 - 8y - 6$$
$$x + 6 = 2(y^2 - 4y)$$
$$x + 6 + 8 = 2(y^2 - 4y + 4)$$
$$x + 14 = 2(y - 2)^2$$
$$x = 2(y - 2)^2 - 14$$

opens right

vertex is $(-14, 2)$

axis of symmetry is $y = 2$

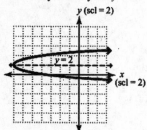

7.
$$x = -y^2 - 2y + 3$$
$$x - 3 = -(y^2 + 2y)$$
$$x - 3 - 1 = -(y^2 + 2y + 1)$$
$$x - 4 = -(y + 1)^2$$
$$x = -(y + 1)^2 + 4$$

opens left

vertex is $(4, -1)$

axis of symmetry is $y = -1$

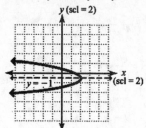

8.
$$x = -3y^2 - 12y - 9$$
$$x + 9 = -3(y^2 + 4y)$$
$$x + 9 - 12 = -3(y^2 + 4y + 4)$$
$$x - 3 = -3(y + 2)^2$$
$$x = -3(y + 2)^2 + 3$$

opens left

vertex is $(3, -2)$

axis of symmetry is $y = -2$

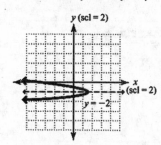

9. a)
$$h = -16t^2 + 32t + 128$$
$$h - 128 = -16(t^2 - 2t)$$
$$h - 128 - 16 = -16(t^2 - 2t + 1)$$
$$h - 144 = -16(t - 1)^2$$
$$h = -16(t - 1)^2 + 144$$

The vertex for the parabola is (1, 144). The maximum height for the rock will be 144 ft.

b) It will take 1 second to reach the maximum height.

c) $0 = -16t^2 + 32t + 128$

$0 = -16(t^2 - 2t - 8)$

$0 = t^2 - 2t - 8$

$0 = (t-4)(t+2)$

$t - 4 = 0$ or $t + 2 = 0$

$t = 4$ $t = \cancel{-2}$

The object will strike the ground after 4 seconds.

10. $(-1, -2)$ and $(-5, 1)$

$d = \sqrt{(-1-(-5))^2 + (-2-1)^2}$

$= \sqrt{(4)^2 + (-3)^2}$

$= \sqrt{16 + 9}$

$= \sqrt{25}$

$= 5$

$\text{Midpoint} = \left(\dfrac{-1+(-5)}{2}, \dfrac{-2+1}{2}\right)$

$= \left(\dfrac{-6}{2}, \dfrac{-1}{2}\right)$

$= \left(-3, -\dfrac{1}{2}\right)$

11. $(2, -5)$ and $(-2, 3)$

$d = \sqrt{(2-(-2))^2 + (-5-3)^2}$

$= \sqrt{(4)^2 + (-8)^2}$

$= \sqrt{16 + 64}$

$= \sqrt{80}$

$= 4\sqrt{5}$

$\text{Midpoint} = \left(\dfrac{2+(-2)}{2}, \dfrac{-5+3}{2}\right)$

$= \left(\dfrac{0}{2}, \dfrac{-2}{2}\right)$

$= (0, -1)$

12. Find the distance from the center to a point on the circle. This is the radius.

$r = \sqrt{(6-(-2))^2 + (-4-2)^2}$

$= \sqrt{(8)^2 + (-6)^2}$

$= \sqrt{64 + 36}$

$= \sqrt{100}$

$= 10$

13. $(x-3)^2 + (y+2)^2 = 25$

Center $(3, -2)$; radius 5

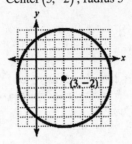

14. $(x+5)^2 + (y-1)^2 = 4$

Center $(-5, 1)$; radius 2

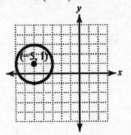

15. $x^2 + y^2 - 4x + 8y + 11 = 0$

$(x^2 - 4x) + (y^2 + 8y) = -11$

$(x^2 - 4x + 4) + (y^2 + 8y + 16) = -11 + 4 + 16$

$(x-2)^2 + (y+4)^2 = 9$

Center $(2, -4)$; radius 3

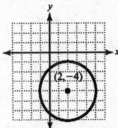

16. $x^2 + y^2 + 10x + 2y + 22 = 0$

$(x^2 + 10x) + (y^2 + 2y) = -22$

$(x^2 + 10x + 25) + (y^2 + 2y + 1) = -22 + 25 + 1$

$(x+5)^2 + (y+1)^2 = 4$

Center $(-5, -1)$; radius 2

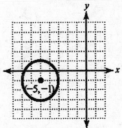

17. $(x-6)^2 + (y+8)^2 = 9^2$

    $(x-6)^2 + (y+8)^2 = 81$

18. $(x-(-3))^2 + (y-(-5))^2 = 10^2$

    $(x+3)^2 + (y+5)^2 = 100$

19. Since the rope is 20 ft. long and that is the distance from the center to any point on the edge of the circle, the radius is 20.
    The equation in standard form is:

    $x^2 + y^2 = 20^2$

    $x^2 + y^2 = 400$

20. Find the distance from the center to a point on the circle. This is the radius.

    $r = \sqrt{(-6-3)^2 + (8-(-4))^2}$

    $ = \sqrt{(-9)^2 + (12)^2}$

    $ = \sqrt{81+144}$

    $ = \sqrt{225}$

    $ = 15$

    The equation in standard form is:

    $(x+6)^2 + (y-8)^2 = 225$

21. $\dfrac{x^2}{49} + \dfrac{y^2}{25} = 1$

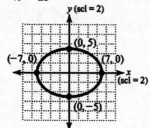

22. $\dfrac{x^2}{9} + \dfrac{y^2}{25} = 1$

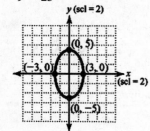

23. $4x^2 + 9y^2 = 36$

    $\dfrac{4x^2}{36} + \dfrac{9y^2}{36} = \dfrac{36}{36}$

    $\dfrac{x^2}{9} + \dfrac{y^2}{4} = 1$

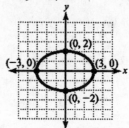

24. $25x^2 + 9y^2 = 225$

    $\dfrac{25x^2}{225} + \dfrac{9y^2}{225} = \dfrac{225}{225}$

    $\dfrac{x^2}{9} + \dfrac{y^2}{25} = 1$

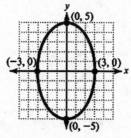

25. $\dfrac{(x+2)^2}{16} + \dfrac{(y-3)^2}{4} = 1$

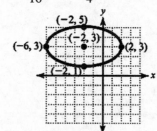

26. $\dfrac{(x-1)^2}{9} + \dfrac{(y+4)^2}{25} = 1$

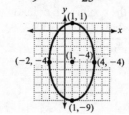

27. $\dfrac{x^2}{25^2} + \dfrac{y^2}{15^2} = 1$

$\dfrac{x^2}{625} + \dfrac{y^2}{225} = 1$

28. $\dfrac{x^2}{1.625^2} + \dfrac{y^2}{2^2} = 1$

$\dfrac{x^2}{1.625^2} + \dfrac{y^2}{4} = 1$

29. $\dfrac{x^2}{25} - \dfrac{y^2}{16} = 1$

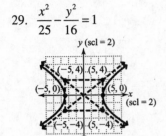

30. $\dfrac{x^2}{36} - \dfrac{y^2}{9} = 1$

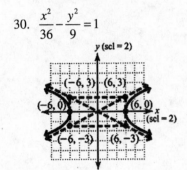

31. $y^2 - 9x^2 = 36$

$\dfrac{y^2}{36} - \dfrac{9x^2}{36} = \dfrac{36}{36}$

$\dfrac{y^2}{36} - \dfrac{x^2}{4} = 1$

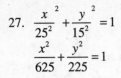

32. $25y^2 - 9x^2 = 225$

$\dfrac{25y^2}{225} - \dfrac{9x^2}{225} = \dfrac{225}{225}$

$\dfrac{y^2}{9} - \dfrac{x^2}{25} = 1$

33. $\begin{cases} y = 2x^2 - 3 \\ 2x - y = -1 \end{cases}$

The first equation is already solved for y.

$2x - \left(2x^2 - 3\right) = -1$

$2x - 2x^2 + 3 = -1$

$0 = 2x^2 - 2x - 4$

$0 = x^2 - x - 2$

$0 = (x - 2)(x + 1)$

$x - 2 = 0$ or $x + 1 = 0$

$\quad x = 2 \qquad\qquad x = -1$

$x = 2: \qquad y = 2 \cdot 2^2 - 3$

$\qquad\qquad\quad y = 2 \cdot 4 - 3$

$\qquad\qquad\quad y = 8 - 3$

$\qquad\qquad\quad y = 5$

$\qquad\qquad\quad (2, 5)$

$x = -1: \qquad y = 2(-1)^2 - 3$

$\qquad\qquad\quad y = 2 \cdot 1 - 3$

$\qquad\qquad\quad y = 2 - 3$

$\qquad\qquad\quad y = -1$

$\qquad\qquad\quad (-1, -1)$

34. $\begin{cases} x^2 + y^2 = 17 \\ x - y = 3 \end{cases}$

Solve the second equation for x. $x = y + 3$

$$(y+3)^2 + y^2 = 17$$
$$y^2 + 6y + 9 + y^2 = 17$$
$$2y^2 + 6y - 8 = 0$$
$$y^2 + 3y - 4 = 0$$
$$(y+4)(y-1) = 0$$

$y + 4 = 0$ or $y - 1 = 0$
     $y = -4$ $y = 1$

$y = -4:$ $x = -4 + 3$
                    $x = -1$
                    $(-1, -4)$

$y = 1:$ $x = 1 + 3$
                    $x = 4$
                    $(4, 1)$

35. $\begin{cases} 25x^2 + 3y^2 = 100 \\ 2x - y = -3 \end{cases}$

Solve the second equation for y. $y = 2x + 3$

$$25x^2 + 3(2x+3)^2 = 100$$
$$25x^2 + 3(4x^2 + 12x + 9) = 100$$
$$25x^2 + 12x^2 + 36x + 27 = 100$$
$$37x^2 + 36x - 73 = 0$$
$$(37x + 73)(x - 1) = 0$$

$37x + 73 = 0$ or $x - 1 = 0$
    $37x = -73$ $x = 1$
    $x = -\dfrac{73}{37}$

$x = -\dfrac{73}{37}:$ $y = 2\left(-\dfrac{73}{37}\right) + 3$

                    $y = -\dfrac{35}{37}$

                    $\left(-\dfrac{73}{37}, -\dfrac{35}{37}\right)$

$x = 1:$ $y = 2 \cdot 1 + 3$
                    $y = 5$
                    $(1, 5)$

36. $\begin{cases} x^2 + y^2 = 64 \\ x^2 - y^2 = 64 \end{cases}$

$\begin{aligned} x^2 + y^2 &= 64 \\ \underline{x^2 - y^2} &\underline{= 64} \\ 2x^2 \quad\;\; &= 128 \\ x^2 \quad\;\; &= 64 \\ x \quad\;\; &= \pm 8 \end{aligned}$

$x = -8:$ $(-8)^2 + y^2 = 64$
                    $64 + y^2 = 64$
                    $y^2 = 0$
                    $y = 0$
                    $(-8, 0)$

$x = 8:$ $(8)^2 + y^2 = 64$
                    $64 + y^2 = 64$
                    $y^2 = 0$
                    $y = 0$
                    $(8, 0)$

37. $\begin{cases} x^2 + y^2 = 25 \\ 25y^2 - 16x^2 = 256 \end{cases}$

$16(x^2 + y^2 = 25)$

$\dfrac{-16x^2 + 25y^2 = 256}{}$

$\begin{aligned} 16x^2 + 16y^2 &= 400 \\ \underline{-16x^2 + 25y^2} &\underline{= 256} \\ 41y^2 &= 656 \\ y^2 &= 16 \\ y &= \pm 4 \end{aligned}$

$y = -4:$ $x^2 + (-4)^2 = 25$
                    $x^2 + 16 = 25$
                    $x^2 = 9$
                    $x = \pm 3$
                    $(-3, -4), (3, -4)$

$x = 4:$ $x^2 + (4)^2 = 25$
                    $x^2 + 16 = 25$
                    $x^2 = 9$
                    $x = \pm 3$
                    $(-3, 4), (3, 4)$

38. $\begin{cases} x^2 + 5y^2 = 36 \\ 4x^2 - 7y^2 = 36 \end{cases}$

$-4\left(x^2 + 5y^2 = 36\right)$

$\underline{\quad 4x^2 - 7y^2 = 36 \quad}$

$-4x^2 - 20y^2 = -144$

$\underline{\quad 4x^2 - 7y^2 \quad = 36 \quad}$

$\qquad -27y^2 = -108$

$\qquad y^2 = 4$

$\qquad y = \pm 2$

$y = 2: \qquad x^2 + 5(2)^2 = 36$

$\qquad\qquad x^2 + 5 \cdot 4 = 36$

$\qquad\qquad x^2 + 20 = 36$

$\qquad\qquad x^2 = 16$

$\qquad\qquad x = \pm 4$

$\qquad\qquad (-4, 2), (4, 2)$

$y = -2: \qquad x^2 + 5(-2)^2 = 36$

$\qquad\qquad x^2 + 5 \cdot 4 = 36$

$\qquad\qquad x^2 + 20 = 36$

$\qquad\qquad x^2 = 16$

$\qquad\qquad x = \pm 4$

$\qquad\qquad (-4, -2), (4, -2)$

39. $\begin{cases} y = x^2 - 2 \\ 4x^2 + 5y^2 = 36 \end{cases}$

The first equation is already solved for y.

$4x^2 + 5\left(x^2 - 2\right)^2 = 36$

$4x^2 + 5\left(x^4 - 4x^2 + 4\right) = 36$

$4x^2 + 5x^4 - 20x^2 + 20 = 36$

$5x^4 - 16x^2 - 16 = 0$

$\cancel{\left(5x^2 + 4\right)}\left(x^2 - 4\right) = 0$

$\qquad\qquad x^2 = 4$

$\qquad\qquad x = \pm 2$

$5x^2 + 4 = 0$ would have yielded an imaginary solution.

$x = -2: \quad y = (-2)^2 - 2$

$\qquad\qquad y = 4 - 2$

$\qquad\qquad y = 2$

$\qquad\qquad (-2, 2)$

$x = 2: \quad y = (2)^2 - 2$

$\qquad\qquad y = 4 - 2$

$\qquad\qquad y = 2$

$\qquad\qquad (2, 2)$

40. $\begin{cases} y = x^2 - 1 \\ 4y^2 - 5x^2 = 16 \end{cases}$

The first equation is already solved for y.

$4\left(x^2 - 1\right)^2 - 5x^2 = 16$

$4\left(x^4 - 2x^2 + 1\right) - 5x^2 = 16$

$4x^4 - 8x^2 + 4 - 5x^2 = 16$

$4x^4 - 13x^2 - 12 = 0$

$\cancel{\left(4x^2 + 3\right)}\left(x^2 - 4\right) = 0$

$\qquad\qquad x^2 = 4$

$\qquad\qquad x = \pm 2$

$4x^2 + 3 = 0$ would have yielded an imaginary solution.

$x = -2: \quad y = (-2)^2 - 1$

$\qquad\qquad y = 4 - 1$

$\qquad\qquad y = 3$

$\qquad\qquad (-2, 3)$

$x = 2: \quad y = (2)^2 - 1$

$\qquad\qquad y = 4 - 1$

$\qquad\qquad y = 3$

$\qquad\qquad (2, 3)$

41. Let x be one of the integers and y be the other integer.

$\begin{cases} x^2 + y^2 = 89 \\ x^2 - y^2 = 39 \end{cases}$

$x^2 + y^2 = 89$

$\underline{\quad x^2 - y^2 = 39 \quad}$

$2x^2 \qquad = 128$

$x^2 \qquad = 64$

$x \qquad = \pm 8$

$x = -8: \quad (-8)^2 + y^2 = 89$

$\qquad\qquad 64 + y^2 = 89$

$\qquad\qquad y^2 = 25$

$\qquad\qquad y = \pm 5$

$\qquad\qquad (-8, -5), (-8, 5)$

$x = 8: \quad (8)^2 + y^2 = 89$

$\qquad\qquad 64 + y^2 = 89$

$\qquad\qquad y^2 = 25$

$\qquad\qquad y = \pm 5$

$\qquad\qquad (8, -5), (8, 5)$

42. Let x be one dimension and let y be the other.
$$\begin{cases} xy = 80 \\ 2x + 2y = 36 \end{cases}$$
Solve the second equation for x. $x = 18 - y$
$$(18 - y)y = 80$$
$$18y - y^2 = 80$$
$$0 = y^2 - 18y + 80$$
$$0 = (y - 10)(y - 8)$$
$$y - 10 = 0 \quad \text{or} \quad y - 8 = 0$$
$$y = 10 \qquad\qquad y = 8$$
If $y = 10$, then $xy = 80$ and $x = 8$.
If $y = 8$, then $xy = 80$ and $x = 10$.
The rug is 8 ft. by 10 ft.

43.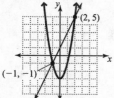

44.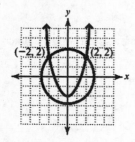

45. $x^2 + y^2 \le 64$

First graph the related equation $x^2 + y^2 = 64$
with a solid line.
Let region 1 be the area inside the circle, and let
region 2 be the area outside the circle.
Region 1: Choose (0, 0).
$$0^2 + 0^2 \le 64$$
$$0 + 0 \le 64$$
$$0 \le 64$$
True, so region 1 is in the solution set.
Region 2: Choose (10, 0).
$$10^2 + 0^2 \le 64$$
$$100 + 0 \le 64$$
$$100 \le 64$$
False, so region 2 is not in the solution set.

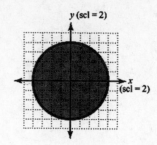

46. $y < 2(x - 3)^2 + 4$

First graph the related equation $y = 2(x - 3)^2 + 4$
with a dotted line.
Let region 1 be the area above the boundary, and
let region 2 be the area below the boundary.
Region 1: Choose (2, 10).
$$10 < 2(2 - 3)^2 + 4$$
$$10 < 2(-1)^2 + 4$$
$$10 < 2(1) + 4$$
$$10 < 2 + 4$$
$$10 < 6$$
False, so region 1 is not in the solution set.
Region 2: Choose $(0, 0)$.
$$0 < 2(0 - 3)^2 + 4$$
$$0 < 2(-3)^2 + 4$$
$$0 < 2(9) + 4$$
$$0 < 18 + 4$$
$$0 < 22$$
True, so region 2 is in the solution set.

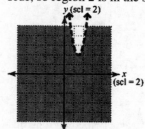

47. $\dfrac{y^2}{25} + \dfrac{x^2}{49} > 1$

First graph the related equation $\dfrac{y^2}{25} + \dfrac{x^2}{49} = 1$ with a dotted line.

Let region 1 be the area inside the ellipse, and let region 2 be the area outside the ellipse.

Region 1: Choose $(0, 0)$.

$\dfrac{0^2}{25} + \dfrac{0^2}{49} > 1$

$0 + 0 > 1$

$0 > 1$

False, so region 1 is not in the solution set.

Region 2: Choose $(0, 6)$.

$\dfrac{6^2}{25} + \dfrac{0^2}{49} > 1$

$\dfrac{36}{25} + 0 > 1$

$\dfrac{36}{25} > 1$

True, so region 2 is in the solution set.

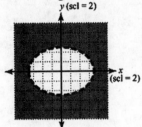

48. $\dfrac{x^2}{25} - \dfrac{y^2}{36} < 1$

First graph the related equation $\dfrac{x^2}{25} - \dfrac{y^2}{36} = 1$ with a dotted line.

Let region 1 be the area to the left of the left branch of the hyperbola, let region 3 be the area to the right of the right branch of the hyperbola, and let region 2 be the area between the two branches.

Region 1: Choose $(-8, 0)$.

$\dfrac{(-8)^2}{25} - \dfrac{0^2}{36} < 1$

$\dfrac{64}{25} - \dfrac{0}{36} < 1$

$\dfrac{64}{25} < 1$

False, so region 1 is not in the solution set.

Region 2: Choose $(0,0)$.

$\dfrac{0^2}{25} - \dfrac{0^2}{36} < 1$

$\dfrac{0}{25} - \dfrac{0}{36} < 1$

$0 < 1$

True, so region 2 is in the solution set.

Region 3: Choose $(8,0)$.

$\dfrac{(8)^2}{25} - \dfrac{0^2}{36} < 1$

$\dfrac{64}{25} - \dfrac{0}{36} < 1$

$\dfrac{64}{25} < 1$

False, so region 3 is not in the solution set.

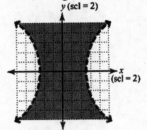

49. $\begin{cases} y \geq x^2 - 3 \\ 2x + y < 2 \end{cases}$

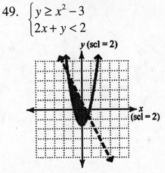

50. $\begin{cases} x + 2y < 4 \\ x^2 + y^2 \leq 25 \end{cases}$

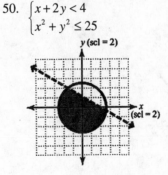

51. $\begin{cases} y > x^2 - 2 \\ y < -x^2 + 1 \end{cases}$

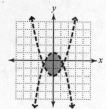

52. $\begin{cases} \dfrac{x^2}{9} + \dfrac{y^2}{25} \le 1 \\ x^2 + y^2 \ge 9 \end{cases}$

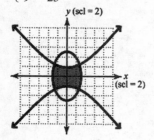

53. $\begin{cases} \dfrac{y^2}{4} - \dfrac{x^2}{9} \le 1 \\ \dfrac{x^2}{9} + \dfrac{y^2}{25} \le 1 \end{cases}$

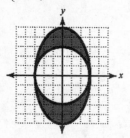

54. $\begin{cases} \dfrac{y^2}{4} + \dfrac{x^2}{16} \le 1 \\ x^2 + y^2 \le 9 \\ y \le x + 1 \end{cases}$

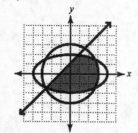

Chapter 12 Practice Test

1. $y = 2(x+1)^2 - 4$; $a = 2$, $h = -1$; and $k = -4$

    The parabola opens upward, has its vertex at $(-1, -4)$, and has its axis of symmetry at $x = -1$.

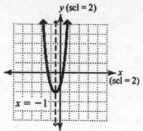

2. $x = -2(y-3)^2 + 1$; $a = -2$, $k = 3$; and $h = 1$

    The parabola opens left, has its vertex at $(1, 3)$, and has its axis of symmetry at $y = 3$.

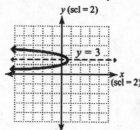

3. $\begin{aligned} x &= y^2 + 4y - 3 \\ x + 3 &= y^2 + 4y \\ x + 3 + 4 &= y^2 + 4y + 4 \\ x + 7 &= (y+2)^2 \\ x &= (y+2)^2 - 7 \end{aligned}$

  $a = 1$, $k = -2$; and $h = -7$

  The parabola opens right, has its vertex at $(-7, -2)$, and has its axis of symmetry at $y = -2$.

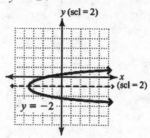

4. $(x_2, y_2) = (2, -3)$ and $(x_1, y_1) = (6, -5)$

$$d = \sqrt{(x_2 - x_1)^2 + (y_2 - y_1)^2}$$
$$= \sqrt{(2 - 6)^2 + (-3 - (-5))^2}$$
$$= \sqrt{(-4)^2 + (2)^2}$$
$$= \sqrt{16 + 4}$$
$$= \sqrt{20}$$
$$= 2\sqrt{5}$$

$$\text{Midpoint} = \left(\frac{x_1 + x_2}{2}, \frac{y_1 + y_2}{2} \right)$$
$$= \left(\frac{6 + 2}{2}, \frac{-5 + (-3)}{2} \right)$$
$$= \left(\frac{8}{2}, \frac{-8}{2} \right)$$
$$= (4, -4)$$

5. $(x + 4)^2 + (y - 3)^2 = 36$

Because $h = -4$, $k = 3$, and $r = 6$, the center is $(-4, 3)$ and the radius is 6.

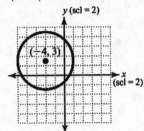

6. $$x^2 + y^2 + 4x - 10y + 20 = 0$$
$$(x^2 + 4x) + (y^2 - 10y) = -20$$
$$(x^2 + 4x + 4) + (y^2 - 10y + 25) = -20 + 4 + 25$$
$$(x + 2)^2 + (y - 5)^2 = 9$$

Because $h = -2$, $k = 5$, and $r = 3$, the center is $(-2, 5)$ and the radius is 3.

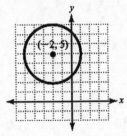

7. The center is $(2, -4)$, so $h = 2$ and $k = -4$. Find the distance from the center to a point on the circle $(-4, 4)$. This is the radius.

$$d = \sqrt{(2 - (-4))^2 + (-4 - 4)^2}$$
$$= \sqrt{(6)^2 + (-8)^2}$$
$$= \sqrt{36 + 64}$$
$$= \sqrt{100}$$
$$= 10$$

The equation in standard form is
$$(x - 2)^2 + (y + 4)^2 = 100$$

8. $$16x^2 + 36y^2 = 576$$
$$\frac{16x^2}{576} + \frac{36y^2}{576} = \frac{576}{576}$$
$$\frac{x^2}{36} + \frac{y^2}{16} = 1$$

$a = 6$, $b = 4$

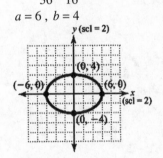

9. $$\frac{(x + 2)^2}{4} + \frac{(y - 1)^2}{25} = 1$$

$a = 2$, $b = 5$, $h = -2$, $k = 1$

10. $$\frac{x^2}{49} - \frac{y^2}{25} = 1$$

$a = 7$, $b = 5$

11. $\dfrac{y^2}{9} - \dfrac{x^2}{16} = 1$

$a = 4$, $b = 3$

12. $\begin{cases} y = (x+1)^2 + 2 \\ 2x + y = 8 \end{cases}$

Solve the second equation for y. $y = 8 - 2x$.

$y = (x+1)^2 + 2$

$8 - 2x = (x+1)^2 + 2$

$8 - 2x = x^2 + 2x + 1 + 2$

$0 = x^2 + 4x - 5$

$0 = (x+5)(x-1)$

$x = -5, 1$

$x = -5:$ $y = 8 - 2(-5)$

$y = 8 + 10$

$y = 18$

$(-5, 18)$

$x = 1:$ $y = 8 - 2(1)$

$y = 8 - 2$

$y = 6$

$(1, 6)$

13. $\begin{cases} 3x - y = 4 \\ x^2 + y^2 = 34 \end{cases}$

Solve the first equation for y. $y = 3x - 4$.

$x^2 + y^2 = 34$

$x^2 + (3x - 4)^2 = 34$

$x^2 + 9x^2 - 24x + 16 = 34$

$10x^2 - 24x - 18 = 0$

$2(5x^2 - 12x - 9) = 0$

$5x^2 - 12x - 9 = 0$

$(5x + 3)(x - 3) = 0$

$x = -\dfrac{3}{5}, 3$

$x = -\dfrac{3}{5}:$ $y = 3\left(-\dfrac{3}{5}\right) - 4$

$y = -\dfrac{29}{5}$

$\left(-\dfrac{3}{5}, -\dfrac{29}{5}\right)$

$x = 3:$ $y = 3(3) - 4$

$y = 9 - 4$

$y = 5$

$(3, 5)$

14. $\begin{cases} x^2 + y^2 = 13 \\ 3x^2 + 4y^2 = 48 \end{cases}$

$-3(x^2 + y^2 = 13)$

$\underline{3x^2 + 4y^2 = 48}$

$-3x^2 - 3y^2 = -39$

$\underline{3x^2 + 4y^2 = 48}$

$y^2 = 9$

$y = \pm 3$

$y = -3:$ $x^2 + (-3)^2 = 13$

$x^2 + 9 = 13$

$x^2 = 4$

$x = \pm 2$

$(-2, -3), (2, -3)$

$y = 3:$ $x^2 + (3)^2 = 13$

$x^2 + 9 = 13$

$x^2 = 4$

$x = \pm 2$

$(2, 3), (-2, 3)$

15. $\begin{cases} x^2 - 2y^2 = 1 \\ 4x^2 + 7y^2 = 64 \end{cases}$

$-4(x^2 - 2y^2 = 1)$

$\underline{4x^2 + 7y^2 = 64}$

$-4x^2 + 8y^2 = -4$

$\underline{4x^2 + 7y^2 = 64}$

$15y^2 = 60$

$y^2 = 4$

$y = \pm 2$

$y = -2:$ $x^2 - 2(-2)^2 = 1$

$x^2 - 8 = 1$

$x^2 = 9$

$x = \pm 3$

$(3, -2), (-3, -2)$

$y = 2:$ $x^2 - 2(2)^2 = 1$

$x^2 - 8 = 1$

$x^2 = 9$

$x = \pm 3$

$(3, 2), (-3, 2)$

16. $y \le -2(x+3)^2 + 2$

First graph the related equation

$y = -2(x+3)^2 + 2$ with a solid line.

Let region 1 be the area above the boundary, and let region 2 be the area below the boundary.

Region 1: Choose $(0,0)$.

$0 \le -2(0+3)^2 + 2$

$0 \le -2(3)^2 + 2$

$0 \le -2(9) + 2$

$0 \le -18 + 2$

$0 \le -16$

False, so region 1 is not in the solution set.

Region 2: Choose $(-2, -6)$.

$-6 < -2(-2+3)^2 + 2$

$-6 < -2(1)^2 + 2$

$-6 < -2 + 2$

$-6 < 0$

True, so region 2 is in the solution set.

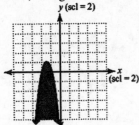

17. $\dfrac{x^2}{4} + \dfrac{y^2}{9} > 1$

First graph the related equation $\dfrac{x^2}{4} + \dfrac{y^2}{9} = 1$ with a dotted line.

Let region 1 be the area inside the ellipse, and let region 2 be the area outside the ellipse.

Region 1: Choose $(0, 0)$.

$\dfrac{0^2}{4} + \dfrac{0^2}{9} > 1$

$0 + 0 > 1$

$0 > 1$

False, so region 1 is not in the solution set.

Region 2: Choose $(0, 4)$.

$\dfrac{0^2}{4} + \dfrac{4^2}{9} > 1$

$0 + \dfrac{16}{9} > 1$

$\dfrac{16}{9} > 1$

True, so region 2 is in the solution set.

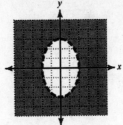

18. $\begin{cases} y \ge x^2 - 4 \\ \dfrac{x^2}{9} + \dfrac{y^2}{16} \le 1 \end{cases}$

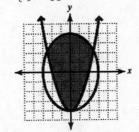

19. $x = 0$: $y = -\dfrac{1}{2} \cdot 0 + 18$

$y = 18$

The height is 18.

$y = 0$: $\quad 0 = -\dfrac{1}{2}x^2 + 18$

$-18 = -\dfrac{1}{2}x^2$

$36 = x^2$

$\pm 6 = x$

The distance across the base is 12.

20. $a = \dfrac{19}{2} = 9.5$ and $b = \dfrac{2}{2} = 1$

$\dfrac{x^2}{a^2} + \dfrac{y^2}{b^2} = 1$

$\dfrac{x^2}{9.5^2} + y^2 = 1$

Chapters 1–12 Cumulative Review

1. False; a positive base raised to a negative exponent yields a reciprocal.

2. True

3. True

4. $-8, -5$

5. i

6. $d = \sqrt{(x_2 - x_1)^2 + (y_2 - y_1)^2}$

7. $6 - (-3) + 2^{-4} = 6 + 3 + \dfrac{1}{16} = 9\dfrac{1}{16}$

8. $\left(5m^3 n^{-2}\right)^{-3} = 5^{-3} m^{3(-3)} n^{-2(-3)}$

    $\qquad\qquad = \dfrac{1}{5^3} m^{-9} n^6$

    $\qquad\qquad = \dfrac{n^6}{125 m^9}$

9. $\sqrt{-20} = \sqrt{-4 \cdot 5} = 2i\sqrt{5}$

10. $ax + bx + ay + by = x(a + b) + y(a + b)$

    $\qquad\qquad\qquad\quad = (a + b)(x + y)$

11. $k^4 - 81 = \left(k^2 + 9\right)\left(k^2 - 9\right)$

    $\qquad\quad = \left(k^2 + 9\right)(k + 3)(k - 3)$

12. $|2x + 1| = |x - 3|$

    $2x + 1 = x - 3 \quad$ or $\quad 2x + 1 = -(x - 3)$

    $\qquad x = -4 \qquad\qquad 2x + 1 = -x + 3$

    $\qquad\qquad\qquad\qquad\quad 3x = 2$

    $\qquad\qquad\qquad\qquad\quad x = \dfrac{2}{3}$

13. $\sqrt{x - 3} = 7$

    $\left(\sqrt{x - 3}\right)^2 = 7^2$

    $\qquad x - 3 = 49$

    $\qquad\quad x = 52$

14. $4x^2 - 2x + 1 = 0$

    $x = \dfrac{-(-2) \pm \sqrt{(-2)^2 - 4(4)(1)}}{2(4)} = \dfrac{2 \pm \sqrt{4 - 16}}{8}$

    $\quad = \dfrac{2 \pm \sqrt{-12}}{8} = \dfrac{2 \pm 2i\sqrt{3}}{8} = \dfrac{1 \pm i\sqrt{3}}{4}$

15. $\qquad 9 + 24x^{-1} + 16x^{-2} = 0$

    $x^2\left(9 + 24x^{-1} + 16x^{-2}\right) = x^2 \cdot 0$

    $\qquad\quad 9x^2 + 24x + 16 = 0$

    $\qquad\qquad\quad (3x + 4)^2 = 0$

    $\qquad\qquad \sqrt{(3x + 4)^2} = 0$

    $\qquad\qquad\qquad 3x + 4 = 0$

    $\qquad\qquad\qquad\qquad x = -\dfrac{4}{3}$

16. $\qquad 9^x = 27$

    $\qquad \left(3^2\right)^x = 3^3$

    $\qquad\quad 3^{2x} = 3^3$

    $\qquad\quad 2x = 3$

    $\qquad\quad x = \dfrac{3}{2} = 1.5$

17. $\log_3(x + 1) - \log_3 x = 2$

    $\qquad\quad \log_3 \dfrac{x + 1}{x} = 2$

    $\qquad\qquad 3^2 = \dfrac{x + 1}{x}$

    $\qquad\qquad 9x = x + 1$

    $\qquad\qquad 8x = 1$

    $\qquad\qquad x = \dfrac{1}{8}$

18. $5m^2 - 3m = 0$

    $m(5m - 3) = 0$

    $m = 0 \quad$ or $\quad 5m - 3 = 0$

    $\qquad\qquad\qquad\quad m = \dfrac{3}{5}$

    For $(-\infty, 0)$, we choose -1.

    $5(-1)^2 - 3(-1) < 0 \qquad$ False

    $\qquad 5 + 3 < 0$

    $\qquad\quad 8 < 0$

    For $\left(0, \dfrac{3}{5}\right)$, we choose $\dfrac{1}{5}$.

    $5\left(\dfrac{1}{5}\right)^2 - 3\left(\dfrac{1}{5}\right) < 0 \qquad$ True

    $\qquad \dfrac{1}{5} - \dfrac{3}{5} < 0$

    $\qquad\quad -\dfrac{2}{5} < 0$

    For $\left(\dfrac{3}{5}, \infty\right)$, we choose 2.

    $5(2)^2 - 3(2) < 0 \qquad$ False

    $\qquad 20 - 6 < 0$

    $\qquad\quad 14 < 0$

    a)

    b) $\left\{ m \,\middle|\, 0 < m < \dfrac{3}{5} \right\}$

    c) $\left(0, \dfrac{3}{5}\right)$

19. a) $(f+g)(x) = (2x+1) + (x^2-1)$
$= x^2 + 2x$

b) $(f-g)(x) = (2x+1) - (x^2-1)$
$= (2x+1) + (-x^2+1)$
$= -x^2 + 2x + 2$

c) $(f \cdot g)(x) = (2x+1)(x^2-1)$
$= 2x^3 - 2x + x^2 - 1$
$= 2x^3 + x^2 - 2x - 1$

d) $(f/g)(x) = \dfrac{2x+1}{x^2-1}; x \neq \pm 1$

e) $(f \circ g)(x) = f[g(x)]$
$= f(x^2-1)$
$= 2(x^2-1) + 1$
$= 2x^2 - 2 + 1$
$= 2x^2 - 1$

f) $(g \circ f)(x) = g[f(x)]$
$= g(2x+1)$
$= (2x+1)^2 - 1$
$= 4x^2 + 4x + 1 - 1$
$= 4x^2 + 4x$

20. $\log_5 125 = 3$

21. $\log_5 x^5 y = \log_5 x^5 + \log_5 y = 5\log_5 x + \log_5 y$

22. Find the distance from the center to a point on the circle. This is the radius.
$r = \sqrt{(2-7)^2 + (4-16)^2}$
$= \sqrt{(-5)^2 + (-12)^2}$
$= \sqrt{25 + 144}$
$= \sqrt{169}$
$= 13$
The equation in standard form is:
$(x-2)^2 + (y-4)^2 = 13^2$
$(x-2)^2 + (y-4)^2 = 169$

23. $\begin{cases} y < -x+2 \\ y \geq x-4 \end{cases}$

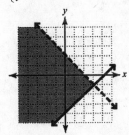

24. $f(x) = \log_3 x$
$y = \log_3 x$
$3^y = x$

y	-1	0	1	2
x	$\dfrac{1}{3}$	1	3	9

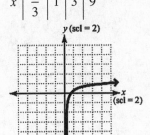

25. $\dfrac{x^2}{4} + \dfrac{y^2}{9} = 1$

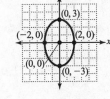

26. a) $V = \pi r^2 h$
$\dfrac{V}{\pi r^2} = \dfrac{\pi r^2 h}{\pi r^2}$
$h = \dfrac{V}{\pi r^2}$

b) $h = \dfrac{V}{\pi r^2}$
$h = \dfrac{15\pi}{\pi r^2}$
$h = \dfrac{15}{r^2}$

27. Let s be pounds of shrimp and t be pounds of tuna consumed.

Translate to a system of equations. $\begin{cases} s+t = 6.3 \\ s = 0.5 + t \end{cases}$

$0.5 + t + t = 6.3 \qquad s = 0.5 + 2.9$
$\qquad 2t = 5.8 \qquad\quad s = 3.4$
$\qquad\quad t = 2.9$

2.9 lbs. of tuna and 3.4 lbs. of shrimp

28. Translate to a system of equations.

Let x be the money market fund, y be the income fund, and z be the growth fund.

$$\begin{cases} x + y + z = 5000 \\ 0.05x + 0.06y + 0.03z = 255 \\ z = x - 500 \end{cases}$$

$z = 1500 - 500$ $1500 + y + 1000 = 5000$

$z = 1000$ $2500 + y = 5000$

$y = 2500$

1500 at 5%, $\$2500$ at 6%, and $\$1000$ at 3%

29. $M = \log \dfrac{I}{I_0}$

$M = \log \dfrac{10^{2.9}}{10^{-4}}$

$M = \log 10^{6.9}$

$M = 6.9$

30. $\dfrac{x^2}{75^2} + \dfrac{y^2}{20^2} = 1$

$\dfrac{x^2}{5625} + \dfrac{y^2}{400} = 1$